Protein Kinases: Molecular Insights

Protein Kinases: Molecular Insights

Edited by Bailey Graham

Syrawood Publishing House
New York

Published by Syrawood Publishing House,
750 Third Avenue, 9th Floor,
New York, NY 10017, USA
www.syrawoodpublishinghouse.com

Protein Kinases: Molecular Insights
Edited by Bailey Graham

International Standard Book Number: 978-1-64740-385-0 (Hardback)

Cataloging-in-publication Data

Protein kinases : molecular insights / edited by Bailey Graham.
p. cm.
Includes bibliographical references and index.
ISBN 978-1-64740-385-0
1. Protein kinases. 2. Molecular biology. I. Graham, Bailey.
QP606.P76 P765 2023
572.792--dc23

TABLE OF CONTENTS

PREFACE

It is often said that books are a boon to mankind. They document every progress and pass on the knowledge from one generation to the other. They play a crucial role in our lives. Thus I was both excited and nervous while editing this book. I was pleased by the thought of being able to make a mark but I was also nervous to do it right because the future of students depends upon it. Hence, I took a few months to research further into the discipline, revise my knowledge and also explore some more aspects. Post this process, I begun with the editing of this book.

Protein kinases are enzymes that use ATP to phosphorylate other proteins, which results in the alteration of their functions. These are signaling enzymes, which are responsible for regulating signal transduction in multiple intracellular processes such as glycogen synthesis, hormone responses, ion transport, proliferation, apoptosis, metabolism, transcription and antibiotic resistance. There are more than 30 families of kinases, of which protein kinase family is the largest. A few examples of protein kinases include AMP-activated protein kinase (AMPK), calcium calmodulin dependent kinases (CAMK), AGC kinase, casein kinase 1 (CK1), CMGC kinases, tyrosine kinases (TKs), and receptor interacting protein kinases (RIPKs). AMPK affects some major physiological functions such as metabolic activity. AMPKs are also critical for regulating lipid metabolism, mitochondrial function, and carbohydrate metabolism. This book includes some of the vital pieces of works being conducted across the world, on various topics related to protein kinases. It will also provide molecular insights on protein kinases to the readers. The book will serve as a valuable source of reference for graduate and postgraduate students.

I thank my publisher with all my heart for considering me worthy of this unparalleled opportunity and for showing unwavering faith in my skills. I would also like to thank the editorial team who worked closely with me at every step and contributed immensely towards the successful completion of this book. Last but not the least, I wish to thank my friends and colleagues for their support.

Editor

1

Discovery of a Novel Class of Covalent Dual Inhibitors Targeting the Protein Kinases BMX and BTK

Michael Forster [1,2,†], Xiaojun Julia Liang [1,2,†], Martin Schröder [3,4], Stefan Gerstenecker [1], Apirat Chaikuad [3,4], Stefan Knapp [3,4,5], Stefan Laufer [1,2,6] and Matthias Gehringer [1,2,*]

1 Department of Pharmaceutical/Medicinal Chemistry, Institute of Pharmaceutical Sciences, Faculty of Sciences, University of Tübingen, 72076 Tübingen, Germany; michael.forster@uni-tuebingen.de (M.F.); xiaojun-julia.liang@uni-tuebingen.de (X.J.L.); stefan.gerstenecker@uni-tuebingen.de (S.G.); stefan.laufer@uni-tuebingen.de (S.L.)
2 Cluster of Excellence iFIT (EXC 2180) 'Image-Guided & Functionally Instructed Tumor Therapies', University of Tübingen, 72076 Tübingen, Germany
3 Structural Genomics Consortium, Goethe University Frankfurt, Buchmann Institute for Molecular Life Sciences, Max-von-Laue-Straße 15, 60438 Frankfurt am Main, Germany; m.schroeder@pharmchem.uni-frankfurt.de (M.S.); chaikuad@pharmchem.uni-frankfurt.de (A.C.); knapp@pharmchem.uni-frankfurt.de (S.K.)
4 Institute of Pharmaceutical Chemistry, Goethe University Frankfurt, Buchmann Institute for Molecular Life Sciences, Max-von-Laue-Straße 9, 60438 Frankfurt am Main, Germany
5 Frankfurt Cancer Institute (FCI) and German Translational Cancer Network (DKTK) Site Frankfurt/Mainz, 60438 Frankfurt am Main, Germany
6 Tübingen Center for Academic Drug Discovery (TüCAD2), 72076 Tübingen, Germany
* Correspondence: matthias.gehringer@uni-tuebingen.de
† These authors contributed equally to this work.

Abstract: The nonreceptor tyrosine TEC kinases are key regulators of the immune system and play a crucial role in the pathogenesis of diverse hematological malignancies. In contrast to the substantial efforts in inhibitor development for Bruton's tyrosine kinase (BTK), specific inhibitors of the other TEC kinases, including the bone marrow tyrosine kinase on chromosome X (BMX), remain sparse. Here we present a novel class of dual BMX/BTK inhibitors, which were designed from irreversible inhibitors of Janus kinase (JAK) 3 targeting a cysteine located within the solvent-exposed front region of the ATP binding pocket. Structure-guided design exploiting the differences in the gatekeeper residues enabled the achievement of high selectivity over JAK3 and certain other kinases harboring a sterically demanding residue at this position. The most active compounds inhibited BMX and BTK with apparent IC_{50} values in the single digit nanomolar range or below showing moderate selectivity within the TEC family and potent cellular target engagement. These compounds represent an important first step towards selective chemical probes for the protein kinase BMX.

Keywords: tyrosine kinases; bone marrow tyrosine kinase on chromosome X; Bruton's tyrosine kinase; Janus kinase 3; covalent inhibitors; chemical probes

1. Introduction

The TEC kinase family constitutes the second largest family of nonreceptor tyrosine kinases including the five members Bruton's tyrosine kinase (BTK), interleukin (IL)-2-inducible T-cell kinase (ITK or TSK/EMT), tyrosine kinase expressed in hepatocellular carcinoma (TEC), bone marrow tyrosine kinase on chromosome X (BMX or ETK) and tyrosine-protein kinase (TXK or RLK) [1]. All TEC family members except TXK commonly harbor an *N*-terminal pleckstrin homology (PH) domain required for membrane association via binding to phosphatidylinositols. The PH domain is followed by a TEC homology (TH) domain which includes a BTK-homology motif and two proline-rich regions in BTK, ITK and TEC, one of which is absent in BMX. In contrast to the other TEC family members, TXK is devoid of the PH and TH domains and membrane targeting is promoted by a palmitoylated cysteine string motif (CC) instead. All Tec family kinases also contain two SRC homology domains, SH3 and SH2, preceding the highly conserved C-terminal kinase domain [2]. This makes TEC kinases closely related to the SRC kinase family, which represents the largest family of nonreceptor tyrosine kinases.

TEC kinases differ in their expression pattern. While TEC and BMX are widely expressed in hematopoietic and endothelial cells, expression of BTK appears to be more restricted, mainly to myeloid and B-cells. On the contrary, ITK and TXK are predominantly expressed in T-lymphocytes [1]. Among the TEC kinases, BTK stands out as the best-studied family member due to its key function in B-cell receptor signaling required for B-cell development. Mutations in the *btk* gene are causative to human X-linked agammaglobulinemia (XLA) and murine X-linked immunodeficiency, which are characterized by a lack of mature B-lymphocytes and defective humoral immunity [3]. BTK is essential for the proliferation and survival of leukemic B-cells making this kinase not only a promising target for treatment of inflammatory and autoimmune diseases but also for B-cell malignancies. Currently, the three BTK inhibitors ibrutinib (**1**, Figure 1), acalabrutinib (**2**) and zanubrutinib (**3**) are approved for the treatment of various B-cell lymphoma and chronic graft-versus-host disease (only ibrutinib) and many more are in preclinical or clinical development [4–6]. Despite different selectivity profiles, these current drugs share a common mode of inhibition by covalently targeting BTK C481 located within the front pocket of the ATP binding site in the proximity of the αD helix. By comparison, a cysteine in this position exists also in other members of the TEC family and few additional kinases (*vide infra*).

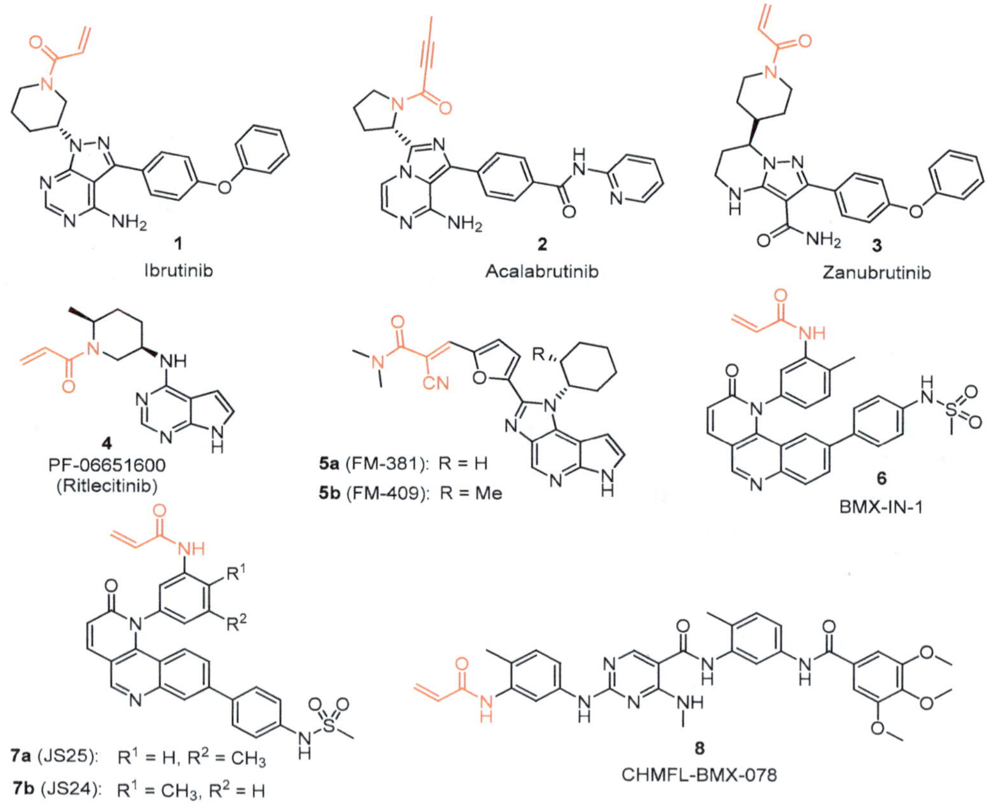

Figure 1. Examples of covalent Bruton's tyrosine kinase (BTK), Janus kinase 3 (JAK3) and bone marrow tyrosine kinase on chromosome X (BMX) inhibitors including the FDA-approved BTK-targeted drugs ibrutinib, acalabrutinib and zanubrutinib. The electrophilic warhead group is highlighted in red.

Contrary to BTK, the effort on inhibitor development for other TEC kinases has remained rather limited despite several lines of evidence suggesting important roles in hematopoiesis as well as their potential as therapeutic targets. For instance, the function of ITK is crucial in T-cell receptor signaling. As such, it is required for T-cell development and response, controlling the production of proinflammatory cytokines [7]. ITK has primarily been considered as a target in the treatment of inflammatory and autoimmune diseases; however, so far, no selective ITK inhibitors have entered clinical development [4]. The role of other TEC kinases, including BTK's closest relative BMX, in health and disease remains less clear. This highlights the need for specific chemical probes [8] to explore the biology of these understudied TEC kinases.

In recent years, targeting protein kinases with covalent inhibitors has become a mature strategy not only for the development of selective chemical probes, but also drugs [4,5]. Although protein kinases do not feature a catalytic nucleophile, a substantial fraction of human protein kinases are known to possess one or several accessible cysteine moieties inside or in close proximity to the ATP binding site [4,9–11]. These cysteines are located at diverse positions with a low degree of conservation. The mechanism of covalent binding typically follows a two-step process with an initial reversible binding event that positions the inhibitor's "warhead" in the spatial proximity of a reactive amino acid residue [12]. High affinity noncovalent interaction along with suitable preorientation between the target kinase and the covalent inhibitor in the first step are essential for efficiency of the subsequent covalent bond formation, and hence on-target specificity [13]. Thus, covalent targeting of a poorly conserved amino acid residue may be employed as an additional selectivity filter complementing reversible recognition. Such a strategy has been very valuable in the context of protein kinase inhibitor design, where achieving selectivity is intrinsically challenging due to the high degree of conservation within this target class, especially in the ATP binding site [4,5]. Besides the possible advantages in terms of selectivity, covalent inhibitors possess other potential benefits. For example, persistent target occupancy enables an increased, time-dependent potency, lowering a hurdle on ATP competition. Such property may also decouple pharmacodynamics from pharmacokinetics, which can be particularly useful in a drug discovery setting [14]. Moreover, initial concerns about the safety of targeted covalent inhibitor drugs have not materialized so far [15]. Beyond drug discovery, the attachment of reporter tags (e.g., fluorescent dyes, affinity labels or click handles) to covalent ligands yields very useful tool compounds to explore various facets of target biology [13].

The presence of a nonconserved cysteine within the solvent-exposed front region at the N-terminus of the αD helix, often referred to as the αD-1 position [9], in all five TEC kinases (C496 in BMX; C481 in BTK) opens an opportunity for irreversible targeting. However, cysteines at the αD-1 position exist also in six other kinases, namely most members of the HER family (EGFR, HER2 and HER4), the Janus kinase JAK3, the SRC kinase BLK and the mitogen activated protein kinase kinase MKK7 (MAP2K7). Currently, the αD-1 cysteine has been targeted by seven approved drugs, including the EGFR/pan-HER inhibitors afatinib, dacomitinib and osimertinib, the HER2 inhibitor neratinib and the aforementioned BTK inhibitors ibrutinib (**1**), acalabrutinib (**2**) and zanubrutinib (**3**) [4,5]. Covalent BTK inhibitors often feature a certain cross-reactivity with other TEC family members. For example, ibrutinib shows strong off-target (re)activity in the entire TEC kinase family, but also for other kinases sharing the αD-1 cysteine [16] while acalabrutinib is more specific [17]. Covalent JAK3 inhibitors have attracted considerable interest [18,19] since the αD-1 cysteine (C909) is the key feature distinguishing JAK3 from the other JAK family members, JAK1, JAK2 and TYK2. For example, the acrylamide-based JAK3 inhibitor PF-06651600 (Ritlecitinib, **4**, Figure 1) [20] is currently being investigated in several phase II and III clinical trials. In our own JAK3 program, we generated highly selective covalent-reversible inhibitors FM-381 (**5a**) and FM-409 (**5b**) [21,22]. The equivalent cysteine in MKK7 (C218) is modified by hypothemicin analogs [4] but also by recently discovered acrylamide-based inhibitors [16,23,24].

As exemplified by BTK, irreversible inhibition represents an excellent strategy for targeting TEC kinases. Nevertheless, the development of inhibitors for other members of this family remains challenging due to the subtly different nature of their binding sites as well as the lack of understanding on these pockets. Design of specific ITK inhibitors, for example, is hindered by lower cysteine reactivity [25] and the larger gatekeeper residue. Only few reports describing the development of selective inhibitors of the less prominent TEC kinases, namely TXK, TEC and BMX, have been published so far. While no inhibitors addressing TXK or TEC as the primary target are known, two structural series of covalent BMX inhibitors targeting C496 have been disclosed. In 2013, Gray and colleagues identified the benzo[*h*] [1,6]naphthyridin-2(1*H*)-one derivative BMX-IN-1 (**6**, Figure 1), a dual covalent BMX/BTK inhibitor [26]. This compound has recently been complemented by a series of analogs from Bernardes and co-workers [27] (preprint on Chemrxiv), which feature increased potency but limited selectivity against JAK3, BLK and within the TEC family. JS25 (**7a**), the most potent BMX inhibitor from this set (IC_{50} = 3.5 nM, k_{inact}/K_I = 19 $\mu M^{-1}s^{-1}$) showed increased cellular target engagement compared to **6** in a NanoBRET™ assay [28] while covalent binding has been demonstrated for the close analog JS24 (**7b**) by mass spectrometry and crystal structure. A structurally distinct series of covalent BMX inhibitors has been published in 2017 by Liu and colleagues [29]. These compounds exemplified by CHMFL-BMX-078 (**8**) showed high affinity for the inactive (nonphosphorylated) kinase while affinity for the active (phosphorylated) kinase was weak (apparent K_d > 10 μM). It is assumed that the latter compound binds BMX and related kinases in a type II binding mode addressing the hydrophobic back-pocket sometimes referred to as "deep pocket" [30], which is only accessible in the DFG-out conformation. However, there is still a need for expanding the scope of novel BMX/TEC family kinase inhibitors with high selectivity or complementary selectivity patterns to study biology. Here we present a novel and structurally unrelated series of covalent BMX/TEC-family kinase inhibitors featuring a distinct selectivity profile among the kinases with a αD-1 cysteine.

2. Results

2.1. Discovery of a New Class of Covalent JAK3/TEC Family Kinase Inhibitors

Our previous study demonstrated the successful use of the tricyclic imidazo [5,4-*d*]pyrrolo[2,3-*b*]pyridine scaffold for the design of selective covalent-reversible JAK3 probes FM-381 (**5a**, see Figure 1) and FM-409 (**5b**) targeting JAK3 C909 via an α-cyano acrylamide warhead [21,22]. Based on this, we aimed to expand the inhibitor series and explore their potentials as irreversible inhibitors for JAK3 and other kinases that harbor a cysteine at the equivalent position. With no success in generating acrylamide-derived irreversible inhibitors based on the aforementioned tricyclic scaffold, we tested alternative hinge binding motifs and synthesized 1*H*-pyrrolo[2,3-*b*]pyridine-derived compound **9a** (Figure 2, left). This derivative retained excellent JAK inhibitory potency (IC_{50} = 0.2 nM) and high selectivity against the other JAK family members and EGFR, demonstrating the hinge binder replacement as a potential strategy for optimizing the parent inhibitor towards other kinases (Table 1).

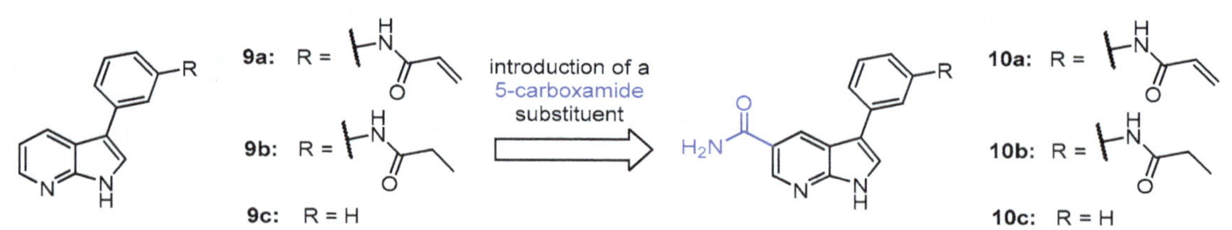

Figure 2. Design of covalent JAK3 inhibitor **10a** from analog **9a** by introduction of a carboxamide group at the 5-postion of the 7-azaindole core.

Table 1. Selectivity of **9a** and **10a** in the JAK family and against EGFR.

Kinase	IC_{50} [nM] [1]	
	9a	**10a**
JAK1	2970	460
JAK2	3580	1120
JAK3	0.2	<0.1
TYK2	5540	1900
EGFR	5250	>10,000

[1] IC_{50} values were determined in a radiometric assay (Reaction Biology HotSpot™ [31]). Data were obtained as five-dose singlicate with 10-fold serial dilution starting at 10 μM. [ATP] = 10 μM.

We hypothesized that the 1*H*-pyrrolo[2,3-*b*]pyridine (7-azaindole) core in **9a** might adopt an inverted binding orientation when compared to **5a,b** or common 7*H*-pyrrolo[2,3-*d*]pyrimidine derived JAK inhibitors like **4** with two hydrogen bonds between the 7-azaindole and the backbone of L905 (Figure 3). On this basis, we introduced a carboxamide group in the 5-position of the 7-azaindole scaffold to establish a third hydrogen bond towards the hinge region [32]. As expected, the resulting compound **10a** (Figure 2, right) exhibited an increased potency for JAK3 (IC_{50} < 0.1 nM) while maintaining a similar degree of selectivity against other JAKs. Interestingly, we observed that this compound demonstrated increased selectivity against EGFR compared to **9a**. In line with covalent binding, nonreactive analogs **9b,c** and **10b,c** showed a strongly decreased potency with JAK3 IC_{50}'s in the micromolar range (not shown). The determined crystal structures of **9a** and **10a** (Figure 4) in complex with JAK3 confirmed the predicted hinge interaction pattern and covalent modification of C909. As expected, the 7-azaindole core in both compounds was anchored to the hinge region via two hydrogen bonds with the L905 backbone NH and carbonyl oxygen atom, respectively. Compound **10a** established an additional hydrogen bond between a carboxamide NH and the E903 backbone carbonyl oxygen atom. The second amide proton of the 5-carboxamide was involved in a presumably weak interaction with the methionine sulfur atom. The warhead amide was present in two flipped conformations (see Figure 4b) and involved in water-mediated hydrogen bonds, most notably with the backbone of C909. The latter interactions may assist in pre-orienting the Michael acceptor to facilitate cysteine addition.

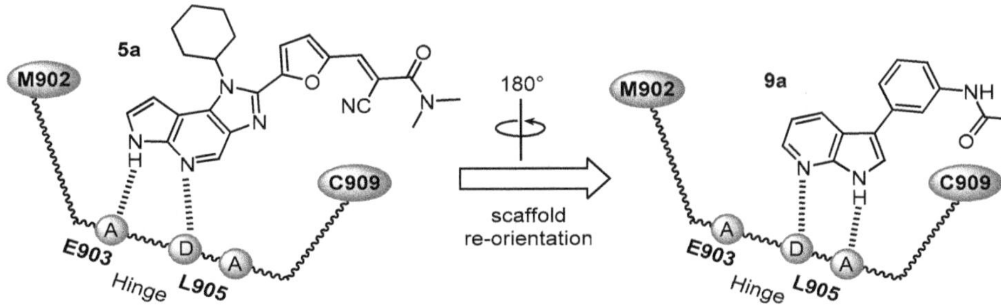

Figure 3. Differences in the hinge binding of different scaffolds exemplified by **5a** and **9a**. Usually, the protein backbone in the hinge region of a kinase provides three potential hydrogen bonding partners (two carbonyl-acceptors (A) and one NH-donor (D)). Depending on the orientation of the hinge binding heterocycle different hydrogen-bonding patterns are possible.

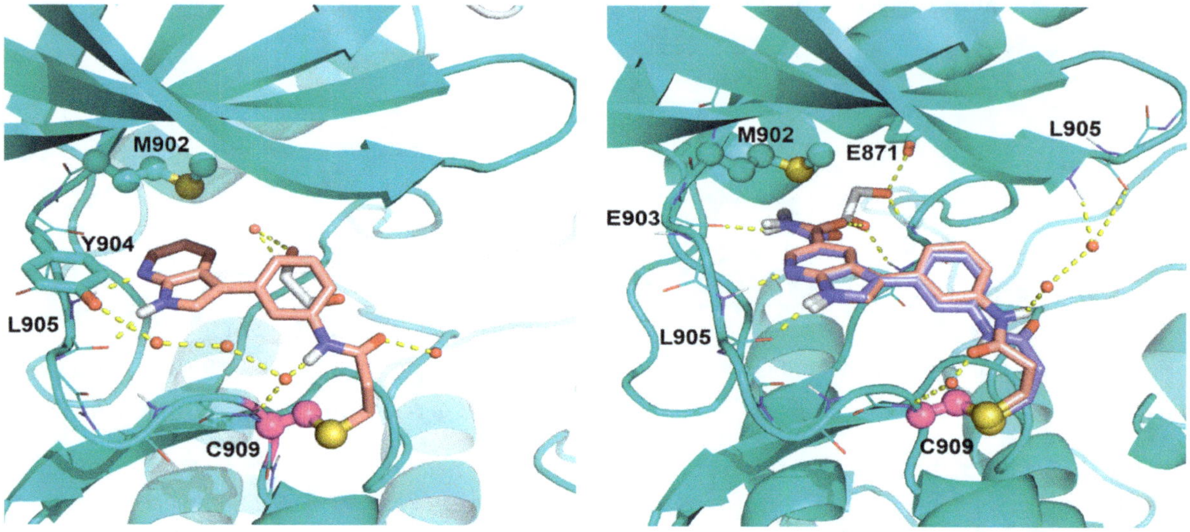

(a) (b)

Figure 4. X-ray crystal structures of compounds **9a** (PDB-code 7APG) and **10a** (PDB-code 7APF) in complex with JAK3. (**a**) Compound **9a** covalently bound to JAK3-C909. An ethylene glycol molecule (depicted in gray) fills the space between the ligand and the DFG motif forming a hydrogen bond to D967 but no direct hydrogen bonds with the ligand. Certain water-mediated hydrogen bonds as well as an alternative pose with a flipped amide conformation were omitted for clarity; (**b**) Compound **10a** covalently bound to JAK3-C909. Two poses with a flipped warhead amide conformation were observed (depicted in salmon and blue). The carbonyl group of the 5-carboxamide moiety forms a hydrogen bond to an ethylene glycol molecule in the back pocket, which is further anchored by hydrogen bonding to E871 in the αC-helix and to the DFG motif. Certain water-mediated hydrogen bonds were omitted for clarity.

To examine whether our replacement strategy of the hinge binding motif for expanding the profile of the JAK3 covalent inhibitors to other kinases succeeded, we characterized the activity of inhibitors **9a** and **10a** against all kinases featuring the αD-1 cysteine. This screening revealed that the compounds strongly inhibited BMX and TXK with no or less significant binding to other kinases at 200 nM (see Table 2). Notably, the observed selectivity pattern did not seem to rely on differences in the gatekeeper moiety which is methionine in JAK3 and MKK7 and a threonine in the HER and TEC family kinases (except ITK).

Table 2. Selectivity of **9a** and **10a** in among the protein kinases with an equivalent cysteine.

Kinase	Gatekeeper Residue	Residual Activity @ 200 nM [1]	
		9a	**10a**
BMX	T	2.7%	1.6%
TXK	T	1.8%	3.2%
BTK	T	18.2%	18.9%
TEC	T	37.1%	42.1%
ITK	**F**	88.1%	90.8%
JAK3	**M**	0.8%	0.5%
MKK7	**M**	37.3%	28.3%
EGFR	T	>100%	95.5%
HER2	T	95.0%	97.0%
HER4	T	66.1%	66.6%
BLK	T	95.8%	82.4%

[1] Residual activities were determined in a radiometric assay (Reaction Biology HotSpot™ [31]). Data were obtained from duplicate measurements and reported as the arithmetic mean of the individual values. [ATP] = 10 μM.

2.2. Evaluation of N-Substitution in the 5-Carboxamide Series

To exploit the steric nature of the gatekeeper moiety as an additional selectivity filter, we introduced *N*-substituents at the 5-carboxamide's nitrogen atom to preserve the triple hydrogen bonding interaction with the hinge region while forcing a steric clash with the bulky methionine gatekeeper in JAK3 or MKK7 and the phenylalanine in ITK. The first analog from this series with a relatively bulky *N*-cyclopentyl substituent (**11a**, Figure 5) did not show substantial activity at 200 nM on any of the kinases harboring an αD-1 cysteine. Nevertheless, we determined the IC_{50} values for JAK3, BMX and TXK to compare acrylamide **11a** with nonreactive propionamide **11b** (Table 3). While **11a** still showed residual activity on BMX, no JAK3 inhibitory activity was observed (IC_{50} > 10 μM). In contrast, analog **11b** did not inhibit any of the latter kinases up to 10 μM.

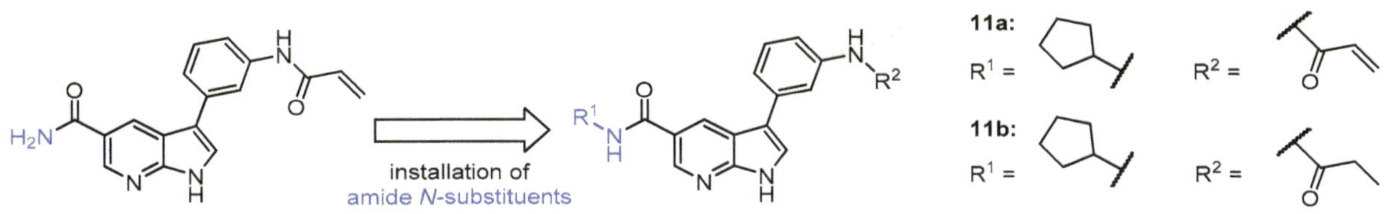

Figure 5. *N*-alkylation of the 5-carboxamide moiety.

Table 3. Inhibitory activity of acrylamides **10a** and **11a** and unreactive analog **11b**.

Kinase	IC_{50} [nM] [1]		
	10a	**11a**	**11b**
BMX	29	539	>10,000
TXK	28	>1000 [2]	>10,000
JAK3	<0.1	>10,000 [3]	n.d.

[1] IC_{50} values were determined in a radiometric assay (Reaction Biology HotSpot™ [31]). Data were obtained as five-dose singlicate with 10-fold serial dilution starting at 1 μM (for **10a/11a**) or 10 μM (**11b**). [ATP] = 10 μM. [2] 72% residual activity at 1 μM. [3] Determined in a JAK3 ELISA assay [33].

To test whether smaller *N*-substituents could prevent the loss in BMX potency, we prepared the corresponding *N*-methyl and *N*-ethyl analogs **11c** and **11d** (Table 4). In addition, we synthesized compounds **11e–g** to evaluate whether a methylene linker would be suitable to direct apolar cyclic moieties of variable size to the hydrophobic pocket behind the threonine gatekeeper often termed "hydrophobic region I" [34]. The binding of these derivatives was assessed using a thermal shift (ΔT_m) assay, which unfortunately revealed a dramatic decrease of ΔT_m for all compounds in relation to the starting point **10a** (Table 4). These results suggested that the applied modification strategy was not suitable to generate potent BMX inhibitors with high selectivity against JAK3.

Table 4. Thermal stabilization of JAK3 and BMX by compound **10a** and *N*-alkylated analogs **11a** and **11c–g**.

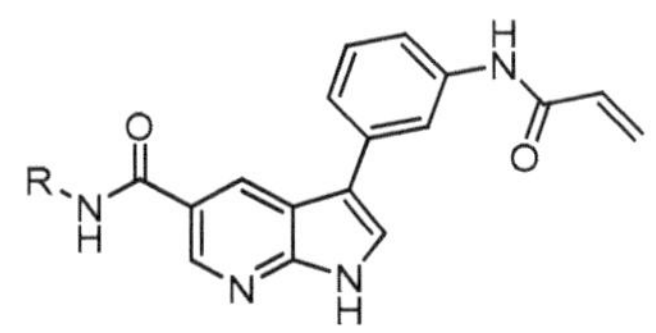

Compound	R	JAK3 ΔT_m [K] [1]	BMX ΔT_m [K] [1]
10a	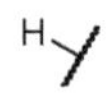	21.07 ± 0.03	10.49 ± 0.09
11a	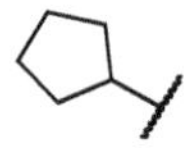	9.12 ± 0.02	5.31 ± 0.18
11c	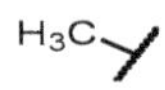	10.27 ± 0.37	7.84 ± 0.04
11d	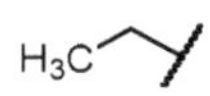	10.35 ± 0.26	7.35 ± 0.19
11e	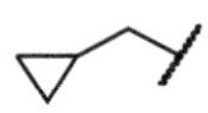	9.64 ± 0.28	6.94 ± 0.07
11f	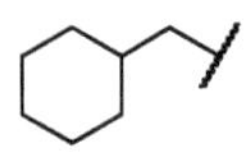	5.54 ± 0.82	3.68 ± 0.13
11g	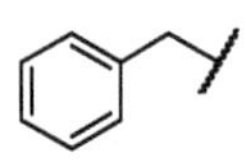	8.29 ± 1.26	6.84 ± 0.23

[1] Mean of at least three independent measurements ± standard deviation.

2.3. *Design and Structure–Activity Exploration of the 5-Acylamino Series*

In search for alternative approaches exploiting the gatekeeper residue as a selectivity filter, molecular modeling studies were performed. Docking simulations suggested that inverting the 5-carboxamide substituent would enable a favorable NH···O hydrogen bond to the hydroxy group of the threonine gatekeeper in BMX. While the third hydrogen bond to the hinge region would not be compatible with this arrangement, the interaction with the gatekeeper was predicted to direct moieties attached at the carbonyl group towards the hydrophobic region I behind the gatekeeper moiety (see Figure 6).

We thus prepared a set of analogs with an *N*-linked amide moiety (compounds **12a–i**, see Figure 6 and Table 5) to validate this hypothesis. Satisfactorily, the compounds from this series showed increased BMX thermal shifts while JAK3 thermal shifts remained in a moderate range suggesting low inhibitory potency on the latter. To confirm these results, the IC_{50} values of a subset of compounds were determined for BMX and JAK3. In agreement, all tested inhibitors exhibited high inhibitory activities for BMX with IC_{50} values in a low nanomolar range. Moreover, all compounds displayed excellent selectivity against JAK3, which was most pronounced for the most active analog **12c** (>9000-fold selectivity).

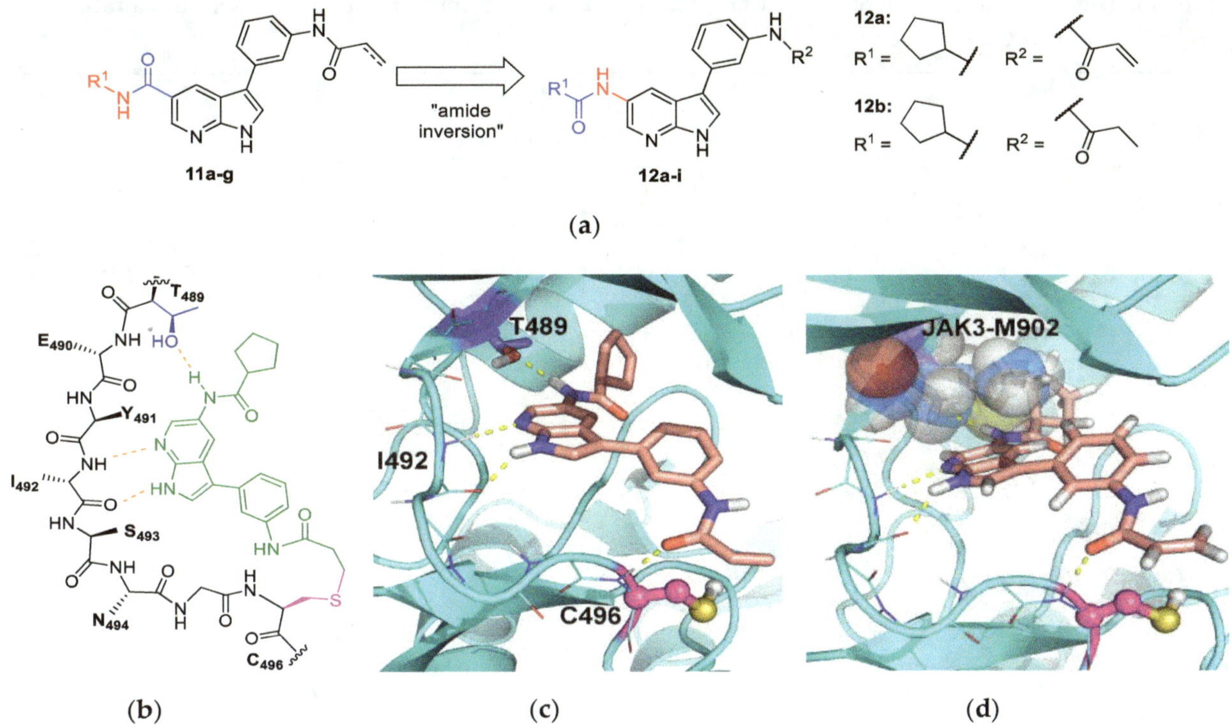

Figure 6. Amide inversion as a strategy to rescue BMX inhibitory activity. (**a**) General strategy of amide inversion and representative analogs **12a** and **12b**; (**b**) Schematic depiction of the expected binding mode of analog **12a**; (**c**) Representative noncovalent docking pose of compound **12a** modelled into the BMX active site using the BMX structure with the PDB-code 3SXR (BMX in complex with dasatinib) as a template. Similar poses were obtained using an alternative crystal structure (PDB-code 3SXS) in which the P-loop region is completely resolved. (**d**) Overlay of the docking pose of **12a** with the crystal structure of JAK3 (7APF) highlighting the clash with the JAK3 gatekeeper moiety. Only the gatekeeper methionine (transparent spheres) of the JAK3 structure is shown for clarity.

We selected potent cyclopentanoic amide **12a** and compound **12c** with a thiophene 2-carboxylic acid amide moiety for further profiling against the kinases that have an equivalent cysteine at the αD-1 position. At a concentration of 200 nM, **12a** strongly inhibited BMX, BTK and TXK while TEC, BLK, HER4 and EGFR were less affected (Table 6). As expected, no inhibition of JAK3, MKK7 and ITK was observed, which may be attributed to their bulkier gatekeeper incapable of forming the predicted hydrogen bond to the inverted amide moiety. In comparison, we observed that compound **12c** had a slightly better potency than **12a**, yet both shared a similar selectivity profile. Moreover, saturated analog **12b** expectedly showed a much weaker BMX inhibition with an IC_{50} value of 582 nM while being devoid any significant activity on other kinases in this set except BTK and TXK (70% and 89% residual activity at 200 nM, respectively). For a better selectivity ranking, we determined IC_{50} values of **12a** and **12c** against all kinases that were significantly inhibited at 200 nM (Table 6). The results confirmed that both compounds were highly potent inhibitors of BMX demonstrated by IC_{50} values of 2 nM and 1 nM, respectively. However, these inhibitors also inhibited BTK with similar potencies, suggesting that the moderate selectivity against BTK observed earlier for the parent compounds **9a** and **10a** was unfortunately compromised. Nevertheless, moderate selectivity of **12a** and **12c** over the other kinases in this test set, which included TXK, TEC, EGFR, HER2, HER4 and BLK, was evident and both had a slightly different selectivity profile. In comparison to **12c**, which exhibited 8–30-fold lower potencies for the tested kinases (>100-fold only for HER2), compound **12a** showed a better selectivity against HER family kinases (72-, 280- and 48-fold against EGFR, HER2 and HER4, respectively), TEC (14-fold) and BLK (24-fold), while selectivity against TXK (6-fold) remained moderate.

Table 5. Thermal stabilization of compounds **12a** and **12c–i** and inhibitory potency of selected inhibitors on JAK3 and BMX.

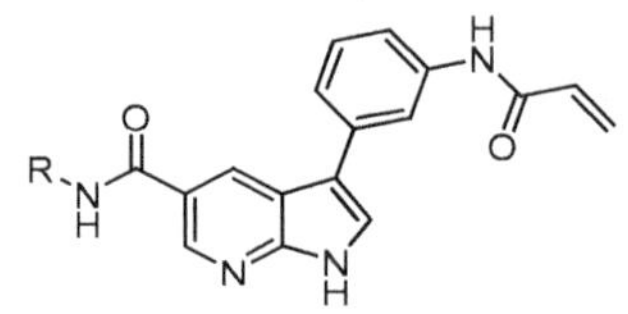

Compound	R	JAK3 ΔT_m [K] [1]	BMX ΔT_m [K] [1]	IC_{50} JAK3 [nM] [2]	IC_{50} BMX [nM] [2]
12a	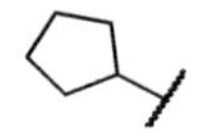	13.80 ± 0.13	11.92 ± 0.03	1340	2
12c	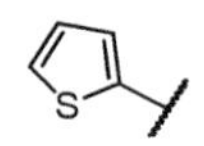	7.36 ± 0.44	11.52 ± 0.08	>10,000	1
12d	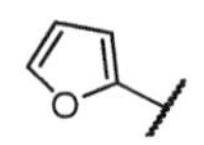	9.93 ± 0.12	10.75 ± 0.09	1080	4
12e	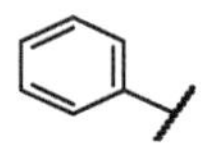	10.88 ± 0.95	11.37 ± 0.03	n.d.	n.d.
12f	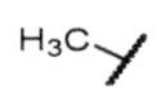	13.44 ± 0.43	10.57 ± 0.12	n.d.	n.d.
12g	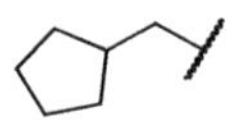	11.89 ± 0.23	11.88 ± 0.07	>10,000	6
12h	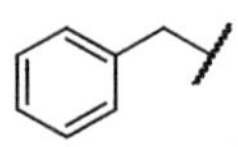	12.1	11.9	n.d.	n.d.
12i	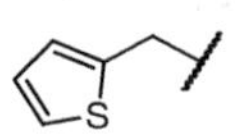	10.1	11.5	n.d.	16

[1] Mean of at least three independent measurements ± standard deviation. [2] IC_{50} values were determined in a radiometric assay (Reaction Biology HotSpot™ [31]). Data were obtained as five-dose singlicate with 10-fold serial dilution starting at 10 μM (JAK3) or 1 μM (**12d,g,i** vs. BMX). For **12a,c** vs. BMX, 5-fold serial dilution starting from 0.5 μM was used. [ATP] to 10 μM.

Table 6. Inhibitory activity of compounds **12a–c** on all protein kinases with an αD-1 cysteine.

Kinase	Gatekeeper Residue	Residual Activity @ 200 nM [1]			IC_{50} [nM] [2]		
		12a	**12b**	**12c**	**12a**	**12b**	**12c**
BMX	T	4.8%	n.d.	0.2%	2.2	582	1.1
TXK	T	5.9%	88.8%	0.3%	13	n.d.	10
BTK	T	2.2%	69.7%	0.8%	2.2	n.d.	<1
TEC	T	25.4%	99.2%	17.2%	31	n.d.	13
ITK	**F**	98.5%	n.d.	94.0%	n.d.	n.d.	n.d
JAK3	**M**	94.5%	n.d.	>100%	1340	n.d.	>10,000
MKK7	**M**	100%	n.d.	>100%	n.d.	n.d.	n.d.
EGFR	T	51.3%	>100%	14.9%	158	n.d.	12
HER2	T	81.9%	>100%	41.1%	612	n.d.	118
HER4	T	15.9%	95.4%	4.7%	106	n.d.	8.5
BLK	T	14.7%	95.1%	8.6%	52	n.d.	30

[1] Residual activities were determined in a radiometric assay (Reaction Biology HotSpot™ [31]). Data were obtained from duplicate measurements and reported as the arithmetic mean of the individual values. [ATP] = 10 μM. [2] IC_{50} values were determined in the above radiometric assay. Data were obtained as five-dose singlicate with 5-fold serial dilution starting at 0.5 μM (**12a,c** vs. BMX, BTK, TEC, TXK and HER4), 1 μM (**12a,c** vs. EGFR and HER2) or 10 μM (**12b** vs. BMX). A 10-fold serial dilution starting at 10 μM was employed for **12a,b** vs. JAK3. [ATP] = 10 μM.

Cellular target engagement was evaluated by a bioluminescence resonance energy transfer (NanoBRET™) assay [28] using HEK293T cells (see Figure 7). Lead compounds **12a** and **12c** showed favorable low nanomolar IC_{50} values against both, BMX and BTK with a slight bias towards BMX. High potency was also observed for compounds **12g,i** and early hit compound **9a** while the close analog **10a** showed decreased potency, which might be a result of lower cell penetration due to the primary amide functionality. As expected, and in line with the data from enzymatic assays, *N*-alkylated analog **11a** and non-reactive control compounds **9b**, **10b**, **11b** and **12b** showed weak or no affinity for both kinases in this cellular model.

(a)

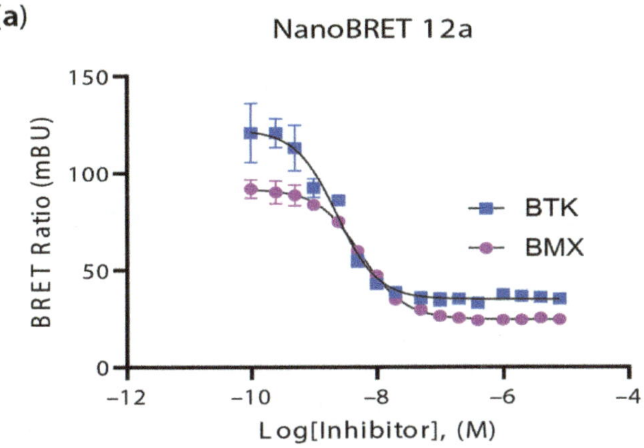

(b)

Compounds	IC_{50} (μM) BMX	IC_{50} (μM) BTK
9a	0.026 ± 0.011	0.015 ± 0.002
9b	>10	9.22 ± 3.70
10a	0.765 ± 0.283	0.283 ± 0.024
10b	>10	>10
11a	6.066 ± 1.830	2.068 ± 0.032
11b	>10	3.705 ± 1.204
12a	0.003 ± 0.001	0.006 ± 0.001
12b	>10	8.207 ± 3.461
12c	0.003 ± 0.001	0.009 ± 0.001
12g	0.003 ± 0.001	0.013 ± 0.002
12i	0.017 ± 0.007	0.037 ± 0.010

Figure 7. (**a**) Exemplified dose–response curve of **12a** titrated versus BTK (blue squares) or BMX (magenta circles) in a NanoBRET™ assay using HEK293T cells. Each point represents the average of triplicates with standard deviation. For dose response curves of the other inhibitors tested, see the supporting information. (**b**) IC_{50} values determined in NanoBRET™ assays in HEK293T cells.

2.4. Validation of Covalent Modification and Binding Mode Prediction

Covalent adduction between BMX and the inhibitors was investigated by mass spectrometry (see Supplementary Table S1). When incubating highly potent acrylamide-derived inhibitors **9a**, **10a** and **12a,c,g,i** at 1.5-fold molar excess with BMX at 4 °C, complete labeling was observed readily after 30 min indicating a high efficiency of covalent bond formation. In contrast, analog **11a**, which was less potent in the activity assay required prolonged exposure (240 min) to approach full target modification. As expected, no modification was detectable for propionamides **9b**, **10b**, **11b** and **12b**, which were employed as negative controls. To examine general reactivity towards thiols, compound **12c** was tested for its stability against glutathione (GSH). In the presence of 5 mM GSH at pH 7.4, **12c** showed a half-life of approx. 10 h which compares favorably to Afatinib (<1 h) tested under the same conditions (see Supplementary Figure S2). It can therefore be concluded that covalent target modification is facilitated by the preceding reversible binding event and not a result of nonspecific thiol addition. To gain a detailed insight in binding interactions, we aimed to crystallize the complex of BMX and the inhibitors, yet unsuccessfully. Thus, covalent docking analyses were instead performed. Suggested covalent binding modes of the two representative compounds **12a** and **12c** are depicted in Figure 8. As expected, the docking poses show the dual hinge interaction pattern along with an additional hydrogen bond between the amide NH and the hydroxy group of T489 orienting the nonpolar substituent at the 5-acylamino group towards the hydrophobic region I.

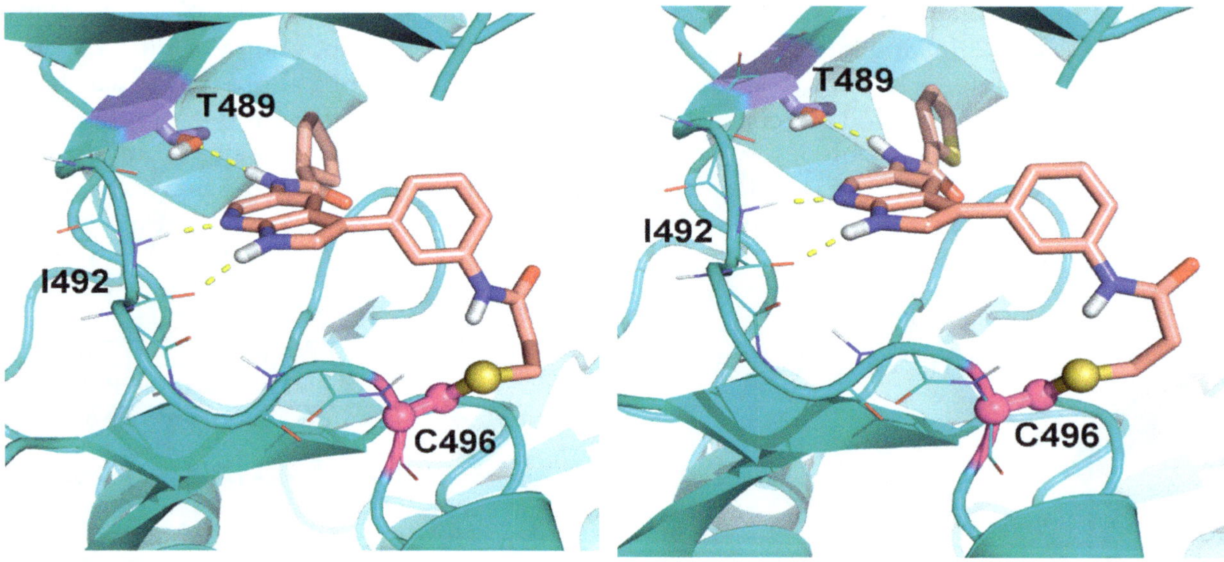

(a) (b)

Figure 8. Covalent docking models of compounds **12a** (panel (**a**)) and **12c** (panel (**b**)). Poses were generated using the BMX crystal structure with the PDB-code 3SXR as the template.

Although we did not experimentally prove selective modification of C496 by tryptic digestion/MS experiments, crystal structure or mutation of the target cysteine, the efficiency and stoichiometry of covalent modification in conjunction with modeling data and the binding modes of analogs **9a** and **10a** in JAK3 strongly support the specific labeling of this residue, which is the only accessible cysteine in or proximal to the BMX ATP binding site. In addition, much lower activity of non-reactive analog **12b** compared to acrylamide **12a** on all enzymes tested complies with a covalent mode of action on other kinases with an equivalent cysteine placement.

2.5. Compound Synthesis

The key transformation in the synthetic route to the inhibitors disclosed herein was the late stage introduction of the entire *N*-aryl acrylamide warhead/linker fragment via a Suzuki coupling. This reaction could be performed under very mild conditions preventing possible side reactions involving the electrophilic Michael acceptor system. The synthesis of the key boronic acid ester **14** was realized as a high yielding two step sequence starting from 3-bromoaniline. The latter was smoothly converted to the pinacol boronate **13** under Miyaura conditions [35] followed by the acylation with acryloyl chloride to deliver the bench stable building block **14** in a good overall yield of 71% (Scheme 1, top).

For the facile derivatization in position 5 of the 7-azaindole scaffold, the corresponding 5-amino and 5-carboxy functionalized key intermediates **18** and **22** were prepared. The synthesis of **18** started with the SEM-protection of commercially available 5-bromo-7-azaindole (**15**) to deliver compound **16**. The carboxy group was then introduced via lithium–halogen-exchange followed by quenching the intermediate aryllithium species with gaseous carbon dioxide at low temperatures. The treatment with elemental bromine in methylene chloride converted acid **17** cleanly to the desired intermediate **18** in excellent yields (Scheme 1, middle).

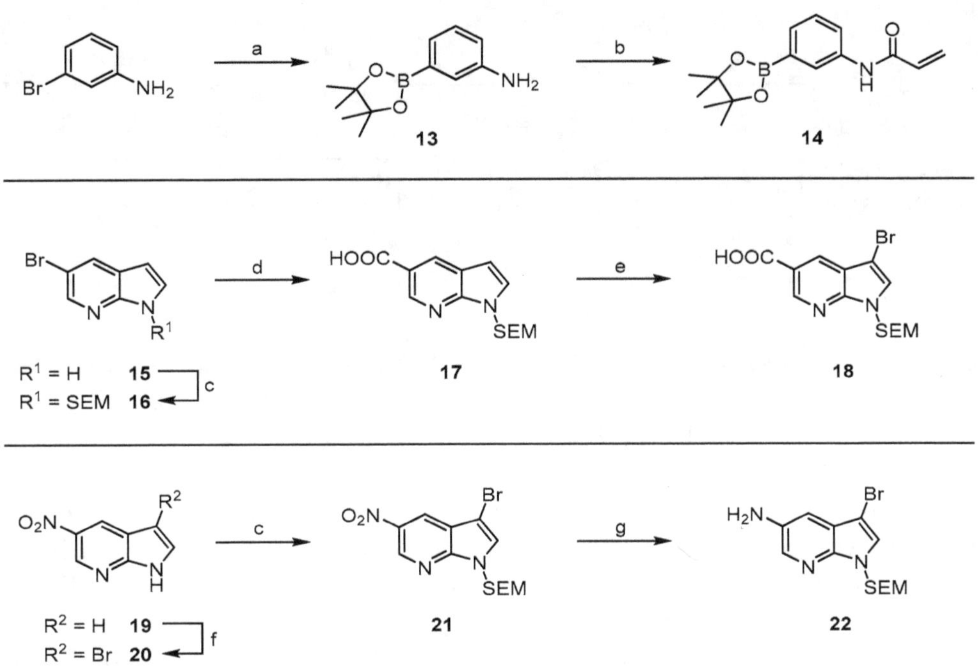

Scheme 1. Synthesis of key building blocks: (a) B_2pin_2, XPhos Pd G4, KOAc, dioxane, 90 °C (84%); (b) acryloyl chloride, Et_3N, DCM, −10°C (85%); (c) NaH, SEM-Cl, DMF, 0°C to rt (56–78%) (d) *n*-BuLi, CO_2, THF. −78 °C (51%); (e) Br_2, DCM, 0°C (89%); (f) NBS, DMF, rt (70%); (g) Zn^0, ammonium formate, EtOH, 50 °C (51%).

The corresponding 5-amino derivative **22** was prepared from 5-nitro-7-azaindole **19**, which is accessible via a previously reported scalable three-step process [36]. Starting from intermediate **19**, the 3-bromo substituent was introduced via electrophilic bromination with NBS (**20**) followed by the SEM-protection of the indole nitrogen atom (**21**). Finally, the nitro group was reduced under mild conditions with zinc and ammonium formate to obtain the key building block **22** (Scheme 1, bottom).

The early hit compounds **9a** and **10a** and their corresponding unsubstituted phenyl analogs **9c** and **10c** were synthesized following slightly modified routes (Schemes 2 and 3). Starting from plain 7-azaindole, the bromination of position 3 was performed with NBS to give compound **23** in almost quantitative yield. In the following step, the indole nitrogen atom was protected with a tosyl group via deprotonation with sodium hydride and reaction with tosyl chloride to obtain intermediate **24**. The latter was coupled with phenylboronic acid or building block **14** under Suzuki conditions to yield the compounds **25** and **26**, respectively. Tosyl deprotection under basic conditions in methanol or *tert*-butanol, respectively, delivered the final compounds **9c** and **9a**.

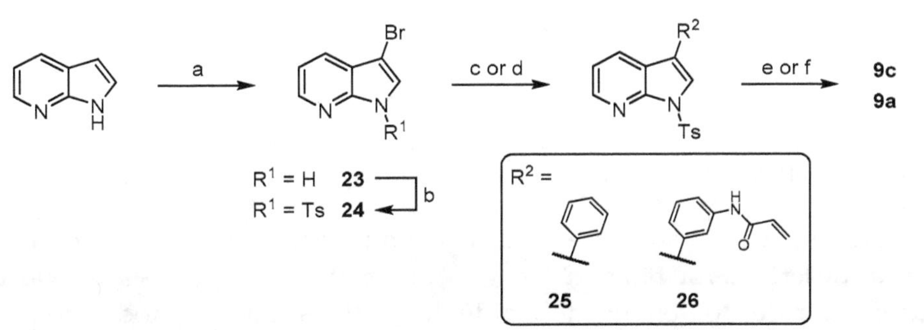

Scheme 2. Synthesis of compounds **9c** and **9a**. (a) NBS, DMF, rt (96%); (b) NaH, TsCl, THF, 0 °C to rt (85%); (c) phenylboronic acid, $Pd(OAc)_2$, XPhos, Na_2CO_3, dioxane/H_2O, 70 °C (82%); (d) **14**, XPhos Pd G3, K_2CO_3, dioxane/H_2O, 50 °C (quant.); (e) for **9c**: KOH, MeOH, 60°C (85%); (f) for **9a**: KOH, *t*BuOH, 50 °C (81%).

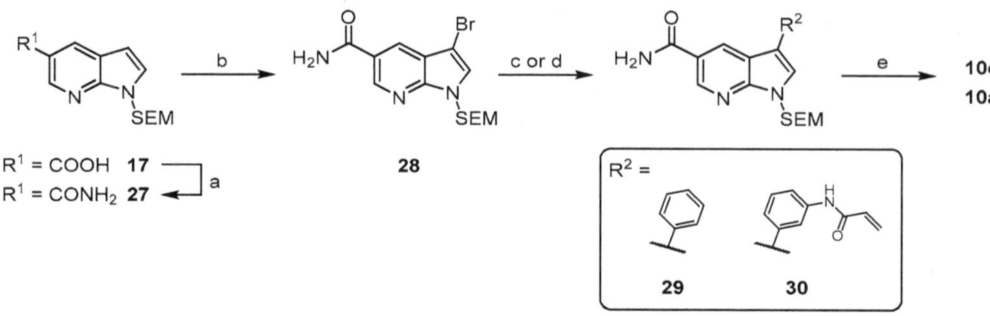

Scheme 3. Synthesis of compounds **10c** and **10a**. (a) CDI, $NH_{3(aq)}$, DMF, rt (72%); (b) Br_2, DCM, 0 °C (68%); (c) phenylboronic acid, tBu$_3$P Pd G3, K_3PO_4, dioxane/H_2O, rt (quant.); (d) **14**, tBu$_3$P Pd G3, K_2CO_3, dioxane/H_2O, rt (93%); (e) TFA, DCM, rt (71–92%).

The corresponding derivatives with an unsubstituted carboxamide in position 5 (**10c** and **10a**) were accessible from the SEM-protected acid **17**. CDI-mediated coupling of the latter with ammonia led to carboxamide **27**. Bromination in position 3 with elemental bromine afforded intermediate **28**, which was then Suzuki-coupled with phenylboronic acid or building block **14**, respectively. The precursors **29** and **30** were finally deprotected under acidic conditions with TFA in DCM to obtain compounds **10c** and **10a** (Scheme 3).

The series of *N*-substituted azaindole 5-carboxamides (compounds **11a** and **11c–g**) was accessible via a three-step procedure starting from key intermediate **18** (Scheme 4). First, amides **31a–g** were prepared via CDI-mediated acid activation and reaction with the corresponding amines. The latter intermediates were then coupled with **14** under mild Suzuki conditions at ambient temperature to give **32a–g** in favorable yields. The final compounds **11a** and **11c–g** were obtained after acid-promoted SEM-cleavage using TFA in DCM at ambient temperature (Scheme 4).

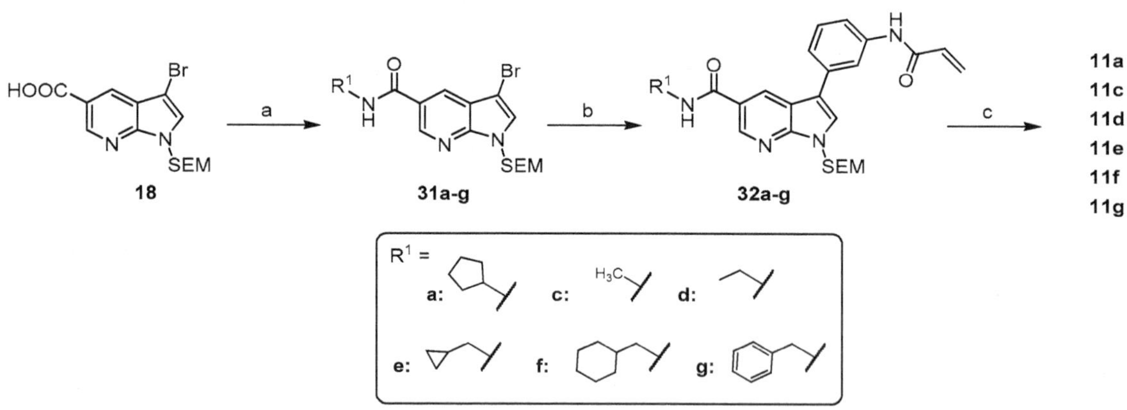

Scheme 4. Synthesis of *N*-alkylated 7-azaindole-5-carboxamides. (a) CDI, corresponding amine, DMF, rt (46–80%); (b) **14**, tBu$_3$P Pd G3, K_3PO_4 or K_2CO_3, dioxane/H_2O, rt (53–79%); (c) TFA, DCM, rt (23–97%).

The final inhibitor series **12a** and **12c–i** featuring an inverted amide functionality was prepared via a similar route as described for **11a** and **11c–g** (Scheme 5). Building block **22** was reacted with the corresponding acid chlorides to deliver the amide derivatives **33a–i** in moderate to excellent yields. The Suzuki coupling with boronate ester **14** was performed under the mild conditions established before to afford precursors **34a–i**. The latter were converted to the final compounds **12a** and **12c–i** under the same acidic conditions as described above (Scheme 5).

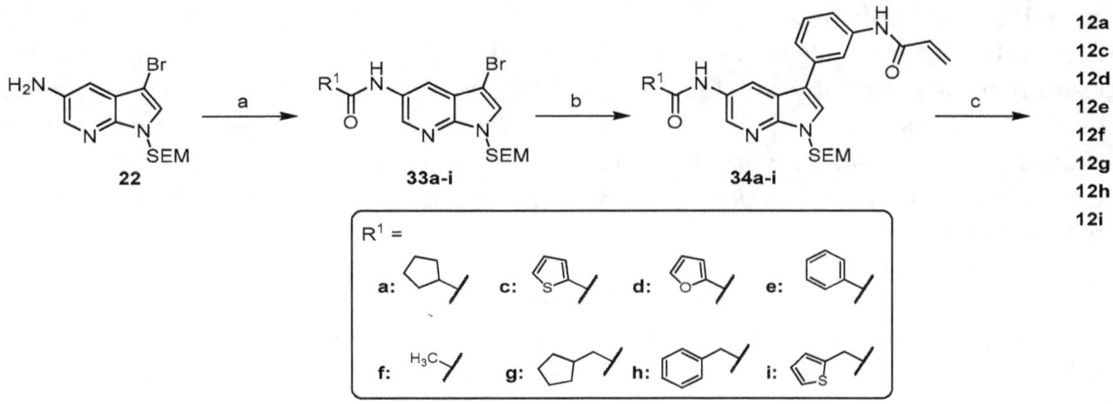

Scheme 5. Synthesis of 5-acylamino-7-azaindoles. (a) corresponding acid chloride, Et_3N, DCM, rt (47–90%); (b) **14**, tBu$_3$P Pd G3, K_3PO_4 or K_2CO_3, dioxane/H_2O, rt (45–73%); (c) TFA, DCM, rt (29–77%).

To access the propionamide analogs of selected inhibitors as negative control compounds, the corresponding acrylamides were hydrogenated in the presence of a Pd/C catalyst. Due to the absence of sensitive functional groups, these conversions usually proceeded smoothly and gave the desired propionamides **9b**, **10b**, **11b** and **12b** in good to excellent yields (Scheme 6).

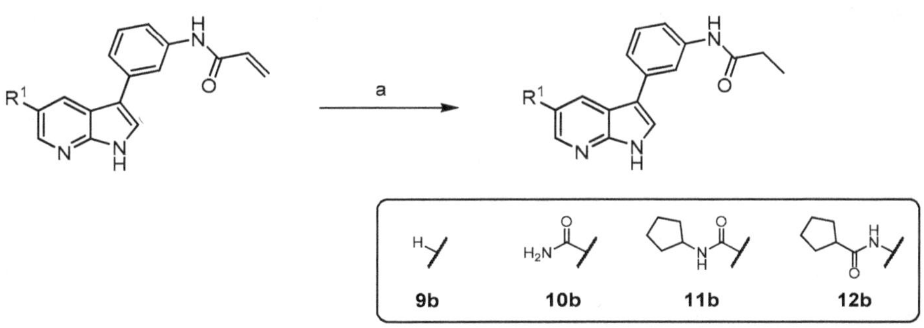

Scheme 6. Synthesis of propionamides **9b**, **10b**, **11b** and **12b** as negative controls. (a) H_2 (1 bar), Pd/C, THF/MeOH, (60–95%).

3. Discussion

Here we describe the discovery of a novel class of covalent inhibitors of the TEC family kinases, most notably BMX and BTK. Development started from compounds **9a** and **10a** belonging to an unprecedented class of covalent JAK3 inhibitors, which possessed potent off-target activity on the TEC kinases BMX and TXK while showing moderate to high selectivity against the other eight protein kinases with an equivalent cysteine placement. Guided by the crystal structures of **9a** and **10a** in complex with JAK3, modification of the hinge binding motif as well as back pocket binding moieties was successfully exploited as a strategy to fine tune the binding profiles of these irreversible inhibitors that target a cysteine located at the αD-1 position. This approach enabled the development of irreversible inhibitors for BMX and BTK, which showed a general preference for kinases with a less bulky threonine gatekeeper residue, enabling high selectivity over JAK3 and MKK7 that harbor a methionine, and ITK that possess a phenylalanine at this position.

The key step in the synthetic access to these inhibitors was the late stage introduction of the *N*-phenyl acrylamide warhead via Suzuki coupling under exceptionally mild conditions. This facilitated broad SAR evaluation and fast library synthesis. Starting from unsubstituted carboxamide **10a**, we introduced *N*-alkyl substituents to induce a clash with bulkier gatekeeper moieties, but this modification led to a drop in potency on both JAK3 and BMX. In contrast, inversion of the carboxamide in the 5-position of the azaindole scaffold retained BMX inhibitory potency, while activity on JAK3 was strongly decreased. Besides the expected steric clash with the methionine gatekeeper moiety in JAK3, molecular modeling suggested inverted amides to form an additional hydrogen bond towards the hydroxy group of threonine gatekeeper moieties, which drives the carbonyl's substituent towards the hydrophobic region I.

The most promising compounds **12a** and **12c** were further profiled against the TEC kinase family and other kinases with an αD-1 cysteine. As expected, low activity was detected on kinases with a non-threonine gatekeeper. However, the promising selectivity pattern observed for starting compounds **9a** and **10a** was partially lost. Most notably, optimization was accompanied by a strong increase in BTK potency with compound **12a** being equipotent on BMX and BTK, while **12c** slightly favored BTK with subnanomolar inhibitory activity in the enzymatic assay. Moderate selectivity was observed against the other TEC kinases and BLK. However, compound **12c** also showed significant off-target activity on HER family kinases while **12a** maintained good selectivity against the latter. Comparison with unreactive analog **12b**, which showed weak activity on all tested kinases underlined the importance of the covalent mode of action. Highly efficient covalent modification of BMX was experimentally shown for several analogs via mass spectrometry while potent cellular BMX and BTK engagement was demonstrated by means of a NanoBRET^{TM} assay.

Targeting BMX has been proposed as a strategy for treatment of various diseases including cardiovascular disorders [37] and certain cancers [26,38,39]. Nevertheless, only three inhibitors have been developed to date: BMX-IN-1 (**6**, see Figure 1), the analog JS25 (**7a**) and the type II inhibitor CHMFL-BMX-078 (**8**). However, these inhibitors still exhibit pronounced affinities to several other kinases with different selectivity profiles. Off-target activities could thus limit their use as a sole tool for biological study. Moreover, their chemical structures suggest unfavorable physicochemical properties due to size and the high portion of sp^2 atoms. The dual covalent BMX/BTK inhibitors presented here thus complement the repository of available BMX inhibitors. It should be mentioned, however, that caution must be taken when comparing covalent inhibitors based on (apparent) IC_{50} data since the latter depend on incubation time. For example, it has recently been demonstrated in the case of MKK7 and ibrutinib that a longer incubation could result in a misled nanomolar inhibitory activity observed for an inhibitor that has only weak reversible binding affinity to the kinase [16]. Nonetheless, determination of meaningful kinetic data, i.e., k_{inact} and K_I, to disentangle contributions of reversible binding and the subsequent bond-forming step remains highly elaborate, thus for BMX such values have only been reported for JS25 and a few related compounds from the same study [27]. Moreover, and in contrast to the aforementioned studies, the compounds presented herein have not been characterized in terms of wider kinome selectivity. Improvements also need to be made concerning the intra-TEC family selectivity. Future efforts will primarily focus on increasing the selectivity against BTK, for example by modifying the residues targeting the hydrophobic region I or by extending the latter towards the DFG-out pocket to generate type II inhibitors [16].

Overall, the presented optimization approach offers a strategy that can be exploited to fine tune existing covalent inhibitors towards unrelated kinases. Compared to previous compounds, the BMX inhibitors discovered here are based on an alternative and readily optimizable chemotype, which may be beneficial for further development of probes that exclusively target BMX. Currently, there is no selective inhibitor or chemical probe available for BMX. However, combinatorial use of our dual irreversible BMX/BTK inhibitors along with other inhibitors with different off-target profiles as a set of chemogenomic compounds would nevertheless allow dissecting biological functions of BMX as well as its role in disease development, which will enable further validation of this kinase as a potential therapeutic target.

4. Materials and Methods

4.1. Chemical Synthesis

4.1.1. General Information

Chemical synthesis was carried out using commonly applied techniques and general procedures. All starting materials and reagents were of commercial quality and were used without further purification. Thin layer chromatography (TLC) was carried out on Merck 60 F254 silica plates (Merck KGaA, Darmstadt, Germany) and were visualized under UV light (254 nm and 366 nm) or developed with an appropriate staining reagent. Preparative column chromatography was carried out with an Interchim PuriFlash 430 or PuriFlash XS420 (Interchim S.A., Montlucon, Allier, France) automated flash chromatography system on normal phase silica gel (Grace Davison Davisil LC60A 20–45 micron (W.R. Grace and Company, Columbia, MD, USA) or Merck Geduran Si60 63–200-micron silica (Merck KGaA, Darmstadt, Germany).

Nuclear magnetic resonance (NMR) spectral analysis was performed on Bruker Avance 200 or Bruker Avance 400 instruments (Bruker Corporation, Billerica, MA, USA). The samples were dissolved in deuterated solvents and chemical shifts are given in relation to tetramethylsilane (TMS). Spectra were calibrated using the residual peaks of the used solvent.

Mass spectrometry (MS) was carried out with an Advion TLC-MS interface (Advion, Ithaca, NY, USA) with electron spray ionization (ESI) in positive and/or negative mode. Instrument settings were as follows: ESI voltage 3.50 kV, capillary voltage 187 V, source voltage 44 V, capillary temperature 250 °C, desolvation gas temperature 250 °C, gas flow 5 L/min nitrogen.

Purities of final compounds were determined via high performance liquid chromatography (HPLC) using an Agilent 1100 Series LC system (Agilent Technologies, Santa Clara, CA, USA) with Phenomenex Luna C8 columns (150 × 4.6 mm, 5 μm) (Phenomenex Inc. Torrance, CA, USA) and detection was performed with a UV DAD at 254 nm and 230 nm wavelength. Elution was carried out with the following gradient: 0.01 M KH_2PO_4, pH 2.30 (solvent A), MeOH (solvent B), 40% B to 85% B in 8 min, 85% B for 5 min, 85% to 40% B in 1 min, 40% B for 2 min, stop time 16 min, flow 1.5 mL/min. All final compounds showed a purity above 95% in the means of peak area at the two different wavelengths.

4.1.2. General Synthetic Procedures

General Procedure A (Suzuki coupling): In a screw-top reaction vial were combined the corresponding boronic acid or ester and the aryl bromide under argon atmosphere. Dioxane and the aqueous base were added subsequently, and the mixture was degassed carefully via several vacuum/argon cycles. Finally, the catalyst system was added to the reaction and the degassing procedure was repeated. The vial was sealed and heated to the indicated temperature until reaction control (TLC or HPLC) showed complete consumption of starting materials. The reaction was cooled to ambient temperature, diluted with EtOAc and washed with brine. The organic phase was dried over Na_2SO_4 and evaporated to dryness. The residue was usually purified via flash chromatography using an appropriate solvent system.

General Procedure B (SEM cleavage): The SEM-protected substrate was dissolved in dry DCM and TFA was added subsequently. The reaction was stirred until TLC indicated complete consumption of the starting material. The majority of TFA was removed under reduced pressure and the residue was taken up in EtOH and was basified with aqueous NH_3 solution. The mixture was stirred at ambient temperature until complete consumption of the hydroxymethyl intermediate (usually overnight) was observed. The volatiles were stripped of under reduced pressure and the residue purified via flash chromatography as indicated.

General Procedure C (amide coupling with CDI): Carboxylic acid **18** was dissolved in dry DMF (ca. 0.1 M) and CDI was added at ambient temperature. The mixture was stirred until gas evolution ceased (usually about 1 h) before the corresponding amine was added to the reaction. Stirring was continued until reaction control indicated complete conversion and the reaction was quenched by addition of water followed by dilution with EtOAc. The organic phase was washed with sat. $NaHCO_3$ (two times) and brine, prior to solvent evaporation and purification by flash chromatography.

General Procedure D (amide coupling with acid chlorides): To a solution of **22** and Et_3N in dry DCM (ca. 0.1 M) was added the corresponding acid chloride under ice-cooling. After complete addition, the cooling bath was removed and stirring was continued at ambient temperature until reaction control (HPLC or TLC) indicated complete conversion. The reaction was quenched by addition of water followed by dilution with EtOAc. The organic phase was washed with sat. $NaHCO_3$, and brine, prior to drying over Na_2SO_4 and evaporation. The residue was purified via flash chromatography with an appropriate eluent system.

4.1.3. Synthesis and Characterization of Intermediates and Final Compounds

3-(4,4,5,5-tetramethyl-1,3,2-dioxaborolan-2-yl)aniline (**13**): In a flame-dried Schlenk flask were suspended 1.47 g dry KOAc (15.0 mmol), 860 mg 3-bromoaniline (5.00 mmol) and 1.27 g bis(pinacolato)diboron (5.00 mmol) in 20 mL dry dioxane. The mixture was carefully degassed via three vacuum/argon cycles and 9 mg XPhos Pd G4 (10 μmol) was added subsequently. The degassing procedure was repeated and then the reaction was heated to 90 °C oil-bath temperature overnight. After cooling to ambient temperature, the mixture was diluted with EtOAc and filtered over a bed of celite. The filtrate was washed with water (three times) and brine, prior to drying over Na_2SO_4 and evaporation. The residue was triturated with heptane and the precipitate isolated by filtration to yield 914 mg (84%) of the product as a slightly brownish solid. ^{1}H NMR (200 MHz, DMSO) δ 7.06–6.97 (m, 1H), 6.95 (d, J = 1.9 Hz, 1H), 6.82 (dt, J = 7.2, 1.1 Hz, 1H), 6.66 (ddd, J = 7.9, 2.4, 1.1 Hz, 1H), 5.02 (br s, 2H), 1.26 (s, 12H) ^{13}C NMR (50 MHz, DMSO) δ 148.0, 128.3, 128.3, 121.9, 120.1, 116.8, 83.2, 24.7 TLC-MS (ESI) m/z: 219.9 $[M + H]^+$ HPLC t_{ret} = 6.97 min.

N-(3-(4,4,5,5-tetramethyl-1,3,2-dioxaborolan-2-yl)phenyl)acrylamide (**14**): To a solution of 891 mg **13** (4.07 mmol) and 715 μL Et_3N (5.09 mmol) in 16 mL dry DCM were dropwise added 368 μL acryloyl chloride (4.48 mmol) at −10 °C. After 1 h, the reaction was quenched by addition of sat. NH_4Cl and transferred to a separatory funnel. The phases were separated and the organic phase was washed with sat $NaHCO_3$ and brine prior to drying over Na_2SO_4 and evaporation. The crude product was purified via flash chromatography (petrol ether/EtOAc + 5% MeOH (10–40%)) to obtain 780 mg (85%) of the title compound as a white solid. ^{1}H NMR (200 MHz, DMSO) δ 10.14 (br s, 1H), 7.98 (d, J = 1.4 Hz, 1H), 7.85 (dt, J = 6.3, 2.4 Hz, 1H), 7.41–7.26 (m, 2H), 6.43 (dd, J = 17.0, 9.7 Hz, 1H), 6.25 (dd, J = 17.0, 2.4 Hz, 1H), 5.75 (dd, J = 9.7, 2.4 Hz, 1H), 1.29 (s, 12H) ^{13}C NMR (50 MHz, $CDCl_3$) δ 163.1, 138.6, 131.8, 129.3, 128.4, 126.8, 125.3, 122.2, 83.7, 24.7 TLC-MS (ESI) m/z: 296.4 $[M + H]^+$ HPLC t_{ret} = 7.45 min.

5-bromo-1-((2-(trimethylsilyl)ethoxy)methyl)-1H-pyrrolo[2,3-b]pyridine (**16**): To an ice-cooled solution of 1.96 g 5-bromo-1*H*-pyrrolo[2,3-*b*]pyridine (10 mmol) in 10 mL dry DMF were added 518 mg sodium hydride (60%wt dispersion in min. oil, 12.9 mmol) and stirring was continued for about 30 min. Subsequently were added 1.95 mL SEM-Cl (10.9 mmol) and stirring was continued until TLC indicated complete conversion. The reaction was diluted with 120 mL EtOAc and transferred to a separatory funnel. The organic phase was successively washed with water (four times) and brine, prior to drying over Na_2SO_4 and evaporation. The residue was purified via flash chromatography (petrol ether/EtOAc (0–15%)) to obtain 2.55 g (78%) as a colorless oil. ^{1}H NMR (200 MHz, $CDCl_3$) δ 8.34 (d, J = 1.9 Hz, 1H), 8.02 (d, J = 1.9 Hz, 1H), 7.35 (d, J = 3.6 Hz, 1H), 6.46 (d, J = 3.6 Hz, 1H), 5.64 (s, 2H), 3.52 (t, J = 8.2 Hz, 2H), 0.89 (t, J = 8.2 Hz, 2H), −0.07 (s, 9H) ^{13}C NMR (50 MHz, $CDCl_3$) δ 146.6, 143.8, 131.1, 129.5, 122.3, 112.3, 100.7, 73.2, 66.5, 17.9, −1.3 TLC-MS (ESI) m/z: 349.3 $[M + Na]^+$ HPLC t_{ret} = 10.84 min.

1-((2-(trimethylsilyl)ethoxy)methyl)-1H-pyrrolo[2,3-b]pyridine-5-carboxylic acid (**17**): Under argon atmosphere were dissolved 2.49 g **16** (7.6 mmol) in 80 mL dry THF in a flame-dried Schlenk tube. At −78 °C, 3.3 mL of a 2.5 M *n*-BuLi solution in hexane (8.36 mmol) was added dropwise. After complete addition, stirring was continued for 30 min before dry CO_2 gas was bubbled through the solution.

After another 30 min, the reaction was quenched with water and slowly warmed to ambient temperature. After dilution with EtOAc, the organic phase was extracted with 0.5 M aqueous NaOH (three times). The combined extracts were acidified with 2M HCl and back-extracted with EtOAc (three times). The combined organic phases were washed with brine, dried over Na_2SO_4 and evaporated to dryness. The crude product was purified via flash chromatography (petrol ether/EtOAc + 5% MeOH + 2% AcOH (25–75%)) to yield 1.14 g (51%) of pure **17** as an off-white semisolid. ^{1}H NMR (200 MHz, $CDCl_3$) δ 10.39 (br s, 1H), 9.10 (s, 1H), 8.65 (s, 1H), 7.43 (d, *J* = 3.5 Hz, 1H), 6.64 (d, *J* = 3.5 Hz, 1H), 5.73 (s, 2H), 3.55 (t, *J* = 8.2 Hz, 2H), 0.90 (t, *J* = 8.2 Hz, 2H), -0.09 (s, 9H) ^{13}C NMR (50 MHz, $CDCl_3$) δ 171.4, 150.0, 145.8, 132.0, 129.8, 120.3, 118.7, 102.8, 73.3, 66.6, 17.8, −1.4 TLC-MS (ESI) *m*/*z*: 291.4 $[M-H]^-$ HPLC t_{ret} = 8.66 min.

3-bromo-1-((2-(trimethylsilyl)ethoxy)methyl)-1H-pyrrolo[2,3-b]pyridine-5-carboxylic acid (**18**): In a 100 mL round-bottomed flask were dissolved 1.14 g **17** (3.89 mmol) in 50 mL dry DCM and the solution was cooled in an ice/water bath. To this were added 684 mg bromine (4.28 mmol) as 0.5 M solution in DCM in small portions until the persistence of yellow coloration was observed. The reaction was quenched by addition of aqueous $Na_2S_2O_3$ solution and extracted with DCM (three times). The combined extracts were washed with sat. NH_4Cl, water and brine, prior to drying over Na_2SO_4 and evaporation. The residue was subjected to flash purification (petrol ether/EtOAc + 5% MeOH + 2% AcOH (25–75%)) to obtain 1.29 g (89%) of **18** as a brownish oil. ^{1}H NMR (200 MHz, $CDCl_3$) δ 11.16 (br s, 1H), 9.10 (s, 1H), 8.59 (s, 1H), 7.48 (s, 1H), 5.70 (s, 2H), 3.56 (t, *J* = 8.2 Hz, 2H), 0.92 (t, *J* = 8.2 Hz, 2H), −0.07 (s, 9H) ^{13}C NMR (50 MHz, $CDCl_3$) δ δ 171.1, 149.1, 146.8, 130.9, 128.7, 119.8, 119.3, 91.8, 73.2, 66.9, 17.8, −1.4 TLC-MS (ESI) *m*/*z*: 369.2 $[M-H]^-$ HPLC t_{ret} = 9.82 min.

3-bromo-5-nitro-1H-pyrrolo[2,3-b]pyridine (**20**): To an ice-cooled suspension of 1.8 g 5-nitro-1*H*-pyrrolo[2,3-*b*]pyridine (11.0 mmol) in 40 mL dry DMF were added 2.36 g *N*-bromosuccinimide (13.2 mmol) as solid in small portions. After complete addition, the cooling bath was removed and the reaction was stirred for 3 h at ambient temperature. The mixture was diluted with water and aqueous $Na_2S_2O_3$ and the precipitate was isolate by filtration. The filter cake was washed with water and dried at 60 °C in a convection oven to yield 1.86 g (70%) of the title compound as a yellow solid. ^{1}H NMR (200 MHz, DMSO-d_6) δ 12.95 (s, 1H), 9.18 (d, *J* = 2.5 Hz, 1H), 8.64 (d, *J* = 2.5 Hz, 1H), 8.09 (d, *J* = 2.6 Hz, 1H). TLC-MS (ESI) *m*/*z*: 239.1 $[M-H]^-$ HPLC t_{ret} = 7.73 min.

3-bromo-5-nitro-1-((2-(trimethylsilyl)ethoxy)methyl)-1H-pyrrolo[2,3-b]pyridine (**21**): To an ice-cooled solution of 2.14 g **20** (8.9 mmol) in 20 mL dry DMF were added 462 mg sodium hydride (60%wt dispersion in min. oil, 11.5 mmol) and stirring was continued for about one h. Subsequently were added 1.7 mL SEM-Cl (9.7 mmol) and stirring was continued until TLC indicated complete conversion. The reaction was diluted with 120 mL EtOAc and transferred to a separatory funnel. The organic phase was successively washed with water (four times) and brine, prior to drying over Na_2SO_4 and evaporation. The residue was purified via flash chromatography (petrol ether/EtOAc (0–20%)) to obtain 1.85 g (56%) **21** as a yellow oil. ^{1}H NMR (200 MHz, $CDCl_3$) δ 9.24 (d, *J* = 2.4 Hz, 1H), 8.73 (d, *J* = 2.4 Hz, 1H), 7.58 (s, 1H), 5.70 (s, 2H), 3.54 (m, 2H), 0.92 (m, 2H), −0.05 (s, 9H). HPLC t_{ret} = 11.56 min.

3-bromo-1-((2-(trimethylsilyl)ethoxy)methyl)-1H-pyrrolo[2,3-b]pyridin-5-amine (**22**): To a solution of 1.85 g **21** (5.0 mmol) in 40 mL abs. EtOH were added 1.63 g finely powdered zinc (24.9 mmol) and 1.57 g ammonium formate (24.9 mmol). The mixture was stirred for 22 h at 50 °C and was then filtered over a celite pad to remove unreacted zinc. The filtrate was concentrated under reduced pressure and taken up in EtOAc. The organic phase was washed with sat NH_4Cl and brine, prior to drying over Na_2SO_4 and evaporation. The residue was purified via flash chromatography (petrol ether/EtOAc 20–60%) to obtain 0.86 g (51%) of the title compound as a brown oil. ^{1}H NMR (200 MHz, $CDCl_3$) δ 7.93 (d, *J* = 2.6 Hz, 1H), 7.27 (s, 1H), 7.14 (d, *J* = 2.6 Hz, 1H), 5.55 (s, 2H), 4.60 (s, 2H), 3.51 (t, 2H), 0.88 (t, 2H), −0.08 (s, 9H). ^{13}C NMR (50 MHz, $CDCl_3$) δ 142.1, 137.5, 134.6, 127.2, 120.2, 112.5, 88.5, 72.9, 66.2, 17.7, −1.6. HPLC t_{ret} = 8.65 min.

3-bromo-1H-pyrrolo[2,3-b]pyridine (**23**): To an ice-cooled solution of 473 mg 1*H*-pyrrolo[2,3-*b*]pyridine (4.0 mmol) in 8 mL DMF were added 726 mg *N*-bromosuccinimide (4.1 mmol) in several portions. After complete addition, the cooling bath was removed and stirring was continued for 4 h. The reaction mixture was poured on sat. $NaHCO_3$/ice and the resulting suspension was stirred for ca. 10 min until a homogenous precipitate was formed. The solids were collected by filtration, washed with water and dried in vacuo to yield 753 mg (96%) of the title compound as a white solid. ^{1}H NMR (200 MHz, $CDCl_3$) δ 12.07 (br s, 1H), 8.29 (dd, J = 4.7, 1.5 Hz, 1H), 7.83 (dd, J = 7.9, 1.5 Hz, 1H), 7.70 (s, 1H), 7.16 (dd, J = 7.9, 4.7 Hz, 1H) ^{13}C NMR (50 MHz, DMSO) δ 147.2, 143.9, 126.4, 125.6, 118.7, 116.3, 87.1 TLC-MS (ESI) *m/z*: 197.1 $[M + H]^+$ HPLC t_{ret} = 6.56 min.

3-bromo-1-tosyl-1H-pyrrolo[2,3-b]pyridine (**24**): To an ice-cooled solution of 197 mg **23** (1.0 mmol) in 5 mL dry THF were added 50 mg sodium hydride (60%wt dispersion in min. oil, 1.25 mmol) and stirring was continued for about 30 min. Subsequently were added 210 mg TsCl (1.1 mmol) and stirring was continued until TLC indicated complete conversion. The reaction was diluted with 25 mL EtOAc and transferred to a separatory funnel. The organic phase was successively washed with water and brine, prior to drying over Na_2SO_4 and evaporation. The residue was triturated with chilled MeOH and filtered to obtain 298 mg (85%) **24** as a white solid. ^{1}H NMR (200 MHz, DMSO) δ 8.49–8.40 (m, 1H), 8.20 (s, 1H), 8.01 (d, J = 8.4 Hz, 2H), 7.96–7.86 (m, 1H), 7.46–7.33 (m, 3H), 2.31 (s, 3H) ^{13}C NMR (50 MHz, DMSO) δ 146.1, 146.0, 145.4, 134.2, 130.2, 128.7, 127.8, 125.7, 121.7, 120.2, 95.1, 21.1 HPLC t_{ret} = 8.86 min.

3-phenyl-1-tosyl-1H-pyrrolo[2,3-b]pyridine (**25**): The preparation was performed following *General Procedure A* from 100 mg **24** (0.25 mmol) and 38 mg phenylboronic acid (0.31 mmol), catalyzed by 1 mg $Pd(OAc)_2$ (5 μmol) and 5 mg XPhos (10 μmol) in 3 mL dioxane and 0.75 mL of an 1 M Na_2CO_3 solution. The reaction was conducted at 70 °C and the crude product was purified via flash chromatography (petrol ether/EtOAc + 5% MeOH (5–30%)) to obtain 72 mg (82%) of the title compound as a colorless semisolid. ^{1}H NMR (200 MHz, $CDCl_3$) δ 8.50 (dd, J = 4.7, 1.2 Hz, 1H), 8.25–8.04 (m, 3H), 7.92 (s, 1H), 7.68–7.56 (m, 2H), 7.55–7.17 (m, 6H), 2.36 (s, 3H) ^{13}C NMR (50 MHz, $CDCl_3$) δ 147.6, 145.3, 145.1, 135.4, 132.6, 129.7, 129.1, 128.9, 128.1, 127.7, 127.4, 122.7, 121.6, 120.3, 119.1, 21.6 TLC-MS (ESI) *m/z*: 403.0 $[M + Na + MeOH]^+$ HPLC t_{ret} = 9.02 min.

N-(3-(1-tosyl-1H-pyrrolo[2,3-b]pyridin-3-yl)phenyl)acrylamide (**26**): The preparation was performed following *General Procedure A* from 53 mg **24** (0.15 mmol) and 51 mg **14** (0.19 mmol), catalyzed by 3 mg XPhos Pd G3 (5 μmol) in 3.6 mL dioxane and 0.9 mL of an 0.5 M K_2CO_3 solution. The reaction was conducted at 50 °C and the crude product was purified via flash chromatography (petrol ether/EtOAc + 5% MeOH (40–70%)) to obtain 63 mg (quant.) of the title compound as a colorless waxy solid. ^{1}H NMR (200 MHz, $CDCl_3$) δ 8.47–8.35 (m, 5H), 8.23–7.95 (m, 5H), 7.85 (s, 1H), 7.57–7.45 (m, 1H), 7.40–7.11 (m, 5H), 6.52–6.24 (m, 2H), 5.74 (dd, J = 8.7, 2.5 Hz, 1H), 2.34 (s, 3H) ^{13}C NMR (50 MHz, $CDCl_3$) δ 164.3, 147.5, 145.5, 145.1, 138.8, 135.2, 133.2, 131.3, 129.8, 129.6, 129.2, 128.0, 127.9, 123.3, 122.8, 121.5, 120.1, 119.3, 119.2, 21.6 TLC-MS (ESI) *m/z*: 416.5 $[M - H]^-$ HPLC t_{ret} = 8.00 min.

3-phenyl-1H-pyrrolo[2,3-b]pyridine (**9c**): In a 25 mL round-bottomed flask were dissolved 72 mg **25** (0.3 mmol) in 6 mL of a 1 M solution of KOH in MeOH at 60 °C oil-bath temperature. The reaction was stirred for 3 h and was then quenched by the addition of sat. NH_4Cl solution. The aqueous phase was extracted with EtOAc (4 × 15 mL) and the combined extracts were washed with brine. After drying over Na_2SO_4 and evaporation, the residue was purified via flash chromatography (DCM/EtOAc (10–80%)) to yield 34 mg (85%) of the final compound as a white solid. ^{1}H NMR (400 MHz, DMSO) δ 11.92 (br s, 1H), 8.35–8.20 (m, 2H), 7.87 (s, 1H), 7.77–7.64 (m, 2H), 7.49–7.36 (m, 2H), 7.25 (t, J = 7.9 Hz, 1H), 7.17–7.08 (m, 1H) ^{13}C NMR (100 MHz, DMSO) δ 149.1, 142.9, 135.0, 128.8, 127.4, 126.2, 125.6, 123.6, 117.3, 116.0, 114.3 TLC-MS (ESI) *m/z*: 193.0 $[M - H]^-$ HPLC t_{ret} = 7.60 min.

N-(3-(1H-pyrrolo[2,3-b]pyridin-3-yl)phenyl)acrylamide (**9a**): To a suspension of 63 mg **26** (0.15 mmol) in 10 mL *t*BuOH were added 561 mg finely powdered KOH (10 mmol) and the mixture was heated to 50 °C oil-bath temperature until TLC indicated complete conversion. The reaction was quenched by the addition of sat. NH_4Cl solution, followed by extraction with EtOAc (3 × 20 mL). The combined extracts were washed with brine and dried over Na_2SO_4. After evaporation of the volatiles, the residue was purified via flash chromatography (DCM/MeOH (2–8%)) to yield 32 mg (81%) of the final compound as a white solid. ^{1}H NMR (200 MHz, DMSO) δ 11.94 (s, 1H), 10.22 (s, 1H), 8.39–8.23 (m, 2H), 8.14 (br s, 1H), 7.86 (d, J = 2.5 Hz, 1H), 7.63–7.50 (m, 1H), 7.47–7.30 (m, 2H), 7.19 (dd, J = 7.9, 4.9 Hz, 1H), 6.48 (dd, J = 16.8, 9.7 Hz, 1H), 6.29 (dd, J = 16.8, 2.1 Hz, 1H), 5.78 (dd, J = 9.7, 2.1 Hz, 1H) ^{13}C NMR (50 MHz, DMSO) δ 163.3, 149.1, 143.0, 139.5, 135.5, 132.0, 129.3, 127.4, 126.9, 123.7, 121.4, 117.2, 117.1, 116.6, 116.1, 114.1 TLC-MS (ESI) m/z: 318.5 $[M + Na + MeOH]^+$ HPLC t_{ret} = 5.42 min.

1-((2-(trimethylsilyl)ethoxy)methyl)-1H-pyrrolo[2,3-b]pyridine-5-carboxamide (**27**): To a solution of 235 mg **17** (0.8 mmol) were added 156 mg CDI (0.96 mmol) and stirring was continued at ambient temperature for 30 min. Then 5 mL conc. NH_3 solution were added and after another hour of stirring the reaction was further diluted with water and transferred to a separatory funnel. The aqueous phase was extracted with EtOAc (3 × 20 mL) and the combined organic phases were backwashed two times with brine. After drying over Na_2SO_4 and evaporation, the residue was purified via flash chromatography (petrol ether/EtOAc + 5% MeOH (40–100%)) to obtain 169 mg (72%) of the title compound as a white solid. ^{1}H NMR (200 MHz, $CDCl_3$) δ 8.82 (s, 1H), 8.40 (s, 1H), 7.38 (d, J = 3.3 Hz, 1H), 6.60 (br s, 2H), 6.55 (d, J = 3.3 Hz, 1H), 5.66 (s, 2H), 3.51 (t, J = 8.2 Hz, 2H), 0.87 (t, J = 8.2 Hz, 2 H), −0.11 (s, 9H) ^{13}C NMR (50 MHz, $CDCl_3$) δ 169.4, 149.5, 142.9, 129.6, 129.1, 122.4, 120.1, 102.3, 73.2, 66.5, 17.8, -1.4 TLC-MS (ESI) m/z: 314.4 $[M + Na]^+$ HPLC t_{ret} = 7.58 min.

3-bromo-1-((2-(trimethylsilyl)ethoxy)methyl)-1H-pyrrolo[2,3-b]pyridine-5-carboxamide (**28**): An ice-cooled solution of 150 mg **27** (0.52 mmol) in 10 mL dry DCM was treated dropwise with a 1 M bromine solution in DCM until a yellow coloration persisted. TLC indicated complete conversion at this point and the reaction was quenched by addition of aqueous Na_2SO_3 solution. After dilution with 40 mL EtOAc, the organic phase was washed with sat. $NaHCO_3$ and brine, prior to drying over Na2SO4 and evaporation. The residue was purified via flash chromatography (petrol ether/EtOAc + 5% MeOH (20–100%)) to obtain 130 mg (68%) of the title compound as a white solid. ^{1}H NMR (200 MHz, $CDCl_3$) δ 8.85 (s, 1H), 8.31 (s, 1H), 7.42 (s, 1H), 6.81–6.45 (m, 2H), 5.64 (s, 2H), 3.52 (t, J = 8.2 Hz, 2H), 0.88 (t, J = 8.2 Hz, 2H), −0.09 (s, 9H) ^{13}C NMR (50 MHz, $CDCl_3$) δ 168.9, 148.4, 144.2, 128.5, 127.8, 123.1, 119.5, 91.3, 73.2, 66.8, 17.8, −1.4 TLC-MS (ESI) m/z: 424.3 $[M + Na + MeOH]^+$ HPLC t_{ret} = 8.77 min.

3-phenyl-1-((2-(trimethylsilyl)ethoxy)methyl)-1H-pyrrolo[2,3-b]pyridine-5-carboxamide (**29**): The preparation was performed following *General Procedure A* from 62 mg **28** (0.17 mmol) and 24 mg phenylboronic acid (0.20 mmol), catalyzed by 2 mg tBu_3P Pd G3 (3 μmol) in 4 mL dioxane and 1 mL of an 0.5 M K_3PO_4 solution. The reaction was conducted at ambient temperature and the crude product was of sufficient purity to be used directly in the next step. Yield: 62 mg (quant.) as a colorless semisolid. ^{1}H NMR (200 MHz, $CDCl_3$/MeOD) δ 8.85 (d, J = 1.6 Hz, 1H), 8.73 (d, J = 1.9 Hz, 1H), 7.67–7.56 (m, 3H), 7.50–7.27 (m, 3 H), 5.71 (s, 2 H), 3.59 (t, J = 8.3 Hz, 2H), 0.93 (t, J = 8.3 Hz, 2H), −0.06 (s, 9H). ^{13}C NMR (50 MHz, $CDCl_3$/MeOD) δ 169.3, 149.6, 143.0, 133.5, 128.9, 128.6, 127.1, 126.8, 126.2, 122.5, 118.3, 117.7, 73.1, 66.5, 17.6, −1.6. TLC-MS (ESI) m/z: 422.2 $[M + Na + MeOH]^+$ HPLC t_{ret} = 9.44 min.

3-(3-acrylamidophenyl)-1-((2-(trimethylsilyl)ethoxy)methyl)-1H-pyrrolo[2,3-b]pyridine-5-carboxamide (**30**): The preparation was performed following *General Procedure A* from 61 mg **28** (0.17 mmol) and 56 mg **14** (0.21 mmol), catalyzed by 3 mg tBu_3P Pd G3 (5 μmol) in 4 mL dioxane and 1 mL of an 0.5 M K_2CO_3 solution. The reaction was conducted at ambient temperature and the crude product was purified via flash chromatography (petrol ether/EtOAc + 5% MeOH (40–100%)). Yield: 67 mg (93%) as colorless a semisolid. ^{1}H NMR (200 MHz, DMSO) δ 10.28 (br s, 1H), 8.87 (s, 1H), 8.78 (s, 1H), 8.18 (br s, 1H), 8.07 (s, 1H), 8.03 (s, 1H), 7.77–7.60 (m, 1 H), 7.45 (d, J = 4.7 Hz, 3H), 6.49 (dd, J = 17.0, 9.8 Hz, 1 H), 6.29 (dd, J = 17.0, 1.7 Hz, 1H), 5.85–5.53 (m, 3H), 3.58 (t, J = 7.8 Hz, 2H), 0.85 (t, J = 7.8 Hz, 2H), −0.10 (s, 9H) ^{13}C NMR (50 MHz, DMSO) δ 167.4, 163.3, 149.2, 143.5, 143.5, 139.7, 134.2, 131.9, 129.5, 127.9, 127.8, 127.0,

123.5, 122.0, 117.6, 117.0, 115.6, 72.7, 65.7, 17.2, −1.4 TLC-MS (ESI) *m*/*z*: 459.5 [M + Na]$^+$ HPLC t_{ret} = 8.53 min.

3-phenyl-1H-pyrrolo[2,3-b]pyridine-5-carboxamide (**10c**): The preparation was carried out following *General Procedure B* starting from 62 mg **29** (0.17 mmol) in 6 mL DCM and 2 mL TFA. Flash purification (DCM/MeOH (8–16%)) afforded 37 mg (92%) of the final compound as a white solid. ^{1}H NMR (200 MHz, DMSO) δ 12.18 (br s, 1H), 8.82 (d, *J* = 1.4 Hz, 1H), 8.77 (d, *J* = 1.4 Hz, 1H), 8.18 (br s, 1H), 7.96 (d, *J* = 2.0 Hz, 1H), 7.79 (s, 1H), 7.76 (s, 1H), 7.54–7.21 (m, 4H). ^{13}C NMR (50 MHz, DMSO) δ 167.7, 150.2, 143.4, 134.5, 128.9, 127.2, 126.5, 126.0, 125.0, 122.4, 116.4, 115.4. TLC-MS (ESI) *m*/*z*: 238.0 [M + H]$^+$ HPLC t_{ret} = 5.19 min.

3-(3-acrylamidophenyl)-1H-pyrrolo[2,3-b]pyridine-5-carboxamide (**10a**): The preparation was carried out following *General Procedure B* starting from 90 mg **30** (0.21 mmol) in 8 mL DCM and 2 mL TFA. Flash purification (DCM/MeOH (6–16%)) afforded 45 mg (71%) of the final compound as a white solid. ^{1}H NMR (400 MHz, DMSO) δ 12.17 (s, 1H), 10.26 (s, 1H), 8.81 (s, 1H), 8.76 (s, 1H), 8.12 (br s, 1H), 7.99 (s, 1H), 7.88 (s, 1H), 7.68 (d, *J* = 7.4 Hz, 1H), 7.50–7.31 (m, 3H), 6.55–6.42 (m, 1H), 6.35–6.22 (m, 1H), 5.83–5.72 (m, 1H) ^{13}C NMR (100 MHz, DMSO) δ 167.7, 163.2, 150.1, 143.2, 139.5, 134.9, 131.9, 129.3, 127.2, 126.7, 124.8, 122.6, 121.9, 117.5, 117.2, 116.4, 115.3 TLC-MS (ESI) *m*/*z*: 329.3 [M + Na]$^+$ HPLC t_{ret} = 3.94 min.

3-bromo-N-cyclopentyl-1-((2-(trimethylsilyl)ethoxy)methyl)-1H-pyrrolo[2,3-b]pyridine-5-carboxamide (**31a**): The amide coupling was performed following *General Procedure C* starting from 111 mg **18** (0.30 mmol) and 58 mg CDI (0.36 mmol) followed by reaction with 59 μL cyclopentylamine (0.60 mmol). The crude product was purified via flash chromatography (hexane/EtOAc (10–40%)). Yield: 105 mg (80%) as a colorless oil. ^{1}H NMR (200 MHz, $CDCl_3$) δ 8.73 (d, *J* = 1.9 Hz, 1H), 8.15 (d, *J* = 1.9 Hz, 1H), 7.36 (s, 1H), 6.60 (d, *J* = 7.2 Hz, 1H), 5.58 (s, 2H), 4.49–4.28 (m, 1H), 3.56–3.37 (m, 2H), 2.15–1.93 (m, 2H), 1.80–1.38 (m, 6H), 0.90–0.76 (m, 2H), −0.11 (s, 9H). ^{13}C NMR (50 MHz, $CDCl_3$) δ 166.2, 148.1, 144.0, 128.1, 126.7, 124.5, 119.1, 90.9, 73.0, 66.6, 51.9, 33.2, 23.9, 23.9, 17.8, −1.4. TLC-MS (ESI) *m*/*z*: 492.3 [M + Na + MeOH]$^+$ HPLC t_{ret} = 10.49 min.

3-bromo-N-methyl-1-((2-(trimethylsilyl)ethoxy)methyl)-1H-pyrrolo[2,3-b]pyridine-5-carboxamide (**31c**): The amide coupling was performed following *General Procedure C* starting from 100 mg **18** (0.27 mmol) and 53 mg CDI (0.32 mmol) followed by addition with 270 μL of a 2 M solution of methylamine in THF (0.54 mmol). The crude product was purified via flash chromatography (hexane/EtOAc (10–50%)). Yield: 60 mg (58%) as a white solid. ^{1}H NMR (200 MHz, $CDCl_3$) δ 8.77 (d, *J* = 1.9 Hz, 1H), 8.21 (d, *J* = 2.0 Hz, 1H), 7.38 (s, 1H), 6.90 (d, *J* = 5.1 Hz, 1H), 5.60 (s, 2H), 3.49 (t, *J* = 8.3 Hz, 2H), 3.01 (d, *J* = 4.7 Hz, 3H), 0.86 (t, *J* = 8.2 Hz, 2H), -0.11 (s, 9H). ^{13}C NMR (50 MHz, $CDCl_3$) δ 167.4, 148.2, 143.9, 128.3, 126.9, 124.2, 119.2, 91.0, 73.1, 66.7, 27.1, 17.8, −1.4 TLC-MS (ESI) *m*/*z*: 405.9 [M + Na]$^+$ HPLC t_{ret} = 9.47 min.

3-bromo-N-ethyl-1-((2-(trimethylsilyl)ethoxy)methyl)-1H-pyrrolo[2,3-b]pyridine-5-carboxamide (**31d**): The amide coupling was performed following *General Procedure C* starting from 150 mg **18** (0.41 mmol) and 86 mg CDI (0.53 mmol) followed by addition with 1 mL of a 2 M solution of ethylamine in MeOH (2.0 mmol). The crude product was purified via flash chromatography (hexane/EtOAc (10–50%)). Yield: 74 mg (46%) as a yellowish solid. ^{1}H NMR (200 MHz, $CDCl_3$) δ 9.02 (d, *J* = 1.8 Hz, 1H), 8.53 (d, *J* = 2.0 Hz, 1H), 7.45 (s, 1H), 7.38 (s, 1H), 5.67 (s, 2H), 3.70 (dd, *J* = 16.2, 10.0 Hz, 2H), 3.53 (t, *J* = 8.3 Hz, 2H), 1.01–0.77 (m, 5H), −0.07 (s, 9H). ^{13}C NMR (50 MHz, $CDCl_3$) δ 166.3, 148.0, 143.6, 128.2, 126.8, 124.2, 119.2, 90.9, 73.0, 66.6, 35.0, 17.7, 14.9, −1.5. HPLC t_{ret} = 9.85 min.

3-bromo-N-(cyclopropylmethyl)-1-((2-(trimethylsilyl)ethoxy)methyl)-1H-pyrrolo[2,3-b]pyridine-5-carboxamide (**31e**): The amide coupling was performed following *General Procedure C* starting from 100 mg **18** (0.27 mmol) and 53 mg CDI (0.32 mmol) followed by reaction with 94 μL cyclopropylmethanamine (1.35 mmol). The crude product was purified via flash chromatography (hexane/EtOAc (20–80%)). Yield: 83 mg (80%) as a white solid. ^{1}H NMR (200 MHz, $CDCl_3$ + MeOD) δ 8.77 (d, *J* = 2.0 Hz, 1H),

8.20 (d, J = 2.0 Hz, 1H), 7.44 (s, 1H), 6.42 (s, 1H), 5.65 (s, 2H), 3.70 (s, 2H), 3.60–3.42 (m, 2H), 3.04–2.85 (m, 1H), 0.99–0.80 (m, 4H), 0.76–0.58 (t, 2H), −0.07 (s, 9H). ^{13}C NMR (50 MHz, $CDCl_3$ + MeOD) δ 168.0, 148.5, 144.0, 128.4, 126.9, 124.1, 119.3, 91.2, 73.1, 67.2, 66.8, 23.4, 17.9, 7.0, −1.3. TLC-MS (ESI) *m/z*: 446.0 $[M + Na]^+$ HPLC t_{ret} = 9.88 min.

3-bromo-N-(cyclohexylmethyl)-1-((2-(trimethylsilyl)ethoxy)methyl)-1H-pyrrolo[2,3-b]pyridine-5-carboxamide (**31f**): The amide coupling was performed following *General Procedure* C starting from 100 mg **18** (0.27 mmol) and 53 mg CDI (0.32 mmol) followed by reaction with 175 μL cyclohexylmethanamine (1.35 mmol). The crude product was purified via flash chromatography (hexane/EtOAc (0–40%)). Yield: 72 mg (53%) as a brown oil. ^{1}H NMR (200 MHz, $CDCl_3$) δ 8.82 (d, J = 2.1 Hz, 1H), 8.26 (d, J = 2.1 Hz, 1H), 7.46 (s, 1H), 6.27 (s, 1H), 5.67 (s, 2H), 3.53 (t, 2H), 3.37 (dd, J = 7.1, 5.4 Hz, 2H), 1.37 –1.06 (m, 4H), 0.99–0.82 (m, 4H), 0.65–0.53 (m, 2H), 0.33 (t, J = 5.2 Hz, 2H), 0.07 (s, 1H), −0.06 (s, 9H). TLC-MS (ESI) *m/z*: 488.0 $[M + Na]^+$ HPLC t_{ret} = 12.44 min.

N-benzyl-3-bromo-1-((2-(trimethylsilyl)ethoxy)methyl)-1H-pyrrolo[2,3-b]pyridine-5-carboxamide (**31g**): The amide coupling was performed following *General Procedure* C starting from 300 mg **18** (0.81 mmol) and 171 mg CDI (1.05 mmol) followed by reaction with 441 μL benzylamine (1.35 mmol). The crude product was purified via flash chromatography (hexane/EtOAc (10–50%)). Yield: 205 mg (55%) as a brown viscous oil. ^{1}H NMR (200 MHz, $CDCl_3$) δ 8.83 (d, J = 2.1 Hz, 1H), 8.26 (d, J = 2.0 Hz, 1H), 7.42 (s, 1H), 7.41–7.27 (m, 5H), 6.73 (t, J = 5.7 Hz, 1H), 5.64 (s, 2H), 4.67 (d, J = 5.6 Hz, 2H), 3.58–3.41 (m, 2H), 0.96–0.81 (m, 2H), −0.07 (s, 9H). TLC-MS (ESI) *m/z*: 481.9 $[M + Na]^+$ HPLC t_{ret} = 10.60 min.

3-(3-acrylamidophenyl)-N-cyclopentyl-1-((2-(trimethylsilyl)ethoxy)methyl)-1H-pyrrolo[2,3-b]pyridine-5-carboxamide (**32a**): The preparation was performed following General Procedure A from 105 mg **31a** (0.24 mmol) and 71 mg **14** (0.26 mmol), catalyzed by 3 mg tBu_3P Pd G3 (5 μmol) in 5.6 mL dioxane and 1.4 mL of a 0.5 M K_3PO_4 solution. The reaction was conducted at ambient temperature and the crude product was purified via flash chromatography (hexane/EtOAc (30–70%)). Yield: 92 mg (78%) as a white solid. ^{1}H NMR (200 MHz, DMSO) δ 10.28 (s, 1H), 8.82 (d, J = 1.9 Hz, 1H), 8.72 (d, J = 1.9 Hz, 1H), 8.44 (d, J = 7.3 Hz, 1H), 8.10–8.00 (m, 2H), 7.71–7.60 (m, 1H), 7.52–7.40 (m, 2H), 6.49 (dd, J = 17.0, 9.8 Hz, 1H), 6.29 (dd, J = 17.0, 2.2 Hz, 1H), 5.79 (dd, J = 9.8, 2.2 Hz, 1H), 5.71 (s, 2H), 4.38–4.18 (m, 1H), 3.58 (t, J = 7.9 Hz, 2H), 2.02–1.84 (m, 2H), 1.76–1.39 (m, 6H), 0.84 (t, J = 7.9 Hz, 2H), −0.10 (s, 9H). ^{13}C NMR (50 MHz, DMSO) δ 165.4, 163.3, 149.0, 143.2, 139.7, 134.3, 131.9, 129.5, 127.8, 127.4, 126.9, 124.2, 121.9, 117.6, 117.5, 116.9, 115.6, 72.7, 65.7, 51.0, 32.2, 23.7, 17.2, −1.4. TLC-MS (ESI) *m/z*: 527.5 $[M + Na]^+$ HPLC t_{ret} = 9.90 min.

3-(3-acrylamidophenyl)-N-methyl-1-((2-(trimethylsilyl)ethoxy)methyl)-1H-pyrrolo[2,3-b]pyridine-5-carboxamide (**32c**): The preparation was performed following General Procedure A from 60 mg **31c** (0.16 mmol) and 64 mg **14** (0.23 mmol), catalyzed by 3 mg tBu_3P Pd G3 (5 μmol) in 4 mL dioxane and 0.9 mL of a 0.5 M K_3PO_4 solution. The reaction was conducted at ambient temperature and the crude product was purified via flash chromatography (petrol ether/EtOAc + 5% MeOH (40–100%)). Yield: 48 mg (69%) as a white solid. The material was carried on directly to the next step. TLC-MS (ESI) *m/z*: 473.0 $[M + Na]^+$ HPLC t_{ret} = 8.97 min.

3-(3-acrylamidophenyl)-N-ethyl-1-((2-(trimethylsilyl)ethoxy)methyl)-1H-pyrrolo[2,3-b]pyridine-5-carboxamide (**32d**): The preparation was performed following General Procedure A from 74 mg **31d** (0.19 mmol) and 64 mg **14** (0.23 mmol), catalyzed by 3 mg tBu_3P Pd G3 (5 μmol) in 4 mL dioxane and 1.1 mL of a 0.5 M K_2CO_3 solution. The reaction was conducted at ambient temperature and the crude product was purified via flash chromatography (petrol ether/EtOAc (40–100%)). Yield: 48 mg (69%) as white solid. The material was carried on directly to the next step. TLC-MS (ESI) *m/z*: 487.2 $[M + Na]^+$ HPLC t_{ret} = 9.11 min.

3-(3-acrylamidophenyl)-N-(cyclopropylmethyl)-1-((2-(trimethylsilyl)ethoxy)methyl)-1H-pyrrolo[2,3-b]pyridine-5-carboxamide (**32e**): The preparation was performed following General Procedure A from 83 mg **31e** (0.20 mmol) and 83 mg **14** (0.30 mmol), catalyzed by 3 mg tBu_3P Pd G3 (5 μmol) in 5 mL dioxane and

1.2 mL of a 0.5 M K_2CO_3 solution. The reaction was conducted at ambient temperature and the crude product was purified via flash chromatography (petrol ether/EtOAc + 5% MeOH (40–100%)). Yield: 78 mg (59%) as a white solid. ^{1}H NMR (200 MHz, $CDCl_3$) δ 8.82 (d, *J* = 2.0 Hz, 1H), 8.68 (d, *J* = 2.0 Hz, 1H), 8.08 (s, 1H), 7.93 (s, 1H), 7.55 (s, 1H), 7.49 (s, 1H), 7.34 (q, *J* = 7.4 Hz, 2H), 6.73 (t, *J* = 5.5 Hz, 1H), 6.55–6.33 (m, 2H), 5.76 (dd, *J* = 8.3, 3.3 Hz, 1H), 5.69 (s, 2H), 3.56 (t, 2H), 3.37 (t, 2H), 1.13 (d, *J* = 9.8 Hz, 1H), 1.01–0.82 (m, 2H), 0.65–0.46 (m, 2H), 0.30 (d, *J* = 5.1 Hz, 2H), −0.07 (s, 9H). TLC-MS (ESI) *m/z*: 513.2 $[M + Na]^+$ HPLC t_{ret} = 8.82 min.

3-(3-acrylamidophenyl)-N-(cyclohexylmethyl)-1-((2-(trimethylsilyl)ethoxy)methyl)-1H-pyrrolo[2,3-b]pyridine-5-carboxamide (**32f**): The preparation was performed following General Procedure A from 72 mg **31f** (0.16 mmol) and 64 mg **14** (0.23 mmol), catalyzed by 3 mg tBu_3P Pd G3 (5 µmol) in 4 mL dioxane and 0.9 mL of a 0.5 M K_2CO_3 solution. The reaction was conducted at ambient temperature and the crude product was purified via flash chromatography (petrol ether/EtOAc (20–60%)). Yield: 57 mg (69%) as a yellowish solid. ^{1}H NMR (200 MHz, $CDCl_3$) δ 8.81 (s, 1H), 8.70 (s, 1H), 8.10 (s, 1H), 7.99 (s, 1H), 7.78–7.30 (m, 4H), 6.74 (s, 1H), 6.42 (s, 1H), 6.37 (s, 1H), 5.77 (s, 1H), 5.68 (s, 2H), 3.55 (t, 2H), 3.34 (t, 2H), 1.73 (d, *J* = 16.2 Hz, 9H), 1.10 (s, 1H), 0.91 (t, *J* = 8.1 Hz, 2H), 0.07 (s, 1H), −0.07 (s, 9H). TLC-MS (ESI) *m/z*: 555.3 $[M + Na]^+$ HPLC t_{ret} = 10.98 min.

3-(3-acrylamidophenyl)-N-benzyl-1-((2-(trimethylsilyl)ethoxy)methyl)-1H-pyrrolo[2,3-b]pyridine-5-carboxamide (**32g**): The preparation was performed following General Procedure A from 200 mg **31g** (0.43 mmol) and 149 mg **14** (0.54 mmol), catalyzed by 8 mg tBu_3P Pd G3 (13 µmol) in 12 mL dioxane and 2.6 mL of a 0.5 M K_2CO_3 solution. The reaction was conducted at ambient temperature and the crude product was purified via flash chromatography (petrol ether/EtOAc + 5% MeOH (40–100%)). Yield: 120 mg (53%) as a brown solid. ^{1}H NMR (200 MHz, $CDCl_3$ + MeOD) δ 8.61 (s, 1H), 8.51 (s, 1H), 8.36 (t, *J* = 5.4 Hz, 1H), 7.78 (s, 1H), 7.36 (s, 1H), 7.30–7.19 (m, 1H), 7.17–6.95 (m, 8H), 6.18–6.10 (m, 2H), 5.51–5.44 (m, 1H), 5.41 (s, 2H), 4.08 (s, 2H), 3.33 (t, *J* = 8.2 Hz, 2H), 0.66 (t, *J* = 8.2 Hz, 2H), −0.31 (s, 9H). ^{13}C NMR (50 MHz, CDCl3 + MeOD) δ 167.4, 164.7, 149.2, 143.1, 138.6, 138.3, 134.1, 130.9, 129.3, 128.3, 127.9, 127.4, 127.3, 127.0, 126.4, 123.3, 122.8, 118.7, 118.3, 118.1, 117.0, 73.0, 66.4, 43.6, 17.5, −1.9. TLC-MS (ESI) *m/z*: 548.9 $[M+Na]^+$ HPLC t_{ret} = 10.08 min.

3-(3-acrylamidophenyl)-N-cyclopentyl-1H-pyrrolo[2,3-b]pyridine-5-carboxamide (**11a**): The preparation was carried out following *General Procedure B* starting from 81 mg **32a** (0.16 mmol) in 7.5 mL DCM and 2.5 mL TFA. Flash purification (DCM/MeOH (4–12%)) afforded 49 mg (82%) of the final compound as a white solid. ^{1}H NMR (400 MHz, DMSO) δ 12.16 (br s, 1H), 10.26 (s, 1H), 8.77 (d, *J* = 1.6 Hz, 1H), 8.70 (d, *J* = 1.6 Hz, 1H), 8.39 (d, *J* = 7.1 Hz, 1H), 8.06–7.95 (m, 1H), 7.88 (d, *J* = 2.4 Hz, 1H), 7.72–7.61 (m, 1H), 7.49–7.34 (m, 2H), 6.48 (dd, *J* = 17.0, 10.1 Hz, 1H), 6.29 (dd, *J* = 17.0, 1.8 Hz, 1H), 5.78 (dd, *J* = 10.1, 1.8 Hz, 1H), 4.34–4.20 (m, 1H), 2.00–1.81 (m, 2H), 1.79–1.64 (m, 2H), 1.63–1.46 (m, 4H). ^{13}C NMR (101 MHz, DMSO) δ 165.6, 163.2, 149.9, 142.9, 139.5, 135.0, 131.9, 129.3, 126.9, 126.7, 124.8, 123.3, 121.8, 117.5, 117.1, 116.3, 115.3, 51.0, 32.1, 23.6. TLC-MS (ESI) *m/z*: 397.0 $[M + Na]^+$ HPLC t_{ret} = 6.94 min.

3-(3-acrylamidophenyl)-N-methyl-1H-pyrrolo[2,3-b]pyridine-5-carboxamide (**11c**): The preparation was carried out following *General Procedure B* starting from 48 mg **32c** (0.11 mmol) in 7 mL DCM and 3 mL TFA. Flash purification (DCM/MeOH (6–16%)) afforded 31 mg (91%) of the final compound as a white solid. ^{1}H NMR (400 MHz, DMSO-d_6) δ 12.18 (s, 1H), 10.27 (s, 1H), 8.76 (d, *J* = 2.0 Hz, 1H), 8.71 (d, *J* = 2.0 Hz, 1H), 8.57 (d, *J* = 4.6 Hz, 1H), 7.98 (s, 1H), 7.89 (d, *J* = 2.6 Hz, 1H), 7.69 (d, *J* = 7.5 Hz, 1H), 7.51–7.37 (m, 2H), 6.49 (dd, *J* = 16.9, 10.1 Hz, 1H), 6.29 (dd, *J* = 16.9, 2.0 Hz, 1H), 5.78 (dd, 1H), 2.83 (d, *J* = 4.4 Hz, 3H) ^{13}C NMR (101 MHz, DMSO-d_6) δ 166.9, 163.7, 150.5., 143.3, 140.0, 135.4, 132.5, 129.8, 127.3, 127.1, 125.4, 123.5, 122.4, 118.0, 117.7, 116.9, 115.8, 26.7. TLC-MS (ESI) *m/z*: 343.0 $[M + Na]^+$. HRMS ESI-TOF $[M + H]^+$ *m/z* calcd. for $C_{18}H_{16}N_4O_2$: 321.1346, found: 321.1351. HPLC t_{ret} = 4.86 min.

3-(3-acrylamidophenyl)-N-ethyl-1H-pyrrolo[2,3-b]pyridine-5-carboxamide (**11d**): The preparation was carried out following *General Procedure B* starting from 68 mg **32d** (0.15 mmol) in 4.8 mL DCM and 1.2 mL TFA. Flash purification (DCM/MeOH (6–10%)) afforded 34 mg (70%) of the final compound as a white solid. ^{1}H NMR (400 MHz, DMSO-d_6) δ 12.16 (s, 1H), 10.25 (s, 1H), 8.77 (d, J = 2.0 Hz, 1H), 8.71 (d, J = 2.0 Hz, 1H), 8.58 (t, J = 5.5 Hz, 1H), 7.98 (d, J = 1.7 Hz, 1H), 7.88 (d, J = 2.6 Hz, 1H), 7.68 (dt, J = 7.1, 2.1 Hz, 1H), 7.49–7.38 (m, 2H), 6.48 (dd, J = 16.9, 10.1 Hz, 1H), 6.29 (dd, J = 17.0, 2.1 Hz, 1H), 5.78 (dd, J = 10.1, 2.1 Hz, 1H), 3.40–3.33 (m, 2H), 1.16 (t, J = 7.2 Hz, 3H). ^{13}C NMR (101 MHz, DMSO-d_6) δ 166.2, 163.7, 150.5, 143.3, 140.0, 135.5, 132.4, 129.8, 127.3, 127.2, 125.4, 123.6, 122.4, 118.0, 117.7, 116.9, 115.8, 34.5, 15.3. TLC-MS (ESI) m/z: 357.0 [M + Na]$^+$. HRMS ESI-TOF [M + H]$^+$ m/z calcd. for $C_{19}H_{18}N_4O_2$: 335,1503, found: 335.1508. HPLC t_{ret} = 5.53 min.

3-(3-acrylamidophenyl)-N-(cyclopropylmethyl)-1H-pyrrolo[2,3-b]pyridine-5-carboxamide (**11e**): The preparation was carried out following *General Procedure B* starting from 78 mg **32e** (0.16 mmol) in 4.2 mL DCM and 1.8 mL TFA. Flash purification (DCM/MeOH (6–16%)) afforded 13 mg (23%) of the final compound as a yellowish solid. ^{1}H NMR (400 MHz, DMSO-d_6) δ 12.18 (s, 1H), 10.30 (s, 1H), 8.80 (d, J = 2.0 Hz, 1H), 8.74 (d, J = 2.1 Hz, 1H), 8.71 (t, J = 5.6 Hz, 1H), 8.00 (d, J = 2.2 Hz, 1H), 7.90 (s, 1H), 7.70 (dt, J = 7.3, 1.9 Hz, 1H), 7.50–7.38 (m, 2H), 6.49 (dd, J = 17.0, 10.1 Hz, 1H), 6.29 (dd, J = 17.0, 2.1 Hz, 1H), 5.79 (dd, J = 10.0, 2.1 Hz, 1H), 3.20 (t, J = 6.2 Hz, 2H), 1.07 (ttd, J = 12.9, 6.8, 3.7 Hz, 1H), 0.51–0.37 (m, 2H), 0.36–0.22 (m, 2H). ^{13}C NMR (101 MHz, DMSO-d_6) δ 165.8, 163.2, 149.9, 142.8, 139.5, 134.9, 131.9, 129.2, 126.8, 126.7, 124.8, 123.0, 121.9, 117.5, 117.1, 116.4, 115.3, 43.5, 11.0, 3.3. TLC-MS (ESI) m/z: 383.1 [M + Na]$^+$. HRMS ESI-TOF [M + H]$^+$ m/z calcd. for $C_{21}H_{20}N_4O_2$: 361.1659, found: 361.1662. HPLC t_{ret} = 6.54 min.

3-(3-acrylamidophenyl)-N-(cyclohexylmethyl)-1H-pyrrolo[2,3-b]pyridine-5-carboxamide (**11f**): The preparation was carried out following *General Procedure B* starting from 30 mg **32f** (0.06 mmol) in 7 mL DCM and 3 mL TFA. Flash purification (DCM/MeOH (5–10%)) afforded 15 mg (64%) of the final compound as a yellowish solid. ^{1}H NMR (400 MHz, DMSO-d_6) δ 12.16 (s, 1H), 10.27 (s, 1H), 8.78 (s, 1H), 8.72 (s, 1H), 8.54 (s, 1H), 8.02 (s, 1H), 7.88 (d, J = 3.0 Hz, 1H), 7.66 (d, J = 7.2 Hz, 1H), 7.44 (s, 2H), 6.49 (ddd, J = 17.2, 10.1, 3.3 Hz, 1H), 6.29 (d, J = 16.8 Hz, 1H), 5.78 (d, J = 10.2 Hz, 1H), 3.16 (q, J = 6.5, 5.0 Hz, 2H), 1.72 (dd, J = 24.7, 11.6 Hz, 4H), 1.61 (d, J = 10.0 Hz, 2H), 1.22 (s, 1H), 1.17 (d, J = 9.2 Hz, 2H), 0.95 (t, J = 12.0 Hz, 2H). ^{13}C NMR (101 MHz, DMSO-d_6) δ 165.9, 163.2, 149.9, 142.8, 139.5, 134.9, 131.9, 129.2, 126.7, 126.6, 124.8, 123.1, 121.8, 117.5, 117.1, 116.3, 115.2, 45.4, 37.4, 30.5, 26.0, 25.4. TLC-MS (ESI) m/z: 425.3 [M + Na]$^+$. HRMS ESI-TOF [M + H]$^+$ m/z calcd. for $C_{24}H_{26}N_4O_2$: 403.2129, found: 403.2133. HPLC t_{ret} = 8.50 min.

3-(3-acrylamidophenyl)-N-benzyl-1H-pyrrolo[2,3-b]pyridine-5-carboxamide (**11g**): The preparation was carried out following *General Procedure B* starting from 80 mg **32g** (0.15 mmol) in 7 mL DCM and 3 mL TFA. Flash purification (DCM/MeOH (6–8%)) afforded 58 mg (97%) of the final compound as a pale red solid. ^{1}H NMR (400 MHz, DMSO-d_6) δ 12.20 (d, J = 2.7 Hz, 1H), 10.26 (s, 1H), 9.18 (t, J = 6.0 Hz, 1H), 8.84 (d, J = 2.0 Hz, 1H), 8.79 (d, J = 2.0 Hz, 1H), 8.00 (t, J = 1.9 Hz, 1H), 7.90 (d, J = 2.6 Hz, 1H), 7.67 (dt, J = 7.7, 1.8 Hz, 1H), 7.52–7.18 (m, 7H), 6.48 (q, J = 16.9, 10.1 Hz, 1H), 6.29 (dd, J = 17.0, 2.1 Hz, 1H), 5.78 (dd, J = 10.1, 2.0 Hz, 1H), 4.54 (d, J = 5.9 Hz, 2H). ^{13}C NMR (101 MHz, DMSO-d_6) δ 166.0, 163.2, 150.1, 142.9, 139.7, 139.5, 134.9, 131.9, 129.3, 128.2, 127.2, 126.9, 126.8, 126.7, 124.9, 122.7, 121.9, 117.5, 117.2, 116.4, 115.4, 42.6. TLC-MS (ESI) m/z: 419.3 [M + Na]$^+$. HRMS ESI-TOF [M + H]$^+$ m/z calcd. for $C_{24}H_{20}N_4O_2$: 397.1659, found: 397.1663. HPLC t_{ret} = 7.20 min.

N-(3-bromo-1-((2-(trimethylsilyl)ethoxy)methyl)-1H-pyrrolo[2,3-b]pyridin-5-yl)cyclopentanecarboxamide (**33a**): The preparation was carried out following *General Procedure D* with 90 mg **22** (0.26 mmol), 111 µL Et_3N (0.79 mmol) and 70 mg cyclopentanecarboxylic acid chloride (0.53 mmol). Flash purification (hexane/EtOAc (10–50%)) afforded 90 mg (78%) of the product as a colorless oily residue. ^{1}H NMR (200 MHz, $CDCl_3$) δ 8.30–8.21 (m, 1H), 8.20–8.15 (m, 1H), 8.11 (br s, 1H), 7.29 (s, 1H), 5.54 (s, 2H), 3.48 (t, J = 8.2 Hz, 2H), 2.82–2.60 (m, 1H), 1.95–1.80 (m, 4H), 1.78–1.43 (m, 4H), 0.86 (t, J = 8.2 Hz, 2H), −0.10 (s, 9H). ^{13}C NMR (50 MHz, $CDCl_3$) δ 175.6, 144.1, 138.3, 129.5, 127.6, 120.0, 119.7, 89.9, 73.0, 66.5, 46.4, 30.7, 26.1, 17.8, −1.4. TLC-MS (ESI) m/z: 492.1 [M + Na + MeOH]$^+$ HPLC t_{ret} = 10.84 min.

N-(3-bromo-1-((2-(trimethylsilyl)ethoxy)methyl)-1H-pyrrolo[2,3-b]pyridin-5-yl)thiophene-2-carboxamide (**33c**): The preparation was carried out following *General Procedure D* with 100 mg **22** (0.29 mmol), 123 μL Et_3N (0.88 mmol) and 41 μL thiophene-2-carbonyl chloride (0.38 mmol). Flash purification (petrol ether/EtOAc + 10% THF (0–50%)) afforded 62 mg (47%) of the product as a white solid. ^{1}H NMR (200 MHz, $CDCl_3$) δ 8.39 (d, *J* = 2.3 Hz, 1H), 8.24 (d, *J* = 2.3 Hz, 1H), 8.22 (s, 1H), 7.71 (dd, *J* = 3.8, 1.1 Hz, 1H), 7.54 (dd, *J* = 5.0, 1.1 Hz, 1H), 7.36 (s, 1H), 7.10 (dd, *J* = 5.0, 3.7 Hz, 1H), 5.59 (s, 2H), 3.52 (t, 2H), 0.89 (t, *J* = 11.1, 2.2, 0.9 Hz, 2H), −0.07 (s, 9H). ^{13}C NMR (50 MHz, $CDCl_3$) δ 161.4, 145.1, 139.5, 139.3, 131.8, 129.6, 129.5, 128.6, 121.5, 120.6, 90.8, 73.8, 67.3, 18.5, −0.7. TLC-MS (ESI) *m/z*: 474.3 $[M + Na]^+$ HPLC t_{ret} = 10.07 min.

N-(3-bromo-1-((2-(trimethylsilyl)ethoxy)methyl)-1H-pyrrolo[2,3-b]pyridin-5-yl)furan-2-carboxamide (**33d**): The preparation was carried out following *General Procedure D* with 100 mg **22** (0.29 mmol), 123 μL Et_3N (0.88 mmol) and 44 μL furane-2-carbonyl chloride (0.38 mmol). Flash purification (petrol ether/EtOAc (40–50%)) afforded 101 mg (80%) of the product as a yellow oil. ^{1}H NMR (200 MHz, $CDCl_3$) δ 8.44 (d, *J* = 2.4 Hz, 1H), 8.33 (d, *J* = 2.4 Hz, 1H), 8.29 (s, 1H), 7.51 (dd, *J* = 1.7, 0.8 Hz, 1H), 7.37 (s, 1H), 7.25 (dd, *J* = 3.3, 1.1 Hz, 1H), 6.56 (dd, *J* = 3.6, 1.8 Hz, 1H), 5.61 (s, 2H), 3.52 (t, 2H), 0.89 (t, 2H), −0.07 (s, 9H). ^{13}C NMR (50 MHz, $CDCl_3$) δ 156.4, 147.5, 144.4, 144.3, 138.0, 128.5, 127.8, 119.9, 119.7, 115.4, 112.6, 89.9, 72.9, 66.4, 17.7, −1.5. TLC-MS (ESI) *m/z*: 458.1 $[M + Na]^+$ HPLC t_{ret} = 9.92 min.

N-(3-bromo-1-((2-(trimethylsilyl)ethoxy)methyl)-1H-pyrrolo[2,3-b]pyridin-5-yl)benzamide (**33e**): The preparation was carried out following *General Procedure D* with 100 mg **22** (0.29 mmol), 123 μL Et_3N (0.88 mmol) and 68 μL benzoyl chloride (0.58 mmol). Flash purification (petrol ether/EtOAc (30–60%)) afforded 91 mg (70%) of the product as a yellow solid. ^{1}H NMR (200 MHz, $CDCl_3$) δ 8.40 (s, 1H), 8.38 (s, 1H), 8.28 (s, 1H), 7.90 (d, *J* = 7.5 Hz, 2H), 7.49 (q, *J* = 13.4, 11.4 Hz, 3H), 7.35 (s, 1H), 5.59 (s, 2H), 3.52 (t, *J* = 8.3 Hz, 2H), 0.89 (t, *J* = 8.0 Hz, 2H), −0.07 (s, 9H). ^{13}C NMR (50 MHz, $CDCl_3$) δ 166.2, 144.3, 138.5, 134.3, 131.9, 129.0, 128.7, 127.8, 127.1, 120.5, 119.7, 89.9, 72.9, 66.4, 17.7, −1.5. TLC-MS (ESI) *m/z*: 468.2 $[M + Na]^+$ HPLC t_{ret} = 10.20 min.

N-(3-bromo-1-((2-(trimethylsilyl)ethoxy)methyl)-1H-pyrrolo[2,3-b]pyridin-5-yl)acetamide (**33f**): The preparation was carried out following *General Procedure D* with 130 mg **22** (0.38 mmol), 160 μL Et_3N (1.14 mmol) and 54 μL acetyl chloride (0.76 mmol). Flash purification (petrol ether/EtOAc (30–80%)) afforded 134 mg (83%) of the product as a yellow solid. ^{1}H NMR (200 MHz, $CDCl_3$) δ 8.25 (d, *J* = 2.3 Hz, 1H), 8.19 (s, 1H), 8.12 (d, *J* = 2.3 Hz, 1H), 7.32 (s, 1H), 5.56 (s, 2H), 3.59–3.40 (t, 2H), 2.19 (s, 3H), 0.99–0.73 (t, 2H), −0.08 (s, 9H). ^{13}C NMR (50 MHz, $CDCl_3$) δ 169.4, 144.3, 138.4, 129.2, 127.9, 120.4, 119.8, 90.0, 73.1, 66.6, 24.3, 17.8, −1.4. TLC-MS (ESI) *m/z*: 406.1 $[M + Na]^+$ HPLC t_{ret} = 9.44 min.

N-(3-bromo-1-((2-(trimethylsilyl)ethoxy)methyl)-1H-pyrrolo[2,3-b]pyridin-5-yl)-2-cyclopentylacetamide (**33g**): The preparation was carried out following *General Procedure D* with 100 mg **22** (0.29 mmol), 123 μL Et_3N (0.88 mmol) and 79 μL 2-cyclopentylacetyl chloride (0.58 mmol). Flash purification (petrol ether/EtOAc (0–40%)) afforded 120 mg (91%) of the product as a brown oil. ^{1}H NMR (200 MHz, $CDCl_3$) δ 8.30 (d, *J* = 2.2 Hz, 1H), 8.27 (d, *J* = 2.1 Hz, 1H), 7.68–7.52 (m, 1H), 7.36 (s, 1H), 5.59 (s, 2H), 3.65–3.36 (m, 2H), 2.36 (d, *J* = 6.5 Hz, 2H), 2.23 (p, *J* = 7.3 Hz, 1H), 2.01–1.75 (m, 4H), 1.72–1.45 (m, 4H), 0.91 (td, *J* = 8.3, 7.7, 4.8 Hz, 2H), −0.06 (d, *J* = 2.5 Hz, 9H). TLC-MS (ESI) *m/z*: 474.2 $[M + Na]^+$ HPLC t_{ret} = 11.32 min.

N-(3-bromo-1-((2-(trimethylsilyl)ethoxy)methyl)-1H-pyrrolo[2,3-b]pyridin-5-yl)-2-phenylacetamide (**33h**): The preparation was carried out following *General Procedure D* with 100 mg **22** (0.29 mmol), 123 μL Et_3N (0.88 mmol) and 77 μL 2-phenylacetyl chloride (0.58 mmol). Flash purification (petrol ether/EtOAc (40–70%)) afforded 111 mg (82%) of the product as a yellow oil. ^{1}H NMR (200 MHz, $CDCl_3$) δ 8.18 (d, *J* = 2.4 Hz, 1H), 8.11 (d, *J* = 2.4 Hz, 1H), 7.45 (d, *J* = 2.5 Hz, 1H), 7.41 (d, *J* = 3.4 Hz, 2H), 7.37 (d, *J* = 2.1 Hz, 2H), 7.34 (s, 2H), 5.57 (s, 2H), 3.77 (s, 2H), 3.49 (t, 2H), 0.88 (t, *J* = 8.2 Hz, 2H), −0.07 (s, 9H). ^{13}C NMR (50 MHz, $CDCl_3$) δ 169.6, 144.4, 138.2, 134.3, 129.5, 129.2, 128.7, 127.7, 127.6, 120.1, 119.5, 89.8, 72.9, 66.4, 44.4, 17.7, −1.5. TLC-MS (ESI) *m/z*: 482.1 $[M + Na]^+$ HPLC t_{ret} = 10.42 min.

N-(3-bromo-1-((2-(trimethylsilyl)ethoxy)methyl)-1H-pyrrolo[2,3-b]pyridin-5-yl)-2-(thiophen-2-yl)acetamide (**33i**): The preparation was carried out following *General Procedure D* with 100 mg **22** (0.29 mmol), 123 μL Et_3N (0.88 mmol) and 72 μL 2-(thiophen-2-yl)acetyl chloride (0.58 mmol). Flash purification (petrol ether/EtOAc (20–50%)) afforded 123 mg (90%) of the product as a brown oil. ^{1}H NMR (200 MHz, $CDCl_3$) δ 8.22 (d, *J* = 2.3 Hz, 1H), 8.12 (d, *J* = 2.3 Hz, 1H), 7.73 (s, 1H), 7.34 (s, 1H), 7.30 (dd, 1H), 7.05 (s, 1H), 7.03 (d, *J* = 1.6 Hz, 1H), 5.57 (s, 2H), 3.97 (s, 2H), 3.49 (t, 2H), 0.88 (t, 2H), −0.08 (s, 9H). ^{13}C NMR (50 MHz, $CDCl_3$) δ 168.6, 144.5, 138.4, 135.6, 128.7, 128.0, 127.9, 127.7, 126.1, 120.4, 119.8, 90.0, 73.1, 66.6, 38.3, 17.9, −1.3. TLC-MS (ESI) *m/z*: 488.1 $[M + Na]^+$ HPLC t_{ret} = 10.15 min.

N-(3-(3-acrylamidophenyl)-1-((2-(trimethylsilyl)ethoxy)methyl)-1H-pyrrolo[2,3-b]pyridin-5-yl)cyclopentane carboxamide (**34a**): The preparation was performed following General Procedure A from 90 mg **33a** (0.21 mmol) and 64 mg **14** (0.24 mmol), catalyzed by 2 mg tBu_3P Pd G3 (4 μmol) in 3.3 mL dioxane and 0.4 mL of a 1.5 M K_2CO_3 solution. The reaction was conducted at ambient temperature and the crude product was purified via flash chromatography (hexane/EtOAc (30–80%)). Yield: 76 mg (73%) as a white solid. ^{1}H NMR (200 MHz, DMSO) δ 10.25 (s, 1H), 10.04 (s, 1H), 8.57 (d, *J* = 2.1 Hz, 1H), 8.49 (d, *J* = 2.1 Hz, 1H), 8.02–7.95 (m, 1H), 7.93 (s, 1H), 7.66–7.56 (m, 1H), 7.43 (t, *J* = 7.7 Hz, 1H), 7.38–7.28 (m, 1H), 6.48 (dd, *J* = 16.9, 9.9 Hz, 1H), 6.28 (dd, *J* = 16.9, 2.2 Hz, 1H), 5.78 (dd, *J* = 9.9, 2.2 Hz, 1H), 5.65 (s, 2H), 3.56 (t, *J* = 8.0 Hz, 2H), 2.91–2.70 (m, 1H), 1.97–1.46 (m, 8H), 0.84 (t, *J* = 8.0 Hz, 2H), −0.10 (s, 9H). ^{13}C NMR (50 MHz, DMSO) δ 174.6, 163.3, 144.8, 139.6, 136.7, 134.8, 131.9, 130.6, 129.4, 127.3, 127.0, 121.8, 121.8, 118.6, 117.4, 117.3, 114.5, 72.6, 65.6, 45.1, 30.1, 25.7, 17.2, −1.4. TLC-MS (ESI) *m/z*: 527.1 $[M + Na]^+$ HPLC t_{ret} = 10.24 min.

N-(3-(3-acrylamidophenyl)-1-((2-(trimethylsilyl)ethoxy)methyl)-1H-pyrrolo[2,3-b]pyridin-5-yl)thiophene-2-carboxamide (**34c**): The preparation was performed following General Procedure A from 90 mg **33c** (0.20 mmol) and 81 mg **14** (0.30 mmol), catalyzed by 3 mg tBu_3P Pd G3 (5 μmol) in 5.0 mL dioxane and 1.2 mL of a 0.5 M K_2CO_3 solution. The reaction was conducted at ambient temperature and the crude product was purified via flash chromatography (petrol ether/EtOAc + 10% THF (20–50%)). Yield: 46 mg (45%) as a white solid. ^{1}H NMR (200 MHz, DMSO-d_6) δ 10.44 (s, 1H), 10.24 (s, 1H), 8.61 (s, 1H), 8.60 (s, 1H), 8.14–7.97 (m, 2H), 7.99 (s, 1H), 7.87 (d, *J* = 5.0 Hz, 1H), 7.63 (d, *J* = 7.4 Hz, 1H), 7.42 (d, *J* = 8.1 Hz, 2H), 7.25 (t, *J* = 4.4 Hz, 1H), 6.48 (dd, *J* = 16.9, 9.9 Hz, 1H), 6.27 (dd, *J* = 16.8, 1.7 Hz, 1H), 5.77 (d, *J* = 10.0 Hz, 1H), 5.68 (s, 2H), 3.59 (t, *J* = 7.9 Hz, 2H), 0.86 (t, *J* = 7.9 Hz, 2H), −0.08 (s, 9H). ^{13}C NMR (50 MHz, -$CDCl_3$ + MeOD) δ 164.8, 161.5, 145.6, 139.2, 138.7, 137.7, 134.7, 131.2, 131.1, 129.5, 129.2, 128.9, 127.9, 127.4, 126.1, 122.4, 121.5, 118.5, 118.3, 118.0, 115.9, 73.2, 66.4, 17.7, −1.7 TLC-MS (ESI) *m/z*: 541.5 $[M + Na]^+$ HPLC t_{ret} = 9.48 min.

N-(3-(3-acrylamidophenyl)-1-((2-(trimethylsilyl)ethoxy)methyl)-1H-pyrrolo[2,3-b]pyridin-5-yl)furan-2-carboxamide (**34d**): The preparation was performed following General Procedure A from 101 mg **33d** (0.23 mmol) and 95 mg **14** (0.35 mmol), catalyzed by 4 mg tBu_3P Pd G3 (7 μmol) in 6.0 mL dioxane and 1.4 mL of a 0.5 M K_2CO_3 solution. The reaction was conducted at ambient temperature and the crude product was purified via flash chromatography (petrol ether/EtOAc + 10% MeOH (40–60%)). Yield: 58 mg (50%) as a white solid. ^{1}H NMR (200 MHz, $CDCl_3$ + MeOD) δ 9.38 (s, 1H), 9.19 (s, 1H), 8.63 (d, *J* = 2.3 Hz, 1H), 8.46 (d, *J* = 2.3 Hz, 1H), 7.85 (s, 1H), 7.65 (d, *J* = 7.0 Hz, 1H), 7.59–7.46 (m, 2H), 7.40–7.23 (m, 2H), 7.22 (dd, *J* = 3.5, 0.8 Hz, 1H), 6.51 (dd, *J* = 3.5, 1.8 Hz, 1H), 6.38 (s, 1H), 6.35 (d, *J* = 3.3 Hz, 1H), 5.69 (dd, *J* = 7.5, 4.3 Hz, 1H), 5.60 (s, 2H), 3.62–3.43 (m, 2H), 0.96–0.78 (m, 2H), −0.12 (s, 9H). ^{13}C NMR (50 MHz, $CDCl_3$ + MeOD) δ 222.3, 221.4, 164.6, 147.5, 145.9, 144.9, 138.9, 137.4, 134.8, 131.3, 129.6, 128.4, 127.5, 126.3, 122.4, 121.2, 118.5, 118.2, 116.0, 115.5, 112.5, 73.2, 66.5, 17.8, −1.5. TLC-MS (ESI) *m/z*: 525.4 $[M + Na]^+$ HPLC t_{ret} = 9.34 min.

N-(3-(3-acrylamidophenyl)-1-((2-(trimethylsilyl)ethoxy)methyl)-1H-pyrrolo[2,3-b]pyridin-5-yl)benzamide (**34e**): The preparation was performed following *General Procedure A* from 91 mg **33e** (0.20 mmol) and 84 mg **14** (0.31 mmol), catalyzed by 4 mg tBu_3P Pd G3 (7 μmol) in 5.0 mL dioxane and 1.2 mL of a 0.5 M K_2CO_3 solution. The reaction was conducted at ambient temperature and the crude product was purified via flash chromatography (petrol ether/EtOAc + 10% MeOH (20–60%)). Yield: 47 mg (45%) as a white solid. ^{1}H NMR (200 MHz, $CDCl_3$ + MeOD) δ 9.65 (s, 1H), 9.47 (s, 1H), 8.57 (d, J = 2.1 Hz, 1H), 8.49 (d, J = 2.1 Hz, 1H), 7.99–7.81 (m, 3H), 7.63–7.25 (m, 6H), 7.23 (s, 1H), 6.32 (s, 1H), 6.29 (d, J = 1.5 Hz, 1H), 5.63 (dd, J = 6.6, 5.1 Hz, 1H), 5.55 (s, 2H), 3.47 (t, 2H), 0.81 (t, J = 8.3 Hz, 2H), −0.17 (s, 9H). ^{13}C NMR (50 MHz, $CDCl_3$ + MeOD) δ 224.0, 167.3, 164.7, 145.7, 138.8, 137.7, 134.8, 134.5, 131.9, 131.2, 129.5, 129.4, 128.6, 127.5, 126.1, 122.3, 121.5, 118.5, 118.3, 118.0, 116.0, 73.2, 66.5, 17.8, −1.6. TLC-MS (ESI) m/z: 535.1 [M + Na]$^+$ HPLC t_{ret} = 9.77 min.

N-(3-(5-acetamido-1-((2-(trimethylsilyl)ethoxy)methyl)-1H-pyrrolo[2,3-b]pyridin-3-yl)phenyl)acrylamide (**34f**): The preparation was performed following *General Procedure A* from 121 mg **33f** (0.32 mmol) and 129 mg **14** (0.47 mmol), catalyzed by 6 mg tBu$_3$P Pd G3 (9 μmol) in 6.0 mL dioxane and 1.9 mL of a 0.5 M K_2CO_3 solution. The reaction was conducted at ambient temperature and the crude product was purified via flash chromatography (DCM/MeOH (0–5%)). Yield: 90 mg (63%) as a white solid. ^{1}H NMR (200 MHz, $CDCl_3$ + MeOD) δ 9.44 (s, 1H), 8.23 (s, 1H), 8.19 (s, 1H), 7.75 (s, 1H), 7.24 (d, J = 9.3 Hz, 2H), 7.04 (dd, J = 6.9, 3.8 Hz, 2H), 6.57 (dt, J = 28.2, 8.8 Hz, 1H), 6.14 (s, 1H), 6.11 (s, 1H), 5.45 (t, J = 5.8 Hz, 1H), 5.32 (s, 2H), 3.26 (t, J = 8.1 Hz, 2H), 1.90 (s, 3H), 0.60 (t, J = 8.2 Hz, 2H), −0.37 (s, 9H). ^{13}C NMR (50 MHz, $CDCl_3$ + MeOD) δ 170.1, 164.7, 145.4, 138.6, 137.0, 134.7, 131.2, 129.5, 129.3, 127.2, 126.0, 122.2, 120.5, 118.4, 118.3, 117.8, 115.7, 73.1, 66.3, 23.4, 17.6, −1.7. TLC-MS (ESI) m/z: 473.1 [M + Na]$^+$ HPLC t_{ret} = 9.18 min.

N-(3-(5-(2-cyclopentylacetamido)-1-((2-(trimethylsilyl)ethoxy)methyl)-1H-pyrrolo[2,3-b]pyridin-3-yl)phenyl) acrylamide (**34g**): The preparation was performed following General Procedure A from 120 mg **33g** (0.27 mmol) and 109 mg **14** (0.40 mmol), catalyzed by 5 mg tBu$_3$P Pd G3 (8 μmol) in 6.0 mL dioxane and 1.6 mL of a 0.5 M K_2CO_3 solution. The reaction was conducted at ambient temperature and the crude product was purified via flash chromatography (DCM/MeOH (0–5%)). Yield: 85 mg (62%) as a white solid. ^{1}H NMR (200 MHz, DMSO-d_6) δ 10.26 (s, 1H), 10.04 (s, 1H), 8.54 (s, 1H), 8.47 (s, 1H), 8.01 (s, 1H), 7.94 (s, 1H), 7.58 (d, J = 6.7 Hz, 1H), 7.53–7.34 (m, 2H), 6.68–6.37 (m, 1H), 6.27 (d, J = 17.8 Hz, 1H), 5.78 (d, J = 12.2 Hz, 1H), 5.65 (s, 2H), 4.12 (t, 2H), 2.34 (s, 2H), 1.67 (d, J = 38.3 Hz, 5H), 1.23 (s, 4H), 0.84 (t, 2H), −0.10 (s, 9H). ^{13}C NMR (50 MHz, $CDCl_3$ + MeOD) δ 172.8, 164.7, 145.3, 138.8, 136.7, 134.8, 131.3, 129.7, 129.5, 127.4, 126.0, 122.2, 120.5, 118.6, 118.3, 117.9, 115.8, 73.2, 66.4, 43.2, 37.2, 32.5, 24.9, 17.8, −1.6. TLC-MS (ESI) m/z: 541.5 [M + Na]$^+$ HPLC t_{ret} = 10.60 min.

N-(3-(5-(2-phenylacetamido)-1-((2-(trimethylsilyl)ethoxy)methyl)-1H-pyrrolo[2,3-b]pyridin-3-yl)phenyl) acrylamide (**34h**): The preparation was performed following General Procedure A from 111 mg **33h** (0.24 mmol) and 98 mg **14** (0.36 mmol), catalyzed by 4 mg tBu$_3$P Pd G3 (7 μmol) in 5.0 mL dioxane and 1.4 mL of a 0.5 M K_2CO_3 solution. The reaction was conducted at ambient temperature and the crude product was purified via flash chromatography (petrol ether/EtOAc + 10% MeOH (20–60%)). Yield: 59 mg (47%) as a white solid. ^{1}H NMR (200 MHz, $CDCl_3$ + MeOD) δ 9.79 (s, 1H), 9.72 (s, 1H), 8.53 (d, J = 2.3 Hz, 1H), 8.37 (s, 1H), 7.85 (s, 1H), 7.50 (s, 1H), 7.47 (s, 1H), 7.31 (d, J = 5.3 Hz, 1H), 7.28–7.18 (m, 2H), 7.24–7.07 (m, 2H), 7.10–6.81 (m, 1H), 6.32 (s, 1H), 6.29 (s, 1H), 6.24 (d, J = 5.8 Hz, 1H), 5.62 (dt, J = 11.8, 6.3 Hz, 1H), 5.54 (s, 2H), 4.22 (s, 2H), 3.44 (t, 2H), 0.78 (t, 2H), −0.19 (s, 9H). ^{13}C NMR (50 MHz, $CDCl_3$ + MeOD) δ 170.8, 164.8, 144.9, 138.6, 136.2, 134.8, 134.5, 131.1, 129.6, 129.3, 129.0, 128.5, 127.1, 126.9, 126.2, 122.3, 120.8, 118.7, 118.3, 118.0, 115.9, 73.2, 66.3, 43.6, 17.6, −1.9. TLC-MS (ESI) m/z: 549.1 [M + Na]$^+$ HPLC t_{ret} = 10.02 min.

N-(3-(5-(2-(thiophen-2-yl)acetamido)-1-((2-(trimethylsilyl)ethoxy)methyl)-1H-pyrrolo[2,3-b]pyridin-3-yl) phenyl)acrylamide (**34i**): The preparation was performed following General Procedure A from 123 mg **33i** (0.26 mmol) and 108 mg **14** (0.39 mmol), catalyzed by 5 mg tBu$_3$P Pd G3 (8 μmol) in 6.0 mL dioxane and 1.6 mL of a 0.5 M K_2CO_3 solution. The reaction was conducted at ambient temperature and the crude product was purified via flash chromatography (DCM/MeOH (0–5%)). Yield: 87 mg (62%) as a white solid. ^{1}H NMR (200 MHz, $CDCl_3$) δ 8.62 (d, J = 2.3 Hz, 1H), 8.49 (d, J = 2.3 Hz, 1H), 8.35 (s, 1H), 8.23 (s, 1H), 8.05 (s, 1H), 7.73 (s, 1H), 7.68 (s, 1H), 7.52 (d, J = 3.3 Hz, 4H), 7.28 (s, 1H), 6.69 (dd, J = 16.6,

1.8 Hz, 1H), 6.55 (dd, J = 16.7, 9.4 Hz, 1H), 5.97 (dd, J = 9.5, 1.8 Hz, 1H), 5.84 (s, 2H), 4.21 (s, 2H), 3.79 (t, J = 8.1 Hz, 2H), 1.16 (t, J = 8.2 Hz, 2H), 0.19 (s, 9H). ^{13}C NMR (50 MHz, $CDCl_3$) δ 168.8, 146.0, 142.7, 141.4, 138.3, 137.7, 135.6, 134.9, 129.4, 128.2, 127.6, 127.4, 126.2, 125.8, 121.2, 118.2, 118.0, 115.9, 92.4, 84.8, 84.3, 73.0, 66.3, 38.0, 17.7, −1.5. TLC-MS (ESI) *m/z*: 555.1 [M + Na]$^+$ HPLC t_{ret} = 9.71 min.

N-(3-(3-acrylamidophenyl)-1H-pyrrolo[2,3-b]pyridin-5-yl)cyclopentanecarboxamide (**12a**): The preparation was carried out following *General Procedure B* starting from 70 mg **34a** (0.14 mmol) in 5 mL DCM and 2 mL TFA. Flash purification (DCM/MeOH (2–8%)) afforded 40 mg (77%) of the final compound as a white solid. ^{1}H NMR (200 MHz, DMSO) δ 11.84 (br s, 1H), 10.24 (s, 1H), 9.97 (s, 1H), 8.62–8.47 (m, 1H), 8.47–8.32 (m, 1H), 7.93 (s, 1H), 7.75 (d, J = 2.2 Hz, 1H), 7.68–7.54 (m, 1H), 7.49–7.26 (m, 2H), 6.49 (dd, J = 16.9, 9.9 Hz, 1H), 6.28 (dd, J = 16.9, 1.6 Hz, 1H), 5.78 (dd, J = 9.9, 1.6 Hz, 1H), 2.91–2.71 (m, 1H), 1.99–1.46 (m, 8H). ^{13}C NMR (50 MHz, DMSO) δ 174.5, 163.3, 145.7, 139.5, 136.8, 135.6, 132.0, 129.7, 129.3, 126.9, 124.4, 121.7, 118.3, 117.3, 117.0, 116.7, 114.3, 45.1, 30.2, 25.8. TLC-MS (ESI) *m/z*: 397.0 [M + Na]$^+$ HPLC t_{ret} = 7.38 min.

N-(3-(3-acrylamidophenyl)-1H-pyrrolo[2,3-b]pyridin-5-yl)thiophene-2-carboxamide (**12c**): The preparation was carried out following *General Procedure B* starting from 46 mg **34c** (0.09 mmol) in 4.2 mL DCM and 1.8 mL TFA. Flash purification (DCM/MeOH (4–10%)) afforded 15 mg (44%) of the final compound as a white solid. ^{1}H NMR (200 MHz, DMSO-d_6) δ 11.94 (s, 1H), 10.42 (s, 1H), 10.27 (s, 1H), 8.55 (s, 2H), 8.06 (d, J = 3.7 Hz, 1H), 7.97 (s, 1H), 7.87 (d, J = 5.1 Hz, 1H), 7.82 (s, 1H), 7.65 (dd, J = 5.8, 2.1 Hz, 1H), 7.41 (s, 1H), 7.39 (s, 1H), 7.25 (t, J = 4.4 Hz, 1H), 6.49 (dd, J = 16.9, 9.8 Hz, 1H), 6.27 (dd, J = 17.0, 1.8 Hz, 1H), 5.77 (dd, J = 9.8, 2.2 Hz, 1H). ^{13}C NMR (50 MHz, DMSO-d_6) δ 163.2, 160.1, 146.2, 139.9, 139.5, 138.0, 135.4, 131.9, 131.7, 129.3, 129.0, 128.7, 128.1, 126.9, 124.6, 121.6, 120.2, 117.2, 116.9, 116.7, 114.3. TLC-MS (ESI) *m/z*: 411.3 [M + Na]$^+$. HRMS ESI-TOF [M + H]$^+$ *m/z* calcd. for $C_{21}H_{16}N_4O_2S$: 389.1067, found: 389.1072. HPLC t_{ret} = 6.44 min.

N-(3-(3-acrylamidophenyl)-1H-pyrrolo[2,3-b]pyridin-5-yl)furan-2-carboxamide (**12d**): The preparation was carried out following *General Procedure B* starting from 58 mg **34d** (0.12 mmol) in 4.2 mL DCM and 1.8 mL TFA. Flash purification (DCM/MeOH (4–10%)) afforded 23 mg (54%) of the final compound as a white solid. ^{1}H NMR (200 MHz, DMSO-d_6) δ 11.92 (s, 1H), 10.34 (s, 1H), 10.22 (s, 1H), 8.57 (s, 2H), 7.96 (s, 2H), 7.86–7.74 (m, 1H), 7.62 (d, J = 5.3 Hz, 1H), 7.50–7.33 (m, 2H), 7.33 (d, J = 3.5 Hz, 1H), 6.72 (dd, J = 3.2, 1.4 Hz, 1H), 6.48 (dd, J = 17.0, 9.8 Hz, 1H), 6.29 (dd, 1H), 5.77 (dd, J = 9.6, 2.0 Hz, 1H). ^{13}C NMR (50 MHz, DMSO-d_6) δ 163.2, 156.5, 147.6, 146.2, 145.6, 139.5, 138.0, 135.4, 131.9, 129.3, 128.4, 126.9, 124.5, 121.6, 120.2, 117.2, 116.9, 116.7, 114.6, 114.2, 112.1. TLC-MS (ESI) *m/z*: 395.2 [M + Na]$^+$. HRMS ESI-TOF [M + H]$^+$ *m/z* calcd. for $C_{21}H_{16}N_4O_3$: 373.1295, found: 373.1300. HPLC t_{ret} = 5.88 min

N-(3-(3-acrylamidophenyl)-1H-pyrrolo[2,3-b]pyridin-5-yl)benzamide (**12e**): The preparation was carried out following *General Procedure B* starting from 47 mg **34e** (0.09 mmol) in 4.2 mL DCM and 1.8 mL TFA. Flash purification (DCM/MeOH (4–10%)) afforded 10 mg (29%) of the final compound as a white solid. ^{1}H NMR (400 MHz, DMSO-d_6) δ 11.92 (d, 1H), 10.39 (s, 1H), 10.23 (s, 1H), 8.62 (d, J = 2.3 Hz, 1H), 8.59 (d, J = 2.2 Hz, 1H), 8.03 (s, 1H), 8.01 (d, J = 1.6 Hz, 1H), 7.96 (s, 1H), 7.81 (d, J = 2.6 Hz, 1H), 7.66–7.61 (m, 1H), 7.60 (d, J = 6.7 Hz, 1H), 7.55 (t, J = 7.2 Hz, 2H), 7.41 (d, J = 6.5 Hz, 2H), 6.47 (dd, J = 16.9, 10.1 Hz, 1H), 6.27 (dd, J = 17.0, 2.1 Hz, 1H), 5.76 (dd, J = 10.0, 2.1 Hz, 1H). ^{13}C NMR (101 MHz, DMSO-d_6) δ 165.5, 163.2, 146.1, 139.5, 138.0, 135.4, 134.6, 131.9, 131.5, 129.2, 129.1, 128.3, 127.5, 126.7, 124.4, 121.6, 120.0, 117.2, 116.9, 116.6, 114.3. TLC-MS (ESI) *m/z*: 405.3 [M + Na]$^+$. HRMS ESI-TOF [M + H]$^+$ *m/z* calcd. for $C_{23}H_{18}N_4O_2$: 383.1503, found: 383.1509. HPLC t_{ret} = 6.74 min.

N-(3-(5-acetamido-1H-pyrrolo[2,3-b]pyridin-3-yl)phenyl)acrylamide (**12f**): The preparation was carried out following *General Procedure B* starting from 90 mg **34f** (0.20 mmol) in 4.2 mL DCM and 1.8 mL TFA. Flash purification (DCM/MeOH (10–16%)) afforded 26 mg (41%) of the final compound as a white solid. ^{1}H NMR (200 MHz, DMSO-d_6) δ 11.85 (s, 1H), 10.22 (s, 1H), 10.03 (s, 1H), 8.46 (s, 1H), 8.39 (s, 1H), 7.95 (s, 1H), 7.76 (s, 1H), 7.60 (d, J = 7.3 Hz, 1H), 7.38 (q, J = 7.7 Hz, 2H), 6.49 (dd, J = 16.9, 9.8 Hz, 1H), 6.28 (d, J = 16.5 Hz, 1H), 5.78 (d, J = 9.7 Hz, 1H), 2.08 (s, 3H). ^{13}C NMR (50 MHz, DMSO-d_6) δ

168.3, 163.2, 145.8, 139.5, 136.8, 135.5, 131.9, 129.4, 129.3, 126.9, 124.4, 121.6, 118.4, 117.2, 116.9, 116.7, 114.2, 23.7. TLC-MS (ESI) *m/z*: 343.1 $[M + Na]^+$. HRMS ESI-TOF $[M + H]^+$ *m/z* calcd. for $C_{18}H_{16}N_4O_2$: 321.1346, found: 321.1352. HPLC t_{ret} = 4.74 min.

N-(3-(5-(2-cyclopentylacetamido)-1H-pyrrolo[2,3-b]pyridin-3-yl)phenyl)acrylamide (**12g**): The preparation was carried out following *General Procedure B* starting from 85 mg **34g** (0.16 mmol) in 7 mL DCM and 3 mL TFA. Flash purification (DCM/MeOH (5–6%)) afforded 27 mg (42%) of the final compound as a white solid. ^{1}H NMR (400 MHz, DMSO-d_6) δ 11.84 (s, 1H), 10.22 (s, 1H), 9.94 (s, 1H), 8.49 (d, *J* = 2.2 Hz, 1H), 8.42 (d, *J* = 2.2 Hz, 1H), 7.96 (s, 1H), 7.76 (d, *J* = 2.4 Hz, 1H), 7.60 (d, *J* = 8.2 Hz, 1H), 7.41 (t, *J* = 7.8 Hz, 1H), 7.35 (d, *J* = 7.7 Hz, 1H), 6.48 (dd, *J* = 16.9, 10.1 Hz, 1H), 6.29 (dd, *J* = 17.0, 2.0 Hz, 1H), 5.78 (dd, *J* = 10.0, 2.0 Hz, 1H), 2.35 (s, 1H), 2.33 (s, 1H), 2.27 (dt, *J* = 15.1, 7.4 Hz, 1H), 1.78 (dq, *J* = 11.8, 6.2 Hz, 2H), 1.62 (pd, *J* = 9.0, 8.3, 5.0 Hz, 2H), 1.59–1.45 (m, 2H), 1.22 (dq, *J* = 11.5, 7.3 Hz, 2H). ^{13}C NMR (101 MHz, DMSO-d_6) δ 171.0, 163.3, 145.9, 139.6, 137.0, 135.7, 132.1, 129.6, 129.3, 126.9, 124.4, 121.8, 118.6, 117.5, 117.1, 116.8, 114.4, 42.5, 36.8, 32.1, 24.7. TLC-MS (ESI) *m/z*: 411.4 $[M + Na]^+$. HRMS ESI-TOF $[M + H]^+$ *m/z* calcd. for $C_{23}H_{24}N_4O_2$: 389.1972, found: 389.1977. HPLC t_{ret} = 7.85 min.

N-(3-(5-(2-phenylacetamido)-1H-pyrrolo[2,3-b]pyridin-3-yl)phenyl)acrylamide (**12h**): The preparation was carried out following *General Procedure B* starting from 59 mg **34h** (0.11 mmol) in 4.2 mL DCM and 1.8 mL TFA. Flash purification (DCM/MeOH (4–10%)) afforded 15 mg (34%) of the final compound as a white solid. ^{1}H NMR (400 MHz, DMSO-d_6) δ 11.86 (s, 1H), 10.28 (s, 1H), 10.22 (s, 1H), 8.51 (d, *J* = 2.3 Hz, 1H), 8.43 (d, *J* = 2.3 Hz, 1H), 7.93 (t, *J* = 1.9 Hz, 1H), 7.76 (d, *J* = 2.3 Hz, 1H), 7.60 (dt, *J* = 8.1, 1.6 Hz, 1H), 7.36 (ddt, *J* = 14.8, 12.3, 7.7 Hz, 6H), 7.28–7.22 (m, 1H), 6.47 (dd, *J* = 17.0, 10.1 Hz, 1H), 6.28 (dd, *J* = 17.0, 2.1 Hz, 1H), 5.77 (dd, *J* = 10.1, 2.1 Hz, 1H), 3.68 (s, 2H). ^{13}C NMR (101 MHz, DMSO-d_6) δ 169.0, 163.1, 145.8, 139.4, 136.6, 135.9, 135.4, 131.9, 129.4, 129.1, 129.1, 128.2, 126.7, 126.4, 124.3, 121.6, 118.3, 117.2, 116.9, 116.6, 114.2, 43.0. TLC-MS (ESI) *m/z*: 419.4 $[M + Na]^+$. HRMS ESI-TOF $[M + H]^+$ *m/z* calcd. for $C_{24}H_{20}N_4O_2$: 397.1659, found: 397.1662. HPLC t_{ret} = 6.98 min.

N-(3-(5-(2-(thiophen-2-yl)acetamido)-1H-pyrrolo[2,3-b]pyridin-3-yl)phenyl)acrylamide (**12i**): The preparation was carried out following *General Procedure B* starting from 87 mg **34i** (0.16 mmol) in 4.2 mL DCM and 1.8 mL TFA. Flash purification (DCM/MeOH (6–8%)) afforded 15 mg (34%) of the final compound as a white solid. ^{1}H NMR (200 MHz, DMSO-d_6) δ 11.88 (s, 1H), 10.33 (s, 1H), 10.23 (s, 1H), 8.49 (s, 1H), 8.43 (s, 1H), 7.94 (s, 1H), 7.77 (s, 1H), 7.61 (d, *J* = 6.8 Hz, 1H), 7.38 (d, *J* = 7.9 Hz, 3H), 7.01 (s, 2H), 6.48 (dd, 1H), 6.27 (d, *J* = 17.2 Hz, 1H), 5.78 (d, *J* = 9.6 Hz, 1H), 3.92 (s, 2H). ^{13}C NMR (50 MHz, DMSO-d_6) δ 168.1, 163.2, 145.9, 139.5, 137.1, 136.6, 135.4, 131.9, 129.2, 129.2, 126.9, 126.6, 126.3, 125.0, 124.5, 121.6, 118.4, 117.2, 116.9, 116.7, 114.2, 37.2. TLC-MS (ESI) *m/z*: 425.2 $[M + Na]^+$. HRMS ESI-TOF $[M + H]^+$ *m/z* calcd. for $C_{22}H_{18}N_4O_2S$: 403.1223, found: 403.1227. HPLC t_{ret} = 6.71 min.

N-(3-(1H-pyrrolo[2,3-b]pyridin-3-yl)phenyl)propionamide (**9b**): To a solution of 17 mg **9a** (0.07 mmol) in THF/MeOH (6 mL each) were added 8 mg Pd/C and the reaction was purged with hydrogen. The mixture was then stirred under hydrogen until HPLC indicated complete conversion. The catalyst was filtered off, the filtrate evaporated and the residue purified via flash chromatography (DCM/MeOH (4–8%)). Yield: 12 mg (69%) as a white solid. ^{1}H NMR (200 MHz, DMSO) δ 11.89 (br s, 1H), 9.90 (s, 1H), 8.37–8.19 (m, 2H), 8.10–7.98 (m, 1H), 7.87–7.75 (m, 1H), 7.56–7.41 (m, 1H), 7.41–7.28 (m, 2H), 7.25–7.06 (m, 1H), 2.35 (q, *J* = 7.4 Hz, 2H), 1.11 (t, *J* = 7.4 Hz, 3H) ^{13}C NMR (50 MHz, DMSO) δ 172.1, 149.1, 142.9, 139.8, 135.3, 129.1, 127.4, 123.6, 120.8, 117.2, 116.8, 116.3, 116.0, 114.2, 29.6, 9.7 TLC-MS (ESI) *m/z*: 266.4 $[M + H]^+$ HPLC t_{ret} = 5.47 min.

3-(3-propionamidophenyl)-1H-pyrrolo[2,3-b]pyridine-5-carboxamide (**10b**): To a solution of 25 mg **10a** (0.08 mmol) in THF/MeOH (5 mL each) were added 8 mg Pd/C and the reaction was purged with hydrogen. The mixture was then stirred under hydrogen until HPLC indicated complete conversion. The catalyst was filtered off, the filtrate evaporated and the residue purified via flash chromatography (DCM/MeOH (6–16%)). Yield: 15 mg (60%) as a white solid. ^{1}H NMR (400 MHz, DMSO) δ 12.15 (s, 1H), 9.95 (s, 1H), 8.81 (s, *J* = 24.3 Hz, 1H), 8.75 (s, 1H), 8.10 (br s, 1H), 7.88 (d, *J* = 19.3 Hz, 2H), 7.61 (d,

J = 7.2 Hz, 1H), 7.46–7.25 (m, 3H), 2.36 (q, J = 7.5 Hz, 2H), 1.11 (t, J = 7.5 Hz, 3H) ^{13}C NMR (100 MHz, DMSO) δ 172.0, 167.7, 150.1, 143.2, 139.8, 134.8, 129.1, 127.2, 124.7, 122.6, 121.3, 117.1, 116.9, 116.4, 115.4, 29.5, 9.6 TLC-MS (ESI) m/z: 331.4 [M + Na]$^+$ HPLC t_{ret} = 3.94 min.

N-cyclopentyl-3-(3-propionamidophenyl)-1H-pyrrolo[2,3-b]pyridine-5-carboxamide (**11b**): To a solution of 20 mg **11a** (0.05 mmol) in THF/MeOH (2 mL each) were added 8 mg Pd/C and the reaction was purged with hydrogen. The mixture was then stirred under hydrogen until HPLC indicated complete conversion. The catalyst was filtered off, the filtrate evaporated and the residue purified via flash chromatography (DCM/MeOH (6–14%)). Yield: 19 mg (95%) as a white solid. ^{1}H NMR (400 MHz, DMSO) δ 12.15 (br s, 1H), 9.98 (s, 1H), 8.76 (d, J = 1.6 Hz, 1H), 8.69 (d, J = 1.6 Hz, 1H), 8.39 (d, J = 7.2 Hz, 1H), 7.94 (s, 1H), 7.86 (d, J = 2.3 Hz, 1H), 7.62–7.55 (m, 1H), 7.43–7.32 (m, 2H), 4.33–4.21 (m, 1H), 2.36 (q, J = 7.5 Hz, 2H), 1.96–1.86 (m, 2H), 1.76–1.65 (m, 2H), 1.62–1.49 (m, 4H), 1.11 (t, J = 7.5 Hz, 3H). ^{13}C NMR (101 MHz, DMSO) δ 172.0, 165.7, 149.9, 142.8, 139.8, 134.8, 129.1, 126.9, 124.7, 123.2, 121.2, 117.2, 116.8, 116.3, 115.4, 50.9, 32.1, 29.5, 23.6, 9.6. TLC-MS (ESI) m/z: 399.1 [M + Na]$^+$ HPLC t_{ret} = 7.15 min.

N-(3-(3-propionamidophenyl)-1H-pyrrolo[2,3-b]pyridin-5-yl)cyclopentanecarboxamide (**12b**): To a solution of 18 mg **12a** (0.05 mmol) in THF/MeOH (2 mL each) were added 5 mg Pd/C and the reaction was purged with hydrogen. The mixture was then stirred under hydrogen until HPLC indicated complete conversion. The catalyst was filtered off, the filtrate evaporated and the residue purified via flash chromatography (DCM/MeOH (3–10%)). Yield: 15 mg (83%) as a white solid. ^{1}H NMR (200 MHz, DMSO) δ 11.82 (br s, 1H), 10.05–9.81 (m, 2H), 8.50 (d, J = 1.7 Hz, 1H), 8.41 (d, J = 1.7 Hz, 1H), 7.90–7.78 (m, 1H), 7.72 (d, J = 1.8 Hz, 1H), 7.63–7.46 (m, 1H), 7.44–7.20 (m, J = 7.7 Hz, 2H), 2.91–2.72 (m, 1H), 2.35 (q, J = 7.4 Hz, 2H), 1.97–1.50 (m, 8H), 1.10 (t, J = 7.4 Hz, 3H). TLC-MS (ESI) m/z: 399.2 [M + Na]$^+$ HPLC t_{ret} = 7.40 min.

4.2. JAK3 Crystal Structure Determination

Recombinant JAK3 kinase domain was expressed and purified as described previously [21,22]. The protein (10 mg/mL) was incubated with the inhibitors at 1:1.2 molar ratio, and the complexes were crystallized using sitting drop vapor diffusion method at 4 °C and the conditions containing 25% PEG 3350, 0.1–0.2 M $MgCl_2$ and 0.1 M MES, pH 5.5. The crystals were cryo-protected using mother liquor supplemented with 20% ethylene glycol, and diffraction data were collected at SLS X06SA. The data were processed and scaled with XDS [40] and Aimless [41], respectively. Molecular replacement using Phaser [42] and the coordinate of JAK3 (pdb id: 5lwm) was performed. Model rebuilding alternated with refinement was performed in COOT [43] and REFMAC [44], respectively. Data collection and refinement statistics are summarized in Table 7.

Table 7. Data collection and refinement statistics for the JAK3-inhibitor complexes.

Complex	JAK3-9a	JAK3-10a
PDB Accession Code	**7APG**	**7APF**
Data Collection		
Resolution [a] (Å)	49.93–2.40 (2.53–2.40)	47.95–1.95 (2.02–1.95)
Spacegroup	$P2_1$	$P2_1$
Cell dimensions	a = 57.5, b = 112.9, c = 102.9 Å $\alpha, \gamma = 90.0°, \beta = 97.2$	a = 63.7, b = 62.5, c = 67.9 Å $\alpha, \gamma = 90.0°, \beta = 101.3°$
No. unique reflections [a]	50,511 (7245)	37,245 (3630)
Completeness [a] (%)	99.2 (98.1)	97.6 (97.7)
I/Σi [a]	6.6 (1.8)	8.2 (2.4)
R_{merge} [a]	0.126 (0.629)	0.108 (0.721)
CC (1/2) [a]	0.994 (0.791)	0.994 (0.823)
Redundancy [a]	4.6 (4.3)	6.0 (6.1)

Table 7. *Cont.*

Complex	JAK3-9a	JAK3-10a
PDB Accession Code	**7APG**	**7APF**
Refinement		
No. atoms in refinement (P/L/O) [b]	8952/100/449	4628/92/352
B factor (P/L/O) [b] ($Å^2$)	40/16/37	33/20/36
R_{fact} (%)	21.2	20.4
R_{free} (%)	25.6	25.5
rms deviation bond [c] (Å)	0.010	0.013
rms deviation angle [c] (°)	1.1	1.3

[a] Values in brackets show the statistics for the highest resolution shells. [b] P/L/O indicate protein, ligand molecules presented in the active sites, and other (water and solvent molecules), respectively. [c] rms indicates root-mean-square.

4.3. Biological Assays

4.3.1. Thermal Shift Assay

The kinase domains of JAK3 and BMX at 2 μM were mixed with the inhibitors at 10 μM, and subsequently SyPRO orange dye (Invitrogen Carlsbad, CA, USA) was added. The thermal shift assay and data evaluation were performed as described previously using a Real-Time PCR Mx3005p machine (Stratagene, La Jolla, CA, USA) [45,46].

4.3.2. Enzymatic Activity Assay

If not mentioned differently, in vitro profiling of compounds was performed at Reaction Biology Corporation using the HotSpot™ assay platform. IC_{50} values were determined as singlicates using five doses with 5- or 10-fold serial dilution starting at 0.5 μM, 1 μM or 10 μM. Further details on the assay can be found on the supplier homepage (http://www.reactionbiology.com; see also Anastassiadis et al. 2011 [31]).

4.3.3. Cellular NanoBRET Assay

Full length BMX and BTK was cloned into pFC32K (Promega, Madison, WI, USA) for expression of a C-terminal NanoLuc fusion. The plasmids were transfected into HEK293T cells cultured in DMEM (Gibco, Waltham, MA, USA) supplemented with 10% fetal bovine serum (Gibco, Waltham, MA, USA) and Penicillin/Streptamycin (Gibco, Waltham, MA, USA). NanoBRET assays were performed using the protocol published previously [47]. Briefly, after transfection and 20 h incubation cells were harvested and subsequently resuspended in OptiMEM (Gibco, Waltham, MA, USA). Cells were aliquoted onto 1534-well plates (Greiner, Kremsmünster, Austria), and inhibitors as well as 0.5 μM Tracer K4 (Promega) for BMX or 0.5 μM Tracer K5 (Promega, Madison, WI, USA) for BTK were added using an ECHO acoustic dispenser. The plates were incubated at 37 °C with 5% CO_2 for 2 h prior to the addition of both NanoBRET NanoGlo substrate (Promega, Madison, WI, USA) and extracellular NanoLuc inhibitor. BRET luminescence (450 nm for donor emission and 610 nm for BRET signal) was measured using PHERAstar FSX plate reader (BMG Labtech, Ortenberg, Germany). Milli-BRET units (mBU) were calculated as a ratio between BRET signal and the overall measured luminescence. A dose–response fitting was applied and IC_{50} values were calculated using the Prism software. The affinity of both tracer molecules towards BMX and BTK was tested and showed similar values (EC_{50} (Tracer 4, **BMX**) = 0.240 μM, EC50 (Tracer 5, **BTK**) = 0.231 μM) indicating no bias of the assay due to the higher affinity of one kinase. Experiments were performed as tripicates and repeated at least three times.

4.4. Mass Spectrometric Investigation of Covalent Binding to BMX

The kinase domain of BMX was diluted to 0.05 mM with buffer containing 20 mM Tris pH 8.0, 200 mM NaCl, 0.5 mM TCEP (pH 7.0) and was subsequently mixed with 0.075 mM inhibitors (ratio of protein to inhibitor of 1:1.5). The mixture was incubated at 4°C. The samples were taken at different time points, the reaction stopped by diluting the sample with a 200-fold excess of 1% formic acid. The denatured protein and the adducts were assessed using electrospray time-of-flight (ESI-TOF) mass spectrometry.

4.5. Determination of Glutathione Reactivity

A total of 5 μL of a 10 mM compound stock solution in DMSO were added to 5 μL of a 4 mM Indoprofen solution in PBS as internal standard. The mixture was diluted with PBS buffer to a total volume of 1 mL of component A. Additionally, a freshly prepared solution of 10 mM GSH in PBS buffer was used as component B. A total of 250 μL of each component was mixed and immediately subjected to HPLC analysis for t^0 measurement. The probes were stored at 40 °C in between the respective injections. $t_{1/2}$ was determined by plotting natural logarithm of AUC/AUC_0 against time.

4.6. Molecular Modeling

All the modelling was performed using the Schrödinger Small-Molecule Drug Discovery Suite 2019-1 (Schrödinger, LLC, New York, NY, USA). Noncovalent docking was performed with Glide in the XP mode after protein preparation using standard settings. Covalent docking was performed with the CovDock module in the pose prediction mode using standard settings. The figures were prepared with PyMOL 1.8.2.0. (Schrödinger, LLC, New York, NY, USA).

Author Contributions: Conceptualization: M.F., A.C., S.K., S.L. and M.G.; chemical synthesis: X.J.L. and M.F.; X-ray crystallography, thermal shift assays, NanoBRET and protein MS experiments: M.S. and A.C.; GSH stability experiments: S.G.; data analysis: M.F., X.J.L., M.S., S.G., A.C. and M.G.; manuscript writing—original draft preparation: M.F. and M.G.; manuscript writing—review and editing, M.F., A.C., X.J.L., M.S., S.G., S.K., S.L. and M.G. All authors have read and agreed to the published version of the manuscript.

Funding: M.G. gratefully acknowledges funding by the Institutional Strategy of the University of Tübingen (ZUK 63, German Research Foundation), the RiSC Program of the State Ministry of Baden-Württemberg for Sciences, Research and Arts, the Max Buchner Research Foundation and the Postdoctoral Fellowship Program of the Baden-Württemberg Stiftung. F.M., S.L. and M.G. are grateful for funding by the German Research Foundation (Deutsche Forschungsgemeinschaft, DFG) under Germany's Excellence Strategy-EXC 2180—390900677. S.L. further acknowledges funding by the Federal Ministry of Education and Research (BMBF) and the Baden-Württemberg Ministry of Science as part of the Excellence Strategy of the German Federal and State Governments. A.C., M.S. and S.K. are grateful for support by the German translational cancer network DKTK, the Frankfurt Cancer Institute (FCI) and the SGC, is a registered charity (no: 1097737) that receives funds from: AbbVie, Bayer AG, Boehringer Ingelheim, Canada Foundation for Innovation, Eshelman Institute for Innovation, Genentech, Genome Canada through Ontario Genomics Institute [OGI-196], EU/EFPIA/OICR/McGill/KTH/Diamond, Innovative Medicines Initiative 2 Joint Undertaking [EUbOPEN grant 875510], Janssen, Merck KGaA (aka EMD in Canada and US), Merck & Co (aka MSD outside Canada and US), Pfizer, São Paulo Research Foundation-FAPESP, Takeda and Wellcome. The authors further acknowledge support by the Open Access Publishing Fund of University of Tübingen.

Acknowledgments: The authors thank Kristine Schmidt for language correction and editing. The authors further thank staff at Swiss Light Source (SLS) for their supports during crystallographic data collection.

Abbreviations

ATP	adenosine triphosphate
AUC	area under the curve
B_2pin_2	Bis(pinacolato)diboron
BLK	B-lymphoid tyrosine kinase
BMX/ETK	bone marrow tyrosine kinase on chromosome X
br s	broad singlet
BRET	bioluminescence resonance energy transfer
BTK	Bruton's tyrosine kinase

CDI	1,1′carbonyldiimidazole
DAD	diode array detector
DCM	dichloromethane
DMEM	Dulbecco's modified Eagle's medium
DMF	*N, N*-dimethylformamide
DMSO	dimethyl sulfoxide
EC_{50}	half maximal effective concentration
EGFR	epidermal growth factor receptor
ESI	electron spray ionization
Et_3N	triethylamine
EtOAc	ethyl acetate
EtOH	ethanol
GSH	glutathione
HER	human epidermal growth factor receptor
HPLC	high performance liquid chromatography
IC_{50}	half maximal inhibitory concentration
ITK	interleukin-2-inducible T-cell kinase
JAK	Janus kinase
KOAc	potassium acetate
LC	liquid chromatography
MAP2K7/MKK7	mitogen activated protein kinase kinase 7
MeOH	methanol
MES	2-(*N*-morpholino)ethanesulfonic acid
MS	mass spectrometry
NaH	sodium hydride
n-BuLi	*n*-Butyllithium
NBS	*N*-bromosuccinimide
NMR	nuclear magnetic resonance spectroscopy
Pd/C	palladium on activated carbon
$Pd(OAc)_2$	palladium(II)acetate
PEG	polyethylene glycol
PH	pleckstrin homology
rt	room temperature
SAR	structure–activity relationship
sat.	saturated
SEM	2-(trimethylsilyl)ethoxymethyl
*t*BuOH	*tert*-Butanol
$t\text{Bu}_3\text{P}$ Pd G3	Mesyl[(tri-t-butylphosphine)-2-(2-aminobiphenyl)]palladium(II)
TCEP	tris(2-carboxyethyl)phosphine
ΔT_m	thermal shift
t_{ret}	retention time
TEC	tyrosine kinase expressed in hepatocellular carcinoma
TFA	trifluoroacetic acid
TH	TEC homology
THF	tetrahydrofurane
TLC	thin layer chromatography
TMS	tetramethylsilane
TOF	time of flight
Tris	tris(hydroxymethyl)aminomethan
TsCl	tosyl chloride
TSK/EMT	interleukin-2-inducible T-cell kinase
TXK/RLK	tyrosine-protein kinase
UV	ultraviolet

XLA	X-linked agammaglobulinemia
XPhos	2-Dicyclohexylphosphino-2′,4′,6′-triisopropyl-1,1′-biphenyl
XPhos Pd G3	(2-Dicyclohexylphosphino-2′,4′,6′-triisopropyl-1,1′-biphenyl)[2-(2′-amino-1,1′-biphenyl)]palladium(II) methanesulfonate
XPhos Pd G4	(2-Dicyclohexylphosphino-2′,4′,6′-triisopropyl-1,1′-biphenyl)[2-(2′-*N*-methylamino-1,1′-biphenyl)]palladium(II) methanesulfonate

References

1. Mano, H. Tec family of protein-tyrosine kinases: An overview of their structure and function. *Cytokine Growth Factor Rev.* **1999**, *10*, 267–280. [CrossRef]
2. Smith, C.I.E.; Islam, T.C.; Mattsson, P.T.; Mohamed, A.J.; Nore, B.F.; Vihinen, M. The Tec family of cytoplasmic tyrosine kinases: Mammalian Btk, Bmx, Itk, Tec, Txk and homologs in other species. *BioEssays* **2001**, *23*, 436–446. [CrossRef] [PubMed]
3. Horwood, D.N.J.; Urbaniak, D.A.M.; Danks, D.L. Tec Family Kinases in Inflammation and Disease. *Int. Rev. Immunol.* **2012**, *31*, 87–103. [CrossRef] [PubMed]
4. Gehringer, M. Covalent kinase inhibitors: An overview. In *Topics in Medicinal Chemistry*; Springer: Berlin/Heidelberg, Germany, 2020; pp. 1–52. [CrossRef]
5. Abdeldayem, A.; Raouf, Y.S.; Constantinescu, S.N.; Moriggl, R.; Gunning, P.T. Advances in covalent kinase inhibitors. *Chem. Soc. Rev.* **2020**, *49*, 2617–2687. [CrossRef] [PubMed]
6. Guo, Y.; Liu, Y.; Hu, N.; Yu, D.; Zhou, C.; Shi, G.; Zhang, B.; Wei, M.; Liu, J.; Luo, L.; et al. Discovery of Zanubrutinib (BGB-3111), a Novel, Potent, and Selective Covalent Inhibitor of Bruton's Tyrosine Kinase. *J. Med. Chem.* **2019**, *62*, 7923–7940. [CrossRef]
7. Lechner, K.S.; Neurath, M.F.; Weigmann, B. Role of the IL-2 inducible tyrosine kinase ITK and its inhibitors in disease pathogenesis. *J. Mol. Med.* **2020**, *98*, 1385–1395. [CrossRef]
8. Arrowsmith, C.H.; Audia, J.E.; Austin, C.; Baell, J.; Bennett, J.; Blagg, J.; Bountra, C.; Brennan, P.E.; Brown, P.J.; Bunnage, M.E.; et al. The promise and peril of chemical probes. *Nat. Chem. Biol.* **2015**, *11*, 536–541. [CrossRef]
9. Chaikuad, A.; Koch, P.; Laufer, S.A.; Knapp, S. The Cysteinome of Protein Kinases as a Target in Drug Development. *Angew. Chem. Int. Ed.* **2018**, *57*, 4372–4385. [CrossRef]
10. Liu, Q.; Sabnis, Y.; Zhao, Z.; Zhang, T.; Buhrlage, S.J.; Jones, L.H.; Gray, N.S. Developing Irreversible Inhibitors of the Protein Kinase Cysteinome. *Chem. Biol.* **2013**, *20*, 146–159. [CrossRef]
11. Leproult, E.; Barluenga, S.; Moras, D.; Wurtz, J.-M.; Winssinger, N. Cysteine Mapping in Conformationally Distinct Kinase Nucleotide Binding Sites: Application to the Design of Selective Covalent Inhibitors. *J. Med. Chem.* **2011**, *54*, 1347–1355. [CrossRef]
12. Gehringer, M.; Laufer, S.A. Emerging and Re-Emerging Warheads for Targeted Covalent Inhibitors: Applications in Medicinal Chemistry and Chemical Biology. *J. Med. Chem.* **2019**, *62*, 5673–5724. [CrossRef] [PubMed]
13. Barf, T.; Kaptein, A. Irreversible Protein Kinase Inhibitors: Balancing the Benefits and Risks. *J. Med. Chem.* **2012**, *55*, 6243–6262. [CrossRef] [PubMed]
14. Gehringer, M. Covalent inhibitors: Back on track? *Future Med. Chem.* **2020**, *12*, 1363–1368. [CrossRef] [PubMed]
15. Baillie, T.A. Approaches to mitigate the risk of serious adverse reactions in covalent drug design. *Expert Opin. Drug Discov.* **2020**. [CrossRef] [PubMed]
16. Schröder, M.; Tan, L.; Wang, J.; Liang, Y.; Gray, N.S.; Knapp, S.; Chaikuad, A. Catalytic Domain Plasticity of MKK7 Reveals Structural Mechanisms of Allosteric Activation and Diverse Targeting Opportunities. *Cell Chem. Biol.* **2020**, *27*, 1285–1295.e4. [CrossRef]
17. Barf, T.; Covey, T.; Izumi, R.; van de Kar, B.; Gulrajani, M.; van Lith, B.; van Hoek, M.; de Zwart, E.; Mittag, D.; Demont, D.; et al. Acalabrutinib (ACP-196): A Covalent Bruton Tyrosine Kinase Inhibitor with a Differentiated Selectivity and In Vivo Potency Profile. *J. Pharmacol. Exp. Ther.* **2017**, *363*, 240–252. [CrossRef] [PubMed]
18. Forster, M.; Gehringer, M.; Laufer, S.A. Recent advances in JAK3 inhibition: Isoform selectivity by covalent cysteine targeting. *Bioorg. Med. Chem. Lett.* **2017**, *27*, 4229–4237. [CrossRef]

19. Gehringer, M.; Forster, M. Covalent janus kinase 3 inhibitors. In *Topics in Medicinal Chemistry*; Springer: Berlin/Heidelberg, Germany, 2020; pp. 1–32. [CrossRef]
20. Telliez, J.-B.; Dowty, M.E.; Wang, L.; Jussif, J.; Lin, T.; Li, L.; Moy, E.; Balbo, P.; Li, W.; Zhao, Y.; et al. Discovery of a JAK3-Selective Inhibitor: Functional Differentiation of JAK3-Selective Inhibition over pan-JAK or JAK1-Selective Inhibition. *ACS Chem. Biol.* **2016**, *11*, 3442–3451. [CrossRef]
21. Forster, M.; Chaikuad, A.; Bauer, S.M.; Holstein, J.; Robers, M.B.; Corona, C.R.; Gehringer, M.; Pfaffenrot, E.; Ghoreschi, K.; Knapp, S.; et al. Selective JAK3 Inhibitors with a Covalent Reversible Binding Mode Targeting a New Induced Fit Binding Pocket. *Cell Chem. Biol.* **2016**, *23*, 1335–1340. [CrossRef]
22. Forster, M.; Chaikuad, A.; Dimitrov, T.; Döring, E.; Holstein, J.; Berger, B.-T.; Gehringer, M.; Ghoreschi, K.; Müller, S.; Knapp, S.; et al. Development, Optimization, and Structure–Activity Relationships of Covalent-Reversible JAK3 Inhibitors Based on a Tricyclic Imidazo[5,4-d]pyrrolo[2,3-b]pyridine Scaffold. *J. Med. Chem.* **2018**, *61*, 5350–5366. [CrossRef]
23. Shraga, A.; Olshvang, E.; Davidzohn, N.; Khoshkenar, P.; Germain, N.; Shurrush, K.; Carvalho, S.; Avram, L.; Albeck, S.; Unger, T.; et al. Covalent Docking Identifies a Potent and Selective MKK7 Inhibitor. *Cell Chem. Biol.* **2019**, *26*, 98–108.e5. [CrossRef] [PubMed]
24. Wolle, P.; Engel, J.; Smith, S.; Goebel, L.; Hennes, E.; Lategahn, J.; Rauh, D. Characterization of Covalent Pyrazolopyrimidine–MKK7 Complexes and a Report on a Unique DFG-in/Leu-in Conformation of Mitogen-Activated Protein Kinase Kinase 7 (MKK7). *J. Med. Chem.* **2019**, *62*, 5541–5546. [CrossRef] [PubMed]
25. Zapf, C.W.; Gerstenberger, B.S.; Xing, L.; Limburg, D.C.; Anderson, D.R.; Caspers, N.; Han, S.; Aulabaugh, A.; Kurumbail, R.; Shakya, S.; et al. Covalent Inhibitors of Interleukin-2 Inducible T Cell Kinase (Itk) with Nanomolar Potency in a Whole-Blood Assay. *J. Med. Chem.* **2012**, *55*, 10047–10063. [CrossRef] [PubMed]
26. Liu, F.; Zhang, X.; Weisberg, E.; Chen, S.; Hur, W.; Wu, H.; Zhao, Z.; Wang, W.; Mao, M.; Cai, C.; et al. Discovery of a Selective Irreversible BMX Inhibitor for Prostate Cancer. *ACS Chem. Biol.* **2013**, *8*, 1423–1428. [CrossRef] [PubMed]
27. Seixas, J.D.; Sousa, B.B.; Marques, M.C.; Guerreiro, A.; Traquete, R.; Rodrigues, T.; Albuquerque, I.S.; Sousa, M.; Lemos, A.R.; Sousa, P.M.F.; et al. Rationally Designed Potent BMX Inhibitors Reveals Mode of Covalent Binding at the Atomic Level. *ChemRxiv* **2020**. [CrossRef]
28. Machleidt, T.; Woodroofe, C.C.; Schwinn, M.K.; Méndez, J.; Robers, M.B.; Zimmerman, K.; Otto, P.; Daniels, D.L.; Kirkland, T.A.; Wood, K.V. NanoBRET—A Novel BRET Platform for the Analysis of Protein–Protein Interactions. *ACS Chem. Biol.* **2015**, *10*, 1797–1804. [CrossRef]
29. Liang, X.; Lv, F.; Wang, B.; Yu, K.; Wu, H.; Qi, Z.; Jiang, Z.; Chen, C.; Wang, A.; Miao, W.; et al. Discovery of 2-((3-Acrylamido-4-methylphenyl)amino)-N-(2-methyl-5-(3,4,5-trimethoxybenzamido)phenyl)-4-(methylamino)pyrimidine-5-carboxamide (CHMFL-BMX-078) as a Highly Potent and Selective Type II Irreversible Bone Marrow Kinase in the X Chromosome (BMX) Kinase Inhibitor. *J. Med. Chem.* **2017**, *60*, 1793–1816. [CrossRef]
30. Backes, A.C.; Müller, G.; Sennhenn, P.C. Design principles of deep pocket-targeting protein kinase inhibitors. In *Protein Kinases as Drug Targets*; John Wiley & Sons, Ltd.: Hoboken, NJ, USA, 2011; pp. 145–193, ISBN 978-3-527-63347-0.
31. Anastassiadis, T.; Deacon, S.W.; Devarajan, K.; Ma, H.; Peterson, J.R. Comprehensive assay of kinase catalytic activity reveals features of kinase inhibitor selectivity. *Nat. Biotechnol.* **2011**, *29*, 1039–1045. [CrossRef]
32. Duan, J.J.-W.; Lu, Z.; Jiang, B.; Yang, B.V.; Doweyko, L.M.; Nirschl, D.S.; Haque, L.E.; Lin, S.; Brown, G.; Hynes, J., Jr.; et al. Discovery of pyrrolo[1,2-b]pyridazine-3-carboxamides as Janus kinase (JAK) inhibitors. *Bioorg. Med. Chem. Lett.* **2014**, *24*, 5721–5726. [CrossRef]
33. Bauer, S.M.; Gehringer, M.; Laufer, S.A. A direct enzyme-linked immunosorbent assay (ELISA) for the quantitative evaluation of Janus Kinase 3 (JAK3) inhibitors. *Anal. Methods* **2014**, *6*, 8817–8822. [CrossRef]
34. Traxler, P. Tyrosine kinase inhibitors in cancer treatment (Part II). *Expert Opin. Ther. Pat.* **1998**, *8*, 1599–1625. [CrossRef]
35. Ishiyama, T.; Murata, M.; Miyaura, N. Palladium(0)-Catalyzed Cross-Coupling Reaction of Alkoxydiboron with Haloarenes: A Direct Procedure for Arylboronic Esters. *J. Org. Chem.* **1995**, *60*, 7508–7510. [CrossRef]
36. Bhat, P.V.; Dere, R.T.; Ravikumar, S.; Hindupur, R.M.; Pati, H.N. Efficient and Scalable Process for Synthesis of 5-Nitro-7-azaindole. *Org. Process Res. Dev.* **2015**, *19*, 1282–1285. [CrossRef]

37. Holopainen, T.; Räsänen, M.; Anisimov, A.; Tuomainen, T.; Zheng, W.; Tvorogov, D.; Hulmi, J.J.; Andersson, L.C.; Cenni, B.; Tavi, P.; et al. Endothelial Bmx tyrosine kinase activity is essential for myocardial hypertrophy and remodeling. *Proc. Natl. Acad. Sci. USA* **2015**, *112*, 13063–13068. [CrossRef] [PubMed]
38. Bagheri-Yarmand, R.; Mandal, M.; Taludker, A.H.; Wang, R.-A.; Vadlamudi, R.K.; Kung, H.-J.; Kumar, R. Etk/Bmx Tyrosine Kinase Activates Pak1 and Regulates Tumorigenicity of Breast Cancer Cells. *J. Biol. Chem.* **2001**, *276*, 29403–29409. [CrossRef]
39. Potter, D.S.; Kelly, P.; Denneny, O.; Juvin, V.; Stephens, L.R.; Dive, C.; Morrow, C.J. BMX Acts Downstream of PI3K to Promote Colorectal Cancer Cell Survival and Pathway Inhibition Sensitizes to the BH3 Mimetic ABT-737. *Neoplasia* **2014**, *16*, 147–157, W9–W16. [CrossRef]
40. Kabsch, W. XDS. *Acta Crystallogr. Sect. D Biol. Crystallogr.* **2010**, *66*, 125–132. [CrossRef]
41. Evans, P.R.; Murshudov, G.N. Scaling and assessment of data quality. *Acta Crystallogr. Sect. D Biol. Crystallogr.* **2013**, *69*, 1204–1214. [CrossRef]
42. McCoy, A.J. Acknowledging Errors: Advanced Molecular Replacement with Phaser. *Methods Mol. Biol.* **2017**, *1607*, 421–453. [CrossRef]
43. Emsley, P. Tools for ligand validation in Coot. *Acta Crystallogr. Sect. D Biol. Crystallogr.* **2017**, *73*, 203–210. [CrossRef]
44. Skubák, P.; Murshudov, G.N.; Pannu, N.S. Direct incorporation of experimental phase information in model refinement. *Acta Crystallogr. Sect. D Biol. Crystallogr.* **2004**, *60*, 2196–2201. [CrossRef] [PubMed]
45. Fedorov, O.; Niesen, F.H.; Knapp, S. Kinase inhibitor selectivity profiling using differential scanning fluorimetry. In *Kinase Inhibitors: Methods and Protocols*; Kuster, B., Ed.; Methods in Molecular Biology; Humana Press: Totowa, NJ, USA, 2012; pp. 109–118, ISBN 978-1-61779-337-0.
46. Fedorov, O.; Marsden, B.; Pogacic, V.; Rellos, P.; Müller, S.; Bullock, A.N.; Schwaller, J.; Sundström, M.; Knapp, S. A systematic interaction map of validated kinase inhibitors with Ser/Thr kinases. *Proc. Natl. Acad. Sci. USA* **2007**, *104*, 20523–20528. [CrossRef] [PubMed]
47. Robers, M.B.; Vasta, J.D.; Corona, C.R.; Ohana, R.F.; Hurst, R.; Jhala, M.A.; Comess, K.M.; Wood, K.V. Quantitative, real-time measurements of intracellular target engagement using energy transfer. In *Systems Chemical Biology: Methods and Protocols*; Ziegler, S., Waldmann, H., Eds.; Methods in Molecular Biology; Springer: New York, NY, USA, 2019; pp. 45–71, ISBN 978-1-4939-8891-4.

2

The 14-3-3 Proteins as Important Allosteric Regulators of Protein Kinases

Veronika Obsilova [1,*] and Tomas Obsil [1,2,*]

[1] Department of Structural Biology of Signaling Proteins, Division BIOCEV, Institute of Physiology of the Czech Academy of Sciences, 25250 Vestec, Czech Republic
[2] Department of Physical and Macromolecular Chemistry, Faculty of Science, Charles University, 12843 Prague, Czech Republic
* Correspondence: veronika.obsilova@fgu.cas.cz (V.O.); obsil@natur.cuni.cz (T.O.)

Abstract: Phosphorylation by kinases governs many key cellular and extracellular processes, such as transcription, cell cycle progression, differentiation, secretion and apoptosis. Unsurprisingly, tight and precise kinase regulation is a prerequisite for normal cell functioning, whereas kinase dysregulation often leads to disease. Moreover, the functions of many kinases are regulated through protein–protein interactions, which in turn are mediated by phosphorylated motifs and often involve associations with the scaffolding and chaperon protein 14-3-3. Therefore, the aim of this review article is to provide an overview of the state of the art on 14-3-3-mediated kinase regulation, focusing on the most recent mechanistic insights into these important protein–protein interactions and discussing in detail both their structural aspects and functional consequences.

Keywords: 14-3-3; kinase; phosphorylation; RAF kinase; ASK1; CaMKK2; PI4KB; LRRK2; PKC

---

1. Introduction

Phosphorylation by kinases governs many key cellular and extracellular processes, such as transcription, cell cycle progression, differentiation, secretion and apoptosis. Tight and precise kinase regulation is a prerequisite for normal cell functioning. Conversely, kinase dysregulation often leads to disease. The function of many kinases is regulated through protein–protein interactions. These interactions are in turn mediated by phosphorylated motifs and often involve associations with the scaffolding and chaperon protein 14-3-3.

14-3-3-mediated regulation has already been demonstrated for more than 30 kinases, and more than 170 kinases contain phosphosites known to conform to 14-3-3-binding sites [1]. However, many details concerning these interactions, especially the exact role of 14-3-3 binding and the mechanism of action of 14-3-3-mediated kinase regulation, remain elusive because only a few 14-3-3:kinase complexes have been structurally characterized so far, mostly using low-resolution approaches [2–4]. Nevertheless, several structures of rapidly accelerated fibrosarcoma B-type (B-RAF) kinase bound to 14-3-3 have been recently solved, thus providing the first glimpse, at atomic resolution, into a structural mechanism of 14-3-3-mediated kinase regulation [5–8].

Considering the above, the aim of this review article is to provide an overview of the current knowledge on 14-3-3-mediated kinase regulation. Although previous studies have shown that 14-3-3 proteins are involved in the regulation of many kinases, this review focuses only on complexes, whose available data enabled more detailed mechanistic insight into the 14-3-3-mediated regulation. Based on these data, we discuss structural aspects and functional consequences of these important protein–protein interactions.

2. 14-3-3 Proteins

The 14-3-3 protein family encompasses structurally similar acidic proteins found in all eukaryotes. Although 14-3-3 proteins were originally characterized as abundant brain proteins in 1967, with no clear function, they were later identified as universal adaptor and chaperon proteins, known for interacting with hundreds of other proteins and for participating in the regulation of almost every major cellular pathway [9–12]. Since their discovery, seven human 14-3-3 isoforms have been identified α/β, γ, ε, ζ/δ, η, τ, σ, where α and δ are the phosphorylated forms of the β and ζ isoforms [13]. Other organisms, such as budding yeast, contain only two isoforms, Bmh1 and Bmh2, whereas plants express 15 different 14-3-3 isoforms [14,15]. In addition, 14-3-3 proteins form a stable homo- or heterodimers, with each protomer consisting of nine tightly packed antiparallel α-helices, which create an amphipathic ligand-binding groove (Figure 1A) [16,17]. The 14-3-3 dimers can be destabilized by the lipid sphingosine and phosphorylation of Ser^{58} at the dimer interface [18,19] and it has been shown that such destabilization induces downregulation of Raf-MAPK and PI3K-Akt signaling [20].

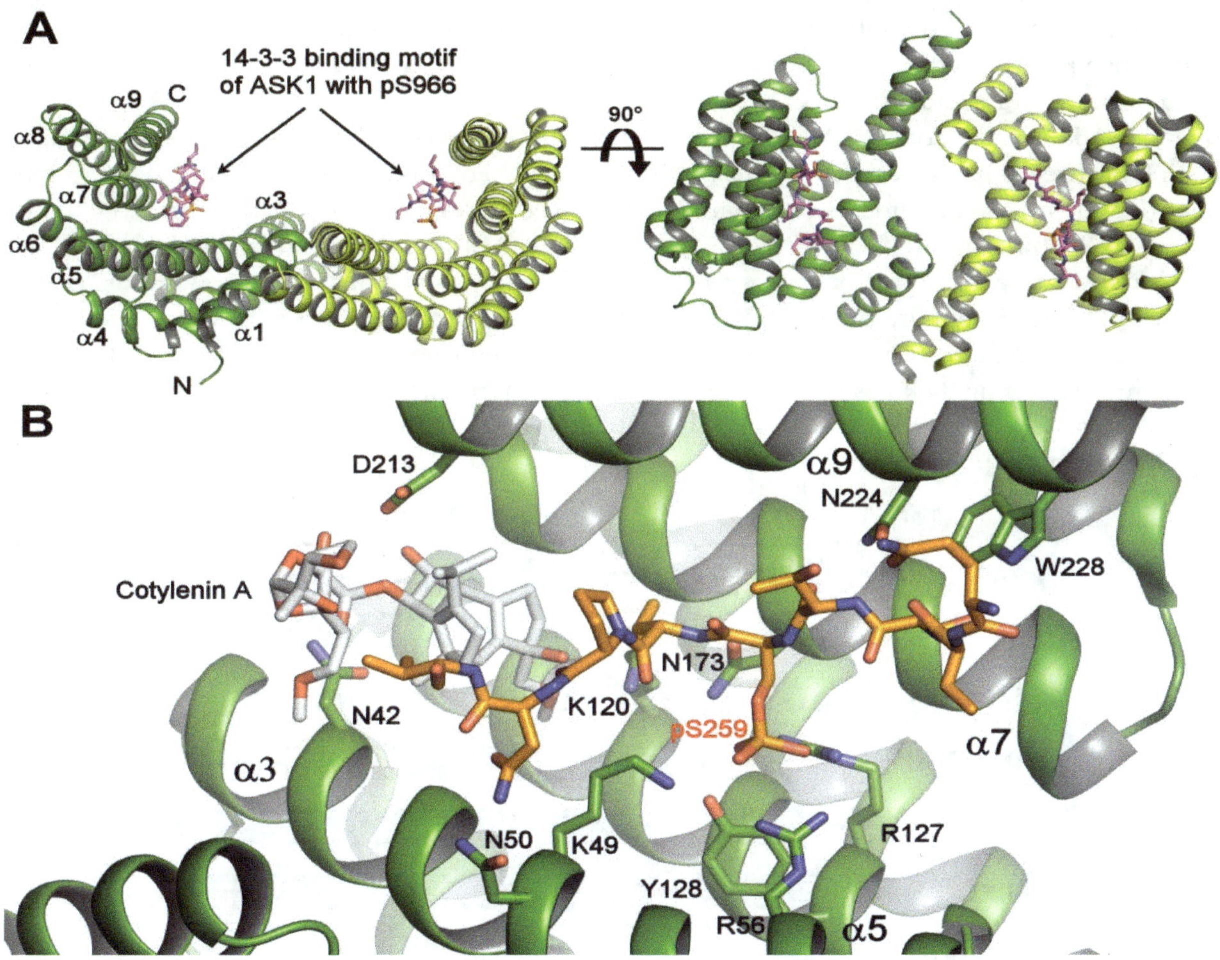

Figure 1. Crystal structure of the 14-3-3 protein with bound phosphopeptides. (**A**) The 14-3-3 binding motif of apoptosis signal-regulating kinase 1 (ASK1) (sequence RSIpS^{966}LPVP) bound to human 14-3-3ζ (PDB ID: 6EJL). The 14-3-3 protein molecule is a dimer with a two-fold symmetry, and each protomer consists of nine antiparallel α-helices and contains an amphipathic groove which is a binding site for the phosphorylated motifs; (**B**) the ternary complex between human 14-3-3σ (shown in green), the 14-3-3 binding motif pSer^{259} of C-RAF (shown in orange), and Cotylenin A (shown in gray) (PDB ID: 4IHL [21]). The 14-3-3σ residues that make polar contacts with the phosphopeptide and Cotylenin are shown as sticks. The Cotylenin A considerably enhances the binding of the C-RAF pSer^{259}-motif to 14-3-3. The figure was prepared with PyMOL.

3. 14-3-3 Protein-Dependent Regulation of Selected Kinases

Phosphorylation by protein kinases governs many key cellular and extracellular processes by participating in various signaling pathways. Therefore, tight kinase regulation is a key prerequisite for normal cell functioning and its dysfunction often leads to disease. Many kinases are regulated in the 14-3-3 protein-dependent manner; however, the structural data available so far enable a detailed understanding of these regulations for only a few kinases. In this review article, we discuss regulation of RAF, ASK1, leucine-rich repeat protein kinase-2 (LRRK2), protein kinase C (PKC), calcium/calmodulin-dependent protein kinase kinase (CaMKK), and phosphatidylinositol-4-kinase-IIIβ (PI4KB), for which the mechanistic insights into their 14-3-3 protein-mediated regulation are available.

3.1. RAF Kinases

RAF kinases are key components of the RAS-RAF-MEK-ERK signaling pathway. These kinases were among the first oncoproteins characterized in the 1980s, and B-RAF is the most frequently mutated protein kinase in various cancers, with V600E standing out as the most common and resistant mutation [22,23]. Selective inhibitors targeting RAF proteins are currently at various stages of clinical trials because RAF activity is strictly regulated, and any misbalance in ERK signaling activation or dysregulation usually leads to cancer.

Mammalian RAF kinases (A-RAF, B-RAF and C-RAF) share the same domain organization, consisting of three conserved regions (CR1-CR3) (Figure 2). The CR1 region contains the RAS-binding domain (RBD) and the cysteine-rich domain (CRD), the CR2 region is rich in Ser/Thr and contains an inhibitory 14-3-3 binding site (Ser^{259} in C-RAF, Ser^{365} in B-RAF), and the third CR3 region comprises the kinase domain and a dominant activatory 14-3-3 binding site (Ser^{621} and Ser^{729} in C-RAF and B-RAF, respectively). C-RAF contains an additional inhibitory 14-3-3 binding site Ser^{233} located between CR1 and CR2 [24]. In the autoinhibited state, the N-terminal region of RAF interacts with the catalytic domain, thus inhibiting its kinase activity [22,25]. Furthermore, the inactive state of RAF is maintained by the 14-3-3 protein dimer, which simultaneously recognizes both phosphorylated motifs bordering the catalytic domain [6,7,24,26].

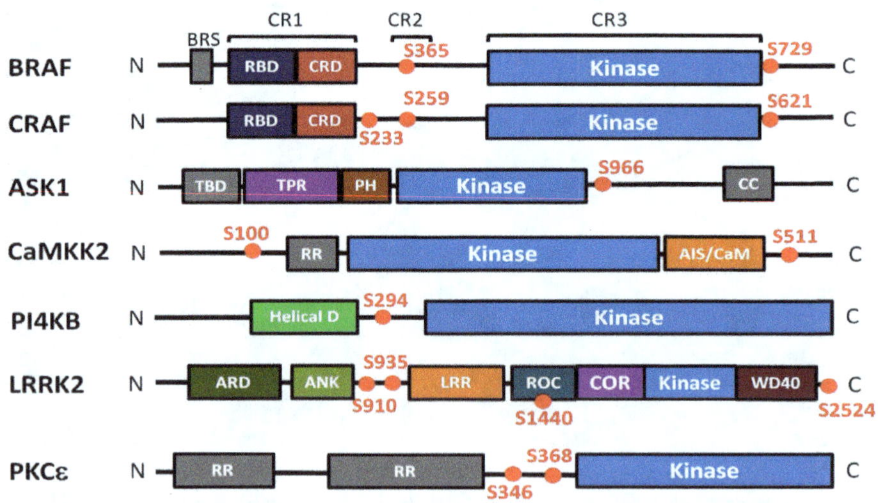

Figure 2. Domain structure and positions of 14-3-3 binding motifs of kinases regulated in the 14-3-3-dependent manner. Kinase domains are shown in blue, the 14-3-3 binding sites are shown as red dots. BRS, B-RAF specific domain; RBD, Ras binding domain; CRD, cysteine rich domain; TBD, TRX-binding domain; TPR, tetratricopeptide repeats domain; PH, pleckstrin homology domain; CC, coiled-coil region; RR, N-terminal regulatory site; AIS/CaM, autoinhibitory segment/CaM-binding domain; Helical, helical domain; ARD, armadillo repeat; ANK, ankyrin repeat; LRR, leucine-rich repeat; ROC, Ras of the complex GTPase domain; COR, C-terminal of ROC; WD40, WD40 or beta-transducin repeat. The kinases differ in their length and they are not shown on the same scale.

Mitogen stimulation induces RAS-GTP binding to the RAF RBD, which anchors RAF to the plasma membrane, followed by dephosphorylation of the 14-3-3 binding sites within the CR2 by PP1 and PP2A phosphatases. This dephosphorylation releases the 14-3-3 protein from this site and enables RAF binding to the plasma membrane through the CRD. As a result, RAF moves closer in proximity to other kinases involved in this process (SRC family kinases and casein kinase 2), which phosphorylate several activating sites within the N-terminal region of RAF.

The key event in RAF activation is the 14-3-3-mediated dimerization of the RAF kinase domain, which promotes the catalytic activity of RAF. RAF dimerization is stabilized by the 14-3-3 dimer, which simultaneously binds two RAF molecules through its C-terminal 14-3-3 binding motifs located within the CR3 C-terminal to the kinase domain [5–8,27]. Activated RAF then phosphorylates MEK, which activates ERK, which, in turn, activates a wide range of targets associated with growth, proliferation, differentiation, survival, and migration [22]. Negative feedback is facilitated by ERK phosphorylation of several distinct inhibitory sites within the N-terminal domains of RAF followed by RAF dissociation from the activated RAS and disruption of RAF dimers.

14-3-3 proteins were identified as key RAF-binding partners, which control RAF activity, in 1994 [28–31]. The crystal structure of the peptide containing the motif around phosphoserine $pSer^{259}$ from the CR2 of C-RAF bound to 14-3-3 was one of the first structures of the 14-3-3:phosphopeptide complexes ever published, showing how 14-3-3 proteins recognize phosphorylated targets [32]. Three 14-3-3 binding sites were identified in C-RAF: inhibitory Ser^{233} between CR1 and CR2, inhibitory Ser^{259} within CR2 and activatory Ser^{621} in CR3 [24,26]. C-RAF phosphorylation by PKA promotes its association with the 14-3-3 protein via Ser^{233} and Ser^{259} [24]. The crystal structures reported by the Ottmann group indicated that the 14-3-3 dimer can simultaneously recognize both phosphorylated motifs Ser^{233} and Ser^{259}, with $pSer^{259}$ functioning as a high affinity binding site and $pSer^{233}$ as a low-affinity binding site [33]. Moreover, both inhibitory 14-3-3 binding sites N-terminal to the kinase domain (Ser^{233} and Ser^{259}), but not the activatory site Ser^{621}, can be stabilized by small molecule compounds targeting cavities at the 14-3-3:C-RAF interface (Figure 1B) [21,34]. Accordingly, such stabilization of the interaction between 14-3-3 and the inhibitory C-RAF 14-3-3 binding motifs may be an effective strategy for the treatment of RAS-connected diseases [21].

The crucial step in both normal and disease-connected RAF signaling is RAF dimerization through its kinase domains, and any destabilization of RAF dimers formation may contribute to RAF inhibition [35,36]. The recently reported cryo-EM and crystal structures of 14-3-3:B-RAF complexes have provided detailed mechanistic insights into how 14-3-3 proteins regulate B-RAF activity both negatively and positively (Figure 3) [5–8].

Negative regulation: In the structure of the autoinhibited B-RAF:MEK1:14-3-3 complex, the 14-3-3 protein simultaneously binds phosphorylated Ser^{365} and Ser^{729} motifs bordering the kinase domain of one B-RAF molecule sequesters the CRD of B-RAF within the central channel of the 14-3-3 dimer and obstructs the dimerization interface of the B-RAF kinase domain (Figure 3A). Both the position of the C-helix and the conformation of the activation segment in the kinase domain correspond to the autoinhibited state. Thus, the 14-3-3 protein appears to keep B-RAF in an inactive state by blocking the membrane recruitment of B-RAF and by preventing B-RAF kinase dimerization through steric occlusion.

Positive regulation: In turn, the structure of the active dimeric B-RAF:MEK:14-3-3 complex revealed that B-RAF dephosphorylation at Ser^{365} induces a structural rearrangement, resulting in an active B-RAF dimer stabilized by the 14-3-3 dimer, which anchors Ser^{729}-containing C-terminal segments of two B-RAF molecules (Figure 3B). Within this complex, the B-RAF kinase domains are oriented in a back-to-back fashion, with the C-helix in an inward position, in line with the active conformation.

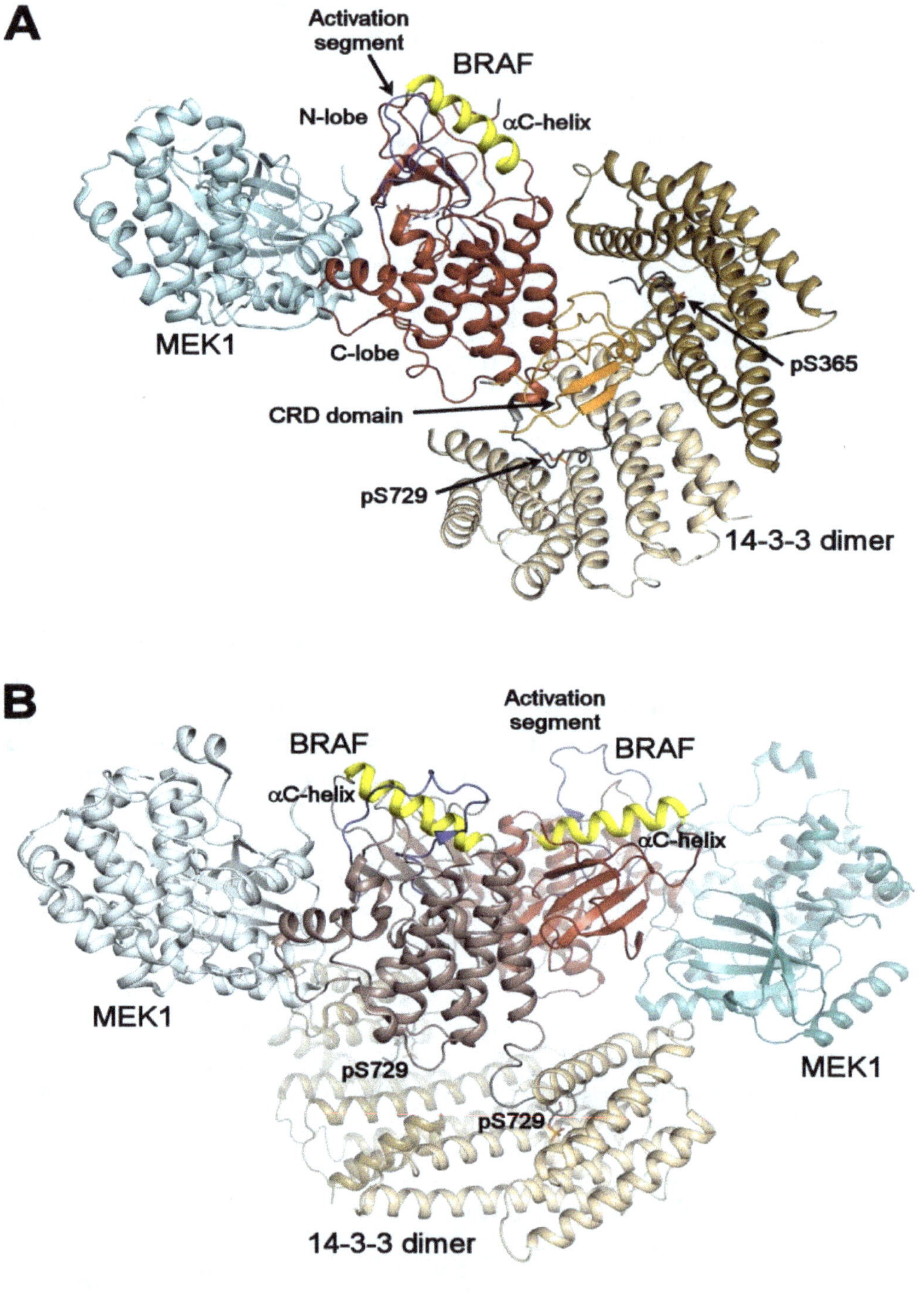

Figure 3. Autoinhibited and active B-RAF:MEK1:14-3-3 complexes. (**A**) structure of the autoinhibited BRAF:MEK1:14-3-3 complex (PDB ID: 6NYB [6,7,24,26]). The 14-3-3 dimer simultaneously binds both $pSer^{365}$ and $pSer^{729}$ motifs (shown in gray) bordering the B-RAF kinase domain (shown in dark red). The CRD domain (shown in orange) is sequestered within the central channel of the 14-3-3 dimer. The position of the αC-helix (shown in yellow) and the activation segment (shown in blue) correspond to the autoinhibited state. The 14-3-3 protein inhibits B-RAF by blocking its membrane localization dimerization through steric occlusion; (**B**) structure of the active B-RAF:MEK1:14-3-3 complex (PDB ID: 6Q0J [6,7,24,26]). Dephosphorylation of the N-terminal motif ($pSer^{365}$) causes structural rearrangement resulting in an active B-RAF dimer stabilized by the 14-3-3 dimer through anchoring C-terminal $pSer^{729}$ motifs of two B-RAF molecules. The B-RAF kinase domains are oriented in the back-to-back fashion with the αC-helix in a position consistent with the active conformation. The figure was prepared with PyMOL.

3.2. Apoptosis Signal-Regulating Kinase 1 (ASK1)

ASK1 belongs to a group of mitogen-activated protein kinase kinase kinases (MAP3Ks). The hierarchical system of the mitogen-activated protein (MAP) kinase cascade is formed by upstream kinases MAP3Ks that phosphorylate intermediate MAP kinases MAP2Ks (MKK3, MKK4 and MKK6) which in turn phosphorylate terminal MAP kinases (p38, JNK or ERK) [37]. Active, phosphorylated MAPKs phosphorylate a wide range of substrates to stimulate cell proliferation, cell death, or inflammation [38]. The aberrantly enhanced ASK1-MAPK signaling leads to many disorders and neurological diseases, such as amyotrophic lateral sclerosis, multiple sclerosis and Parkinson's, Alzheimer´s, and Huntington´s disease. Due to the complexity of the signaling processes in which ASK1 participates, mechanistic and structural insights into the ASK1 signaling pathways are necessary for developing inhibitors as therapeutical targets against these devastating diseases [39].

Human ASK1 contains a thioredoxin binding domain (TBD) at the N-terminus, the central regulatory region (CRR) with a tumor necrosis factor receptor associated factor (TRAF)-binding region, a catalytic domain (CD), in the middle of the molecule, and C-terminal coiled-coiled (CC) and sterile-alpha motif (SAM) domains, which are involved in ASK1 homo- and hetero-oligomerization with the closely related ASK2 [40–43].

Under normal, non-stress conditions, ASK1/ASK2 hetero-oligomers interact with several other proteins, including thioredoxin (TRX) and 14-3-3, forming a high molecular mass complex known as the ASK1 signalosome [44]. Federspiel et al. have shown that the ASK1 signalosome contains ASK2 at a 1:1 stoichiometric ratio with ASK1 and 14-3-3 proteins [45]. The 14-3-3 protein binds to the phosphorylated motifs containing $pSer^{966}$ and $pSer^{964}$ in ASK1 and ASK2, respectively, located C-terminally to the CD domains, suppressing both the autokinase and the transkinase activities of ASK1 (Figure 2) [46,47]. Both 14-3-3 and TRX are physiological inhibitors of ASK1.

Under stress conditions, 14-3-3 and TRX dissociate from ASK1. Then, the N-terminal parts of ASK1/ASK2 homo-oligomerize, and TRAF 2 and 6 bind to the CRR domain(s). After autophosphorylation of ASK1 Thr^{838} within the activation loop, ASK1 is activated, phosphorylating its downstream targets [46,48–50]. The crystal structure of the ASK1-CRR domain showed that a conserved surface of CRR, formed by the pleckstrin homology (PH) domain and the tetratricopeptide repeat (TPR) region, is necessary for ASK1 activity. CRR has a dual role: on the one hand, it stimulates ASK1 activity through the PH domain; on the other hand, it aids ASK1 autoinhibition by delivering the TBD and CD domains near each other [51].

The closeness of the 14-3-3 binding motif (Ser^{966}) to ASK1-CD suggests that the 14-3-3 protein modulates the enzyme activity or accessibility of its active site, as previously shown for other enzymes regulated by 14-3-3 [52,53]. The low-resolution structural characterization of the complex between ASK1-CD and 14-3-3 indicated that the 14-3-3 protein might inhibit ASK1 by modulating the structure of its active site, by sterically blocking the phosphorylation of Thr^{838} within the activation loop and by blocking the interactions between ASK1 and its substrates [2]. Moreover, ASK1-CD forms a stable dimer in solution, and its complex with the 14-3-3 protein, which is also dimeric, is conformationally heterogeneous, lacking a clearly defined interface, presumably due to the absence of another protein, such as ASK2 [46,54]. The SAXS-based structural model of the ASK1:14-3-3ζ complex also suggest that the $pSer^{966}$ in one protomer of the ASK1 dimer binds to one protomer of the 14-3-3 dimer, leaving the other protomers of 14-3-3 and ASK1 ligand-free [2].

Because the 14-3-3 protein is a physiological inhibitor of ASK1, the ASK1:14-3-3 and ASK2:14-3-3 complexes are potential targets for therapeutic intervention as an alternative or complimentary strategy in suppressing the ASK1 activity. Recent advances in the development of small molecule compounds aimed at stabilizing 14-3-3 protein-kinase interactions [21,55–58] have demonstrated the feasibility of this approach.

3.3. *Calcium/Calmodulin-Dependent Protein Kinase Kinases (CaMKK)*

CaMKKs are Ser/Thr protein kinases that function as upstream elements of the CaMK signaling cascade where they phosphorylate and activate two downstream CaMKs, CaMKI and CaMKIV [59–62]. Upstream extracellular signals, including insulin, hormones (insulin, thyroid hormone, growth hormone), glucose and amino acids, trigger the increase in the intracellular concentration of free Ca^{2+}, which is followed by the activation of Ca^{2+}/CaM targets such as CaMKs. In the inactive state, the kinase activity of CaMKs is blocked by the auto-inhibitory segment (AIS), which interacts with the kinase domain (KD) and prevents substrate binding and/or structurally modulates the catalytic site [63]. This autoinhibition is relieved by Ca^{2+}/CaM binding to the region overlapping with the AIS. CaMKI participates in the regulation of hippocampal memory formation, neuronal migration, cell survival and synaptogenesis, whereas CaMKIV is involved in the control of protein synthesis and gene expression programs in response to nutrients and hormones [64,65].

Mammalian CaMKKs include two isoforms: CaMKK1 and CaMKK2. These isoforms have a similar domain organization, that is, the catalytic kinase domain is followed by the C-terminal AIS, which partly overlaps with the Ca^{2+}/CaM-binding region [62,66]. However, these CaMKK isoforms exhibit different biochemical properties because CaMKK1 is strictly regulated in a Ca^{2+}/CaM-dependent manner, whereas CaMKK2 also shows considerable Ca^{2+}/CaM-independent activity. In addition, CaMKK2 is an important upstream activator of the histone deacetylase SIRT 1 and AMP-activated protein kinase (AMPK) [67,68], a key regulator of energy homeostasis, inflammation, and autophagy whose dysregulation has been implicated in chronic diseases such as obesity, diabetes and cancer [60,69].

Both CaMKK isoforms are partly inhibited upon phosphorylation by PKA at multiple sites in a process involving 14-3-3 protein binding. CaMKK1 possesses five PKA phosphorylation sites (Ser^{52}, Ser^{74}, Thr^{108}, Ser^{458}, and Ser^{475}), whereas CaMKK2 contains only four such sites (Ser^{100}, Thr^{145}, Ser^{495}, and Ser^{511}). Two sites phosphorylated by PKA are responsible for 14-3-3 protein binding; the first site is located at the N-terminus (Ser^{74} in CaMKK1; Ser^{100} in CaMKK2), and the second site is on the C-terminus (Ser^{475} in CaMKK1; Ser^{511} in CaMKK2) (Figure 2) [70–73]. 14-3-3 protein recruitment by CaMKK1 partly suppresses its catalytic activity, whereas CaMKK2 activity remains unaltered, suggesting different regulation mechanisms for these isoforms [3,70,71,74]. Moreover, it has recently been shown that phosphorylation of CaMKK2 at Ser^{495} is involved in regulating VEGF-induced AMPK activation, a pathway shown to regulate angiogenesis [75].

Low-resolution structural analysis of the CaMKK2:14-3-3 complex indicated that the formation of the complex affects the structure of several regions of CaMKK2 outside the 14-3-3 binding motifs, including the regulatory region within the N-terminal extension and the C-lobe of the kinase domain, which are apparently located outside the central channel of the 14-3-3γ dimer [3]. Nevertheless, previous studies on both CaMKK isoforms suggested that the main role of 14-3-3 protein binding in CaMKKs regulation is to slow down their dephosphorylation, especially at the serine residue located within the Ca^{2+}/CaM-binding region, whose phosphorylation has been shown to block Ca^{2+}/CaM binding. This, in turn, keeps CaMKKs in their PKA-mediated inhibited states [3,70,74].

Furthermore, the crystal structure of the N-terminal 14-3-3 binding motif of CaMKK2 (Ser^{100}) bound to 14-3-3 revealed that it adopts an unusual conformation within the 14-3-3 ligand binding groove (PDB ID: 6EWW) [3]. The N-terminal motif (sequence RKLpS^{100}LQE) does not conform to any canonical 14-3-3 binding motif because it contains a Gln residue and not a Pro residue at position +2 relative to the phosphorylated residue pSer100 [3]. In this crystal structure, the asymmetric unit contained four copies of the phosphopeptide:14-3-3 complex and in three of them the side chain of Gln^{102} interacts with $pSer^{100}$, likely forcing the direction of the polypeptide chain to change and thus mimicking the role of the Pro residue. Due to the abrupt change in the direction of the polypeptide chain, the part of the 14-3-3 ligand binding groove known as "the fusicoccin binding cavity" remains empty. As such, the interaction between CaMKK2 and 14-3-3 may be stabilized by small-molecule compounds, as previously described for other 14-3-3 complexes [34,57,58,76]. Indeed, the recently published study confirmed that fusicoccins can stabilize protein–protein interactions between CaMKK2 and 14-3-3 [56]. On the other hand, structure of the diphosphorylated pSer^{100}-pSer^{511} peptide (PDB ID: 6EF5) did not show the interaction between Gln^{102} and pSer^{100}, thus suggesting that the proline-like conformation of the polypeptide chain can also be formed without that interaction [74].

3.4. *Phosphatidylinositol 4-Kinases*

Phosphoinositides (PIPs) are phosphorylated derivatives of phosphatidylinositol (PI). Seven PIPs have been described in eukaryotes with the combination of 1-3 phosphates in positions 3, 4, and 5 of the inositol ring. PIPs are critically involved in numerous cellular processes, such as signal transduction and vesicular trafficking, modulating lipid distribution and metabolism, regulating ion channels, pumps, and transporters and controlling both endocytic and exocytic processes [77,78]. Among the PIPs, phosphatidylinositol 4,5-bisphosphate (PI(4,5)P_2) stands out as the most abundant anionic lipid. Its synthesis is primarily mediated by the PIP5K group of phosphatidylinositol kinases, which phosphorylate phosphatidylinositol 4-phosphate (PI(4)P) at position 5 of the inositol ring. PI(4,5)P_2 phosphorylation by PI3K yields PI(3,4,5)P_3, which is a potent signal for survival and proliferation, whereas PI(4,5)P_2 hydrolysis by phospholipase C yields two second messengers: inositol trisphosphate (cytosolic) and diacylglycerol (membrane-bound). Together, these second messengers initiate several downstream signaling cascades [79].

Phosphoinositide 4-kinases (PI4K), which synthesize PI(4)P, are classified into two groups: type II and type III (the missing type I kinase is a historical artefact). Type II kinases are not similar to any other lipid kinase. In contrast, type III kinases are typical PI4Ks because they resemble PI3Ks. Type III phosphatidylinositol-4-kinases are hijacked by various +RNA viruses, such as viruses from the *Picornaviridae*, *Flaviviridae*, and *Coronaviridae* families. For this reason, these enzymes are attractive targets for pharmaceutical intervention [80,81].

Phosphatidylinositol-4-kinase-IIIβ (PI4KB) is a class III PI4K kinase with a domain organization very similar to PI3K. This kinase has an N-terminal helical domain and a C-terminal catalytic domain. The catalytic domain consists of N- and C-lobes, with an internal linker in the N-lobe [78]. Several binding partners of PI4KB have been identified, including the adaptor protein Acyl-CoA-binding protein 3 (ACBD3, often also abbreviated as ACBP) [82], the small GTPase Rab11 [83] and the 14-3-3 protein [84]. The ACBD3 interacts with the PI4KB region preceding the helical domain and is important for enterovirus replication via its interactions with viral proteins and PI4KB. Rab11 interacts with the helical domain of PI4KB, and 14-3-3 protein recognizes the phosphorylated motif around Ser^{294} located within the flexible linker that connects the helical domain to the catalytic domain (Figure 2) [4,85–87]. The 14-3-3 binding site of PI4KB is conserved across the mammalian and yeast species [88], and its phosphorylation is mediated by protein kinase D (PKD) [89]. The PKD-mediated phosphorylation of PI4KB is associated with the Golgi compartment and stimulates the lipid kinase activity of PI4KB [84,90]. However, this activation is not caused by the 14-3-3 protein, whose recruitment has no direct effect on PI4KB activity or on its nucleocytoplasmic shuttling, instead increasing the stability of PI4KB and protecting it against dephosphorylation [4,84]. In addition, 14-3-3 may also regulate PI4KB through 14-3-3 dimers, which recruit other 14-3-3 binding proteins [91].

The crystal structure of yeast 14-3-3 protein with a bound synthetic phosphopeptide derived from the human PI4KB lipid kinase revealed interactions typical of 14-3-3:phosphopeptide complexes [87]. Recently, biophysical characterization of the complex between full-length PI4KB and 14-3-3 has shown that the complex is tight and formed in a 2:2 molar stoichiometry. Low-resolution structural analysis by SAXS indicated that the complex is compact but flexible, especially within the disordered loops connecting the 14-3-3 binding site to the rest of the enzyme, thus allowing multiple orientations between PI4KB and the 14-3-3 protein dimer. The resulting models have also shown unrestricted access to the active site of PI4KB and a small binding interface between 14-3-3 and PI4KB [4].

Interestingly, SAXS-based analysis of the ternary 14-3-3:PI4KB:Rab11 complex highlighted a compact arrangement in a 2:1:1 stoichiometry, that is, one 14-3-3 dimer binds one molecule of phosphorylated PI4KB, which binds one molecule of Rab11 [92]. The orientation of PI4KB with respect to the 14-3-3 dimer is different from that of the binary 14-3-3:PI4KB complex, but the active site of PI4KB is also fully exposed. The disordered C-terminal tail of Rab11 adopts an extended conformation, with no direct contact with 14-3-3, which suggested that 14-3-3 does not interfere with Rab11 binding to the membrane via its C-terminus. The 14-3-3 dimer bound to PI4KB may also play an additional role, as suggested by Wortzel et al. [93], who showed that the ERK1c kinase phosphorylated at Ser^{343} interacts with the 14-3-3:PI4KB complex, thus enabling ERK1c translocation to the Golgi, where ERK1c is phosphorylated by MEK1b and induces mitotic Golgi fragmentation. Therefore, the 14-3-3 protein not only stabilizes PI4KB, protecting it against dephosphorylation, but also enables crosstalk between lipid signaling pathways and the ERK cascade.

3.5. Leucine-Rich Repeat Protein Kinase-2 (LRRK2)

LRRK2 is a large multidomain Ser/Thr protein kinase of the Roco family with a characteristic ROC (Ras of the complex)-COR (C-terminus of ROC) bi-domain [94]. These two domains interact with each other, but ROC has a GTPase activity, whereas COR functions as a dimerization module [95]. LRRK2 mutations, either in the kinase or in the Roco-type GTPase domains, are the most prevalent cause of familial Parkinson´s disease (PD) [96]. Therefore, developing new strategies for the treatment of PD treatment requires understanding the regulatory mechanism of this protein [97].

LRRK2 is phosphorylated on multiple sites in vivo, and the phosphorylation of two or more residues (Ser^{910}, Ser^{935}, Ser^{1444} in the ROC domain, and Thr^{2524} at the very C-terminus) triggers its association with the 14-3-3 protein. Among these phosphorylated residues, $pSer^{910}$, $pSer^{935}$, and $pSer^{1444}$ account for the tightest binding (Figure 2) [98–102]. In addition, several PD-associated LRRK2 mutants suppress phosphorylation on these sites, thereby also reducing 14-3-3 protein binding. Accordingly, Muda et al. [101] demonstrated that disrupting the interaction between the LRRK2 ROC domain and 14-3-3 by introducing the $R^{1441}H$ and $S^{1444}A$ mutations increases LRRK2 kinase activity, whereas in vitro LRRK2 WT phosphorylation by PKA at Ser1444 in the presence of 14-3-3 decreases LRRK2 activity.

The autophosphorylation site Thr^{2524} located at the very C-terminus of LRRK2 in the WD40 domain has been recently identified as another 14-3-3 binding motif resembling the C-terminal canonical mode III [100]. Deletion of the C-terminal segment abolished both the kinase activity and 14-3-3 protein binding, thus highlighting the importance of the C-terminus for LRRK2 function, as corroborated by the recently reported cryo-EM structure of the LRRK2 construct with ROC-COR-kinase-WD40 domains. This structure revealed that the C-terminal segment is located near the N-lobe of the kinase domain and the adjacent COR domain [103].

Because the kinase activity of LRRK2 is enhanced in PD and reduced upon binding to 14-3-3, stabilizing the interaction between LRRK2 and 14-3-3 using small-molecule compounds is another potential therapeutic strategy for PD. The structure of the doubly phosphorylated LRRK2 peptide containing both $pSer^{910}$ and $pSer^{935}$ bound to 14-3-3 reported by the Ottmann group suggested the "drugability" of this interface, especially at the site Ser^{935}, which contains a pocket targetable by compounds such as the fungal toxin Fusicoccin A [55,99].

Notwithstanding the advances described above, the role of 14-3-3 protein in regulating LRRK2 is far from being fully understood. More specifically, the mechanism whereby 14-3-3 protein suppresses LRRK2 kinase activity remains unclear and may involve interference with the dimerization of the ROC domain, modulation of the interaction between ROC and the adjacent COR domain or even interaction of the C-terminal segment with the kinase and COR domains [100,101,103]. Hopefully, the recently reported cryo-TM and cryo-EM structures of LRRK2 will soon be followed by structural studies aimed at clarifying the role of 14-3-3 in LRRK2 function [103,104].

3.6. Protein Kinase C (PKC)

The PKC family consists of evolutionary conserved calcium and phospholipid-dependent Ser/Thr protein kinases implicated in multiple signal transduction networks [105,106]. In total, nine mammalian PKC genes are divided into three subfamilies: conventional (PKC α, β and γ), novel (PKC δ, ε, θ, η) and atypical (PKC ζ and ι) PKCs. All PKCs share a common architecture of the N-terminal regulatory element connected by a flexible link to a C-terminal catalytic kinase domain [107]. Variable regions within the structure (V1–V5) define the isoform-binding specificity and enable the signal diversity of PKCs [108].

In particular, PKCε controls cytokinesis, the final stage of the cell cycle during which cells divide into daughter cells. This process involves the PKCε:14-3-3 interaction [109,110]. In mitotic cells, 14-3-3 proteins activate the phosphorylated PKCε in the absence of lipids. The selective inhibition of the 14-3-3:PKCε complex formation leads to defects in the completion of cytokinesis. This delay is connected with the prolonged localization of Ras Homolog Family Member A (RhoA) at the midbody, the enhanced RhoA activity, and the delay in actomyosin ring dissociation. In addition, incomplete cytokinesis can promote tumorigenesis [111]. Consequently, the regulatory mechanism of this process is still intensively studied.

Three upstream activators, glycogen synthase kinase 3 (GSK3), p38 MAP kinase and PKC itself, phosphorylate PKCε residues Ser^{346}, Ser^{350}, and Ser^{368}, respectively [110]. All three sites are located in the V3 hinge region that connects the regulatory N-terminal part to the C-terminal kinase domain and conserved among vertebrates. Furthermore, motifs surrounding Ser^{346} and Ser^{368} were identified as 14-3-3 binding sites, and their phosphorylation induces the formation of the PKCε:14-3-3 complex in a 1:2 stoichiometry (Figure 2). The site containing Ser^{346} is a canonical mode I motif, whereas the second site somewhat resembles mode II motif for its lack of a Pro residue at +2 position with respect to $pSer^{368}$.

The Ser^{368} site is a gatekeeper (dominant) site, as shown by structural analysis combined with calorimetry measurements. However, high affinity binding requires the presence of both phosphorylated motifs [112]. Interestingly, simultaneous phosphorylation of the third site, Ser^{350}, substantially reduced the binding affinity of the phosphopeptide containing the V3 hinge region. Thus, the phosphorylation status of this site may function as an additional modulator of the 14-3-3-mediated activation of PKCε [112].

Functionally, 14-3-3 protein binding has been shown to activate PKCε in the absence of lipids [110]. This suggests that 14-3-3 stabilizes the active open conformation of PKCε and promotes its accumulation at the midbody by relieving or blocking an inactive conformation in which the active site of the kinase domain is occupied by the N-terminal pseudosubstrate region.

4. Conclusions and Challenges

In this review article, we demonstrated that 14-3-3 proteins are crucial regulators of many physiologically relevant kinases. However, the data available so far enable a detailed understanding of 14-3-3 protein-mediated regulations for only a few kinases. A *bona fide* exception is the B-RAF kinase. The recently solved high-resolution structures of B-RAF:14-3-3 complexes have shed light into the role of 14-3-3 binding. Accordingly, we hope that these structures will stimulate further structural studies of other 14-3-3:kinase complexes towards a more complete understanding of all nuances of these protein–protein interactions and their roles in regulating kinase functions. Concomitantly, 14-3-3 proteins are physiological "inhibitors" of several kinases whose dysregulation is involved in cancer and neurodegenerative diseases (e.g., RAF, ASK1, LRRK2). As such, these complexes are potential targets for therapeutic interventions. In fact, recent advances in the development of small molecule compounds aimed at stabilizing 14-3-3 protein–protein interactions have demonstrated the feasibility of this approach as an alternative or complimentary strategy in suppressing the activity of these kinases.

Acknowledgments: The authors thank Carlos V. Melo for proofreading the article.

Abbreviations

ACBD3	Acyl-CoA-binding protein 3
AMPK	AMP-activated protein kinase
ASK1	Apoptosis signal-regulating kinase 1
CaMKK	Calcium/calmodulin-dependent protein kinase kinase
GSK3	Glycogen synthase kinase 3
LRRK2	Leucine-rich repeat protein kinase-2
MAP	Mitogen-activated protein
MAPK	Mitogen-activated protein kinase
MAP2K	Mitogen-activated protein kinase kinase
MAP3K	Mitogen-activated protein kinase kinase kinase
PD	Parkinson's disease
PI	Phosphatidylinositol
PIP	Phosphoinositides
PI4K	Phosphoinositide 4-kinases
PI4KB	Phosphatidylinositol-4-kinase-IIIβ
PKC	Protein kinase C
RAF	Rapidly Accelerated Fibrosarcoma
RhoA	Ras Homolog Family Member A

References

1. Tinti, M.; Madeira, F.; Murugesan, G.; Hoxhaj, G.; Toth, R.; Mackintosh, C. ANIA: ANnotation and Integrated Analysis of the 14-3-3 interactome. *Database* **2014**, *2014*, bat085. [CrossRef] [PubMed]
2. Petrvalska, O.; Kosek, D.; Kukacka, Z.; Tosner, Z.; Man, P.; Vecer, J.; Herman, P.; Obsilova, V.; Obsil, T. Structural Insight into the 14-3-3 Protein-dependent Inhibition of Protein Kinase ASK1 (Apoptosis Signal-regulating kinase 1). *J. Biol. Chem.* **2016**, *291*, 20753–20765. [CrossRef] [PubMed]
3. Psenakova, K.; Petrvalska, O.; Kylarova, S.; Lentini Santo, D.; Kalabova, D.; Herman, P.; Obsilova, V.; Obsil, T. 14-3-3 protein directly interacts with the kinase domain of calcium/calmodulin-dependent protein kinase kinase (CaMKK2). *Biochim. Biophys. Acta* **2018**, *1862*, 1612–1625. [CrossRef] [PubMed]
4. Chalupska, D.; Eisenreichova, A.; Rozycki, B.; Rezabkova, L.; Humpolickova, J.; Klima, M.; Boura, E. Structural analysis of phosphatidylinositol 4-kinase IIIbeta (PI4KB)—14-3-3 protein complex reveals internal flexibility and explains 14-3-3 mediated protection from degradation in vitro. *J. Struct. Biol.* **2017**, *200*, 36–44. [CrossRef]
5. Kondo, Y.; Ognjenovic, J.; Banerjee, S.; Karandur, D.; Merk, A.; Kulhanek, K.; Wong, K.; Roose, J.P.; Subramaniam, S.; Kuriyan, J. Cryo-EM structure of a dimeric B-Raf:14-3-3 complex reveals asymmetry in the active sites of B-Raf kinases. *Science* **2019**, *366*, 109–115. [CrossRef]
6. Park, E.; Rawson, S.; Li, K.; Kim, B.W.; Ficarro, S.B.; Pino, G.G.; Sharif, H.; Marto, J.A.; Jeon, H.; Eck, M.J. Architecture of autoinhibited and active BRAF-MEK1-14-3-3 complexes. *Nature* **2019**, *575*, 545–550. [CrossRef]
7. Liau, N.P.D.; Wendorff, T.J.; Quinn, J.G.; Steffek, M.; Phung, W.; Liu, P.; Tang, J.; Irudayanathan, F.J.; Izadi, S.; Shaw, A.S.; et al. Negative regulation of RAF kinase activity by ATP is overcome by 14-3-3-induced dimerization. *Nat. Struct. Mol. Biol.* **2020**, *27*, 134–141. [CrossRef]
8. Liau, N.P.D.; Venkatanarayan, A.; Quinn, J.G.; Phung, W.; Malek, S.; Hymowitz, S.G.; Sudhamsu, J. Dimerization Induced by C-Terminal 14-3-3 Binding Is Sufficient for BRAF Kinase Activation. *Biochemistry* **2020**, *59*, 3982–3992. [CrossRef]
9. Aitken, A.; Amess, B.; Howell, S.; Jones, D.; Martin, H.; Patel, Y.; Robinson, K.; Toker, A. The role of specific isoforms of 14-3-3 protein in regulating protein kinase activity in the brain. *Biochem. Soc. Trans.* **1992**, *20*, 607–611. [CrossRef]
10. Bridges, D.; Moorhead, G.B. 14-3-3 proteins: A number of functions for a numbered protein. *Sci. STKE* **2005**, *2005*, re10. [CrossRef]
11. Aitken, A. Post-translational modification of 14-3-3 isoforms and regulation of cellular function. *Semin. Cell Dev. Biol.* **2011**, *22*, 673–680. [CrossRef] [PubMed]

12. Sluchanko, N.N. Association of Multiple Phosphorylated Proteins with the 14-3-3 Regulatory Hubs: Problems and Perspectives. *J. Mol. Biol.* **2018**, *430*, 20–26. [CrossRef] [PubMed]
13. Aitken, A.; Howell, S.; Jones, D.; Madrazo, J.; Patel, Y. 14-3-3 alpha and delta are the phosphorylated forms of raf-activating 14-3-3 beta and zeta. In vivo stoichiometric phosphorylation in brain at a Ser-Pro-Glu-Lys MOTIF. *J. Biol. Chem.* **1995**, *270*, 5706–5709. [CrossRef] [PubMed]
14. van Hemert, M.J.; van Heusden, G.P.; Steensma, H.Y. Yeast 14-3-3 proteins. *Yeast* **2001**, *18*, 889–895. [CrossRef]
15. Sehnke, P.C.; Rosenquist, M.; Alsterfjord, M.; DeLille, J.; Sommarin, M.; Larsson, C.; Ferl, R.J. Evolution and isoform specificity of plant 14-3-3 proteins. *Plant. Mol. Biol.* **2002**, *50*, 1011–1018. [CrossRef]
16. Xiao, B.; Smerdon, S.J.; Jones, D.H.; Dodson, G.G.; Soneji, Y.; Aitken, A.; Gamblin, S.J. Structure of a 14-3-3 protein and implications for coordination of multiple signalling pathways. *Nature* **1995**, *376*, 188–191. [CrossRef]
17. Liu, D.; Bienkowska, J.; Petosa, C.; Collier, R.J.; Fu, H.; Liddington, R. Crystal structure of the zeta isoform of the 14-3-3 protein. *Nature* **1995**, *376*, 191–194. [CrossRef]
18. Ma, Y.; Pitson, S.; Hercus, T.; Murphy, J.; Lopez, A.; Woodcock, J. Sphingosine activates protein kinase A type II by a novel cAMP-independent mechanism. *J. Biol. Chem.* **2005**, *280*, 26011–26017. [CrossRef]
19. Woodcock, J.M.; Murphy, J.; Stomski, F.C.; Berndt, M.C.; Lopez, A.F. The dimeric versus monomeric status of 14-3-3zeta is controlled by phosphorylation of Ser58 at the dimer interface. *J. Biol. Chem.* **2003**, *278*, 36323–36327. [CrossRef]
20. Woodcock, J.M.; Coolen, C.; Goodwin, K.L.; Baek, D.J.; Bittman, R.; Samuel, M.S.; Pitson, S.M.; Lopez, A.F. Destabilisation of dimeric 14-3-3 proteins as a novel approach to anti-cancer therapeutics. *Oncotarget* **2015**, *6*, 14522–14536. [CrossRef]
21. Molzan, M.; Kasper, S.; Roglin, L.; Skwarczynska, M.; Sassa, T.; Inoue, T.; Breitenbuecher, F.; Ohkanda, J.; Kato, N.; Schuler, M.; et al. Stabilization of physical RAF/14-3-3 interaction by cotylenin A as treatment strategy for RAS mutant cancers. *ACS Chem. Biol.* **2013**, *8*, 1869–1875. [CrossRef] [PubMed]
22. Lavoie, H.; Therrien, M. Regulation of RAF protein kinases in ERK signalling. *Nat. Rev. Mol. Cell Biol.* **2015**, *16*, 281–298. [CrossRef] [PubMed]
23. Tate, J.G.; Bamford, S.; Jubb, H.C.; Sondka, Z.; Beare, D.M.; Bindal, N.; Boutselakis, H.; Cole, C.G.; Creatore, C.; Dawson, E.; et al. COSMIC: The Catalogue Of Somatic Mutations In Cancer. *Nucleic Acids Res.* **2019**, *47*, D941–D947. [CrossRef] [PubMed]
24. Dumaz, N.; Marais, R. Protein kinase A blocks Raf-1 activity by stimulating 14-3-3 binding and blocking Raf-1 interaction with Ras. *J. Biol. Chem.* **2003**, *278*, 29819–29823. [CrossRef]
25. Stanton, V.P., Jr.; Nichols, D.W.; Laudano, A.P.; Cooper, G.M. Definition of the human raf amino-terminal regulatory region by deletion mutagenesis. *Mol. Cell Biol.* **1989**, *9*, 639–647. [CrossRef]
26. Tzivion, G.; Luo, Z.; Avruch, J. A dimeric 14-3-3 protein is an essential cofactor for Raf kinase activity. *Nature* **1998**, *394*, 88–92. [CrossRef]
27. Luo, Z.; Tzivion, G.; Belshaw, P.J.; Vavvas, D.; Marshall, M.; Avruch, J. Oligomerization activates c-Raf-1 through a Ras-dependent mechanism. *Nature* **1996**, *383*, 181–185. [CrossRef]
28. Fantl, W.J.; Muslin, A.J.; Kikuchi, A.; Martin, J.A.; MacNicol, A.M.; Gross, R.W.; Williams, L.T. Activation of Raf-1 by 14-3-3 proteins. *Nature* **1994**, *371*, 612–614. [CrossRef]
29. Freed, E.; Symons, M.; Macdonald, S.G.; McCormick, F.; Ruggieri, R. Binding of 14-3-3 proteins to the protein kinase Raf and effects on its activation. *Science* **1994**, *265*, 1713–1716. [CrossRef]
30. Irie, K.; Gotoh, Y.; Yashar, B.M.; Errede, B.; Nishida, E.; Matsumoto, K. Stimulatory effects of yeast and mammalian 14-3-3 proteins on the Raf protein kinase. *Science* **1994**, *265*, 1716–1719. [CrossRef]
31. Fu, H.; Xia, K.; Pallas, D.C.; Cui, C.; Conroy, K.; Narsimhan, R.P.; Mamon, H.; Collier, R.J.; Roberts, T.M. Interaction of the protein kinase Raf-1 with 14-3-3 proteins. *Science* **1994**, *266*, 126–129. [CrossRef] [PubMed]
32. Petosa, C.; Masters, S.C.; Bankston, L.A.; Pohl, J.; Wang, B.; Fu, H.; Liddington, R.C. 14-3-3zeta binds a phosphorylated Raf peptide and an unphosphorylated peptide via its conserved amphipathic groove. *J. Biol. Chem.* **1998**, *273*, 16305–16310. [CrossRef] [PubMed]
33. Molzan, M.; Ottmann, C. Synergistic binding of the phosphorylated S233- and S259-binding sites of C-RAF to one 14-3-3zeta dimer. *J. Mol. Biol.* **2012**, *423*, 486–495. [CrossRef] [PubMed]
34. Molzan, M.; Schumacher, B.; Ottmann, C.; Baljuls, A.; Polzien, L.; Weyand, M.; Thiel, P.; Rose, R.; Rose, M.; Kuhenne, P.; et al. Impaired binding of 14-3-3 to C-RAF in Noonan syndrome suggests new approaches in diseases with increased Ras signaling. *Mol. Cell Biol.* **2010**, *30*, 4698–4711. [CrossRef] [PubMed]

35. Freeman, A.K.; Ritt, D.A.; Morrison, D.K. The importance of Raf dimerization in cell signaling. *Small Gtpases* **2013**, *4*, 180–185. [CrossRef] [PubMed]
36. Durrant, D.E.; Morrison, D.K. Targeting the Raf kinases in human cancer: The Raf dimer dilemma. *Br. J. Cancer* **2018**, *118*, 3–8. [CrossRef]
37. Ichijo, H.; Nishida, E.; Irie, K.; ten Dijke, P.; Saitoh, M.; Moriguchi, T.; Takagi, M.; Matsumoto, K.; Miyazono, K.; Gotoh, Y. Induction of apoptosis by ASK1, a mammalian MAPKKK that activates SAPK/JNK and p38 signaling pathways. *Science* **1997**, *275*, 90–94. [CrossRef]
38. Johnson, G.L.; Lapadat, R. Mitogen-activated protein kinase pathways mediated by ERK, JNK, and p38 protein kinases. *Science* **2002**, *298*, 1911–1912. [CrossRef]
39. Takenaka, S.; Fujisawa, T.; Ichijo, H. Apoptosis signal-regulating kinase 1 (ASK1) as a therapeutic target for neurological diseases. *Expert Opin. Ther. Targets* **2020**, 1061–1064. [CrossRef]
40. Psenakova, K.; Hexnerova, R.; Srb, P.; Obsilova, V.; Veverka, V.; Obsil, T. The redox-active site of thioredoxin is directly involved in apoptosis signal-regulating kinase 1 binding that is modulated by oxidative stress. *FEBS J.* **2020**, *287*, 1626–1644. [CrossRef]
41. Tobiume, K.; Saitoh, M.; Ichijo, H. Activation of apoptosis signal-regulating kinase 1 by the stress-induced activating phosphorylation of pre-formed oligomer. *J. Cell Physiol.* **2002**, *191*, 95–104. [CrossRef] [PubMed]
42. Bunkoczi, G.; Salah, E.; Filippakopoulos, P.; Fedorov, O.; Muller, S.; Sobott, F.; Parker, S.A.; Zhang, H.; Min, W.; Turk, B.E.; et al. Structural and functional characterization of the human protein kinase ASK1. *Structure* **2007**, *15*, 1215–1226. [CrossRef] [PubMed]
43. Trevelyan, S.J.; Brewster, J.L.; Burgess, A.E.; Crowther, J.M.; Cadell, A.L.; Parker, B.L.; Croucher, D.R.; Dobson, R.C.J.; Murphy, J.M.; Mace, P.D. Structure-based mechanism of preferential complex formation by apoptosis signal-regulating kinases. *Sci. Signal.* **2020**, *13*. [CrossRef] [PubMed]
44. Zhang, L.; Chen, J.; Fu, H. Suppression of apoptosis signal-regulating kinase 1-induced cell death by 14-3-3 proteins. *Proc. Natl. Acad. Sci. USA* **1999**, *96*, 8511–8515. [CrossRef] [PubMed]
45. Federspiel, J.D.; Codreanu, S.G.; Palubinsky, A.M.; Winland, A.J.; Betanzos, C.M.; McLaughlin, B.; Liebler, D.C. Assembly Dynamics and Stoichiometry of the Apoptosis Signal-regulating Kinase (ASK) Signalosome in Response to Electrophile Stress. *Mol. Cell Proteom.* **2016**, *15*, 1947–1961. [CrossRef]
46. Cockrell, L.M.; Puckett, M.C.; Goldman, E.H.; Khuri, F.R.; Fu, H. Dual engagement of 14-3-3 proteins controls signal relay from ASK2 to the ASK1 signalosome. *Oncogene* **2010**, *29*, 822–830. [CrossRef]
47. Goldman, E.H.; Chen, L.; Fu, H. Activation of apoptosis signal-regulating kinase 1 by reactive oxygen species through dephosphorylation at serine 967 and 14-3-3 dissociation. *J. Biol. Chem.* **2004**, *279*, 10442–10449. [CrossRef]
48. Saitoh, M.; Nishitoh, H.; Fujii, M.; Takeda, K.; Tobiume, K.; Sawada, Y.; Kawabata, M.; Miyazono, K.; Ichijo, H. Mammalian thioredoxin is a direct inhibitor of apoptosis signal-regulating kinase (ASK) 1. *EMBO J.* **1998**, *17*, 2596–2606. [CrossRef]
49. Fujino, G.; Noguchi, T.; Matsuzawa, A.; Yamauchi, S.; Saitoh, M.; Takeda, K.; Ichijo, H. Thioredoxin and TRAF family proteins regulate reactive oxygen species-dependent activation of ASK1 through reciprocal modulation of the N-terminal homophilic interaction of ASK1. *Mol. Cell Biol.* **2007**, *27*, 8152–8163. [CrossRef]
50. Liu, H.; Nishitoh, H.; Ichijo, H.; Kyriakis, J.M. Activation of apoptosis signal-regulating kinase 1 (ASK1) by tumor necrosis factor receptor-associated factor 2 requires prior dissociation of the ASK1 inhibitor thioredoxin. *Mol. Cell Biol.* **2000**, *20*, 2198–2208. [CrossRef]
51. Weijman, J.F.; Kumar, A.; Jamieson, S.A.; King, C.M.; Caradoc-Davies, T.T.; Ledgerwood, E.C.; Murphy, J.M.; Mace, P.D. Structural basis of autoregulatory scaffolding by apoptosis signal-regulating kinase 1. *Proc. Natl. Acad. Sci. USA* **2017**, *114*, E2096–E2105. [CrossRef] [PubMed]
52. Obsil, T.; Ghirlando, R.; Klein, D.C.; Ganguly, S.; Dyda, F. Crystal structure of the 14-3-3zeta:serotonin N-acetyltransferase complex. A role for scaffolding in enzyme regulation. *Cell* **2001**, *105*, 257–267. [CrossRef]
53. Alblova, M.; Smidova, A.; Docekal, V.; Vesely, J.; Herman, P.; Obsilova, V.; Obsil, T. Molecular basis of the 14-3-3 protein-dependent activation of yeast neutral trehalase Nth1. *Proc. Natl. Acad. Sci. USA* **2017**, *114*, E9811–E9820. [CrossRef] [PubMed]
54. Obsil, T.; Obsilova, V. Structural aspects of protein kinase ASK1 regulation. *Adv. Biol. Regul.* **2017**, *66*, 31–36. [CrossRef]
55. Kaplan, A.; Ottmann, C.; Fournier, A.E. 14-3-3 adaptor protein-protein interactions as therapeutic targets for CNS diseases. *Pharm. Res.* **2017**, *125*, 114–121. [CrossRef]

56. Lentini Santo, D.; Petrvalska, O.; Obsilova, V.; Ottmann, C.; Obsil, T. Stabilization of Protein-Protein Interactions between CaMKK2 and 14-3-3 by Fusicoccins. *ACS Chem. Biol.* **2020**, *15*, 3060–3071. [CrossRef]
57. Stevers, L.M.; Sijbesma, E.; Botta, M.; MacKintosh, C.; Obsil, T.; Landrieu, I.; Cau, Y.; Wilson, A.J.; Karawajczyk, A.; Eickhoff, J.; et al. Modulators of 14-3-3 Protein-Protein Interactions. *J. Med. Chem.* **2018**, *61*, 3755–3778. [CrossRef]
58. Wurtele, M.; Jelich-Ottmann, C.; Wittinghofer, A.; Oecking, C. Structural view of a fungal toxin acting on a 14-3-3 regulatory complex. *EMBO J.* **2003**, *22*, 987–994. [CrossRef]
59. Marcelo, K.L.; Means, A.R.; York, B. The Ca^{2+}/Calmodulin/CaMKK2 Axis: Nature's Metabolic CaMshaft. *Trends Endocrinol. Metab.* **2016**, *27*, 706–718. [CrossRef]
60. Racioppi, L.; Means, A.R. Calcium/calmodulin-dependent protein kinase kinase 2: Roles in signaling and pathophysiology. *J. Biol. Chem.* **2012**, *287*, 31658–31665. [CrossRef]
61. Soderling, T.R. The Ca-calmodulin-dependent protein kinase cascade. *Trends Biochem. Sci.* **1999**, *24*, 232–236. [CrossRef]
62. Tokumitsu, H.; Wayman, G.A.; Muramatsu, M.; Soderling, T.R. Calcium/calmodulin-dependent protein kinase kinase: Identification of regulatory domains. *Biochemistry* **1997**, *36*, 12823–12827. [CrossRef] [PubMed]
63. Goldberg, J.; Nairn, A.C.; Kuriyan, J. Structural basis for the autoinhibition of calcium/calmodulin-dependent protein kinase I. *Cell* **1996**, *84*, 875–887. [CrossRef]
64. Wayman, G.A.; Kaech, S.; Grant, W.F.; Davare, M.; Impey, S.; Tokumitsu, H.; Nozaki, N.; Banker, G.; Soderling, T.R. Regulation of axonal extension and growth cone motility by calmodulin-dependent protein kinase I. *J. Neurosci.* **2004**, *24*, 3786–3794. [CrossRef] [PubMed]
65. Lin, F.; Marcelo, K.L.; Rajapakshe, K.; Coarfa, C.; Dean, A.; Wilganowski, N.; Robinson, H.; Sevick, E.; Bissig, K.D.; Goldie, L.C.; et al. The camKK2/camKIV relay is an essential regulator of hepatic cancer. *Hepatology* **2015**, *62*, 505–520. [CrossRef]
66. Edelman, A.M.; Mitchelhill, K.I.; Selbert, M.A.; Anderson, K.A.; Hook, S.S.; Stapleton, D.; Goldstein, E.G.; Means, A.R.; Kemp, B.E. Multiple Ca^{2+}-calmodulin-dependent protein kinase kinases from rat brain. Purification, regulation by Ca^{2+}-calmodulin, and partial amino acid sequence. *J. Biol. Chem.* **1996**, *271*, 10806–10810. [CrossRef]
67. Anderson, K.A.; Ribar, T.J.; Lin, F.; Noeldner, P.K.; Green, M.F.; Muehlbauer, M.J.; Witters, L.A.; Kemp, B.E.; Means, A.R. Hypothalamic CaMKK2 contributes to the regulation of energy balance. *Cell Metab.* **2008**, *7*, 377–388. [CrossRef]
68. Wen, L.; Chen, Z.; Zhang, F.; Cui, X.; Sun, W.; Geary, G.G.; Wang, Y.; Johnson, D.A.; Zhu, Y.; Chien, S.; et al. Ca^{2+}/calmodulin-dependent protein kinase kinase beta phosphorylation of Sirtuin 1 in endothelium is atheroprotective. *Proc. Natl. Acad. Sci. USA* **2013**, *110*, E2420–E2427. [CrossRef]
69. Gao, G.; Widmer, J.; Stapleton, D.; Teh, T.; Cox, T.; Kemp, B.E.; Witters, L.A. Catalytic subunits of the porcine and rat 5′-AMP-activated protein kinase are members of the SNF1 protein kinase family. *Biochim. Biophys. Acta* **1995**, *1266*, 73–82. [CrossRef]
70. Davare, M.A.; Saneyoshi, T.; Guire, E.S.; Nygaard, S.C.; Soderling, T.R. Inhibition of calcium/calmodulin-dependent protein kinase kinase by protein 14-3-3. *J. Biol. Chem.* **2004**, *279*, 52191–52199. [CrossRef]
71. Ichimura, T.; Taoka, M.; Hozumi, Y.; Goto, K.; Tokumitsu, H. 14-3-3 Proteins directly regulate Ca^{2+}/calmodulin-dependent protein kinase kinase alpha through phosphorylation-dependent multisite binding. *FEBS Lett.* **2008**, *582*, 661–665. [CrossRef] [PubMed]
72. Matsushita, M.; Nairn, A.C. Inhibition of the Ca^{2+}/calmodulin-dependent protein kinase I cascade by cAMP-dependent protein kinase. *J. Biol. Chem.* **1999**, *274*, 10086–10093. [CrossRef]
73. Wayman, G.A.; Tokumitsu, H.; Soderling, T.R. Inhibitory cross-talk by cAMP kinase on the calmodulin-dependent protein kinase cascade. *J. Biol. Chem.* **1997**, *272*, 16073–16076. [CrossRef] [PubMed]
74. Langendorf, C.G.; O'Brien, M.T.; Ngoei, K.R.W.; McAloon, L.M.; Dhagat, U.; Hoque, A.; Ling, N.X.Y.; Dite, T.A.; Galic, S.; Loh, K.; et al. CaMKK2 is inactivated by cAMP-PKA signaling and 14-3-3 adaptor proteins. *J. Biol. Chem.* **2020**. [CrossRef] [PubMed]
75. Spengler, K.; Zibrova, D.; Woods, A.; Langendorf, C.G.; Scott, J.W.; Carling, D.; Heller, R. Protein kinase A negatively regulates VEGF-induced AMPK activation by phosphorylating CaMKK2 at serine 495. *Biochem. J.* **2020**, *477*, 3453–3469. [CrossRef] [PubMed]

76. De Boer, A.H.; de Vries-van Leeuwen, I.J. Fusicoccanes: Diterpenes with surprising biological functions. *Trends Plant. Sci.* **2012**, *17*, 360–368. [CrossRef]
77. Balla, T. Phosphoinositides: Tiny lipids with giant impact on cell regulation. *Physiol. Rev.* **2013**, *93*, 1019–1137. [CrossRef]
78. Burke, J.E. Structural Basis for Regulation of Phosphoinositide Kinases and Their Involvement in Human Disease. *Mol. Cell* **2018**, *71*, 653–673. [CrossRef]
79. Di Paolo, G.; De Camilli, P. Phosphoinositides in cell regulation and membrane dynamics. *Nature* **2006**, *443*, 651–657. [CrossRef]
80. Balla, A.; Balla, T. Phosphatidylinositol 4-kinases: Old enzymes with emerging functions. *Trends Cell Biol.* **2006**, *16*, 351–361. [CrossRef]
81. Boura, E.; Nencka, R. Phosphatidylinositol 4-kinases: Function, structure, and inhibition. *Exp. Cell Res.* **2015**, *337*, 136–145. [CrossRef] [PubMed]
82. Sasaki, J.; Ishikawa, K.; Arita, M.; Taniguchi, K. ACBD3-mediated recruitment of PI4KB to picornavirus RNA replication sites. *EMBO J.* **2012**, *31*, 754–766. [CrossRef] [PubMed]
83. De Graaf, P.; Zwart, W.T.; van Dijken, R.A.; Deneka, M.; Schulz, T.K.; Geijsen, N.; Coffer, P.J.; Gadella, B.M.; Verkleij, A.J.; van der Sluijs, P.; et al. Phosphatidylinositol 4-kinasebeta is critical for functional association of rab11 with the Golgi complex. *Mol. Biol. Cell* **2004**, *15*, 2038–2047. [CrossRef] [PubMed]
84. Hausser, A.; Link, G.; Hoene, M.; Russo, C.; Selchow, O.; Pfizenmaier, K. Phospho-specific binding of 14-3-3 proteins to phosphatidylinositol 4-kinase III beta protects from dephosphorylation and stabilizes lipid kinase activity. *J. Cell Sci.* **2006**, *119*, 3613–3621. [CrossRef]
85. Demmel, L.; Beck, M.; Klose, C.; Schlaitz, A.L.; Gloor, Y.; Hsu, P.P.; Havlis, J.; Shevchenko, A.; Krause, E.; Kalaidzidis, Y.; et al. Nucleocytoplasmic shuttling of the Golgi phosphatidylinositol 4-kinase Pik1 is regulated by 14-3-3 proteins and coordinates Golgi function with cell growth. *Mol. Biol. Cell* **2008**, *19*, 1046–1061. [CrossRef]
86. Burke, J.E.; Inglis, A.J.; Perisic, O.; Masson, G.R.; McLaughlin, S.H.; Rutaganira, F.; Shokat, K.M.; Williams, R.L. Structures of PI4KIIIbeta complexes show simultaneous recruitment of Rab11 and its effectors. *Science* **2014**, *344*, 1035–1038. [CrossRef]
87. Eisenreichova, A.; Klima, M.; Boura, E. Crystal structures of a yeast 14-3-3 protein from Lachancea thermotolerans in the unliganded form and bound to a human lipid kinase PI4KB-derived peptide reveal high evolutionary conservation. *Acta Cryst. F Struct. Biol. Commun.* **2016**, *72*, 799–803. [CrossRef]
88. Johnson, C.; Crowther, S.; Stafford, M.J.; Campbell, D.G.; Toth, R.; MacKintosh, C. Bioinformatic and experimental survey of 14-3-3-binding sites. *Biochem. J.* **2010**, *427*, 69–78. [CrossRef]
89. Hausser, A.; Storz, P.; Martens, S.; Link, G.; Toker, A.; Pfizenmaier, K. Protein kinase D regulates vesicular transport by phosphorylating and activating phosphatidylinositol-4 kinase IIIbeta at the Golgi complex. *Nat. Cell Biol.* **2005**, *7*, 880–886. [CrossRef]
90. Szivak, I.; Lamb, N.; Heilmeyer, L.M. Subcellular localization and structural function of endogenous phosphorylated phosphatidylinositol 4-kinase (PI4K92). *J. Biol. Chem.* **2006**, *281*, 16740–16749. [CrossRef]
91. Valente, C.; Turacchio, G.; Mariggio, S.; Pagliuso, A.; Gaibisso, R.; Di Tullio, G.; Santoro, M.; Formiggini, F.; Spano, S.; Piccini, D.; et al. A 14-3-3gamma dimer-based scaffold bridges CtBP1-S/BARS to PI(4)KIIIbeta to regulate post-Golgi carrier formation. *Nat. Cell Biol.* **2012**, *14*, 343–354. [CrossRef] [PubMed]
92. Chalupska, D.; Rozycki, B.; Humpolickova, J.; Faltova, L.; Klima, M.; Boura, E. Phosphatidylinositol 4-kinase IIIbeta (PI4KB) forms highly flexible heterocomplexes that include ACBD3, 14-3-3, and Rab11 proteins. *Sci. Rep.* **2019**, *9*, 567. [CrossRef] [PubMed]
93. Wortzel, I.; Hanoch, T.; Porat, Z.; Hausser, A.; Seger, R. Mitotic Golgi translocation of ERK1c is mediated by a PI4KIIIbeta-14-3-3gamma shuttling complex. *J. Cell Sci.* **2015**, *128*, 4083–4095. [CrossRef] [PubMed]
94. Morrison, D.K. The 14-3-3 proteins: Integrators of diverse signaling cues that impact cell fate and cancer development. *Trends Cell Biol.* **2009**, *19*, 16–23. [CrossRef] [PubMed]
95. Marin, I.; van Egmond, W.N.; van Haastert, P.J. The Roco protein family: A functional perspective. *FASEB J.* **2008**, *22*, 3103–3110. [CrossRef] [PubMed]
96. Zimprich, A.; Biskup, S.; Leitner, P.; Lichtner, P.; Farrer, M.; Lincoln, S.; Kachergus, J.; Hulihan, M.; Uitti, R.J.; Calne, D.B.; et al. Mutations in LRRK2 cause autosomal-dominant parkinsonism with pleomorphic pathology. *Neuron* **2004**, *44*, 601–607. [CrossRef] [PubMed]

97. Taylor, M.; Alessi, D.R. Advances in elucidating the function of leucine-rich repeat protein kinase-2 in normal cells and Parkinson's disease. *Curr. Opin. Cell Biol.* **2020**, *63*, 102–113. [CrossRef] [PubMed]
98. Nichols, R.J.; Dzamko, N.; Morrice, N.A.; Campbell, D.G.; Deak, M.; Ordureau, A.; Macartney, T.; Tong, Y.; Shen, J.; Prescott, A.R.; et al. 14-3-3 binding to LRRK2 is disrupted by multiple Parkinson's disease-associated mutations and regulates cytoplasmic localization. *Biochem. J.* **2010**, *430*, 393–404. [CrossRef]
99. Stevers, L.M.; de Vries, R.M.; Doveston, R.G.; Milroy, L.G.; Brunsveld, L.; Ottmann, C. Structural interface between LRRK2 and 14-3-3 protein. *Biochem. J.* **2017**, *474*, 1273–1287. [CrossRef]
100. Manschwetus, J.T.; Wallbott, M.; Fachinger, A.; Obergruber, C.; Pautz, S.; Bertinetti, D.; Schmidt, S.H.; Herberg, F.W. Binding of the Human 14-3-3 Isoforms to Distinct Sites in the Leucine-Rich Repeat Kinase 2. *Front. Neurosci.* **2020**, *14*, 302. [CrossRef]
101. Muda, K.; Bertinetti, D.; Gesellchen, F.; Hermann, J.S.; von Zweydorf, F.; Geerlof, A.; Jacob, A.; Ueffing, M.; Gloeckner, C.J.; Herberg, F.W. Parkinson-related LRRK2 mutation R1441C/G/H impairs PKA phosphorylation of LRRK2 and disrupts its interaction with 14-3-3. *Proc. Natl. Acad. Sci. USA* **2014**, *111*, E34–E43. [CrossRef] [PubMed]
102. Li, X.; Wang, Q.J.; Pan, N.; Lee, S.; Zhao, Y.; Chait, B.T.; Yue, Z. Phosphorylation-dependent 14-3-3 binding to LRRK2 is impaired by common mutations of familial Parkinson's disease. *PLoS ONE* **2011**, *6*, e17153.
103. Deniston, C.K.; Salogiannis, J.; Mathea, S.; Snead, D.M.; Lahiri, I.; Matyszewski, M.; Donosa, O.; Watanabe, R.; Bohning, J.; Shiau, A.K.; et al. Structure of LRRK2 in Parkinson's disease and model for microtubule interaction. *Nature* **2020**. [CrossRef] [PubMed]
104. Watanabe, R.; Buschauer, R.; Bohning, J.; Audagnotto, M.; Lasker, K.; Lu, T.W.; Boassa, D.; Taylor, S.; Villa, E. The In Situ Structure of Parkinson's Disease-Linked LRRK2. *Cell* **2020**, *182*, 1508–1518.e16. [CrossRef] [PubMed]
105. Nishizuka, Y. The role of protein kinase C in cell surface signal transduction and tumour promotion. *Nature* **1984**, *308*, 693–698. [CrossRef] [PubMed]
106. Rosse, C.; Linch, M.; Kermorgant, S.; Cameron, A.J.; Boeckeler, K.; Parker, P.J. PKC and the control of localized signal dynamics. *Nat. Rev. Mol. Cell Biol.* **2010**, *11*, 103–112. [CrossRef] [PubMed]
107. Newton, A.C. Protein kinase C: Poised to signal. *Am. J. Physiol. Endocrinol. Metab.* **2010**, *298*, E395–E402. [CrossRef]
108. Newton, A.C. Protein kinase C: Structure, function, and regulation. *J. Biol. Chem.* **1995**, *270*, 28495–28498. [CrossRef]
109. Glotzer, M. The molecular requirements for cytokinesis. *Science* **2005**, *307*, 1735–1739. [CrossRef]
110. Saurin, A.T.; Durgan, J.; Cameron, A.J.; Faisal, A.; Marber, M.S.; Parker, P.J. The regulated assembly of a PKCepsilon complex controls the completion of cytokinesis. *Nat. Cell Biol.* **2008**, *10*, 891–901. [CrossRef]
111. Fujiwara, T.; Bandi, M.; Nitta, M.; Ivanova, E.V.; Bronson, R.T.; Pellman, D. Cytokinesis failure generating tetraploids promotes tumorigenesis in p53-null cells. *Nature* **2005**, *437*, 1043–1047. [CrossRef] [PubMed]
112. Kostelecky, B.; Saurin, A.T.; Purkiss, A.; Parker, P.J.; McDonald, N.Q. Recognition of an intra-chain tandem 14-3-3 binding site within PKCepsilon. *EMBO Rep.* **2009**, *10*, 983–989. [CrossRef] [PubMed]

3

A769662 Inhibits Insulin-Stimulated Akt Activation in Human Macrovascular Endothelial Cells Independent of AMP-Activated Protein Kinase

Anastasiya Strembitska [1], Sarah J. Mancini [1], Jonathan M. Gamwell [1,2], Timothy M. Palmer [1,3], George S. Baillie [1] and Ian P. Salt [1,*]

1 Institute of Cardiovascular and Medical Sciences, College of Medical, Veterinary and Life Sciences, University of Glasgow, Glasgow G12 8QQ, UK; a.strembitska.1@research.gla.ac.uk (A.S.); sarah.mancini@glasgow.ac.uk (S.J.M.); jonathan.gamwell@rdm.ox.ac.uk (J.M.G.); tim.palmer@hyms.ac.uk (T.M.P.); george.baillie@glasgow.ac.uk (G.S.B.)
2 Radcliffe Department of Medicine, University of Oxford, Oxford OX3 9DU, UK
3 Centre for Atherothrombosis and Metabolic Disease, Hull York Medical School, University of Hull, Hull HU6 7RX, UK
* Correspondence: ian.salt@glasgow.ac.uk

Abstract: Protein kinase B (Akt) is a key enzyme in the insulin signalling cascade, required for insulin-stimulated NO production in endothelial cells (ECs). Previous studies have suggested that AMP-activated protein kinase (AMPK) activation stimulates NO synthesis and enhances insulin-stimulated Akt activation, yet these studies have largely used indirect activators of AMPK. The effects of the allosteric AMPK activator A769662 on insulin signalling and endothelial function was therefore examined in cultured human macrovascular ECs. Surprisingly, A769662 inhibited insulin-stimulated NO synthesis and Akt phosphorylation in human ECs from umbilical veins (HUVECs) and aorta (HAECs). In contrast, the AMPK activators compound **991** and AICAR had no substantial inhibitory effect on insulin-stimulated Akt phosphorylation in ECs. Inhibition of AMPK with SBI-0206965 had no effect on the inhibition of insulin-stimulated Akt phosphorylation by A769662, suggesting the inhibitory action of A769662 is AMPK-independent. A769662 decreased IGF1-stimulated Akt phosphorylation yet had no effect on VEGF-stimulated Akt signalling in HUVECs, suggesting that A769662 attenuates early insulin/IGF1 signalling. The effects of A769662 on insulin-stimulated Akt phosphorylation were specific to human ECs, as no effect was observed in the human cancer cell lines HepG2 or HeLa, as well as in mouse embryonic fibroblasts (MEFs). A769662 inhibited insulin-stimulated Erk1/2 phosphorylation in HAECs and MEFs, an effect that was independent of AMPK in MEFs. Therefore, despite being a potent AMPK activator, A769662 has effects unlikely to be mediated by AMPK in human macrovascular ECs that reduce insulin sensitivity and eNOS activation.

Keywords: AMP-activated protein kinase; protein kinase B; Akt; insulin signalling; A769662; endothelial function

---

1. Introduction

Endothelial cells (ECs) are essential for modulation of vascular homeostasis and signal transduction [1], including the production and regulation of vascular tone, modulation of inflammatory responses and maintenance of an anti-atherogenic phenotype of vascular smooth muscle cells (VSMCs) [1]. Being the downstream target of phosphoinositide (PI) 3-kinase (PI3K), protein kinase B (Akt) is one of the key kinases regulating cell survival, cell-cycle progression and metabolism [2].

Under physiological conditions, insulin [3], VEGF [4] or IGF1 [5] stimulation rapidly leads to increased Akt activity by increasing Ser473 and Thr308 phosphorylation.

In healthy subjects, insulin acts as a vasodilator, stimulating calcium-independent NO synthesis in cultured human aortic ECs (HAECs) through Akt-mediated phosphorylation of endothelial nitric oxide synthase (eNOS) Ser1177 and Ser615 [6]. However, in murine models of diabetes and people with diabetes, impaired insulin-stimulated blood flow and NO bioavailability have been demonstrated [7]. Decreased NO bioavailability leads to endothelial dysfunction in animal models and human subjects, increasing risk of cardiovascular events and pro-inflammatory signalling in the vasculature [8].

AMP-activated protein kinase (AMPK) is a heterotrimeric serine/threonine protein kinase sensitive to intracellular AMP levels which acts as a master regulator of energy metabolism [9,10]. AMPK is activated in response to an increase in the intracellular (AMP + ADP):ATP ratio and simultaneous AMPKα Thr172 phosphorylation by the upstream kinases liver kinase B1 or calcium/calmodulin-dependent protein kinase kinase β [9,10]. Activated AMPK inhibits anabolic processes which consume ATP [9,10], reduces inflammation [11,12], inhibits endothelial cell proliferation [13], stimulates mitochondrial biogenesis [14], and increases insulin sensitivity [9,15], making it an attractive pharmacological target for diabetes and cardiovascular pathologies [10]. There are 12 differentially expressed AMPK isoforms, each composed of catalytic $\alpha1/\alpha2$, and regulatory $\beta1/\beta2$ and $\gamma1/\gamma2/\gamma3$ subunits [9,10], allowing a specialised cellular and systemic response to different metabolic stimuli [10].

Previous studies have reported that AMPK activation improved insulin sensitivity and energy homeostasis, and attenuated inflammatory signalling in insulin-sensitive tissues, such as muscle [14], liver [16,17] and adipose tissue [11,18] as well as ECs [12,15]. Several AMPK activators have been shown to increase NO synthesis in HAECs in an AMPK-dependent manner [10]. Nevertheless, it is unclear whether all AMPK activators improve insulin sensitivity and vascular function, as previous studies have largely been conducted using a variety of compounds that activate AMPK by altering cellular nucleotide ratios, including rosiglitazone, resveratrol, metformin and canagliflozin or mimic AMP, such as AICAR (5-amino-4-imidazolecarboxamide ribonucleoside) [12,19–22].

In the last 15 years, there has been considerable effort to develop AMPK-selective small-molecule activators, resulting in the development of A769662 [23], the discovery of salicylate as an AMPK activator and, more recently, the development of compound **991**—a compound 5–10-fold more potent than A769662 [24]. A769662 is an allosteric activator that binds complexes containing AMPK$\beta1$, also inhibiting AMPKα Thr172 dephosphorylation [24,25]. Since being first described as an AMPK activator, some AMPK-independent effects of A769662 have been reported [26–29], yet it remains a commonly utilised pharmacological tool for selective AMPK activation. The effects of A769662 on insulin signalling and insulin-stimulated Akt/eNOS axis activation in primary human ECs were therefore determined.

2. Results

2.1. Insulin-Stimulated Signalling and NO Production are Reduced by A769662 in Human Endothelial Cells

Stimulation of HUVECs with concentrations of insulin above 0.1 μM robustly stimulated phosphorylation of Akt at Ser473 and Thr308 (Supplementary Figure S1), similar to concentrations of insulin previously demonstrated to be required for insulin-stimulated NO synthesis in cultured endothelial cells [30]. All subsequent experiments were therefore conducted using 1 μM insulin. To examine the effect of A769662 on HUVEC insulin signalling, cells were preincubated in the presence or absence of A769662 prior to stimulation with insulin and NO synthesis and the extent of Akt phosphorylation assessed. Insulin modestly increased NO synthesis, yet insulin-stimulated NO synthesis was lost upon preincubation with A769662 (Figure 1A).

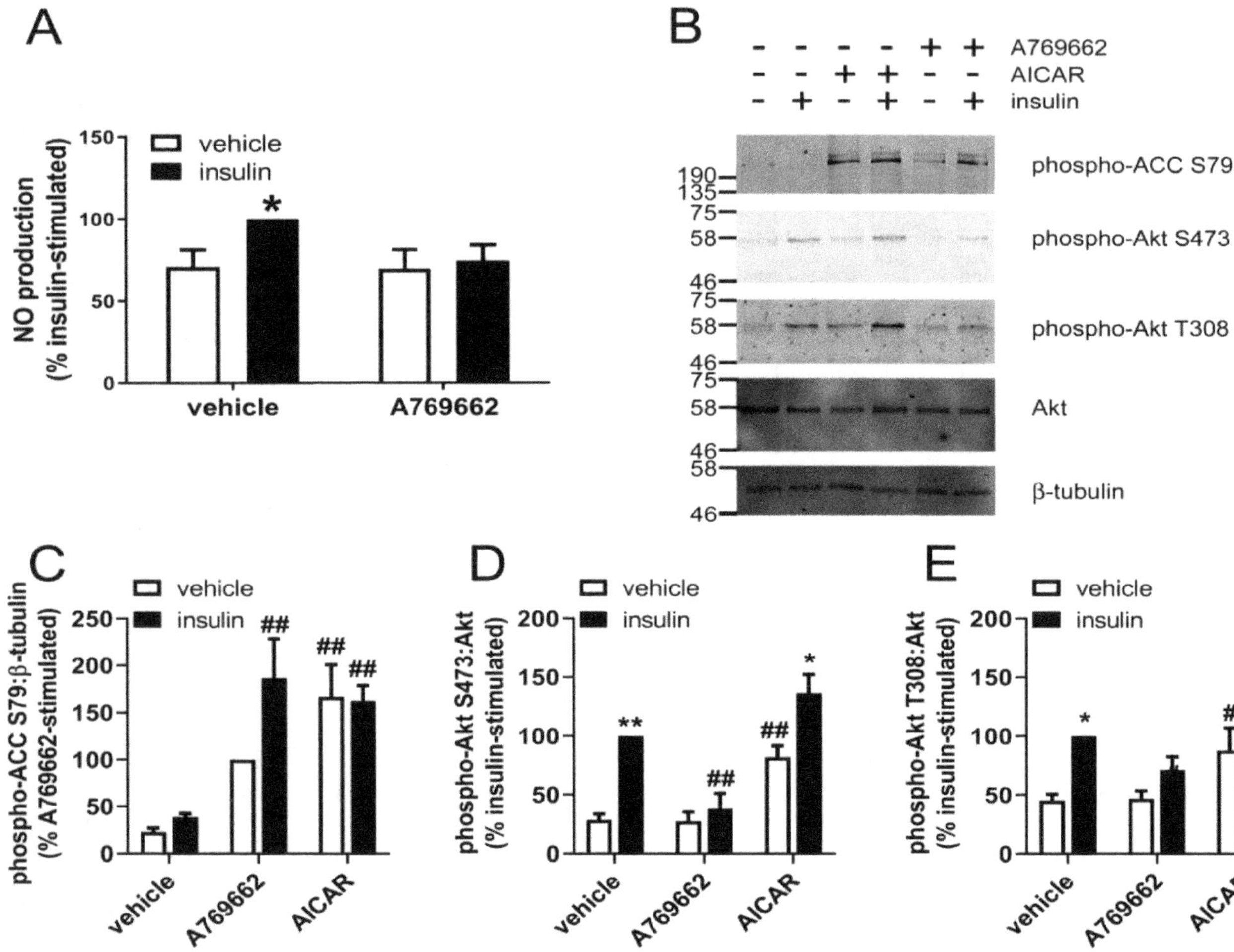

Figure 1. The effect of A769662 on insulin-stimulated NO production and Akt phosphorylation in HUVECs. (**A**) HUVECs were stimulated with A769662 (50 μM, 45 min) prior to insulin (1 μM, 15 min), conditioned media collected and NO production assessed. Data (mean ± SEM) shown from three independent replicates. (**B–E**) HUVECs were stimulated with A769662 (50 μM, 45 min) or AICAR (2 mM, 45 min) prior to insulin (51 μM, 15 min) and cell lysates prepared. Proteins were resolved by SDS-PAGE and immunoblotted with the antibodies indicated. (**B**) Representative immunoblots from three biological replicates with molecular weight markers indicated. Phospho-Akt Thr308 protein levels were assessed by stripping and re-probing the membranes. Densitometric quantification of (**C**) ACC, (**D**) Akt Ser473 and (**E**) Akt Thr308 phosphorylation normalised to β-tubulin or Akt (mean ± SEM). * $p < 0.05$, ** $p < 0.01$, relative to absence of insulin. # $p < 0.05$, ## $p < 0.01$ relative to absence of AMPK activator.

A769662 (50 μM, 45 min) increased AMPK activity, assessed by immunoblotting of AMPK-specific ACC (acetyl CoA carboxylase) Ser79 phosphorylation (Figure 1B,C), yet insulin had no effect on ACC phosphorylation. The inhibition of NO production was associated with markedly reduced insulin-stimulated phosphorylation of Akt at Ser473 and Thr308 (Figure 1D,E). In contrast, the AMPK activator AICAR, which is converted to the AMP mimetic ZMP in cells, increased basal and insulin-stimulated Akt Ser473 and Thr308 phosphorylation (Figure 1D,E), despite activating AMPK to a similar degree as assessed by ACC Ser79 phosphorylation (Figure 1C). Furthermore, the direct AMPK activator compound **991**, which allosterically activates AMPK at the same site as A769662 [31], had no effect on basal or insulin-stimulated Akt Thr308 phosphorylation, and only modestly reduced insulin-stimulated Akt Ser473 phosphorylation by 10% despite activating AMPK to a similar extent (Figure 2).

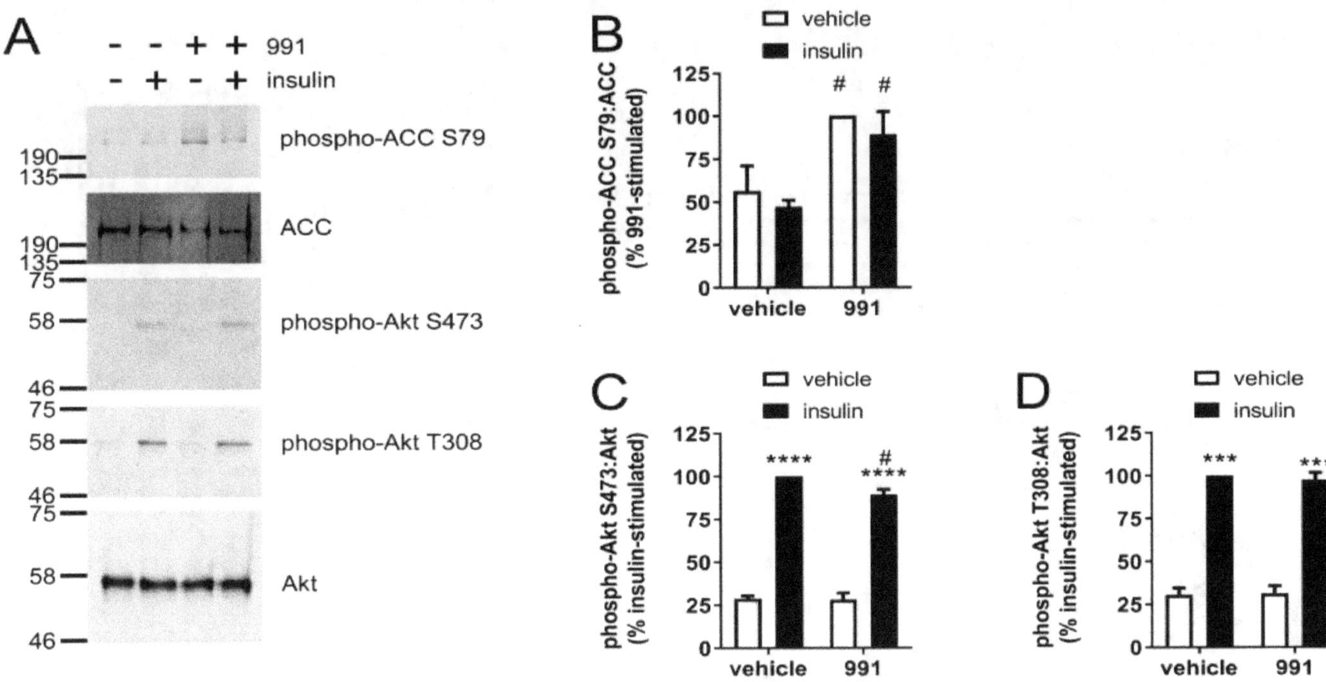

Figure 2. The effect of compound **991** on insulin-stimulated Akt phosphorylation in HUVECs. HUVECs were stimulated with compound **991** (5 μM, 45 min) prior to insulin (1 μM, 15 min) and cell lysates prepared. Proteins were resolved by SDS-PAGE and immunoblotted with the antibodies indicated. (**A**) Representative immunoblots from three biological replicates with molecular weight markers indicated. Phospho-Akt Thr308 protein levels were assessed by stripping and re-probing the membranes. Densitometric quantification of (**B**) ACC Ser79, (**C**) Akt Ser473 and (**D**) Akt Thr308 phosphorylation normalised to ACC or Akt (mean ± SEM). *** $p < 0.001$, **** $p < 0.0001$ relative to absence of insulin. # $p < 0.05$ relative to absence of compound **991**.

To examine the relationship between AMPK activation and inhibition of insulin-stimulated Akt phosphorylation by A769662, the concentration dependence of either effect of A769662 was assessed. Significant A769662-mediated stimulation of ACC phosphorylation was achieved with 50–100 μM A769662 (Figure 3A,B). A769662 decreased both insulin-stimulated Akt Ser473 and Thr308 phosphorylation in HUVECs in a concentration-dependent manner (Figure 3A), whereby 100 μM A769662 significantly inhibited Akt S473 phosphorylation and the statistical significance of insulin-stimulated Akt phosphorylation at either site was lost at concentrations above 10 μM A769662 (Figure 3C,D).

Furthermore, the time dependence of the inhibitory effect of A769662 on insulin-stimulated Akt phosphorylation was assessed. A769662 (50 μM) rapidly stimulated ACC phosphorylation in HUVECs within 5 min, an effect that was sustained for at least 2 h (Figure 3E,F). The inhibitory effect of A769662 on insulin-stimulated Akt Ser473 phosphorylation occurred similarly rapidly within 5 min and was sustained for 1 h (Figure 3E,G).

To determine whether the inhibition of insulin-stimulated Akt phosphorylation in ECs was AMPK-dependent, similar experiments were conducted after prior incubation in the SBI-0206965, which has recently been described as a selective inhibitor of AMPK [32]. Preincubation with 30 μM SBI-0206965 completely inhibited A769662-stimulated ACC Ser79 phosphorylation (Figure 4A,B), yet had no effect on the inhibition of insulin-stimulated Akt phosphorylation at Ser473 or Thr308 (Figure 4A,C,D), further indicating that the inhibitory effect of A769662 was AMPK-independent.

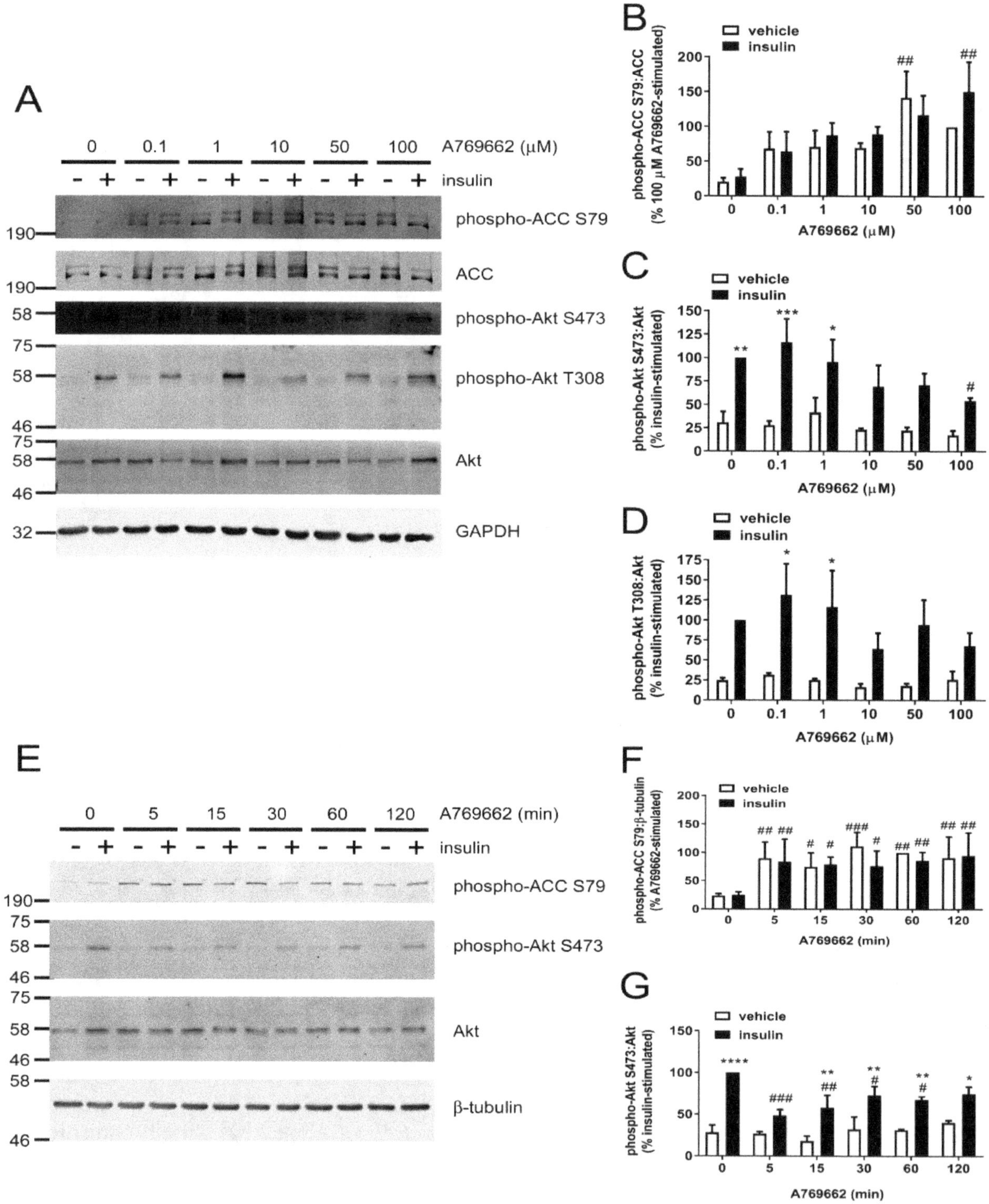

Figure 3. A769662 decreases Akt phosphorylation in a concentration- and time-dependent manner in HUVECs. HUVECs were stimulated with (**A–D**) the indicated concentrations of A769662 for 45 min or (**E–G**) 50 μM A769662 for the indicated durations prior to insulin (1 μM, 15 min). Cell lysates were prepared and immunoblotted with the antibodies indicated. (**A**,**E**) Representative immunoblots from three independent biological replicates with molecular weight markers indicated. Total ACC protein level was assessed by stripping and re-probing the membranes. Densitometric quantification of (**B**,**F**) ACC Ser79, (**C**,**G**) Akt Ser473 and (**D**) Akt Thr308 phosphorylation (mean ± SEM). * $p < 0.05$, ** $p < 0.01$, *** $p < 0.001$, **** $p < 0.0001$ relative to absence of insulin. # $p < 0.05$, ## $p < 0.01$, ### $p < 0.001$ relative to absence of A769662.

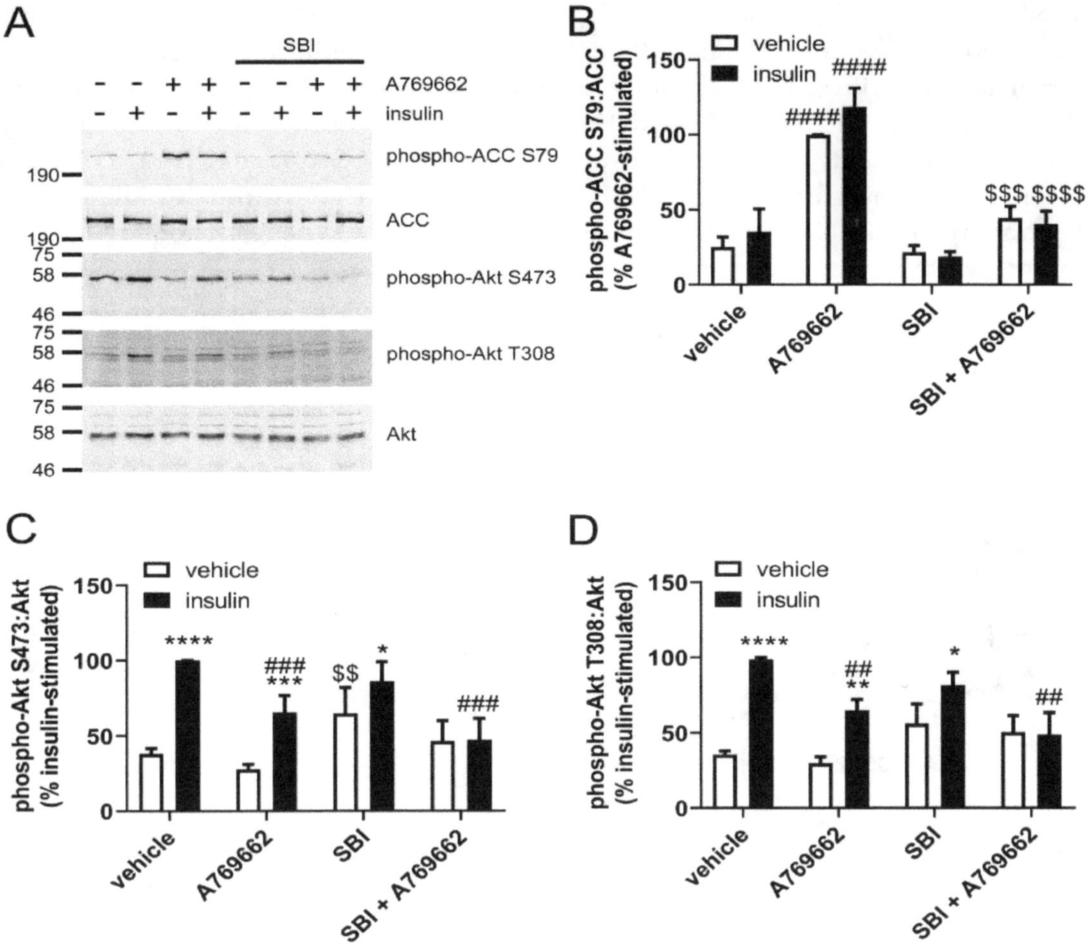

Figure 4. SBI-0206965 inhibits A769662-stimulated AMPK activation without altering inhibition of insulin-stimulated Akt phosphorylation in HUVECs. HUVECs were stimulated with A769662 (50 μM, 45 min) prior to insulin (1 μM, 15 min) after preincubation in the presence or absence of SBI-0206965 (SBI, 30 μM, 30 min) and cell lysates prepared. Proteins were resolved by SDS-PAGE and immunoblotted with the antibodies indicated. (**A**) Representative immunoblots from six biological replicates with molecular weight markers indicated. Densitometric quantification of (**B**) ACC, (**C**) Akt Ser473 and (**D**) Akt Thr308 phosphorylation normalised to total ACC or Akt (mean ± SEM). * $p < 0.05$, ** $p < 0.01$, *** $p < 0.001$, **** $p < 0.0001$ relative to absence of insulin. ## $p < 0.01$, ### $p < 0.001$, #### relative to absence of A769662. $$ $p < 0.01$, $$$ $p < 0.001$, $$$$ $p < 0.0001$ relative to absence of SBI-0206965.

2.2. Insulin Signalling and Insulin-Stimulated NO Production are Significantly Decreased in A769662-Treated HAECs

To determine whether this inhibition of insulin signalling was conserved in ECs from other regions of the vasculature, insulin-stimulated NO synthesis and signalling were assessed in HAECs. A769662 (50 μM) markedly inhibited insulin-stimulated NO synthesis (Figure 5A), an effect associated with reduced insulin-stimulated phosphorylation of Akt Ser473 (Figure 5B,C). To examine whether insulin signalling through an alternative pathway independent of PI3K was influenced by A769662, insulin-stimulated Erk1/2 phosphorylation was assessed. Intriguingly, A769662 significantly inhibited insulin-stimulated Erk1/2 phosphorylation (Figure 5B,D).

To examine whether the inhibitory action of A769662 on insulin-stimulated Akt phosphorylation was observed in non-endothelial cell lines, similar experiments were conducted in the HeLa tumour and HepG2 hepatoma cell lines. Preincubation with A769662 (50 μM, 45 min) or compound **991** (5 μM, 60 min) significantly increased ACC Ser79 phosphorylation in both HeLa (Supplementary Figure S2) and HepG2 cells (Supplementary Figure S3) to a similar extent. Unlike HUVECs, preincubation with A769662 had no statistically significant effect on basal or insulin-stimulated Akt Ser473 phosphorylation in either cell line, although the statistical significance of the effect of insulin was lost upon preincubation with A769662 in HeLa cells (Supplementary Figure S2C,D). Compound **991** did not affect insulin-stimulated Akt Ser473 and Thr308 phosphorylation in HepG2 cells (Supplementary Figure S3C,D) yet intriguingly did increase insulin-stimulated Akt Ser473 and Thr308 phosphorylation in HeLa cells (Supplementary Figure S2C,D). The inhibition of insulin-stimulated Akt Ser473 phosphorylation by preincubation with A769662 therefore seems to be restricted to ECs, and it is not observed in insulin-sensitive human cell lines.

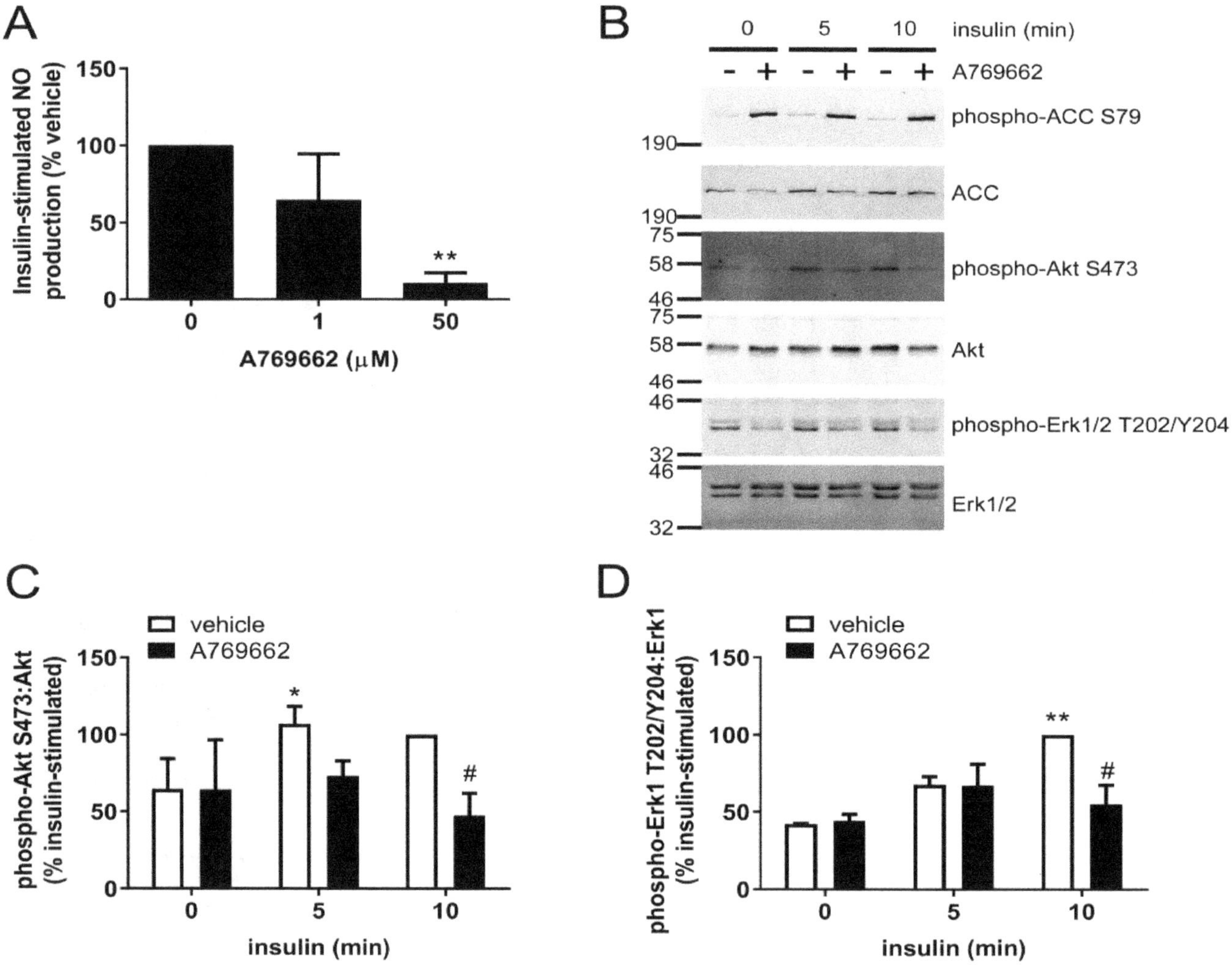

Figure 5. The effect of A769662 on insulin-stimulated NO synthesis and signalling in HAECs. (**A**) HAECs were stimulated with the indicated concentrations of A769662 prior to insulin stimulation (1 μM, 15 min). Conditioned media was collected and NO production assessed. (**B–D**) HAECs were incubated with 50 μM A769662 for 45 min prior to insulin stimulation (1 μM) for the indicated durations. Cell lysates were prepared and immunoblotted with the antibodies indicated. (**B**) Representative immunoblots from three independent biological replicates with molecular weight markers indicated. (**C,D**) Densitometric quantification (mean ± SEM) of (**C**) Akt Ser473 or (**D**) Erk1 Thr202/Tyr204 phosphorylation normalised to total Akt or Erk1 levels respectively (mean ± SEM). * $p < 0.05$, ** $p < 0.01$ relative to absence of insulin. # $p < 0.05$, relative to absence of A769662.

2.3. *A769662 Inhibits Insulin-Stimulated Erk1/2 Phosphorylation in an AMPK-Independent Manner*

To further assess the AMPK-dependence of the inhibition of insulin-stimulated Akt and Erk1/2 phosphorylation by A769662, SV40-immortalised wild-type (WT) or AMPK knockout (KO) mouse embryonic fibroblasts (MEFs) [33] were stimulated with insulin after prior incubation in the presence or absence of A769662 (Figure 6). A769662 (100 μM, 30 min) robustly stimulated ACC phosphorylation in WT MEFs and ACC phosphorylation was undetectable in KO MEFs (Figure 6A). Insulin stimulated Akt Ser473 and Erk1/2 phosphorylation to a similar extent in cells from either genotype, yet A769662 had no effect on insulin-stimulated Akt Ser473 phosphorylation in either genotype (Figure 6B). In contrast, insulin-stimulated Erk1/2 phosphorylation was significantly inhibited by preincubation with A769662 in cells of either genotype (Figure 6C), indicating an AMPK-independent effect.

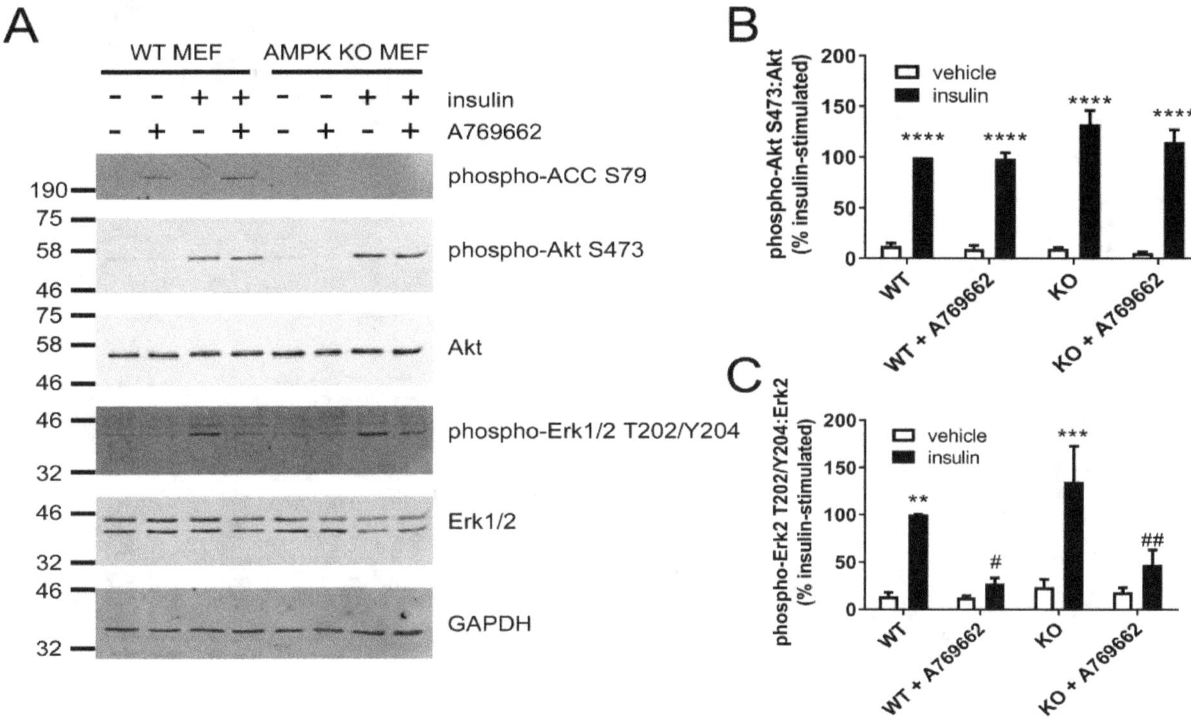

Figure 6. The effect of A769662 on insulin-stimulated Akt and Erk1/2 phosphorylation in wild-type and AMPK knockout MEFs. Wild-type (WT) and AMPK knockout (KO) MEFs were pre-incubated with A769662 (100 μM, 30 min) prior to insulin stimulation (1 μM, 15 min). Cell lysates were prepared, proteins resolved by SDS-PAGE and immunoblotted with the antibodies indicated. (**A**) Representative immunoblots from four independent biological replicates with molecular weight markers shown. (**B**,**C**) Densitometric quantification of (**B**) Akt Ser473 or (**C**) Erk2 Thr202/Tyr204 phosphorylation normalised to total Akt or Erk2 levels respectively (mean ± SEM). ** $p < 0.01$, *** $p < 0.001$, **** $p < 0.0001$ relative to absence of insulin. # $p < 0.05$, ## $p < 0.01$ relative to absence of A769662.

2.4. *A769662 Inhibits IGF1-Stimulated Akt Ser473 Phosphorylation but Has No Effect on VEGF Signalling*

To examine whether A769662 inhibits Akt phosphorylation in response to growth factors other than insulin in ECs, the effect of A769662 on IGF-1 and VEGF were assessed in HUVECs. VEGF (10 ng/mL, 10 min) stimulated a significant increase in Erk1 Thr202/Tyr204 phosphorylation, yet there was only a trend towards an increase in Akt Ser473 phosphorylation (Figure 7). In contrast, IGF1 (25 ng/mL, 10 min) significantly stimulated Akt Ser473 phosphorylation and tended to increase Erk1 Thr202/Tyr204 phosphorylation (Figure 7). Preincubation with A769662 significantly increased ACC Ser79 phosphorylation (Figure 7B) and modestly inhibited IGF1-stimulated Akt phosphorylation (Figure 7C). Although previous studies from our laboratory have shown VEGF stimulated AMPK activity after 5 min [34,35], stimulation with VEGF for 10 min only tended to increase ACC Ser79 phosphorylation in HUVECs (Figure 7B). This disparity may reflect the difference in incubation time, as in previous studies, VEGF stimulated transient ACC Ser79 phosphorylation that reached a maximum at 5 min and decreased rapidly after this point [34]. Furthermore, in contrast to the inhibitory action of A769662 on insulin-stimulated Erk1/2 phosphorylation observed in HAECs (Figure 5), IGF-1 significantly stimulated Erk1/2 phosphorylation only in the presence of A769662 (Figure 7D).

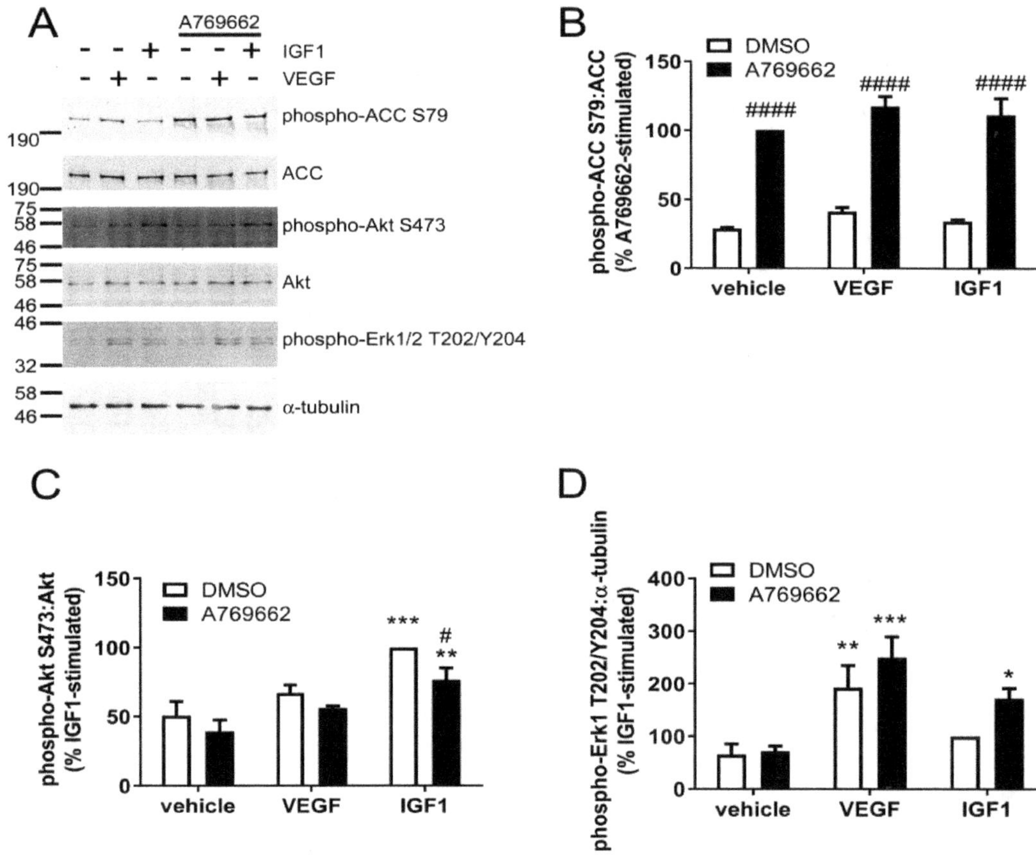

Figure 7. A769662 tends to attenuate IGF1-stimulated Ser473 Akt phosphorylation but has no effect on VEGF signalling in HUVECs. HUVECs were stimulated with A769662 (50 μM 45 min) prior to VEGF (10 ng/mL, 10 min) or IGF1 (25 ng/mL, 10 min). Cell lysates were prepared, proteins resolved by SDS-PAGE and immunoblotted with the antibodies indicated. (**A**) Representative immunoblots from four independent biological replicates with molecular weight markers indicated. Total ACC protein level was assessed by stripping and re-probing the membranes. (**B**–**D**) Densitometric quantification of (**B**) ACC Ser79, (**C**) Akt Ser473 or (**D**) Erk1 Thr202/Tyr204 phosphorylation normalised to total ACC, Akt or α-tubulin levels respectively (mean ± SEM). * $p < 0.05$, ** $p < 0.01$, *** $p < 0.001$ relative to absence of IGF1/VEGF. # $p < 0.05$, #### $p < 0.0001$ relative to absence of A769662.

2.5. AMPK Complexes Containing α1 and β1 Isoforms Contribute the Majority of Total Cellular AMPK Activity in HAECs

A769662 and compound **991** selectively activate AMPK complexes containing the β1 regulatory subunit [31]. It has previously been demonstrated that HepG2 cells principally express AMPKβ1, whereas HeLa cells express both AMPKβ1 and AMPKβ2 [36], yet the proportion of AMPKβ1/β2 complexes in endothelial cells has not been reported. The activities of AMPK complexes containing specific AMPKβ and α isoforms was therefore assessed in HAECs. Complexes containing AMPKβ1 accounted for approximately 60% of the total cellular AMPK activity and AMPKβ2 the remaining 40%

in HAECs. As previously reported, complexes containing AMPKα1 represent the majority (~95%) of total cellular AMPK activity in HAECs [21] (Figure 8).

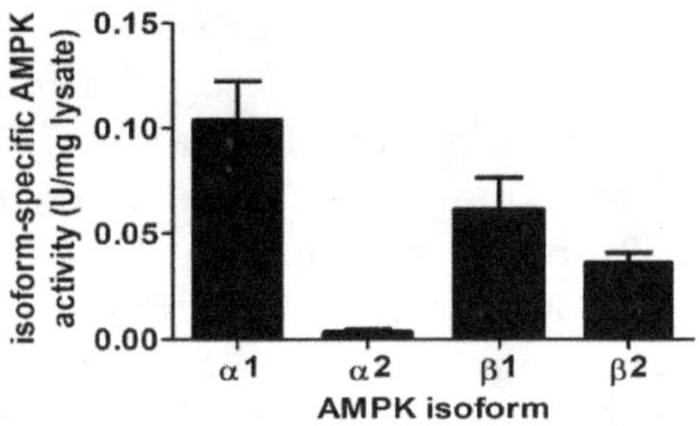

Figure 8. Activities of complexes containing specific AMPK isoforms in HAECs. AMPK was immunoprecipitated from HAEC lysates (100 μg) and AMPK activity assessed in immunoprecipitates by incorporation of ^{32}P from [γ-^{32}P]ATP into SAMS peptide. AMPK activity is presented as U/mg lysate protein (1 U = 1 nmol ^{32}P incorporated into SAMS peptide/min). Results shown are mean ± SEM activity from three independent experiments.

3. Discussion

This study demonstrates that A769662 inhibited the effects of insulin on Akt phosphorylation and NO synthesis in ECs, but had no substantive effect on Akt phosphorylation in MEFs, HeLa or HepG2 cells. This action of A769662 was not attenuated when AMPK activity was inhibited, and was not recapitulated by the alternative AMPK activators compound **991** and AICAR, suggesting this is an AMPK-independent action of A769662. In addition, A769662 inhibited insulin-stimulated Erk1 phosphorylation in ECs and MEFs, an effect that was still apparent in MEFs lacking AMPK activity. Taken together, these data suggest an EC-specific action of A769662 on early insulin signalling that is independent of AMPK.

Previous studies have demonstrated multiple beneficial effects of AMPK activation on insulin signalling, lipid and plasma glucose levels [17,23,37,38]. In ECs, AMPK activation has been reported to stimulate NO synthesis and angiogenesis while inhibiting pro-inflammatory signalling and reactive oxygen species synthesis [10]. Most of these studies have used AMPK activators that either indirectly activate AMPK through the inhibition of mitochondrial ATP synthesis, such as metformin, berberine and resveratrol or mimic AMP, such as AICAR [39]. As a consequence, these AMPK activators also have numerous other effects that are not mediated by AMPK. In contrast, A769662 and compound **991** are direct allosteric activators of AMPK that selectively or show a bias toward activating AMPK complexes containing the β1 regulatory subunit isoform [31]. Few studies have examined the endothelial effects of A769662, although it has been reported to inhibit antioxidant gene expression [40,41], viability and proliferation [13,32] and proinflammatory signalling [12,42] in human ECs. Furthermore, compound **991** has been also reported to inhibit proinflammatory signalling in ECs [12].

It is, therefore, surprising that A769662 inhibited insulin-stimulated NO synthesis in HUVECs and HAECs, as many studies report that AMPK activation stimulates NO synthesis [10]. Indeed, A769662 stimulates activating eNOS Ser1177 phosphorylation in hearts and AICAR and resveratrol have been reported to improve impaired insulin-mediated vascular responses in rodents [43–45]. Endothelial AMPK activation is not always associated with eNOS phosphorylation or NO synthesis, however [46,47], and the lack of insulin-stimulated NO synthesis in A769662-stimulated HAECs and HUVECs is likely to be a consequence of reduced insulin-stimulated Akt phosphorylation, since Akt is required for the activation of eNOS by insulin [48]. A769662 also decreased IGF1-stimulated Akt phosphorylation. Insulin and IGF1 both activate Akt [49], differing only in the initial step, as they bind to the insulin receptor (IR) and IGF1 receptor (IGF1R) respectively. In addition, IGF1R and IR share enough structural similarity to allow interchangeable binding of IGF1 and insulin [49,50]. Although IGF1 has a higher affinity for IGF1R, IGF1 may still bind IR and vice-versa. This may explain why A769662 seems to have a more modest effect on IGF1-stimulated Akt Ser473 phosphorylation, as IGF1 signalling may recruit distinct populations of IR substrate (IRS) proteins compared to insulin [51]. On the other hand, unlike insulin and IGF1, VEGF signalling does not utilise IRSs to recruit PI3K and was unaffected by A769662 preincubation [52]. As both Akt Ser473 and Thr308 phosphorylation were inhibited by A769662 in ECs, this suggests that A769662 does not simply inhibit the kinases that phosphorylate those sites, mammalian target of rapamycin complex 2 (mTORC2) or phosphoinositide-dependent protein kinase-1 (PDK1) respectively [2]. In addition, A769662 also inhibited insulin-stimulated Erk1/2 phosphorylation in ECs and MEFs, which is activated by a pathway separate to that of Akt after insulin receptor activation [53]. Importantly, A769662 also inhibited insulin-stimulated Erk1/2 phosphorylation in MEFs lacking AMPK, demonstrating that this effect is AMPK-independent. A769662 has previously been demonstrated to have no direct effect on the activity of Akt or PDK1, although it did inhibit the Erk1 kinase, mitogen-activated protein kinase kinase-1 (MKK1), in vitro [54]. In contrast, IGF-1-stimulated Erk1 phosphorylation was accentuated by A769662 in HUVECs, such that direct inhibition of MKK1 by A769662 may not simply underlie the inhibition of insulin-stimulated Erk1/2 phosphorylation. Taken together, these data indicate that A769662 inhibits at the level of the insulin/IGF-1 receptor or a receptor-associated protein, thereby inhibiting both pathways.

The inhibition of insulin-stimulated Akt phosphorylation by A769662 in ECs was not recapitulated by either compound 991 or AICAR, both of which activate AMPK to a similar degree. Compound **991** allosterically activates AMPK at a similar site to A769662 [31], whereas AICAR is phosphorylated to the AMP-mimetic, ZMP within cells [39]. Indeed, AICAR increased basal and insulin-stimulated Akt phosphorylation in ECs, in agreement with previous reports [15,21]. As both Akt Ser473 and Thr308 phosphorylation increased, this likely reflects stimulation of PI3K or earlier signalling events by AICAR, independent of AMPK activation. Given that the inhibition of Akt phosphorylation was still observed in ECs in which AMPK activity had been completely inhibited by SBI-0206965 and the lack of effect of compound **991** despite the similar mechanism by which compound **991** activates AMPK, these data indicate an AMPK-independent action of A769662 on insulin-stimulated Akt phosphorylation, although it occurs with a similar concentration to that required for AMPK activation.

Intriguingly, the inhibitory effect of A769662 on insulin-stimulated Akt activation was limited to ECs, as it was not observed in MEFs, HeLa or HepG2 cells, although the statistical significance of stimulation by insulin was lost in HeLa cells preincubated with A769662. A769662 (100 μM) has been demonstrated previously to inhibit basal Akt Ser473 phosphorylation in prostate cancer cell lines, yet stimulated Akt Ser473 phosphorylation at a lower concentration (50 μM) [55]. Indeed, previous studies have shown no effect of A769662 on insulin-stimulated Akt activation in rat adult cardiomyocytes, human myotubes or L6 cells [29,56,57]. In contrast to this neutral effect reported in striated muscle cells, A769662 has also been reported to stimulate glucose uptake in muscle cells by increasing PI3K association with IRS1, suggesting an AMPK-independent effect that increased Akt activity [58]. In addition, high concentrations of A769662 stimulated Akt phosphorylation in a manner sensitive to the PI3K inhibitor, wortmannin in CHO cells expressing the δ-opioid receptor [59]. It is clear, therefore, that A769662 influences Akt differentially in different cell types, providing further evidence that these effects are not mediated by AMPK.

In addition, we demonstrate that preincubation of HeLa cells, but not HepG2 cells or ECs, with compound **991** increased insulin-stimulated Akt phosphorylation. These data argue for a cell-type specific/selective effect of A769662 on insulin signalling, whereby only A769662 markedly reduces insulin signalling in ECs and insulin-stimulated Erk1/2 phosphorylation in ECs and MEFs. This cell-type selectivity could be related to differential expression of IR/IGF1R between cell types. HUVECs have been previously reported to have approximately 400,000 IGF1R/cell and 40,000 IR/cell [30]. Furthermore, according to the updated version of Human Protein Atlas (https://www.proteinatlas.org/) [60], HepG2 and HeLa cells have been reported to express 21.9 and 27.5 IGF1R transcripts per million (TPM), respectively and 15.9 and 1.9 IR TPM respectively [61]. Higher IR levels could preserve insulin-stimulated Akt and Erk1 phosphorylation by increasing the number of activated receptors and signal intensity. The high relative IR expression levels in HepG2 cells may therefore explain why A769662 had no effect on attenuation of insulin signalling in this cell line, whereas in HUVECs and HAECs, which have lower IR expression levels, there was such a marked effect of A769662.

It is unlikely that the differential effects of A769662 in the different cell types can be explained by differences in AMPK isoform expression, due to the selectivity of A769662 and compound **991** [31]. Both compounds would be expected to activate AMPK complexes containing β1, and complexes containing AMPKβ1 were found to contribute ~60% of the total cellular AMPK activity in HAECs. It has previously been reported that HepG2 cells principally express AMPKβ1, whereas HeLa cells express both AMPKβ1 and AMPKβ2 [36]. This further indicates that the inhibition of insulin signalling in ECs by A769662 is an AMPK-independent effect and unrelated to differential actions on specific pools of AMPK within ECs, as all the cell types investigated express abundant AMPKβ1 levels. As AMPK-independent effects of A769662, including inhibition of 26S proteasome activity in MEFs, voltage-gated Na^+ channels in rat neurons and the Na^+/K^+ ATPase in L6 cells [26,28,29] have been reported previously, inhibition of insulin signalling in ECs may similarly be considered AMPK-independent. In addition, A769662 was recently reported to promote vasorelaxation in rabbit and rat arteries by reducing cytosolic Ca^{2+} levels, by an endothelium-dependent yet AMPK-independent manner [62].

In conclusion, A769662 decreases insulin-stimulated NO synthesis and Akt Ser473 phosphorylation in HUVECs and HAECs in a manner likely to be independent of AMPK. Furthermore, A769662 decreases insulin-stimulated Erk1/2 phosphorylation in a manner that is AMPK-independent in MEFs and HAECs. Taken together, these data demonstrate that caution should be exercised when interpreting data obtained using A769662 as a tool in cultured human endothelial cells.

4. Materials and Methods

4.1. Materials

Cryopreserved HUVECs, HAECs and MV2 medium were purchased from Promocell (Heidelberg, Germany). SV40-immortalised wild-type and AMPKα1 and AMPKα2 knock-out MEFs were kindly provided by Dr. B. Viollet (Institut Cochin, Paris, France) and have been described previously [33]. HeLa and HepG2 cells were obtained from ATCC (Manassas, VA, USA). IGF1, VEGF, SBI-0206965 and porcine insulin were purchased from Sigma Aldrich (St. Louis, MO, USA). AICAR was purchased from Toronto Research Chemicals Inc. (Ontario, ON, Canada). Compound **991** was synthetised by MRC Technology. A769662 and mouse anti-β-tubulin (#11307) antibodies were purchased from Abcam (Cambridge, UK). Rabbit anti-phospho-Akt Thr308 (#13038), anti-phospho-Akt Ser473 (#4058), anti-ERK1/2 (#9102), anti-phospho-ACC Ser79 (#3661), anti-ACC (#3676), anti-α-tubulin (#2144) and mouse anti-Akt (#2920) and anti-phospho-p44/42 Erk1/2 (Thr202/Tyr204) (#9106) antibodies were from New England Biolabs UK (Hitchin, UK). Sheep anti-AMPKα1, anti-AMPKα2, anti-AMPKβ1 and anti-AMPKβ2 antibodies used for immunoprecipitation of AMPK complexes containing specific subunit isoforms were a kind gift from Professor D.G. Hardie (University of Dundee) and have been described previously [39,63,64]. IRdye680 or 800-labelled donkey anti-mouse IgG (#926-32212) and anti-rabbit IgG (#926-68023 and #926-32213) antibodies were from LI-COR Biosciences (Lincoln, UK). Medium 199, RPMI 1640 and DMEM (4.5 g/L glucose) were from Life Technologies (Paisley, UK).

4.2. Cell Culture and Experimental Design

HAECs and HUVECs were cultured in MV2 medium supplemented with EC growth factor mix, 5% (v/v) serum (Promocell). HUVECs and HAECs were utilised between passages 3 and 6. MEFs and HeLa cells were cultured in DMEM supplemented with 10% (v/v) foetal calf serum. HepG2 cells were cultured in RPMI 1640 supplemented with 10% (v/v) foetal calf serum, 1 mM sodium pyruvate. Once cells reached 90–100% confluence, HUVECs/HAECs were incubated in serum-free Medium 199 and HeLa cells and HepG2 cells were incubated with serum-free DMEM for 2 h and then with AMPK activators prior to stimulation with insulin (1 μM), VEGF (10 ng/mL) or IGF1 (25 ng/mL). Cells were placed on ice and washed with PBS prior to lysis in Triton X-100-based lysis buffer (50 mM Tris-HCl, pH 7.4 at 4 °C, 50 mM NaF, 1 mM $Na_4P_2O_7$, 1 mM EDTA, 1 mM EGTA, 1% (*v*/*v*) Triton-X-100, 250 mM mannitol, 1 mM DTT, 1 mM Na_3VO_4, 0.1 mM benzamidine, 0.1 mM PMSF, 5 μg/mL SBTI). Cell lysates were scraped into microcentrifuge tubes and incubated on ice for 20 min, centrifuged (5 min, 21,910× *g*, 4 °C) and the subsequent supernatants stored at −20 °C. Lysate protein concentrations were determined using Bradford or BCA methods, as previously described [6,21].

4.3. SDS-Polyacrylamide Gel Electrophoresis and Immunoblotting

Cell lysate proteins were resolved by SDS-PAGE and immunoblotted with antibodies diluted in 50% (*v*/*v*) LI-COR blocking buffer in TBS containing 0.1% (*v*/*v*) Tween-20 as described previously [13]. Proteins were visualised using infrared dye-labelled secondary antibodies on a LI-COR Odyssey infrared imaging system and analysed using the ImageJ software for densitometric quantification of band intensity. In some cases, indicated in the figure legends, immunoblots were stripped in 0.2 M NaOH for 10 min, washed multiple times with Tris-buffered saline (TBS) until the pH returned to 7.2, and then blocked in TBS supplemented with 5% (*w*/*v*) milk powder prior to probing with primary antibodies. In all other cases, immunoblots for phospho- and total protein levels of ACC, Akt and Erk1/2 were obtained on immunoblots conducted concurrently in parallel.

4.4. NO Assay

HAECs and HUVECs were cultured in 6-well plates until 100% confluent and serum-starved for 2 h in Medium 199 and then in the presence of Krebs-Ringer-Hepes (KRH) buffer for 20 min. The medium was replaced and cells incubated in the presence or absence of stimuli in KRH for a further 15–20 min. Samples of medium (50 μL) were taken at different intervals and 200 μL methanol added to each. Samples were centrifuged (21,910× *g*, 4 °C, 20 min), and supernatants were stored at −20 °C. NO concentration was determined using a Sievers 280A NO Meter (Sievers, Boulder, CO, USA) as described previously [21].

4.5. AMPK Assay

AMPK specific isoforms were immunoprecipitated from HAECs using sheep anti-AMPKα1, anti-AMPKα2 [63], anti-AMPKβ1 [39] or anti-AMPKβ2 [64] antibodies bound to protein G-Sepharose (1 μg antibody and 5 μL packed volume protein G-Sepharose/immunoprecipitation) in IP buffer (50 mM Tris-HCl (pH 7.4 at 4 °C), 150 mM NaCl, 50 mM NaF, 5 mM $Na_4P_2O_7$, 1 mM EDTA, 1 mM EGTA, 1% (*v*/*v*) Triton-X-100, 1% (*v*/*v*) glycerol, 1 mM DTT, 0.1 mM benzamidine, 1 mM PMSF, 5 μg/mL SBTI, 1 mM Na_3VO_4). Immunoprecipitates were then washed into HBD buffer (50 mM HEPES-NaOH (pH 7.4), 0.02% (*v*/*v*) Brij-35, 1 mM DTT) and assayed for AMPK activity using SAMS substrate peptide as described previously [12,35].

4.6. Statistical Analysis

All data is expressed as a relative change (mean ± SEM) from control baseline or normalised to group of reference for each experiment. Prism software (GraphPad Software, San Diego, CA, USA) was used to perform one or two-way ANOVA (with Tukey's or Dunnett's post hoc multiple comparison tests), where appropriate, using $p < 0.05$ as significant.

Author Contributions: Conceptualisation, A.S. and I.P.S.; Methodology, A.S. and I.P.S.; Investigation, A.S., S.J.M., J.M.G. and I.P.S.; Validation, A.S., S.J.M. and J.M.G.; Formal Analysis, A.S., S.J.M. and I.P.S.; Writing—Original Draft Preparation, A.S. and I.P.S.; Writing—Review & Editing, I.P.S., T.M.P. and G.S.B.; Supervision, I.P.S.; Funding Acquisition I.P.S. and T.M.P. All listed authors approved the final version of the manuscript.

References

1. Zhao, Y.; Vanhoutte, P.M.; Leung, S.W.S. Vascular nitric oxide: Beyond eNOS. *J. Pharmacol. Sci.* **2015**, *12*, 83–94. [CrossRef] [PubMed]
2. Manning, B.D.; Toker, A. AKT/PKB Signaling: Navigating the Network. *Cell* **2017**, *169*, 381–405. [CrossRef] [PubMed]
3. Jiang, Z.Y.; He, Z.; King, B.L.; Kuroki, T.; Opland, D.M.; Suzuma, K.; Suzuma, I.; Ueki, K.; Kulkarni, R.N.; Kahn, C.R. Characterization of multiple signaling pathways of insulin in the regulation of vascular endothelial growth factor expression in vascular cells and angiogenesis. *J. Biol. Chem.* **2003**, *278*, 31964–31971. [CrossRef] [PubMed]
4. Takahashi, S.; Mendelsohn, M.E. Synergistic activation of endothelial nitric-oxide synthase (eNOS) by HSP90 and Akt. Calcium-independent eNOS activation involves formation of an HSP90-Akt-CaM-bound eNOS complex. *J. Biol. Chem.* **2003**, *278*, 30821–30827. [CrossRef] [PubMed]
5. Michell, B.J.; Griffiths, J.E.; Mitchelhill, K.I.; Rodriguez-Crespo, I.; Tiganis, T.; Bozinovski, S.; Montellano, P.R.O.; Kemp, B.E.; Pearson, R.B. The Akt kinase signals directly to endothelial nitric oxide synthase. *Curr. Biol.* **1999**, *9*, 845–848. [CrossRef]
6. Ritchie, S.A.; Kohlhaas, C.F.; Boyd, A.R.; Yalla, K.C.; Walsh, K.; Connell, J.M.; Salt, I.P. Insulin-stimulated phosphorylation of endothelial nitric oxide synthase at serine-615 contributes to nitric oxide synthesis. *Biochem. J.* **2010**, *426*, 85–90. [CrossRef] [PubMed]
7. Reynolds, L.J.; Credeur, D.P.; Manrique, C.; Padilla, J.; Fadel, P.J.; Thyfault, J.P. Obesity, type 2 diabetes, and impaired insulin-stimulated blood flow: Role of skeletal muscle NO synthase and endothelin-1. *J. Appl. Physiol.* **2017**, *122*, 38–47. [CrossRef] [PubMed]
8. Tabit, C.E.; Chung, W.B.; Vita, J.A. Endothelial dysfunction in diabetes mellitus: Molecular mechanisms and clinical implications. *Rev. Endocr. Metab. Disord.* **2010**, *11*, 61–74. [CrossRef]
9. Day, E.A.; Ford, R.J.; Steinberg, G.R. AMPK as a Therapeutic Target for Treating Metabolic Diseases. *Trends Endocrinol. MeTab.* **2017**, *28*, 545–560. [CrossRef]
10. Salt, I.P.; Hardie, D.G. AMP-Activated Protein Kinase: An Ubiquitous Signaling Pathway with Key Roles in the Cardiovascular System. *Circ. Res.* **2017**, *120*, 1825–1841. [CrossRef]
11. Mancini, S.J.; White, A.D.; Bijland, S.; Rutherford, C.; Graham, D.; Richter, E.A.; Viollet, B.; Touyz, R.M.; Palmer, T.M.; Salt, I.P. Activation of AMP-activated protein kinase rapidly suppresses multiple pro-inflammatory pathways in adipocytes including IL-1 receptor-associated kinase-4 phosphorylation. *Mol. Cell. Endocrinol.* **2017**, *440*, 44–56. [CrossRef] [PubMed]
12. Mancini, S.J.; Boyd, D.; Katwan, O.J.; Strembitska, A.; Almabrouk, T.A.; Kennedy, S.; Palmer, T.M.; Salt, I.P. Canagliflozin inhibits interleukin-1β-stimulated cytokine and chemokine secretion in vascular endothelial cells by AMP-activated protein kinase-dependent and independent mechanisms. *Sci. Rep.* **2018**, *8*, 5276. [CrossRef] [PubMed]
13. Peyton, K.J.; Liu, X.M.; Yu, Y.; Yates, B.; Durante, W. Activation of AMP-Activated Protein Kinase Inhibits the Proliferation of Human Endothelial Cells. *J. Pharmacol. Exp. Ther.* **2012**, *342*, 827–834. [CrossRef] [PubMed]
14. Li, C.; Reif, M.M.; Craige, S.M.; Kant, S.; Keaney, J.F. Endothelial AMPK activation induces mitochondrial biogenesis and stress adaptation via eNOS-dependent mTORC1 signaling. *Nitric Oxide-Biol. Chem.* **2016**, *55–56*, 45–53. [CrossRef] [PubMed]
15. Ido, Y.; Carling, D.; Ruderman, N. Hyperglycemia-induced apoptosis in human umbilical vein endothelial cells: Inhibition by the AMP-activated protein kinase activation. *Diabetes* **2002**, *51*, 159–167. [CrossRef] [PubMed]
16. Li, Y.; Xu, S.; Mihaylova, M.M.; Zheng, B.; Hou, X.; Jiang, B.; Park, O.; Luo, Z.; Lefai, E.; Shyy, J.Y.J.; et al. AMPK phosphorylates and inhibits SREBP activity to attenuate hepatic steatosis and atherosclerosis in diet-induced insulin-resistant mice. *Cell MeTab.* **2011**, *13*, 376–388. [CrossRef] [PubMed]
17. Ford, R.J.; Fullerton, M.D.; Pinkosky, S.L.; Day, E.A.; Scott, J.W.; Oakhill, J.S.; Bujak, A.L.; Smith, B.K.; Crane, J.D.; Blümer, R.M.; et al. Metformin and salicylate synergistically activate liver AMPK, inhibit lipogenesis and improve insulin sensitivity. *Biochem. J.* **2015**, *468*, 125–132. [CrossRef] [PubMed]
18. Mottillo, E.P.; Desjardins, E.M.; Crane, J.D.; Smith, B.K.; Green, A.E.; Ducommun, S.; Henriksen, T.I.; Rebalka, I.A.; Razi, A.; Sakamoto, K.; et al. Lack of Adipocyte AMPK Exacerbates Insulin Resistance and Hepatic Steatosis through Brown and Beige Adipose Tissue Function. *Cell Metab.* **2016**, *24*, 118–129. [CrossRef]

19. Boyle, J.G.; Logan, P.J.; Ewart, M.A.; Reihill, J.A.; Ritchie, S.A.; Connell, J.M.; Cleland, S.J.; Salt, I.P. Rosiglitazone Stimulates Nitric Oxide Synthesis in Human Aortic Endothelial Cells via AMP-activated Protein Kinase. *J. Biol. Chem.* **2008**, *283*, 11210–11217. [CrossRef]
20. Csiszar, A.; Labinskyy, N.; Podlutsky, A.; Kaminski, P.M.; Wolin, M.S.; Zhang, C.; Mukhopadhyay, P.; Pacher, P.; Hu, F.; De Cabo, R.; et al. Vasoprotective effects of resveratrol and SIRT1: Attenuation of cigarette smoke-induced oxidative stress and proinflammatory phenotypic alterations. *Am. J. Physiol. Heart Circ. Physiol.* **2008**, *294*, H2721–H2735. [CrossRef]
21. Morrow, V.A.; Foufelle, F.; Connell, J.M.C.; Petrie, J.R.; Gould, G.W.; Salt, I.P. Direct activation of AMP-activated protein kinase stimulates nitric-oxide synthesis in human aortic endothelial cells. *J. Biol. Chem.* **2003**, *278*, 31629–31639. [CrossRef] [PubMed]
22. Davis, B.J.; Xie, Z.; Viollet, B.; Zou, M.H. Activation of the AMP-activated kinase by antidiabetes drug metformin stimulates nitric oxide synthesis in vivo by promoting the association of heat shock protein 90 and endothelial nitric oxide synthase. *Diabetes* **2006**, *55*, 496–505. [CrossRef] [PubMed]
23. Cool, B.; Zinker, B.; Chiou, W.; Kifle, L.; Cao, N.; Perham, M.; Dickinson, R.; Adler, A.; Gagne, G.; Iyengar, R.; et al. Identification and characterization of a small molecule AMPK activator that treats key components of type 2 diabetes and the metabolic syndrome. *Cell MeTab.* **2006**, *3*, 403–416. [CrossRef] [PubMed]
24. Bultot, L.; Jensen, T.E.; Lai, Y.C.; Madsen, A.L.; Collodet, C.; Kviklyte, S.; Deak, M.; Yavari, A.; Foretz, M.; Ghaffari, S.; et al. Benzimidazole derivative small-molecule 991 enhances AMPK activity and glucose uptake induced by AICAR or contraction in skeletal muscle. *Am. J. Physiol. Endocrinol. MeTab.* **2016**, *311*, E706–E719. [CrossRef] [PubMed]
25. Ducommun, S.; Ford, R.J.; Bultot, L.; Deak, M.; Bertrand, L.; Kemp, B.E.; Steinberg, G.R.; Sakamoto, K. Enhanced activation of cellular AMPK by dual-small molecule treatment: AICAR and A769662. *Am. J. Physiol. Endocrinol. MeTab.* **2014**, *306*, E688–E696. [CrossRef] [PubMed]
26. Asiedu, M.N.; Han, C.; Dib-Hajj, S.D.; Waxman, S.G.; Price, T.J.; Dussor, G. The AMPK activator A769662 blocks voltage-gated sodium channels: Discovery of a novel pharmacophore with potential utility for analgesic development. *PLoS ONE* **2017**, *12*, e0169882. [CrossRef] [PubMed]
27. Vlachaki Walker, J.M.; Robb, J.L.; Cruz, A.M.; Malhi, A.; Weightman Potter, P.G.; Ashford, M.L.; McCrimmon, R.J.; Ellacott, K.L.; Beall, C. AMP-activated protein kinase (AMPK) activator A-769662 increases intracellular calcium and ATP release from astrocytes in an AMPK-independent manner. *Diabetes Obes. MeTab.* **2017**, *19*, 997–1005. [CrossRef]
28. Moreno, D.; Knecht, E.; Viollet, B.; Sanz, P. A769662, a novel activator of AMP-activated protein kinase, inhibits non-proteolytic components of the 26S proteasome by an AMPK-independent mechanism. *FEBS Lett.* **2008**, *582*, 2650–2654. [CrossRef]
29. Benziane, B.; Björnholm, M.; Lantier, L.; Viollet, B.; Zierath, J.R.; Chibalin, A.V. AMP-activated protein kinase activator A-769662 is an inhibitor of the Na^+-K^+-ATPase. *Am. J. Physiol. Cell Physiol.* **2009**, *297*, C1554–C1567. [CrossRef]
30. Zeng, G.; Quon, M.J. Insulin-stimulated production of nitric oxide is inhibited by Wortmannin: Direct measurement in vascular endothelial cells. *J. Clin. Investig.* **1996**, *98*, 894–898. [CrossRef]
31. Xiao, B.; Sanders, M.J.; Carmena, D.; Bright, N.J.; Haire, L.F.; Underwood, E.; Patel, B.R.; Heath, R.B.; Walker, P.A.; Hallen, S.; et al. Structural basis of AMPK regulation by small molecule activators. *Nat Commun.* **2013**, *4*, 1–17. [CrossRef] [PubMed]
32. Dite, T.A.; Langendorf, C.G.; Hoque, A.; Galic, S.; Rebello, R.J.; Ovens, A.J.; Lindqvist, L.M.; Ngoei, K.R.; Ling, N.X.; Furic, L.; et al. AMP-activated protein kinase selectively inhibited by the type II inhibitor SBI-0206965. *J. Biol. Chem.* **2018**, *293*, 8874–8885. [CrossRef] [PubMed]
33. Jørgensen, S.B.; Viollet, B.; Andreelli, F.; Frøsig, C.; Birk, J.B.; Schjerling, P.; Vaulont, S.; Richter, E.A.; Wojtaszewski, J.F. Knockout of the $\alpha 2$ but Not α_1 5′-AMP-activated Protein Kinase Isoform Abolishes 5-Aminoimidazole-4-carboxamide-1-β-4-ribofuranoside- but Not Contraction-induced Glucose Uptake in Skeletal Muscle. *J. Biol. Chem.* **2004**, *279*, 1070–1079. [CrossRef] [PubMed]
34. Reihill, J.A.; Ewart, M.A.; Salt, I.P. The role of AMP-activated protein kinase in the functional effects of vascular endothelial growth factor-A and -B in human aortic endothelial cells. *Vasc. Cell.* **2011**, *3*, 9. [CrossRef] [PubMed]

35. Heathcote, H.R.; Mancini, S.J.; Strembitska, A.; Jamal, K.; Reihill, J.A.; Palmer, T.M.; Gould, G.W.; Salt, I.P. Protein kinase C phosphorylates AMP-activated protein kinase α1 Ser487. *Biochem. J.* **2016**, *473*, 4681–4697. [CrossRef] [PubMed]
36. Scott, J.W.; Galic, S.; Graham, K.L.; Foitzik, R.; Ling, N.X.; Dite, T.A.; Issa, S.M.; Langendorf, C.G.; Weng, Q.P.; Thomas, H.E.; et al. Inhibition of AMP-Activated Protein Kinase at the Allosteric Drug-Binding Site Promotes Islet Insulin Release. *Chem. Biol.* **2015**, *22*, 705–711. [CrossRef] [PubMed]
37. Lee, Y.S.; Kim, W.S.; Kim, K.H.; Yoon, M.J.; Cho, H.J.; Shen, Y.; Ye, J.M.; Lee, C.H.; Oh, W.K.; Kim, C.T.; et al. Berberine, a natural plant product, activates AMP-activated protein kinase with beneficial metabolic effects in diabetic and insulin-resistant states. *Diabetes* **2006**, *55*, 2256–2264. [CrossRef] [PubMed]
38. Zang, M.; Zuccollo, A.; Hou, X.; Nagata, D.; Walsh, K.; Herscovitz, H.; Brecher, P.; Ruderman, N.B.; Cohen, R.A. AMP-activated protein kinase is required for the lipid-lowering effect of metformin in insulin-resistant human HepG2 cells. *J. Biol. Chem.* **2004**, *279*, 47898–47905. [CrossRef]
39. Hawley, S.A.; Ross, F.A.; Chevtzoff, C.; Green, K.A.; Evans, A.; Fogarty, S.; Towler, M.C.; Brown, L.J.; Ogunbayo, O.A.; Evans, A.M.; et al. Use of cells expressing γ subunit variants to identify diverse mechanisms of AMPK activation. *Cell MeTab.* **2010**, *11*, 554–565. [CrossRef]
40. Liu, X.; Peyton, K.J.; Shebib, A.R.; Wang, H.; Korthuis, R.J.; Durante, W. Activation of AMPK stimulates heme oxygenase-1 gene expression and human endothelial cell survival. *Am. J. Physiol. Heart Circ. Physiol.* **2011**, *300*, H84–H93. [CrossRef]
41. Dang, Y.; Ling, S.; Duan, J.; Ma, J.; Ni, R.; Xu, J.W. Bavachalcone-Induced Manganese Superoxide Dismutase Expression through the AMP-Activated Protein Kinase Pathway in Human Endothelial Cells. *Pharmacology* **2015**, *95*, 105–110. [CrossRef] [PubMed]
42. Rutherford, C.; Speirs, C.; Williams, J.J.L.; Ewart, M.A.; Mancini, S.J.; Hawley, S.A.; Delles, C.; Viollet, B.; Costa-Pereira, A.P.; Baillie, G.S.; et al. Phosphorylation of Janus kinase 1 (JAK1) by AMP-activated protein kinase (AMPK) links energy sensing to anti-inflammatory signaling. *Sci. Signal.* **2016**, *9*, ra109. [CrossRef] [PubMed]
43. Kim, A.S.; Miller, E.J.; Wright, T.M.; Li, J.; Qi, D.; Atsina, K.; Zaha, V.; Sakamoto, K.; Young, L.H. A small molecule AMPK activator protects the heart against ischemia-reperfusion injury. *J. Mol. Cell. Cardiol.* **2011**, *51*, 24–32. [CrossRef] [PubMed]
44. Bradley, E.A.; Zhang, L.; Genders, A.J.; Richards, S.M.; Rattigan, S.; Keske, M.A. Enhancement of insulin-mediated rat muscle glucose uptake and microvascular perfusion by 5-aminoimidazole-4-carboxamide-1-β-d-ribofuranoside. *Cardiovasc. Diabetol.* **2015**, *14*, 91. [CrossRef] [PubMed]
45. Liu, Z.; Jiang, C.; Zhang, J.; Liu, B.; Du, Q. Resveratrol inhibits inflammation and ameliorates insulin resistant endothelial dysfunction via regulation of AMP-activated protein kinase and sirtuin 1 activities. *J. Diabetes* **2016**, *8*, 324–335. [CrossRef]
46. Stahmann, N.; Woods, A.; Spengler, K.; Heslegrave, A.; Bauer, R.; Krause, S.; Viollet, B.; Carling, D.; Heller, R. Activation of AMP-activated protein kinase by vascular endothelial growth factor mediates endothelial angiogenesis independently of nitric-oxide synthase. *J. Biol. Chem.* **2010**, *285*, 10638–10652. [CrossRef]
47. Mount, P.F.; Hill, R.E.; Fraser, S.A.; Levidiotis, V.; Katsis, F.; Kemp, B.E.; Power, D.A. Acute renal ischemia rapidly activates the energy sensor AMPK but does not increase phosphorylation of eNOS-Ser^{1177}. *Am. J. Physiol. Physiol.* **2005**, *289*, F1103–F1115. [CrossRef]
48. Montagnani, M.; Chen, H.; Barr, V.A.; Quon, M.J. Insulin-stimulated Activation of eNOS Is Independent of Ca2+ but Requires Phosphorylation by Akt at Ser1179. *J. Biol. Chem.* **2001**, *276*, 30392–30398. [CrossRef]
49. Van Heemst, D. Insulin, IGF-1 and longevity. *Aging Dis.* **2010**, *1*, 147–157. [CrossRef]
50. Varewijck, A.J.; Janssen, J.A. Insulin and its analogues and their affinities for the IGF1 receptor. *Endocr. Relat. Cancer* **2012**, *19*, F63–F75. [CrossRef]
51. Cai, W.; Sakaguchi, M.; Kleinridders, A.; Gonzalez-Del Pino, G.; Dreyfuss, J.M.; O'Neill, B.T.; Ramirez, A.K.; Pan, H.; Winnay, J.N.; Boucher, J.; et al. Domain-dependent effects of insulin and IGF-1 receptors on signalling and gene expression. *Nat. Commun.* **2017**, *8*, 14892. [CrossRef] [PubMed]
52. Olsson, A.K.; Dimberg, A.; Kreuger, J.; Claesson-Welsh, L. VEGF receptor signalling—In control of vascular function. *Nat. Rev. Mol. Cell. Biol.* **2006**, *7*, 359–371. [CrossRef] [PubMed]
53. King, G.L.; Park, K.; Li, Q. Selective insulin resistance and the development of cardiovascular diseases in diabetes: The 2015 Edwin Bierman Award Lecture. *Diabetes* **2016**, *65*, 1462–1471. [CrossRef] [PubMed]

54. Göransson, O.; McBride, A.; Hawley, S.A.; Ross, F.A.; Shpiro, N.; Foretz, M.; Viollet, B.; Hardie, D.G.; Sakamoto, K. Mechanism of action of A-769662, a valuable tool for activation of AMP-activated protein kinase. *J. Biol. Chem.* **2007**, *282*, 32549–32560. [CrossRef] [PubMed]
55. Choudhury, Y.; Yang, Z.; Ahmad, I.; Nixon, C.; Salt, I.P.; Leung, H.Y. AMP-activated protein kinase (AMPK) as a potential therapeutic target independent of PI3K/Akt signaling in prostate cancer. *Oncoscience* **2014**, *1*, 446. [CrossRef] [PubMed]
56. Timmermans, A.D.; Balteau, M.; Gélinas, R.; Renguet, E.; Ginion, A.; de Meester, C.; Sakamoto, K.; Balligand, J.L.; Bontemps, F.; Vanoverschelde, J.L.; et al. A-769662 potentiates the effect of other AMP-activated protein kinase activators on cardiac glucose uptake. *Am. J. Physiol. Heart Circ. Physiol.* **2014**, *306*, H1619–H1630. [CrossRef] [PubMed]
57. Green, C.J.; Pedersen, M.; Pedersen, B.K.; Scheele, C. Elevated NF-κB activation is conserved in human myocytes cultured from obese type 2 diabetic patients and attenuated by AMP-activated protein kinase. *Diabetes* **2011**, *60*, 2810–2819. [CrossRef]
58. Treebak, J.T.; Birk, J.B.; Hansen, B.F.; Olsen, G.S.; Wojtaszewski, J.F.P. A-769662 activates AMPK beta1-containing complexes but induces glucose uptake through a PI3-kinase-dependent pathway in mouse skeletal muscle. *Am. J. Physiol. Cell Physiol.* **2009**, *297*, C1041–C1052. [CrossRef]
59. Olianas, M.C.; Dedoni, S.; Onali, P. Signalling pathways mediating phosphorylation and inactivation of glycogen synthase kinase-3β by the recombinant human δ-opioid receptor stably expressed in Chinese hamster ovary cells. *Neuropharmacology* **2011**, *60*, 1326–1336. [CrossRef]
60. Human Protein Atlas. Available online: https://www.proteinatlas.org/ENSG00000017427-IGF1/cell (accessed on 6 June 2018).
61. Uhlen, M.; Oksvold, P.; Fagerberg, L.; Lundberg, E.; Jonasson, K.; Forsberg, M.; Zwahlen, M.; Kampf, C.; Wester, K.; Hober, S.; et al. Towards a knowledge-based Human Protein Atlas. *Nat. Biotechnol.* **2010**, *28*, 1248–1250. [CrossRef]
62. Huang, Y.; Smith, C.A.; Chen, G.; Sharma, B.; Miner, A.S.; Barbee, R.W.; Ratz, P.H. The AMP-Dependent Protein Kinase (AMPK) activator A-769662 causes arterial relaxation by reducing cytosolic free calcium independently of an increase in AMPK phosphorylation. *Front. Pharmacol.* **2017**, *8*, 1–14. [CrossRef] [PubMed]
63. Woods, A.; Salt, I.; Scott, J.; Hardie, D.G.; Carling, D. The alpha1 and alpha2 isoforms of the AMP-activated protein kinase have similar activities in rat liver but exhibit differences in substrate specificity in vitro. *FEBS Lett.* **1996**, *397*, 347–351. [CrossRef]
64. Durante, P.E.; Mustard, K.J.; Park, S.H.; Winder, W.W.; Hardie, D.G. Effects of endurance training on activity and expression of AMP-activated protein kinase isoforms in rat muscles. *Am. J. Physiol. Endocrinol. Metab.* **2002**, *283*, E178–E186. [CrossRef] [PubMed]

4

Implications of AMPK in the Formation of Epithelial Tight Junctions

Pascal Rowart [1], Jingshing Wu [2], Michael J. Caplan [2] and François Jouret [1,3,*]

1 Groupe Interdisciplinaire de Génoprotéomique Appliquée (GIGA), Cardiovascular Sciences, University of Liège (ULiège), Avenue de L'Hôpital 11, 4000 Liège, Belgium; pascal.rowart@uliege.be
2 Department of Cellular and Molecular Physiology, Yale School of Medicine, New Haven, CT 06520, USA; jingshing.wu@yale.edu (J.W.); michael.caplan@yale.edu (M.J.C.)
3 Division of Nephrology, Centre Hospitalier Universitaire de Liège (CHU of Liège), University of Liège (CHU ULiège), 13-B4000 Liège, Belgium
* Correspondence: francois.jouret@chuliege.be

Abstract: Tight junctions (TJ) play an essential role in the epithelial barrier. By definition, TJ are located at the demarcation between the apical and baso-lateral domains of the plasma membrane in epithelial cells. TJ fulfill two major roles: (i) TJ prevent the mixing of membrane components; and (ii) TJ regulate the selective paracellular permeability. Disruption of TJ is regarded as one of the earliest hallmarks of epithelial injury, leading to the loss of cell polarity and tissue disorganization. Many factors have been identified as modulators of TJ assembly/disassembly. More specifically, in addition to its role as an energy sensor, adenosine monophosphate-activated protein kinase (AMPK) participates in TJ regulation. AMPK is a ubiquitous serine/threonine kinase composed of a catalytic α-subunit complexed with regulatory β-and γ-subunits. AMPK activation promotes the early stages of epithelial TJ assembly. AMPK phosphorylates the adherens junction protein afadin and regulates its interaction with the TJ-associated protein zonula occludens (ZO)-1, thereby facilitating ZO-1 distribution to the plasma membrane. In the present review, we detail the signaling pathways up-and down-stream of AMPK activation at the time of Ca^{2+}-induced TJ assembly.

Keywords: AMPK; tight junctions; epithelial cells; ZO-1; par complex; MDCK; nectin-afadin; adherent junctions

---

The correct establishment and maintenance of cell-cell contacts and cell polarity in multicellular organisms are crucial for normal cell physiology and tissue homeostasis [1,2]. Epithelial cells form barriers that protect and separate multicellular organisms from the external environment. Such compartmentalization provides different internal and external environments with specialized functions [3,4]. In addition to its compartmentalization role, epithelial cell membrane integrity plays a major role in the defense against pathological organisms, as well as against disease development. Hence, the loss of cell polarity and membrane disruption are observed in cancer, acute kidney injury, apoptosis, and infection, as well as at the initial stages of some central nervous system neoplasia [5–9].

To maintain these two distinct regions and to protect the organism, epithelial cells are sealed together by a junctional complex formed by four main components located along the apico-basal axis, as follows: (1) TJ; (2) adherens junctions (AJ); (3) gap junctions; and (4) desmosomes. Disturbances in the formation and maintenance of TJ are observed in many pathological conditions, including cancer [10,11]. For this reason, a better understanding of their regulation may lead to novel targeted therapies [12]. Many factors have been identified as modulators of the assembly and disassembly of TJ. AMP-Activated protein kinase (AMPK) has emerged as one of these. Indeed, epithelial cells are involved in electrolyte and fluid transport across the apical and basal membrane, which consume a major part of an epithelial cell's internal energy currency, namely its stores of

Adenosine triphosphate (ATP) [13]. AMPK is the main energy sensor in all eukaryotic cells that regulate their levels of ATP. During cellular stresses such as hypoxia, starvation, glucose deprivation, or muscle contraction, the ratio of ADP/ATP or AMP/ATP will change. To restore the cellular energy balance, AMPK will promote catabolic pathways and inhibit anabolic pathways [14]. Interestingly enough, the depletion of ATP results in the rapid dislocation of cellular tight junctions (TJ), whereas ATP repletion induces a recovery of tight-junction integrity [15]. As an example, in case of kidney ischemia, TJ disassembly between proximal tubule cells allows the paracellular backleak of the ultrafiltrate into the interstitium, which in turn aggravates the renal hypoperfusion [16]. Zhang et al. were the first to demonstrate a role of AMPK in the regulation of epithelial tight junction assembly and disassembly. Note that the role of AMPK in TJ regulation appears to be independent of intracellular [ATP] levels. These observations opened new investigations into the mechanisms through which AMPK serves at the crossroads between the regulation of cellular energy and TJ homeostasis [17].

1. TJ Are Multiprotein Complexes

The concept of TJ (also known as *zonula occludens*) emerged in 1963 with Farquhar and Palade's experiments demonstrating, by electron microscopy, the regular occurrence in various rat and pig epithelia of a characteristic junctional complex whose components bear a relationship to each other and to the lumen of the organ [18]. Briefly, an individual TJ strand is associated laterally with another TJ strand on adjacent cells to form paired strands, where the extracellular space at this region is completely obliterated [19]. TJ are the most apical constituent of the junctional complex with AJ immediately underneath [20]. TJ can be thought of as gaskets that define and seal the most apical border of cell-cell contacts. For several years following their morphological observation, TJ were further investigated and major progress was achieved by Stevenson et al. in 1986. They explored the biochemical structure of TJ and identified ZO1 as one of their protein components [21], which has gone on to serve as a canonical marker for the assessment of TJ formation.

TJ have two main roles. First, they demarcate the apical and basolateral domains of polarized cells by acting like a fence [22]. TJ prevent membrane proteins from diffusing freely between the two membrane compartments. This function is absolutely required in order for apical and basolateral domains to maintain their distinct lipid and protein compositions [23]. Second, they create a physiological and structural paracellular barrier that regulates the selective passage and exchange of molecules [24–27]. In many situations, various materials are selectively transported across cellular sheets, and this occurs either by direct transcellular transport or by paracellular flux through TJ. The selective passage of these components through TJ is mediated by aqueous pores whose structures have yet to be fully defined [28]. In addition to these functions, TJ are connected to signaling pathways that communicate with the cell cytoplasm and subcellular components [29]. To be able to perform all of these diverse functions, TJ must possess a complicated architecture based upon a multiprotein complex that is composed of more than 40 proteins that are classified either as transmembrane proteins or as cytoplasmic proteins bound to the actin cytoskeleton [30] (Figure 1).

1.1. Transmembrane Proteins

The three main transmembrane proteins are claudins [31], occludins [32], and Junctional Adhesion Molecules (JAMs) [33]. Both claudins and occludins contain four transmembrane regions (Tetraspans) with their amino-and carboxyl-terminal ends directed to the cytoplasm [34]. There are two isoforms of occludin generated by alternative splicing [35], whereas claudins comprise more than 24 members [36]. The extended C-terminus of claudins and occludins have been shown to be essential for interactions with the soluble cytoskeletal ZO proteins. These interactions mediate the association of the ZO proteins with the plasma membrane, which is an obligate step in the formation of TJ [37]. In vivo studies revealed that claudins are more necessary for the structural integrity of TJ than occludins. In occludin knock-out (KO) mice, TJ were morphologically intact [38] in comparison to claudin-KO mice, which showed the disappearance of TJ strands [39,40]. These results indicated that claudin is required for the formation of TJ strands, and suggested that occludin is rather required for TJ stability. The last category of transmembrane proteins are JAMs. They have a single transmembrane domain and their extracellular portion is folded into two immunoglobulin-like domains. Three isoforms are currently known: JAM-A [33], JAM-B, and JAM-C [41,42]. The three JAMs are co-distributed in epithelial cells with the ZO-1 protein. Evidence of a role for JAMs in TJ formation is supported by several studies. Overexpression of JAM proteins enhances the recruitment of TJ components and leads to the increased accumulation of ZO-1 and occludin [43].

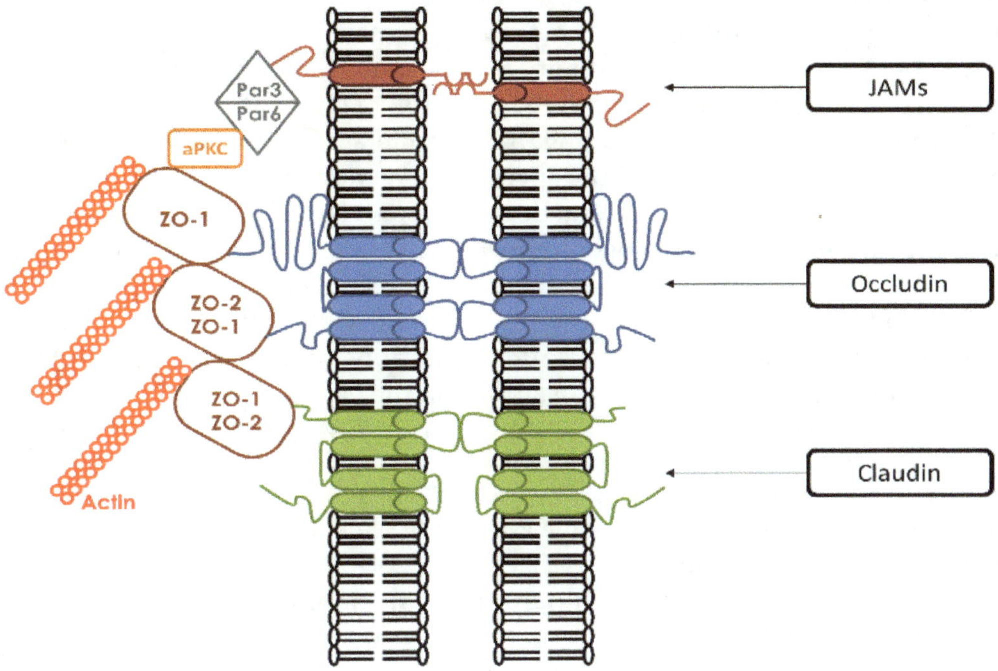

Figure 1. Molecular components of tight junctions. They are composed of three families of transmembrane proteins that include Occludins, Claudins, and Junctional Adhesion Molecule (JAMs). Every transmembrane protein is associated with cytoplasmic adaptor proteins such as the zonula occludens proteins ZO-1 and ZO-2. These interactions are mediated through their cytosolic tails. Other interactions occur between ZO proteins and additional cytoplasmic proteins (Par6, Par3, and aPKC). ZO proteins are also connected with the actin cytoskeleton.

1.2. Cytoplasmic Proteins

The "cytoplasmic plaque of TJ" serves as a link between the transmembrane TJ proteins and the actin cytoskeleton [44]. Most of the cytoplasmic proteins can attach to the TJ plaque via PDZ-domains. A PDZ domain is a common structural domain that interacts with stereotypical sequences embedded within the C-terminal regions of transmembrane proteins. PDZ domain-containing proteins can interact with other PDZ domain-containing proteins and through these multiplexed associations can anchor TJ membrane proteins to the cytoskeleton. PDZ domains are implicated in a variety of signaling mechanisms [45]. The two most important PDZ-containing proteins identified at the TJ plaque are the zonula-occludens-(ZO)1 proteins, which belong to the membrane-associated guanylate kinase family (MAGUK), and the partitioning defective proteins (Par), members of the Par3/aPKC/Par6 polarity complex.

The MAGUK family includes three structurally related proteins: ZO-1, ZO-2, and ZO-3. ZO-1 was the first to be associated with TJ [21,46]. They all share a similar structural organization, with an N-terminal region containing three PDZ domains. In vitro as well as in vivo analyses showed that the first PDZ domain (PDZ1) of the three ZO proteins has binding affinities for the C-terminal domains of claudins [37]. This PDZ-dependent interaction with ZO proteins promotes the proper targeting of claudins to the TJ. Furthermore, ZO-1/ZO-2 knock-down (KD) cells show disruptions in claudin localization associated with barrier dysfunction [47–49]. The second PDZ domain (PDZ2) is responsible for dimerization with other ZO proteins [50]. The third PDZ domain (PDZ3) is associated with the interaction with JAM-A [43]. Surprisingly, these three PDZ domains are not sufficient for the recruitment of ZO proteins to TJ [51]. In addition to these three PDZ domains, ZO proteins also have other regions that are required for their recruitment to TJ. These include SH3 and GUK domains, which can interact with afadin and occludin, respectively [52].

Given the fact that occludin-deficient cells are able to form normal TJ, with the appropriate distribution of ZO-1 [53], alternative interactions must necessarily be involved. One possibility would be the interaction of ZO proteins with α-catenin, a cytoplasmic actin-binding protein that associates with the β-catenin/E-cadherin complex at AJ [54,55]. However, α-catenin-deficient cells are able to recruit ZO-1 to the plasma membrane, indicating that this interaction is not critical. The final hypothesis focuses on the nectin/afadin complex and the specific interaction of the proline-rich regions of afadin with the SH3 domain of ZO proteins. Interaction between afadin and ZO-1 during the formation of cell-cell junctions in MDCK cells has been reported [56]. This interaction is principally observed before TJ are formed. During and after the formation of TJ, ZO proteins appear to be dissociated from afadin, and afadin becomes associated with nectin at AJ [57–60]. The association between TJ components and the AJ complex (α-catenin/β-catenin/E-cadherin and nectin/afadin) thus appears to catalyze the deposition of TJ proteins to the cell surface in the early steps of TJ formation. In summary, ZO proteins are essential for TJ formation, as well as for the linking of TJ membrane proteins to the actin cytoskeleton [51,61].

Besides the MAGUK family, members of the Par family play key roles in TJ assembly. Observations on the asymmetric divisions occurring in the *C. elegans* zygote led to the discovery of six Par proteins by Kemphues et al. in 1988, which are essential for the partitioning of early determinants and the development of embryonic polarity [62]. Only Par3 (Also known as Bazooka) and Par6 were found to be colocalized in *C. elegans* embryos. They both contain a PDZ domain and are able to bind to each other [63]. Par6 contains both N-terminal and C-terminal regions and three conserved domains for their interactions with other members of this complex. Its first domain PB1 (Phox/Bem 1) is located at the N-terminal region and is essential for the interaction of Par6 with atypical protein kinase C (aPKC). The second is Cdc42/Rac interaction binding (CRIB) and can be directly modulated by the cell division control protein 42 (Cdc42). The third one is a PDZ domain located at the C-terminal region. Accumulating evidence showed that Par3 and Par6 function together with aPKC [64]. The PB1 domain of Par6 binds the PB1 domain of aPKC to form a heterodimer. Par3 also contains N-terminal and C-terminal regions separated by three central PDZ domains. This tripartite Par3/aPKC/Par6 is known as the "Par complex" and is conserved from worms to vertebrates [65,66]. This interaction is a membrane targeting signal. In the Par complex, Par3 associates with the Par6/aPKC heterodimer by a PDZ-PDZ domain interaction at the onset of epithelial polarization [67]. Several molecules, such as nectin and JAMs, can bind the PDZ1 domain of Par3.

2. AMPK Is a Key Regulator of Energy Balance

Each accomplish energy-requiring tasks through the hydrolysis of ATP into ADP, which serves as their immediate source of energy [68]. Maintaining an adequate supply of energy is an essential requirement for survival, which means that ATP levels must be kept at a sufficient concentration. The main sensor of cellular energy status is the AMP-activated protein kinase (AMPK). When ATP levels fall, its main function is to switch off anabolic and biosynthetic pathways that consume ATP and to switch on catabolic pathways that produce ATP [69] (Figure 2). When the overall energy levels in cells decrease due to increased demands or decreased availability of substrates, AMPK gets activated through a combination of phosphorylation by upstream kinases and/or direct activation by AMP and ADP [70,71].

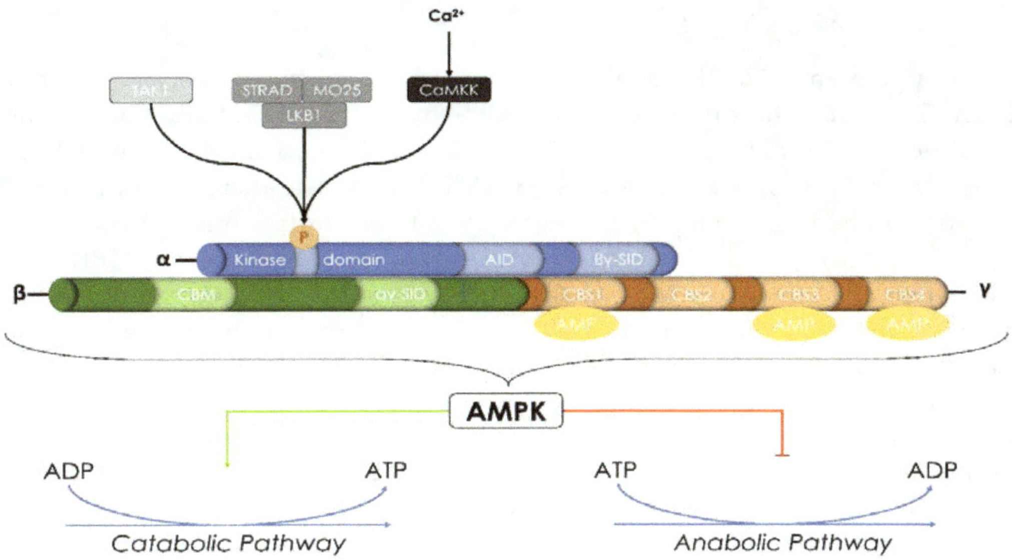

Figure 2. Structure and activation of AMP-activated protein kinase (AMPK). AMPK is a heterotrimeric α-β-γ serine/threonine kinase. It is made up of a catalytic α-subunit complexed with regulatory β-and γ-subunits. It can be activated through the phosphorylation of Thr-172 by two main upstream kinases: Ca^{2+}-Calmodulin Kinase Kinase (CaMKK) and Liver Kinase B1 (LKB1). Transforming growth factor-β-activated kinase (TAK1) was also described as a new AMPK regulatory kinase. In addition to its activation by upstream kinases, AMPK can also be allosterically activated by AMP. Once activated, AMPK responds to changes in the level of Adenosine triphosphate (ATP) by switching off either anabolic and biosynthetic pathways consuming ATP or switching on catabolic pathways that produce ATP.

2.1. AMPK: Structure and Regulation

AMPK is a heterotrimeric serine/threonine kinase. It is made up of a catalytic α-subunit complexed with regulatory β-and γ-subunits [72]. There are 12 unique heterotrimeric combinations of AMPK. In mammals, the α-subunit is encoded by two isoforms, and the β-and γ-subunits are encoded by two and three isoforms, respectively (α1, α2, β1, β2, γ1, γ2, and γ3). All these isoforms have differential tissue-specific expression and activity [73–75].

The α-subunit possesses an *N*-terminal kinase domain that mediates its catalytic activity and a *C*-terminal subunit-interacting domain that plays a role in the interaction with β-and γ-subunits (βγ-subunit interacting domain-βγ-SID) [76]. The α1-subunit is expressed in many organs (kidney, heart, brain, spleen, liver, lung, and skeletal muscle), unlike α2-subunit, which is essentially expressed primarily in skeletal muscle [77]. In addition to their different tissue/organ expression, α1- and α2-subunits are differentially expressed within the cell. Indeed, α1 is predominantly expressed in the cytosol, whereas α2 is localized to the nucleus in periods of high energy demand [78]. The β-subunit contains a central glycogen-binding domain CBM (carbohydrate-binding module) that permits the interaction of AMPK with glycogen particles and a C-terminal region essential for the assembly of the α β γ complex [79]. The γ-subunit contains four cystathionine-β-synthase (CBS) tandem sequence repeats that fold to form two "Bateman domains" and can bind AMP or ATP to regulate the AMPK activation [80] (Figure 2).

Phosphorylation of Thr-172 in the α-subunit catalytic loop is the main pathway that produces the activation of AMPK. Nevertheless, there are three major mechanisms responsible for the AMPK activation: (i) upstream kinases [71]; (ii) the increase of [AMP] and/or [ADP] [81,82]; and (iii) direct binding to the γ-subunit of AMP for the allosteric activation of AMPK [83]. Beside its activation by allosteric AMP binding and upstream kinases, AMPK has been reported to have an autophosphorylation ability at β-subunit Thr-148 [84,85].

2.2. AMPK: Upstream Kinase and Substrates

Two major upstream AMPK-regulatory kinases have been discovered that are both serine/threonine kinases. The first one is the liver-kinase-B1 (LKB1) and the second one is the Ca^{2+}/calmodulin-dependent kinase kinase (CaMKKβ). LKB1 and CaMKK can activate AMPK in response to energy stress as signaled by elevated AMP levels or to increases of intracellular [Ca^{2+}] levels in an AMP-independent manner, respectively. Other studies have also demonstrated that transforming growth factor-β-activated kinase (TAK1) may represent a third AMPK kinase.

LKB1 phosphorylates and activates AMPK in vitro following increased cellular [AMP] levels [86]. LKB1 activity requires the binding of the scaffolding-related adaptor mouse protein 25 (MO25) and the pseudokinase STe-20 Related ADaptor (STRAD) via the formation of the holoenzyme complex [86]. In cells lacking the expression of LKB1, the activation of AMPK in response to the increase of the AMP/ATP ratio is abolished, suggesting that LKB1 is required for the AMPK phosphorylation when [AMP] increases in the cell [86]. Other studies demonstrated that, in certain circumstances, AMPK can be activated, even in the absence of LKB1 [87,88]. Hence, CaMKK emerged as another main AMPK kinase [89]. In contrast to LKB1, the AMPK phosphorylation by CaMKK does not require a disturbance of the ATP/AMP ratio, but rather an increase in intracellular Ca^{2+} [90]. The addition of the Ca^{2+} ionophore A23187 activates AMPK-via the phosphorylation of Thr-172-approximately 10-fold more in cells expressing a kinase-inactive mutant of LKB1 compared to wild-type cells. Conversely, the AMPK activation by Ca^{2+} ionophore A23187 was abolished by the CaMKK inhibitor STO-609 [91]. Other studies confirmed this result and observed that the overexpression of CaMKK increases AMPK activity, whereas the inhibition of CaMKK reduces AMPK activity [92]. These results suggest a physiological role of LKB1 and CaMKK as AMPK regulatory kinases in mammalian cells [92].

AMPK is a modulator of several pathways [93–95]. For example, AMPK negatively regulates two enzymes involved in lipid synthesis: HMGCR (3-hydroxy-3-methylglutaryl coenzyme A reductase) [96] and ACC (acetyl-CoA carboxylase) [97]. AMPK also exerts a potent effect on glucose metabolism. Glucose uptake is facilitated through the translocation of glucose transporter 4 (GLUT4) to the cell membrane and also through the regulation of *GLUT4* gene expression in response to AMPK activation [98]. AMPK also regulates glycogen metabolism. AMPK activation phosphorylates and decreases the activation of glycogen synthase (GS), thus reducing glycogen synthesis [99]. Thus, AMPK activation reduces the production of stored forms of metabolic energy and diminishes the activity of energy utilizing pathways, while it increases the capacity of cells to import energetic precursors and to produce ATP. Kishton et al. showed that AMPK actively restrained aerobic glycolysis in cells through the inhibition of mTORC1, while promoting oxidative metabolism and mitochondrial Complex I activity producing ATP [100]. The inhibition of AMPK-related kinase 5 (ARK5), an upstream regulator of AMPK, leads to a collapse of cellular ATP levels. Proteomics highlighted the down-regulation of multiple subunits of complexes I, III, and IV of the mitochondrial respiratory chain following the depletion of ARK5 [101]. These studies suggest a role of AMPK in the production of ATP by the mitochondrial respiratory chain. AMPK facilitates the assembly of TJ.

3. AMPK and ZO-1

Zhang et al. [17] demonstrated in 2006 that AMPK could regulate the assembly of epithelial TJ in the MDCK cell line. The authors used a Ca^{2+} switch-based model described in 1978 [102] to decipher the role of AMPK in TJ assembly. Cell-cell adhesion, as well as TJ integrity in polarized epithelial cells, is rapidly lost when the Ca^{2+} is removed from the extracellular medium. On the other hand, the re-addition of Ca^{2+} into the culture medium induces the rapid assembly of cell-cell contacts and subsequent TJ formation. Depletion of Ca^{2+} from the medium causes ZO-1 to translocate from the cell periphery to the cytoplasm. Upon the re-addition of Ca^{2+}, ZO-1 moves back to the TJ. With this model, the authors showed that AMPK is phosphorylated during the Ca^{2+}-induced TJ assembly, while the total amount of AMPK remains unchanged. They also examined the AMPK activity by measuring the phosphorylated form of ACC, one of the principle AMPK substrates. They found an eight-fold increase in pACC following a Ca^{2+}-switch. It is important to note that such AMPK phosphorylation and activation was not attributable to changes in cellular [ATP] levels during a Ca^{2+}-switch. To evaluate the potential effect of AMPK in TJ formation, the authors monitored the time course of ZO-1 relocation to cell-cell junctions with or without the chemical AMPK activator AICAR (which acts an AMP mimic) added at the time of the Ca^{2+} switch. The amount of ZO-1 relocated to TJ in the presence of AICAR during the Ca^{2+} switch was higher in comparison to a classic Ca^{2+} switch. Furthermore, the addition of AICAR in Ca^{2+}-depleted medium was sufficient to activate AMPK and to accelerate ZO-1 relocation to TJ [17]. They also measured the paracellular flux of 70-kDa dextran in MDCK monolayers during the Ca^{2+}-switch in the presence or absence of AICAR. The presence of AICAR led to a slight but statistically significant decrease in the dextran flux rate across the monolayers. Similarly to Zhang et al., Peng et al. also measured the paracellular flux by measuring inulin in intestinal epithelial cells with or without AICAR. Incubation with AICAR led to a significant decrease in the flux rate across the cell monolayers. These effects were also abolished by the AMPK inhibitor, Compound C. These studies indicated that the backleak effect is decreased by the activation of AMPK, independently of [ADP/ATP] changes. In MDCK cells expressing dominant negative AMPK, the early initiation of TJ assembly was compromised. Still, normal-appearing TJ could eventually form in AMPK-deficient cells over time, suggesting that AMPK activation supports the initiation of TJ formation, but other factors, including extracellular Ca^{2+}, are required for the long-term stabilization of TJ. Shortly after the publication of the Zhang et al. study, Zheng et al. confirmed these findings and in addition showed that the activation of AMPK in response to the initiation of junction formation requires LKB1. They also generated MDCK cell lines expressing a kinase-dead mutant form of AMPKα1 and monitored the effect of Ca^{2+}-switch on TER, a measurement for the paracellular barrier function and integrity of TJ [103]. Expression of kinase-dead AMPK significantly decreased the peak level of TER, meaning that the formation of functional TJ is suppressed in the absence of AMPK [104].

4. AMPK and Afadin-Nectin System

In the above-detailed model, the nectin-afadin system is required for the deposition of junction components induced by AMPK activation. The involvement of the nectin-afadin complex in cell adhesion has been described in AJ formation [105]. Still, afadin KD cells are not able to induce the relocalization of ZO-1 and occludin at TJ sites following the addition of AICAR in low Ca^{2+} medium. AMPK therefore appears to be connected to afadin in the Ca^{2+}-independent TJ formation. Furthermore, immunoprecipitation between afadin and AMPK revealed that afadin is a direct substrate of AMPK. The addition of Compound-C inhibited the phosphorylation of afadin, whereas the afadin signal was increased without the AMPK inhibitor [106]. Since afadin directly binds to ZO-1 [56], the authors investigated whether AMPK activation increases the interaction between afadin and ZO-1, thereby facilitating the assembly of TJ. Immunoprecipitation, with or without AICAR during a Ca^{2+} switch, revealed an enhanced interaction between these two proteins after Ca^{2+}-switch and even more in the case of AICAR exposure. These results suggested that AMPK activation might facilitate TJ assembly by phosphorylating afadin and inducing its association with ZO-1.

5. AMPK Effectors

Recent research investigated the potential of AMPK effectors to preserve the epithelial architecture. The first study focused on the multimodular polarity scaffold protein GIV (G-alpha interacting vesicle associated protein) [107]. This protein has been demonstrated to regulate cell polarity and morphogenesis [108], as well as cell-cell junction formation through its ability to bind Par3 [109] and the Cadherin-catenin complex [110]. A role for AMPK-mediated phosphorylation of GIV at serine 245 when [ATP] levels decreased was suggested. The phosphorylation of GIV at ps 245 triggered its localization to TJ by increasing its ability to bind TJ-associated microtubules and AJ-localized protein complexes. The addition of Compound-C inhibited AMPK-mediated phosphorylation of GIV and induced the destabilization of TJ and the reduction of TER. On the other hand, metformin (an AMPK activator) and AICAR triggered GIV phosphorylation and stabilized TJ, with subsequent enhanced TER [107].

In 2005, butyrate emerged as a new candidate to promote enhanced intestinal barrier function as reflected by increases in TER in vitro [111]. A few years later, Peng and co-workers explored whether the effect of butyrate on the intestinal epithelial barrier is related to AMPK. Using a model of Ca^{2+} switch, with or without the addition of butyrate, they demonstrated that the amount of pAMPK, as well as pACC, increased after the treatment with butyrate in a time-dependent manner. The addition of AICAR in the culture medium increased TER, whereas Compound-C abolished this effect. Butyrate also promoted a faster relocalization of ZO-1 and occludin at the cell periphery and tightened the intestinal barrier [112]. These results further support a role for AMPK activation in TJ formation. Additional studies using sodium butyrate further support the AMPK activation by CaMKK due to the increasing of the intracellular concentration of Ca^{2+}. Furthermore, the AMPK activation also increases the phosphorylation of PKCβ, a key player in TJ regulation [65,113]. This study underscores the putative interplay between the AMPK and PKC family in the formation of TJ [114].

Another model involving porcine intestinal epithelial cells investigated the effect of L-glutamine (Gln) in the preservation of TJ. Gln was described as a critically important nutrient for the maintenance of intestinal mucosal barrier integrity in humans and animals [115]. Indeed, the depletion of Gln results in a decreased abundance of TJ-associated proteins and increased intestinal paracellular permeability, whereas the addition of Gln resulted in increased TER, enhanced TJ-protein abundance, and the localization of TJ proteins to the plasma membrane [116]. TJ proteins, such as claudin-1, claudin-4, and ZO-1, are localized more abundantly at TJ sites in the presence of Gln. In addition, the abundance of pAMPK was further enhanced by the addition of Gln. The beneficial effect of Gln, as well as the phosphorylation of AMPK, was abrogated in a low Ca^{2+} medium and with the use of STO-609 (CaMKK inhibitor). Gln increased the intracellular [ATP] levels, but these were not affected by STO-609, meaning that AMPK acts as a TJ regulator via the CaMKK pathway in a model of Ca^{2+}-induced TJ formation [117]. Our recent work further supports a similar role for CaMKK in the activation of AMPK during a Ca^{2+} switch. This study found that the pharmacological inhibition of CaMKK or the direct inhibition of AMPK by Compound-C hampered AMPK phosphorylation and ZO-1 relocation to the TJ during a Ca^{2+} switch in MDCK cells, whereas the inactivation of LKB1 by shRNA did not significantly influence these processes [118].

Park et al. focused their research on the beneficial effect of Theaflavins (TFs), a polyphenol pigment in black tea, known to have anti-hyperglycemic, antioxidant, and anti-inflammatory effects. They previously found that TFs induced AMPK activation. By measuring the fluorescein transport across epithelial cells, they observed a decrease in its transport by pre-treatment with TFs, thereby revealing a reduction of paracellular permeability. Moreover, the TJ-related proteins claudin-1, occludin, and ZO-1 were significantly increased at TJ. Furthermore, Compound-C restored the fluorescein transport and inhibited the action of TFs. The authors highlighted the AMPK-mediated expression of claudin-1, occludin, and ZO-1 at TJ in intestinal cells by TFs [119]. Another natural agent, i.e., Forskolin, has been demonstrated to have beneficial effects on the AMPK-mediated TJ formation. This compound increased the phosphorylation and activation of AMPK with a comparable effect to the addition of 2 Deoxy-D-Glucose (2-DG) in a placenta epithelial cell culture. Forskolin treatment markedly enhanced the assembly of TJ strands, with higher ZO-1 relocation at TJ, while only weak ZO-1 staining was observed in control cells. The authors also used dominant negative AMPK transfected cells or Compound-C to inhibit AMPK activation and examined whether or not Forskolin-induced TJ assembly is mediated by AMPK activation. In both cases, the Forskolin effect was abrogated, resulting in lower ZO-1 relocation at TJ [120]. Thus, the activation of AMPK by Forskolin may enhance TJ formation.

AMPKα-null *Drosophila* die before reaching adulthood, while the transgenic expression of wild-type AMPK in AMPKα-null mutants allowed them to successfully develop into adults. A detailed examination of the embryonic epithelial structure of AMPKα-null mutants revealed a major disorganization of apico-basal polarity. They also assessed whether AMPK is necessary for cell polarity in mammalian cells. 2-DG treatment of unpolarized epithelial cells, such as the LS174T line, induced major changes in cell shape with the formation of a polarized actin cytoskeleton and a brush-border-like structure. Interestingly, this actin polarization was suppressed by the AMPK-specific inhibitor Compound C [121]. Furthermore, the authors have shown that AMPKα mutation in *Drosophila* embryos leads to the abnormal distribution of epithelial polarity markers. The consequent loss of polarity along with over-proliferative aberration could promote cancers [122]. In addition, AMPK has been shown to suppress tumorigenesis and the Warburg effect [123]. Therefore, one may speculate that AMPK-mediated TJ strengthening may help inhibit adenocarcinoma and tumorigenesis. Additional studies involving AMPK KO mice showed higher intestinal permeability when compared with WT mice, as indicated by decreased TER and increased paracellular FITC-dextran permeability, indicating a leaking gut. To investigate the integrity of TJ, ZO-1 immunofluorescence staining was analyzed and ZO-1 labeled at the tip of villi was impaired in AMPK KO mice [124].

Another role of AMPK in TJ formation and maintenance could be hypothetically linked to dietary methionine restriction (MR). MR has been found to modify the protein composition of TJ complexes in epithelial cells [125]. In addition, the stimulation of *S*-adenosyl-l-methionine, a key intermediate of methionine metabolism, led to the consumption of both Met and ATP [126] and AMPK activation [127]. These two studies may suggest a new role of AMPK in TJ maintenance in the case of MR. Along with other nutritional regimens modulating AMPK, Zinc has the potential to function as a TJ modifier and selective enhancer of epithelial barrier function [128] by regulating claudin-3 and occludin [129]. Since the rapid activation of AMPK was observed after exposure of neurons to Zinc, one may speculate an interplay between zinc-induced TJ formation and the AMPK pathway. Interestingly, Zinc-induced AMPK activation was mediated by LKB1 in the absence of changes in intracellular AMP levels or CaMKKβ activation [130].

6. AMPK and Co-Culture Models

The involvement of AMPK in TJ formation has also been demonstrated in several models of direct cell-cell co-culture. Tang et al. investigated the role of lymphocytes in the modulation of the epithelial barrier since lymphocytes are recruited by epithelial cells during infection. To mimic an infection state, they used a direct co-culture of MDCK cells with lymphocytes and a Ca^{2+} switch model to measure the TJ formation. The time course of ZO-1 relocation after Ca^{2+} switch was accelerated in the co-culture compared to MDCK alone. Furthermore, lymphocytes drastically increased AMPK phosphorylation in comparison to MDCK alone after a Ca^{2+}-switch. To link the increased AMPK phosphorylation and TJ formation, the authors used Compound-C and MDCK expressing an shRNA directed against AMPKα1. In both cases, the beneficial effect of lymphocytes was abolished and a slower TJ assembly was observed, thereby confirming the requirement for AMPK in the TJ formation [131]. Similar experiments were performed in a co-culture model of mesenchymal stromal cells (MSC) and MDCK cells. Bone marrow-derived MSC can modulate epithelial TJ at the time of their Ca^{2+}-induced assembly. The relocation of ZO-1 to MDCK cell-cell contacts was indeed significantly accelerated in the presence of MSC compared to a MDCK cell culture alone. Furthermore, AMPK activation and activity were also enhanced in the co-culture model. The addition of Compound-C or STO-609 abolished this AMPK activation and ZO-1 relocation. On the other hand, the co-culture of MSC with MDCK expressing an shRNA directed against LKB1 did not suppress the AMPK activity and ZO-1 relocation. This work further supports a role for CaMKK in the activation of AMPK and ZO-1 protein relocation at TJ during a Ca^{2+}-switch, independently of LKB1 activity [118].

Patkee et al. worked on metformin and its role in airway epithelial TJ in a model of co-culture with *P. aeruginosa* (a respiratory pathogen) to mimic an infection and TJ disruption with higher glucose permeability. The addition of *P. aeruginosa* into the culture of airway epithelial cells produced a significant decrease in TER. Metformin treatment attenuated the fall in TER produced by *P. aeruginosa*. AICAR pre-treatment also attenuated the *P. aeuringosa*-induced reduction of TER. On the other hand, this increasing TER was prevented by pre-treatment with the AMPK inhibitor Compound-C. To explain this increased TER phenomenon, the authors investigated the effect of *P. aeruginosa* and metformin on the abundance of TJ proteins. They found a decline in claudin-1 and occludin abundance in the co-culture with *P. aeruginosa*. The addition of metformin enhanced the expression of these two TJ proteins. These data indicate a potential AMPK-dependence that may be responsible for metformin's ability to increase the airway epithelial barrier function [132].

7. Conclusions

TJ are key constituents of polarized epithelial cells. It is well established that the presence of TJ is indispensable for tissue compartmentalization and cellular homeostasis. Their disruption represents one of the earliest markers of epithelial injury and diseases. Accumulating evidence demonstrates that AMPK is a key factor in the formation of TJ via several signaling pathways (Figure 3). Further investigations concerning the impact of AMPK on epithelial maintenance in baseline conditions and in diseased conditions may lead to innovative therapies.

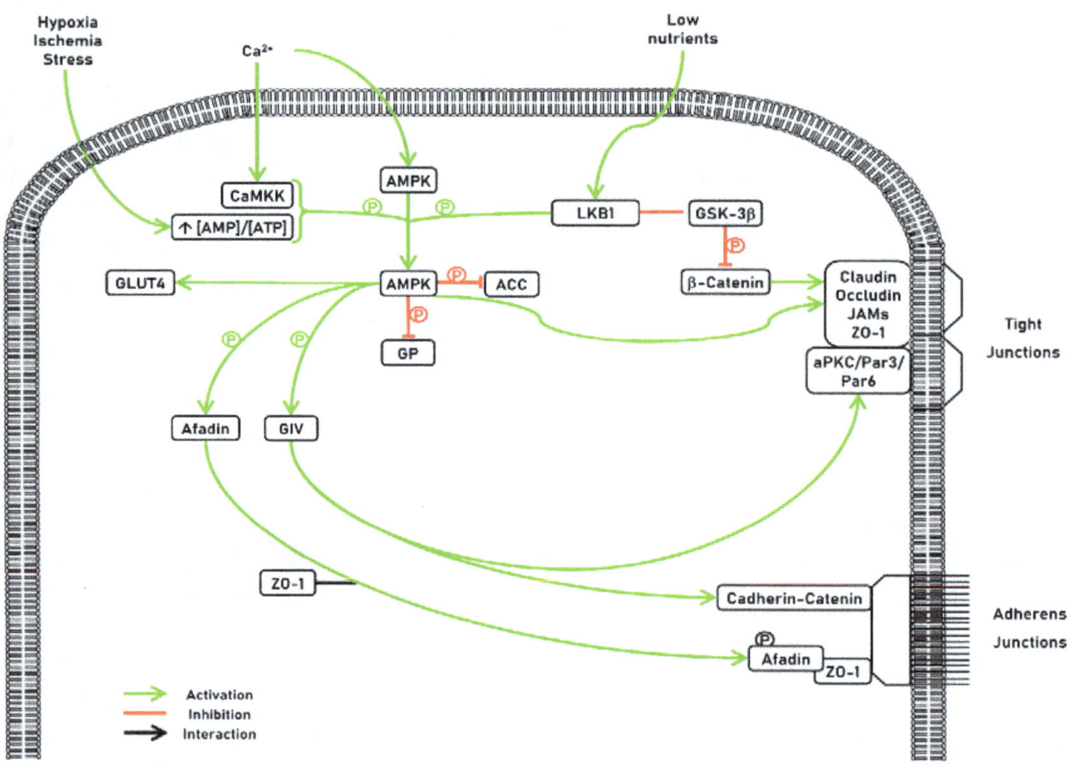

Figure 3. Representative schema of AMPK activators and substrates.

AMPK is involved in tight junctions (TJ) formation. AMPK can be activated by two mains upstream kinases: Ca^{2+}-Calmodulin Kinase Kinase (CaMKK) and Liver Kinase B1 (LKB1). Once activated, AMPK can have several effects. First, AMPK can modulate lipid metabolism by targeting the fatty acid synthesis pathway by the phosphorylation and inhibition of Acetyl CoA Carboxylase (ACC). Second, AMPK also exerts a potent effect on glucose metabolism trough the translocation of glucose transporter 4 (GLUT4) to the cell membrane. Third, AMPK also regulates glycogen metabolism. AMPK activation phosphorylates and decreases the activation of glycogen synthase (GS), thus reducing glycogen synthesis. AMPK is involved in TJ formation. Activated AMPK phosphorylates afadin and induces its association with ZO-1. AMPK also phosphorylates G-alpha interacting vesicle associated protein (GIV), which regulates cell polarity and morphogenesis, as well as cell-cell junction formation through its ability to bind Par3 and the Cadherin-catenin complex.

Author Contributions: Conceptualization, P.R., F.J. and M.J.C.; Writing-Original Draft Preparation, P.R.; Writing-Review & Editing, P.R., J.W., M.J.C. and F.J.; Supervision, F.J. and M.J.C.

Acknowledgments: F.J. is a Fellow of the Fonds National de la Recherche Scientifique (FNRS) and received support from the University of Liège (Fonds Spéciaux à la Recherche, Fonds Léon Fredericq) and the FNRS (Research Credits 2013 and 2016).

References

1. Drubin, D.G.; Nelson, W.J. Origins of Cell Polarity. *Cell* **1996**, *84*, 335–344. [CrossRef]
2. Yeaman, C.; Grindstaff, K.K.; Nelson, W.J. New Perspectives on Mechanisms Involved in Generating Epithelial Cell Polarity. *Physiol. Rev.* **1999**, *79*, 73–98. [CrossRef] [PubMed]
3. Royer, C.; Lu, X. Epithelial cell polarity: A major gatekeeper against cancer? *Cell Death Differ.* **2011**, *18*, 1470–1477. [CrossRef] [PubMed]
4. Marchiando, A.M.; Graham, W.V.; Turner, J.R. Epithelial Barriers in Homeostasis and Disease. *Annu. Rev. Pathol. Mech. Dis.* **2010**, *5*, 119–144. [CrossRef] [PubMed]
5. Wang, Y.; Lu, X. *Cell Polarity: A Key Defence Mechanism Against Infection and Cancer Cell Invasion?* Springer: Cham, Switzerland, 2015.
6. Frisch, S.M.; Francis, H. Disruption of Epithelial Cell-Matrix Interactions Induces Apoptosis. *J. Cell Biol.* **1994**, *124*, 619–626. [CrossRef] [PubMed]
7. Klezovitch, O.; Fernandez, T.E.; Tapscott, S.J.; Vasioukhin, V. Loss of cell polarity causes severe brain dysplasia in Lgl1 knockout mice. *Genes Dev.* **2004**, *18*, 559–571. [CrossRef] [PubMed]
8. Basile, D.P.; Anderson, M.D.; Sutton, T.A. Pathophysiology of acute kidney injury. *Compr. Physiol.* **2012**, *2*, 1303–1353. [CrossRef] [PubMed]
9. Epstein, F.H.; Fish, E.M.; Molitoris, B.A. Alterations in Epithelial Polarity and the Pathogenesis of Disease States. *N. Engl. J. Med.* **1994**, *330*, 1580–1588. [CrossRef] [PubMed]
10. Fakhoury, M.; Negrulj, R.; Mooranian, A.; Al-Salami, H. Inflammatory bowel disease: Clinical aspects and treatments. *J. Inflamm. Res.* **2014**, *7*, 113–120. [CrossRef] [PubMed]
11. Wang, X.; Tully, O.; Ngo, B.; Zitin, M.; Mullin, J.M. Epithelial tight junctional changes in colorectal cancer tissues. *Sci. World J.* **2011**, *11*, 826–841. [CrossRef] [PubMed]
12. Förster, C. Tight junctions and the modulation of barrier function in disease. *Histochem. Cell Biol.* **2008**, *130*, 55–70. [CrossRef] [PubMed]
13. Caplan, M.J.; Seo-Mayer, P.; Zhang, L. Epithelial junctions and polarity: Complexes and kinases. *Curr. Opin. Nephrol. Hypertens.* **2008**, *17*, 506–512. [CrossRef] [PubMed]
14. Li, J.; Zhong, L.; Wang, F.; Zhu, H. Dissecting the role of AMP-activated protein kinase in human diseases. *Acta Pharm. Sin. B* **2017**, *7*, 249–259. [CrossRef] [PubMed]
15. Canfield, P.E.; Geerdes, A.M.; Molitoris, B.A. Effect of reversible ATP depletion on tight-junction integrity in LLC-PK1 cells. *Am. J. Physiol.* **1991**, *261*, F1038–F1045. [CrossRef] [PubMed]
16. Kwon, O.; Nelson, W.J.; Sibley, R.; Huie, P.; Scandling, J.D.; Dafoe, D.; Alfrey, E.; Myers, B.D. Backleak, Tight Junctions, and Cell–Cell Adhesion in Postischemic Injury to the Renal Allograft. *J. Clin. Investig.* **1998**, *101*, 2054–2064. [CrossRef] [PubMed]
17. Zhang, L.; Li, J.; Young, L.H.; Caplan, M.J. AMP-activated protein kinase regulates the assembly of epithelial tight junctions. *Proc. Natl. Acad. Sci. USA* **2006**, *103*, 17272–17277. [CrossRef] [PubMed]
18. Farquhar, M.G.; Palade, G.E. Junctional complexes in various epithelia. *J. Cell Biol.* **1963**, *17*, 375–412. [CrossRef] [PubMed]
19. Cereijido, M.; Valdés, J.; Shoshani, L.; Contreras, R.G. Role of tight junctions in establishing and maintaining cell polarity. *Annu. Rev. Physiol.* **1998**, *60*, 161–177. [CrossRef] [PubMed]
20. Campbell, H.K.; Maiers, J.L.; DeMali, K.A. Interplay between tight junctions & adherens junctions. *Exp. Cell Res.* **2017**, *358*, 39–44. [CrossRef] [PubMed]

21. Stevenson, B.R.; Siliciano, J.D.; Mooseker, M.S.; Goodenough, D.A. Identification of ZO-I: A High Molecular Weight Polypeptide Associated with the Tight Junction (Zonula Occludens) in a Variety of Epithelia. *J. Cell Biol.* **1986**, *103*, 755–766. [CrossRef] [PubMed]
22. Mandel, L.J.; Bacallao, R.; Zampighi, G. Uncoupling of the molecular "fence" and paracellular "gate" functions in epithelial tight junctions. *Nature* **1993**, *361*, 552–555. [CrossRef] [PubMed]
23. Dragsten, P.R.; Blumenthal, R.; Handler, J.S. Membrane asymmetry in epithelia: Is the tight junction a barrier to diffusion in the plasma membrane? *Nature* **1981**, *294*, 718–722. [CrossRef] [PubMed]
24. Van Meer, G.; Simons, K.; Simons, K. The function of tight junctions in maintaining differences in lipid composition between the apical and the basolateral cell surface domains of MDCK cells. *EMBO J.* **1986**, *5*, 1455–1464. [PubMed]
25. Van Itallie, C.M. The Molecular Physiology of Tight Junction Pores. *Physiology* **2004**. [CrossRef] [PubMed]
26. Anderson, J.M.; Van Itallie, C.M. Physiology and function of the tight junction. *Cold Spring Harb. Perspect. Biol.* **2009**, *1*, a002584. [CrossRef] [PubMed]
27. Fanning, S.; Mitic, L.L.; Anderson, J.M. Transmembrane proteins in the tight junction barrier. *J. Am. Soc. Nephrol.* **1999**, *10*, 1337–1345. [PubMed]
28. Tsukita, S.; Furuse, M.; Itoh, M. Multifunctional strands in tight junctions. *Nat. Rev. Mol. Cell Biol.* **2001**, *2*, 285–293. [CrossRef] [PubMed]
29. Matter, K.; Balda, M.S. Signalling to and from tight junctions. *Nat. Rev. Mol. Cell Biol.* **2003**, *4*, 225–236. [CrossRef] [PubMed]
30. Gonzalezmariscal, L. Tight junction proteins. *Prog. Biophys. Mol. Biol.* **2003**, *81*, 1–44. [CrossRef]
31. Furuse, M.; Hirase, T.; Itoh, M.; Nagafuchi, A.; Yonemura, S.; Tsukita, S.; Tsukita, S. Occludin: A novel integral membrane protein localizing at tight junctions. *J. Cell Biol.* **1993**, *123*, 1777–1788. [CrossRef] [PubMed]
32. Furuse, M.; Fujita, K.; Hiiragi, T.; Fujimoto, K.; Tsukita, S. Claudin-1 and -2: Novel integral membrane proteins localizing at tight junctions with no sequence similarity to occludin. *J. Cell Biol.* **1998**, *141*, 1539–1550. [CrossRef] [PubMed]
33. Martìn-Padura, I.; Lostaglio, S.; Schneemann, M.; Williams, L.; Romano, M.; Fruscella, P.; Panzeri, C.; Stoppacciaro, A.; Ruco, L.; Villa, A.; et al. Junctional adhesion molecule, a novel member of the immunoglobulin superfamily that distributes at intercellular junctions and modulates monocyte transmigration. *J. Cell Biol.* **1998**, *142*, 117–127. [CrossRef] [PubMed]
34. Tsukita, S.; Furuse, M. Occludin and claudins in tight-junction strands: Leading or supporting players? *Trends Cell Biol.* **1999**, *9*, 268–273. [CrossRef]
35. Muresan, Z.; Paul, D.L.; Goodenough, D.A. Occludin 1B, a variant of the tight junction protein occludin. *Mol. Biol. Cell* **2000**, *11*, 627–634. [CrossRef] [PubMed]
36. Morita, K.; Furuse, M.; Fujimoto, K.; Tsukita, S. Claudin multigene family encoding four-transmembrane domain protein components of tight junction strands. *Proc. Natl. Acad. Sci. USA* **1999**, *96*, 511–516. [CrossRef] [PubMed]
37. Itoh, M.; Furuse, M.; Morita, K.; Kubota, K.; Saitou, M.; Tsukita, S. Direct binding of three tight junction-associated MAGUKs, ZO-1, ZO-2, and ZO-3, with the COOH termini of claudins. *J. Cell Biol.* **1999**, *147*, 1351–1363. [CrossRef] [PubMed]
38. Saitou, M.; Furuse, M.; Sasaki, H.; Schulzke, J.D.; Fromm, M.; Takano, H.; Noda, T.; Tsukita, S. Complex phenotype of mice lacking occludin, a component of tight junction strands. *Mol. Biol. Cell* **2000**, *11*, 4131–4142. [CrossRef] [PubMed]
39. Gow, A.; Southwood, C.M.; Li, J.S.; Pariali, M.; Riordan, G.P.; Brodie, S.E.; Danias, J.; Bronstein, J.M.; Kachar, B.; Lazzarini, R.A. CNS myelin and sertoli cell tight junction strands are absent in Osp/claudin-11 null mice. *Cell* **1999**, *99*, 649–659. [CrossRef]
40. Gow, A.; Davies, C.; Southwood, C.M.; Frolenkov, G.; Chrustowski, M.; Ng, L.; Yamauchi, D.; Marcus, D.C.; Kachar, B. Deafness in Claudin 11-null mice reveals the critical contribution of basal cell tight junctions to stria vascularis function. *J. Neurosci.* **2004**, *24*, 7051–7062. [CrossRef] [PubMed]
41. Aurrand-Lions, M.; Johnson-Leger, C.; Wong, C.; Du Pasquier, L.; Imhof, B.A. Heterogeneity of endothelial junctions is reflected by differential expression and specific subcellular localization of the three JAM family members. *Blood* **2001**, *98*, 3699–3707. [CrossRef] [PubMed]
42. Aurrand-Lions, M.A.; Duncan, L.; Du Pasquier, L.; Imhof, B.A. Cloning of JAM-2 and JAM-3: An emerging junctional adhesion molecular family? *Curr. Top. Microbiol. Immunol.* **2000**, *251*, 91–98. [PubMed]

43. Bazzoni, G.; Martinez-Estrada, O.M.; Orsenigo, F.; Cordenonsi, M.; Citi, S.; Dejana, E. Interaction of junctional adhesion molecule with the tight junction components ZO-1, cingulin, and occludin. *J. Biol. Chem.* **2000**, *275*, 20520–20526. [CrossRef] [PubMed]
44. Schneeberger, E.E.; Lynch, R.D. The tight junction: A multifunctional complex. *Am. J. Physiol. Cell Physiol.* **2004**, *286*, C1213–C1228. [CrossRef] [PubMed]
45. Guillemot, L.; Paschoud, S.; Pulimeno, P.; Foglia, A.; Citi, S. The cytoplasmic plaque of tight junctions: A scaffolding and signalling center. *Biochim. Biophys. Acta Biomembr.* **2008**, *1778*, 601–613. [CrossRef] [PubMed]
46. Furuse, M. Molecular basis of the core structure of tight junctions. *Cold Spring Harb. Perspect. Biol.* **2010**, *2*, a002907. [CrossRef] [PubMed]
47. Fanning, A.S.; Van Itallie, C.M.; Anderson, J.M. Zonula occludens-1 and -2 regulate apical cell structure and the zonula adherens cytoskeleton in polarized epithelia. *Mol. Biol. Cell* **2012**, *23*, 577–590. [CrossRef] [PubMed]
48. Ikenouchi, J.; Umeda, K.; Tsukita, S.; Furuse, M.; Tsukita, S. Requirement of ZO-1 for the formation of belt-like adherens junctions during epithelial cell polarization. *J. Cell Biol.* **2007**, *176*, 779–786. [CrossRef] [PubMed]
49. Van Itallie, C.M.; Anderson, J.M. Claudin interactions in and out of the tight junction. *Tissue Barriers* **2013**, *1*, e25247. [CrossRef] [PubMed]
50. Giepmans, B.N.; Moolenaar, W.H. The gap junction protein connexin43 interacts with the second PDZ domain of the zona occludens-1 protein. *Curr. Biol.* **1998**, *8*, 931–934. [CrossRef]
51. Umeda, K.; Ikenouchi, J.; Katahira-Tayama, S.; Furuse, K.; Sasaki, H.; Nakayama, M.; Matsui, T.; Tsukita, S.; Furuse, M.; Tsukita, S. ZO-1 and ZO-2 independently determine where claudins are polymerized in tight-junction strand formation. *Cell* **2006**, *126*, 741–754. [CrossRef] [PubMed]
52. Schmidt, A.; Utepbergenov, D.I.; Mueller, S.L.; Beyermann, M.; Schneider-Mergener, J.; Krause, G.; Blasig, I.E. Occludin binds to the SH3-hinge-GuK unit of zonula occludens protein 1: Potential mechanism of tight junction regulation. *Cell. Mol. Life Sci.* **2004**, *61*, 1354–1365. [CrossRef] [PubMed]
53. Saitou, M.; Fujimoto, K.; Doi, Y.; Itoh, M.; Fujimoto, T.; Furuse, M.; Takano, H.; Noda, T.; Tsukita, S. Occludin-deficient embryonic stem cells can differentiate into polarized epithelial cells bearing tight junctions. *J. Cell Biol.* **1998**, *141*, 397–408. [CrossRef] [PubMed]
54. Itoh, M.; Nagafuchi, A.; Moroi, S.; Tsukita, S. Involvement of ZO-1 in cadherin-based cell adhesion through its direct binding to alpha catenin and actin filaments. *J. Cell Biol.* **1997**, *138*, 181–192. [CrossRef] [PubMed]
55. Rajasekaran, A.K.; Hojo, M.; Huima, T.; Rodriguez-Boulan, E. Catenins and zonula occludens-1 form a complex during early stages in the assembly of tight junctions. *J. Cell Biol.* **1996**, *132*, 451–463. [CrossRef] [PubMed]
56. Yamamoto, T.; Harada, N.; Kano, K.; Taya, S.; Canaani, E.; Matsuura, Y.; Mizoguchi, A.; Ide, C.; Kaibuchi, K. The Ras target AF-6 interacts with ZO-1 and serves as a peripheral component of tight junctions in epithelial cells. *J. Cell Biol.* **1997**, *139*, 785–795. [CrossRef] [PubMed]
57. Fukuhara, A.; Irie, K.; Nakanishi, H.; Takekuni, K.; Kawakatsu, T.; Ikeda, W.; Yamada, A.; Katata, T.; Honda, T.; Sato, T.; et al. Involvement of nectin in the localization of junctional adhesion molecule at tight junctions. *Oncogene* **2002**, *21*, 7642–7655. [CrossRef] [PubMed]
58. Yokoyama, S.; Tachibana, K.; Nakanishi, H.; Yamamoto, Y.; Irie, K.; Mandai, K.; Nagafuchi, A.; Monden, M.; Takai, Y. alpha-catenin-independent recruitment of ZO-1 to nectin-based cell-cell adhesion sites through afadin. *Mol. Biol. Cell* **2001**, *12*, 1595–1609. [CrossRef] [PubMed]
59. Yamada, A.; Fujita, N.; Sato, T.; Okamoto, R.; Ooshio, T.; Hirota, T.; Morimoto, K.; Irie, K.; Takai, Y. Requirement of nectin, but not cadherin, for formation of claudin-based tight junctions in annexin II-knockdown MDCK cells. *Oncogene* **2006**, *25*, 5085–5102. [CrossRef] [PubMed]
60. Ooshio, T.; Kobayashi, R.; Ikeda, W.; Miyata, M.; Fukumoto, Y.; Matsuzawa, N.; Ogita, H.; Takai, Y. Involvement of the interaction of afadin with ZO-1 in the formation of tight junctions in Madin-Darby canine kidney cells. *J. Biol. Chem.* **2010**, *285*, 5003–5012. [CrossRef] [PubMed]
61. Fanning, A.S.; Jameson, B.J.; Jesaitis, L.A.; Anderson, J.M. The tight junction protein ZO-1 establishes a link between the transmembrane protein occludin and the actin cytoskeleton. *J. Biol. Chem.* **1998**, *273*, 29745–29753. [CrossRef] [PubMed]

62. Kemphues, K.J.; Priess, J.R.; Morton, D.G.; Cheng, N.S. Identification of genes required for cytoplasmic localization in early C. elegans embryos. *Cell* **1988**, *52*, 311–320. [CrossRef]
63. Hung, T.J.; Kemphues, K.J. PAR-6 is a conserved PDZ domain-containing protein that colocalizes with PAR-3 in Caenorhabditis elegans embryos. *Development* **1999**, *126*, 127–135. [PubMed]
64. Tabuse, Y.; Izumi, Y.; Piano, F.; Kemphues, K.J.; Miwa, J.; Ohno, S. Atypical protein kinase C cooperates with PAR-3 to establish embryonic polarity in Caenorhabditis elegans. *Development* **1998**, *125*, 3607–3614. [PubMed]
65. Chen, J.; Zhang, M. The Par3/Par6/aPKC complex and epithelial cell polarity. *Exp. Cell Res.* **2013**, *319*, 1357–1364. [CrossRef] [PubMed]
66. Assémat, E.; Bazellières, E.; Pallesi-Pocachard, E.; Le Bivic, A.; Massey-Harroche, D. Polarity complex proteins. *Biochim. Biophys. Acta Biomembr.* **2008**, *1778*, 614–630. [CrossRef] [PubMed]
67. Izumi, Y.; Hirose, T.; Tamai, Y.; Hirai, S.; Nagashima, Y.; Fujimoto, T.; Tabuse, Y.; Kemphues, K.J.; Ohno, S. An atypical PKC directly associates and colocalizes at the epithelial tight junction with ASIP, a mammalian homologue of Caenorhabditis elegans polarity protein PAR-3. *J. Cell Biol.* **1998**, *143*, 95–106. [CrossRef] [PubMed]
68. Carling, D. AMPK signalling in health and disease. *Curr. Opin. Cell Biol.* **2017**, *45*, 31–37. [CrossRef] [PubMed]
69. Carling, D.; Viollet, B. Beyond energy homeostasis: The expanding role of AMP-activated protein kinase in regulating metabolism. *Cell Metab.* **2015**, *21*, 799–804. [CrossRef] [PubMed]
70. Kurumbail, R.G.; Calabrese, M.F. Structure and Regulation of AMPK. In *AMP-activated Protein Kinase*; Springer: Cham, Switzerland, 2016; pp. 3–22.
71. Garcia, D.; Shaw, R.J. AMPK: Mechanisms of Cellular Energy Sensing and Restoration of Metabolic Balance. *Mol. Cell* **2017**, *66*, 789–800. [CrossRef] [PubMed]
72. Hardie, D.G. AMP-activated protein kinase: An energy sensor that regulates all aspects of cell function. *Genes Dev.* **2011**, *25*, 1895–1908. [CrossRef] [PubMed]
73. Hardie, D.; Scott, J.; Pan, D.; Hudson, E. Management of cellular energy by the AMP-activated protein kinase system. *FEBS Lett.* **2003**, *546*, 113–120. [CrossRef]
74. Ross, F.A.; MacKintosh, C.; Hardie, D.G. AMP-activated protein kinase: A cellular energy sensor that comes in 12 flavours. *FEBS J.* **2016**, *283*, 2987–3001. [CrossRef] [PubMed]
75. Steinberg, G.R.; Kemp, B.E. AMPK in Health and Disease. *Physiol. Rev.* **2009**, *89*, 1025–1078. [CrossRef] [PubMed]
76. Calabrese, M.F.; Rajamohan, F.; Harris, M.S.; Caspers, N.L.; Magyar, R.; Withka, J.M.; Wang, H.; Borzilleri, K.A.; Sahasrabudhe, P.V.; Hoth, L.R.; et al. Structural Basis for AMPK Activation: Natural and Synthetic Ligands Regulate Kinase Activity from Opposite Poles by Different Molecular Mechanisms. *Structure* **2014**, *22*, 1161–1172. [CrossRef] [PubMed]
77. Moffat, C.; Harper, M.E. Metabolic Functions of AMPK: Aspects of Structure and of Natural Mutations in the Regulatory Gamma Subunits. *IUBMB Life* **2010**, *62*, 739–745. [CrossRef] [PubMed]
78. Salt, I.; Celler, J.W.; Hawley, S.A.; Prescott, A.; Woods, A.; Carling, D.; Hardie, D.G. AMP-activated protein kinase: Greater AMP dependence, and preferential nuclear localization, of complexes containing the alpha2 isoform. *Biochem. J.* **1998**, *334*, 177–187. [CrossRef] [PubMed]
79. Hudson, E.R.; Pan, D.A.; James, J.; Lucocq, J.M.; Hawley, S.A.; Green, K.A.; Baba, O.; Terashima, T.; Hardie, D.G. A novel domain in AMP-activated protein kinase causes glycogen storage bodies similar to those seen in hereditary cardiac arrhythmias. *Curr. Biol.* **2003**, *13*, 861–866. [CrossRef]
80. Kemp, B.E. Bateman domains and adenosine derivatives form a binding contract. *J. Clin. Investig.* **2004**, *113*, 182–184. [CrossRef] [PubMed]
81. Hardie, D.; Hawley, S. AMP-activated protein kinase: The energy charge hypothesis revisited. *Bioessays* **2001**, *23*, 1112–1119. [CrossRef] [PubMed]
82. Xiao, B.; Sanders, M.J.; Underwood, E.; Heath, R.; Mayer, F.V.; Carmena, D.; Jing, C.; Walker, P.A.; Eccleston, J.F.; Haire, L.F.; Saiu, P.; et al. Structure of mammalian AMPK and its regulation by ADP. *Nature* **2011**, *472*, 230–233. [CrossRef] [PubMed]
83. Gowans, G.J.; Hawley, S.A.; Ross, F.A.; Hardie, D.G. AMP is a true physiological regulator of AMP-activated protein kinase by both allosteric activation and enhancing net phosphorylation. *Cell Metab.* **2013**, *18*, 556–566. [CrossRef] [PubMed]

84. Oligschlaeger, Y.; Miglianico, M.; Chanda, D.; Scholz, R.; Thali, R.F.; Tuerk, R.; Stapleton, D.I.; Gooley, P.R.; Neumann, D. The Recruitment of AMP-activated Protein Kinase to Glycogen Is Regulated by Autophosphorylation. *J. Biol. Chem.* **2015**, *290*. [CrossRef] [PubMed]
85. Mitchelhill, K.I.; Michell, B.J.; House, C.M.; Stapleton, D.; Dyck, J.; Gamble, J.; Ullrich, C.; Witters, L.A.; Kemp, B.E. Posttranslational Modifications of the 5′-AMP-activated Protein Kinase 1 Subunit. *J. Biol. Chem.* **1997**, *272*, 24475–24479. [CrossRef] [PubMed]
86. Hawley, S.A.; Boudeau, J.; Reid, J.L.; Mustard, K.J.; Udd, L.; Mäkelä, T.P.; Alessi, D.R.; Hardie, D.G. Complexes between the LKB1 tumor suppressor, STRAD alpha/beta and MO25 alpha/beta are upstream kinases in the AMP-activated protein kinase cascade. *J. Biol.* **2003**, *2*, 28. [CrossRef] [PubMed]
87. Lizcano, J.M.; Göransson, O.; Toth, R.; Deak, M.; Morrice, N.A.; Boudeau, J.; Hawley, S.A.; Udd, L.; Mäkelä, T.P.; Hardie, D.G.; Alessi, D.R. LKB1 is a master kinase that activates 13 kinases of the AMPK subfamily, including MARK/PAR-1. *EMBO J.* **2004**, *23*, 833–843. [CrossRef] [PubMed]
88. Shaw, R.J.; Kosmatka, M.; Bardeesy, N.; Hurley, R.L.; Witters, L.A.; DePinho, R.A.; Cantley, L.C. The tumor suppressor LKB1 kinase directly activates AMP-activated kinase and regulates apoptosis in response to energy stress. *Proc. Natl. Acad. Sci. USA* **2004**, *101*, 3329–3335. [CrossRef] [PubMed]
89. Hawley, S.A.; Selbert, M.A.; Goldstein, E.G.; Edelman, A.M.; Carling, D.; Hardie, D.G. 5′-AMP activates the AMP-activated protein kinase cascade, and Ca^{2+}/calmodulin activates the calmodulin-dependent protein kinase I cascade, via three independent mechanisms. *J. Biol. Chem.* **1995**, *270*, 27186–27191. [CrossRef] [PubMed]
90. Hurley, R.L.; Anderson, K.A.; Franzone, J.M.; Kemp, B.E.; Means, A.R.; Witters, L.A. The Ca^{2+}/calmodulin-dependent protein kinase kinases are AMP-activated protein kinase kinases. *J. Biol. Chem.* **2005**, *280*, 29060–29066. [CrossRef] [PubMed]
91. Hawley, S.A.; Pan, D.A.; Mustard, K.J.; Ross, L.; Bain, J.; Edelman, A.M.; Frenguelli, B.G.; Hardie, D.G. Calmodulin-dependent protein kinase kinase-β is an alternative upstream kinase for AMP-activated protein kinase. *Cell Metab.* **2005**, *2*, 9–19. [CrossRef] [PubMed]
92. Woods, A.; Dickerson, K.; Heath, R.; Hong, S.-P.; Momcilovic, M.; Johnstone, S.R.; Carlson, M.; Carling, D. Ca^{2+}/calmodulin-dependent protein kinase kinase-beta acts upstream of AMP-activated protein kinase in mammalian cells. *Cell Metab.* **2005**, *2*, 21–33. [CrossRef] [PubMed]
93. Lage, R.; Dié Guez, C.; Vidal-Puig, A.; Ló Pez, M. AMPK: A metabolic gauge regulating whole-body energy homeostasis. *Trends Mol. Med.* **2008**, *14*, 539–549. [CrossRef] [PubMed]
94. Brown, M.S.; Brunschede, G.Y.; Goldstein, J.L. Inactivation of 3-hydroxy-3-methylglutaryl coenzyme A reductase in vitro. An adenine nucleotide-dependent reaction catalyzed by a factor in human fibroblasts. *J. Biol. Chem.* **1975**, *250*, 2502–2509. [PubMed]
95. Marcinko, K.; Steinberg, G.R. The role of AMPK in controlling metabolism and mitochondrial biogenesis during exercise. *Exp. Physiol. Exp Physiol* **2014**, *9912*, 1581–1585. [CrossRef] [PubMed]
96. Beg, Z.H.; Allmann, D.W.; Gibson, D.M. Modulation of 3-hydroxy-3-methylglutaryl coenzyme A reductase activity with cAMP and wth protein fractions of rat liver cytosol. *Biochem. Biophys. Res. Commun.* **1973**, *54*, 1362–1369. [CrossRef]
97. Carlson, C.A.; Kim, K.H. Regulation of hepatic acetyl coenzyme A carboxylase by phosphorylation and dephosphorylation. *J. Biol. Chem.* **1973**, *248*, 378–380. [CrossRef]
98. Holmes, B.F.; Kurth-Kraczek, E.J.; Winder, W.W. Chronic activation of 5′-AMP-activated protein kinase increases GLUT-4, hexokinase, and glycogen in muscle. *J. Appl. Physiol.* **1999**, *87*, 1990–1995. [CrossRef] [PubMed]
99. Carling, D.; Grahame Hardie, D. The substrate and sequence specificity of the AMP-activated protein kinase. Phosphorylation of glycogen synthase and phosphorylase kinase. *Biochim. Biophys. Acta Mol. Cell Res.* **1989**, *1012*, 81–86. [CrossRef]
100. Kishton, R.J.; Barnes, C.E.; Nichols, A.G.; Cohen, S.; Gerriets, V.A.; Siska, P.J.; Macintyre, A.N.; Goraksha-Hicks, P.; de Cubas, A.A.; Liu, T.; et al. AMPK Is Essential to Balance Glycolysis and Mitochondrial Metabolism to Control T-ALL Cell Stress and Survival. *Cell Metab.* **2016**, *23*, 649–662. [CrossRef] [PubMed]
101. Liu, L.; Ulbrich, J.; Müller, J.; Wüstefeld, T.; Aeberhard, L.; Kress, T.R.; Muthalagu, N.; Rycak, L.; Rudalska, R.; Moll, R.; et al. Deregulated MYC expression induces dependence upon AMPK-related kinase 5. *Nature* **2012**, *483*, 608–612. [CrossRef] [PubMed]

102. Cereijido, M.; Robbins, E.S.; Dolan, W.J.; Rotunno, C.A.; Sabatini, D.D. Polarized monolayers formed by epithelial cells on a permeable and translucent support. *J. Cell Biol.* **1978**, *77*, 853–880. [CrossRef] [PubMed]
103. Matter, K.; Balda, M.S. Functional analysis of tight junctions. *Methods* **2003**, *30*, 228–234. [CrossRef]
104. Zheng, B.; Cantley, L.C. Regulation of epithelial tight junction assembly and disassembly by AMP-activated protein kinase. *Proc. Natl. Acad. Sci. USA* **2007**, *104*, 819–822. [CrossRef] [PubMed]
105. Mandai, K.; Nakanishi, H.; Satoh, A.; Obaishi, H.; Wada, M.; Nishioka, H.; Itoh, M.; Mizoguchi, A.; Aoki, T.; Fujimoto, T.; et al. Afadin: A novel actin filament-binding protein with one PDZ domain localized at cadherin-based cell-to-cell adherens junction. *J. Cell Biol.* **1997**, *139*, 517–528. [CrossRef] [PubMed]
106. Zhang, L.; Jouret, F.; Rinehart, J.; Sfakianos, J.; Mellman, I.; Lifton, R.P.; Young, L.H.; Caplan, M.J. AMP-activated protein kinase (AMPK) activation and glycogen synthase kinase-3β (GSK-3β) inhibition induce Ca^{2+}-independent deposition of tight junction components at the plasma membrane. *J. Biol. Chem.* **2011**, *286*, 16879–16890. [CrossRef] [PubMed]
107. Aznar, N.; Patel, A.; Rohena, C.C.; Dunkel, Y.; Joosen, L.P.; Taupin, V.; Kufareva, I.; Farquhar, M.G.; Ghosh, P. AMP-activated protein kinase fortifies epithelial tight junctions during energetic stress via its effector GIV/Girdin. *Elife* **2016**, *5*, e20795. [CrossRef] [PubMed]
108. Bhandari, D.; Lopez-Sanchez, I.; To, A.; Lo, I.-C.; Aznar, N.; Leyme, A.; Gupta, V.; Niesman, I.; Maddox, A.L.; Garcia-Marcos, M.; et al. Cyclin-dependent kinase 5 activates guanine nucleotide exchange factor GIV/Girdin to orchestrate migration–proliferation dichotomy. *Proc. Natl. Acad. Sci. USA* **2015**, *112*, E4874–E4883. [CrossRef] [PubMed]
109. Sasaki, K.; Kakuwa, T.; Akimoto, K.; Koga, H.; Ohno, S. Regulation of epithelial cell polarity by PAR-3 depends on Girdin transcription and Girdin-Gαi3 signaling. *J. Cell Sci.* **2015**, *128*, 2244–2258. [CrossRef] [PubMed]
110. Houssin, E.; Tepass, U.; Laprise, P. Girdin-mediated interactions between cadherin and the actin cytoskeleton are required for epithelial morphogenesis in Drosophila. *Development* **2015**, *142*, 1777–1784. [CrossRef] [PubMed]
111. Schumann, A.; Nutten, S.; Donnicola, D.; Comelli, E.M.; Mansourian, R.; Cherbut, C.; Corthesy-Theulaz, I.; Garcia-Rodenas, C. Neonatal antibiotic treatment alters gastrointestinal tract developmental gene expression and intestinal barrier transcriptome. *Physiol. Genom.* **2005**, *23*, 235–245. [CrossRef] [PubMed]
112. Peng, L.; Li, Z.-R.; Green, R.S.; Holzman, I.R.; Lin, J. Butyrate enhances the intestinal barrier by facilitating tight junction assembly via activation of AMP-activated protein kinase in Caco-2 cell monolayers. *J. Nutr.* **2009**, *139*, 1619–1625. [CrossRef] [PubMed]
113. 113. Mullin, J.M.; Laughlin, K.V.; Ginanni, N.; Marano, C.W.; Clarke, H.M.; Peralta Soler, A. Increased Tight Junction Permeability Can Result from Protein Kinase C Activation/Translocation and Act as a Tumor Promotional Event in Epithelial Cancers. *Ann. N. Y. Acad. Sci.* **2006**, *915*, 231–236. [CrossRef]
114. Miao, W.; Wu, X.; Wang, K.; Wang, W.; Wang, Y.; Li, Z.; Liu, J.; Li, L.; Peng, L. Sodium Butyrate Promotes Reassembly of Tight Junctions in Caco-2 Monolayers Involving Inhibition of MLCK/MLC2 Pathway and Phosphorylation of PKCβ2. *Int. J. Mol. Sci.* **2016**, *17*, 1696. [CrossRef] [PubMed]
115. Wu, G. Functional Amino Acids in Growth, Reproduction, and Health. *Adv. Nutr.* **2010**, *1*, 31–37. [CrossRef] [PubMed]
116. Li, N.; Lewis, P.; Samuelson, D.; Liboni, K.; Neu, J. Glutamine regulates Caco-2 cell tight junction proteins. *Am. J. Physiol. Liver Physiol.* **2004**, *287*, G726–G733. [CrossRef] [PubMed]
117. Wang, B.; Wu, Z.; Ji, Y.; Sun, K.; Dai, Z.; Wu, G. L-Glutamine Enhances Tight Junction Integrity by Activating CaMK Kinase 2–AMP-Activated Protein Kinase Signaling in Intestinal Porcine Epithelial Cells. *J. Nutr.* **2016**, *146*, 501–508. [CrossRef] [PubMed]
118. Rowart, P.; Erpicum, P.; Krzesinski, J.-M.; Sebbagh, M.; Jouret, F. Mesenchymal Stromal Cells Accelerate Epithelial Tight Junction Assembly via the AMP-Activated Protein Kinase Pathway, Independently of Liver Kinase B1. *Stem Cells Int.* **2017**, *2017*. [CrossRef] [PubMed]
119. Park, H.-Y.; Kunitake, Y.; Hirasaki, N.; Tanaka, M.; Matsui, T. Theaflavins enhance intestinal barrier of Caco-2 Cell monolayers through the expression of AMP-activated protein kinase-mediated Occludin, Claudin-1, and ZO-1. *Biosci. Biotechnol. Biochem.* **2015**, *79*. [CrossRef] [PubMed]
120. Egawa, M.; Kamata, H.; Kushiyama, A.; Sakoda, H.; Fujishiro, M.; Horike, N.; Yoneda, M.; Nakatsu, Y.; Ying, G.; Jun, Z.; et al. Long-term Forskolin Stimulation Induces AMPK Activation and Thereby Enhances

Tight Junction Formation in Human Placental Trophoblast BeWo Cells. *Placenta* **2008**, *29*, 1003–1008. [CrossRef] [PubMed]

121. Lee, J.H.; Koh, H.; Kim, M.; Kim, Y.; Lee, S.-H.; Karess, R.E.; Shong, M.; Kim, J.-M.J.; Kim, J.-M.J.; Chung, J. Energy-dependent regulation of cell structure by AMP-activated protein kinase. *Nature* **2007**, *447*, 1017–1020. [CrossRef] [PubMed]
122. Bilder, D. Epithelial polarity and proliferation control: Links from the Drosophila neoplastic tumor suppressors. *Genes Dev.* **2004**, *18*, 1909–1925. [CrossRef] [PubMed]
123. Faubert, B.; Boily, G.; Izreig, S.; Griss, T.; Samborska, B.; Dong, Z.; Dupuy, F.; Chambers, C.; Fuerth, B.J.; Viollet, B.; et al. AMPK Is a Negative Regulator of the Warburg Effect and Suppresses Tumor Growth In Vivo. *Cell Metab.* **2013**, *17*, 113–124. [CrossRef] [PubMed]
124. Sun, X.; Yang, Q.; Rogers, C.J.; Du, M.; Zhu, M.-J. AMPK improves gut epithelial differentiation and barrier function via regulating Cdx2 expression. *Cell Death Differ.* **2017**, *24*, 819–831. [CrossRef] [PubMed]
125. Mullin, J.M.; Skrovanek, S.M.; Ramalingam, A.; DiGuilio, K.M.; Valenzano, M.C. Methionine restriction fundamentally supports health by tightening epithelial barriers. *Ann. N. Y. Acad. Sci.* **2016**, *1363*, 59–67. [CrossRef] [PubMed]
126. Thomas, D.; Surdin-Kerjan, Y. Metabolism of sulfur amino acids in Saccharomyces cerevisiae. *Microbiol. Mol. Biol. Rev.* **1997**, *61*, 503–532. [PubMed]
127. Ogawa, T.; Tsubakiyama, R.; Kanai, M.; Koyama, T.; Fujii, T.; Iefuji, H.; Soga, T.; Kume, K.; Miyakawa, T.; Hirata, D.; et al. Stimulating S-adenosyl-l-methionine synthesis extends lifespan via activation of AMPK. *Proc. Natl. Acad. Sci. USA* **2016**, *113*, 11913–11918. [CrossRef] [PubMed]
128. Wang, X.; Valenzano, M.C.; Mercado, J.M.; Zurbach, E.P.; Mullin, J.M. Zinc Supplementation Modifies Tight Junctions and Alters Barrier Function of CACO-2 Human Intestinal Epithelial Layers. *Dig. Dis. Sci.* **2013**, *58*, 77–87. [CrossRef] [PubMed]
129. Miyoshi, Y.; Tanabe, S.; Suzuki, X.T. Cellular zinc is required for intestinal epithelial barrier maintenance via the regulation of claudin-3 and occludin expression. *Am. J. Physiol.* **2016**, *311*, G105–G116. [CrossRef] [PubMed]
130. Eom, J.-W.; Lee, J.-M.; Koh, J.-Y.; Kim, Y.-H. AMP-activated protein kinase contributes to zinc-induced neuronal death via activation by LKB1 and induction of Bim in mouse cortical cultures. *Mol. Brain* **2016**, *9*, 14. [CrossRef] [PubMed]
131. Tang, X.X.; Chen, H.; Yu, S.; Zhang, L.; Caplan, M.J.; Chan, H.C. Lymphocytes accelerate epithelial tight junction assembly: Role of AMP-activated protein kinase (AMPK). *PLoS ONE* **2010**, *5*, e12343. [CrossRef] [PubMed]
132. Patkee, W.R.A.; Carr, G.; Baker, E.H.; Baines, D.L.; Garnett, J.P. Metformin prevents the effects of Pseudomonas aeruginosa on airway epithelial tight junctions and restricts hyperglycaemia-induced bacterial growth. *J. Cell. Mol. Med.* **2016**, *20*, 758–764. [CrossRef] [PubMed]

5

AMPK: An Epigenetic Landscape Modulator

Brendan Gongol [1,2], Indah Sari [2,†], Tiffany Bryant [2,†], Geraldine Rosete [2,†] and Traci Marin [1,3,*]

1 Department of Medicine, University of California, San Diego, CA 92093, USA; brengong@gmail.com
2 Department of Cardiopulmonary Sciences, School of Allied Health Professions, Loma Linda University, Loma Linda, CA 92350, USA; isari@llu.edu (I.S.); tprescott@llu.edu (T.B.); grosete@llu.edu (G.R.)
3 Department of Health Sciences, Victor Valley College, Victorville, CA 92395, USA
* Correspondence: traci.marin@vvc.edu
† These authors contributed equally to this work.

Abstract: Activated by AMP-dependent and -independent mechanisms, AMP-activated protein kinase (AMPK) plays a central role in the regulation of cellular bioenergetics and cellular survival. AMPK regulates a diverse set of signaling networks that converge to epigenetically mediate transcriptional events. Reversible histone and DNA modifications, such as acetylation and methylation, result in structural chromatin alterations that influence transcriptional machinery access to genomic regulatory elements. The orchestration of these epigenetic events differentiates physiological from pathophysiological phenotypes. AMPK phosphorylation of histones, DNA methyltransferases and histone post-translational modifiers establish AMPK as a key player in epigenetic regulation. This review focuses on the role of AMPK as a mediator of cellular survival through its regulation of chromatin remodeling and the implications this has for health and disease.

Keywords: AMPK; epigenetics; chromatin remodeling; histone modification; DNA methylation

---

1. Introduction

Epigenetic regulation gives rise to a spectrum of cellular phenotypes observed in a single organism independent of primary DNA sequence. Such regulation is hereditable and stable, as occurs in the determination of cell type, but also transient, producing a particular phenotypic outcome to ensure survival [1–3]. Influencing gene expression, epigenetics promotes organismal adaption by offering substantial functional variability in response to environmental stimuli [1–3]. This regulation occurs, in part, through nucleosomal remodeling as a result of histone, DNA, and DNA-binding protein modifications that include: phosphorylation, acetylation, *O*-GlcNAcylation, ribosylation, and methylation. Such signature modifications or marks characterize nucleosome remodeling and determine the degree of gene activation or silencing. At a fundamental level, stressors such as nutrient deprivation or heightened physical activity trigger dynamic epigenetic markings that orchestrate adaptive gene regulation to improve survivability [4–7].

AMP-activated protein kinase (AMPK), a master regulator of energy homeostasis and a key mediator of adaptation and cell survival, is activated by conditions that produce energy deprivation such as hypoxia, exercise, nutrient starvation, and infection [8]. AMPK functions as a heterotrimeric serine/threonine protein kinase composed of a catalytic α-subunit, a scaffolding β-subunit, and a regulatory γ-subunit. Both α and β-subunits exist in two isoforms ($\alpha 1$, $\alpha 2$, and $\beta 1$, $\beta 2$), while the regulatory γ-subunit exists in three isoforms ($\gamma 1$, $\gamma 2$, $\gamma 3$). The combination of α, β, and γ isoforms can form 12 different heterotrimeric complexes, each demonstrating unique functions and subcellular or tissue-specific distributions [9–11]. Once activated, AMPK phosphorylates targets with well-defined consensus sequences to regulate bioenergetics by modulating metabolic pathways that promote ATP production and limit energy expenditure [12]. Such roles of AMPK in cellular homeostasis are facilitated by gene regulation as a result of chromatin conformational dynamics. To that end, AMPK orchestrates these changes by phosphorylating several histones and proteins involved in nucleosome remodeling that enhance mitochondrial biogenesis and function [13]. Additionally, during acute metabolic stressors, such as fasting or exercise, AMPK associates with chromatin at promoters of genes involved in lipid and glucose metabolism [14].

Chronic AMPK activation plays a role in cellular and organismal inheritance, evidenced by its multi-isoform 2R-ohnologue characteristics, which are often evolutionarily conserved in gene coding regions to support basic survival functions and to increase the possibility for complex tissue diversity and adaptation [15–17]. Such adaptation for survival early in life initiates epigenetic programming that correlates with AMPK activation and determines predisposition to disease. For example, insults during development, such as placental insufficiency or maternal metabolic disorders, increase the probability of adverse metabolic disorders in later stages of life, such as diabetes mellitus, metabolic syndrome, insulin resistance, hypertension, vascular disease, and cancer [18]. Furthermore, metabolic disorders, such as maternal obesity, are often associated with reduced AMPK expression or activity and concomitant loss of its epigenetic mechanisms associated with adaptation and survival [19,20]. Although the mechanistic basis for this regulation, as well as its implications for inheritance, is still under exploration, these collective observations point to a fundamentally important role of AMPK as an epigenetic regulator.

2. Histone Modification

Histones are nuclear-localized, primarily positively charged proteins. However, a number of histone splice variants have been described with broad cellular functions and localizations [21,22]. Within the nucleus, histones are packaged into an octamer consisting of two H2A, H2B, H3, and H4 [23]. These octameric histone complexes associate via charge–charge interactions with DNA and make up a nucleosome, which is linked to other nucleosomes by histone H1. Given their tight association with DNA, histones serve as mediators between stress response signaling cascades and the regulation of nucleosomal structure that ultimately influence gene expression and cellular survival [24]. AMPK, both directly and indirectly, regulates the post-translational modification (PTM) status of histones that play a major role in the regulation of nucleosome structure.

2.1. Histone Phosphorylation

Bioinformatic analysis has identified several histones that contain an AMPK phosphorylation consensus sequence. These include: H1FX, H2AFX, H2AFY, H2AFY2, H2AFZ, H2BFM, and H3F3B [25], suggesting that AMPK plays an important role in epigenetic regulation through histone phosphorylation. Ultimately, such histone phosphorylation promotes cross-talk between epigenetic regulators to facilitate spatio-temporal nucleosome structural changes that influence transcriptional machinery [24]. For example, activation by glucose deprivation or UV radiation results in the colocalization of AMPK and phosphorylating serine 36 in H2B ($H2B^{S36}$) throughout the *carnitine palmitoyltransferase 1C* (*CPT1C*) and *cyclin-dependent kinase inhibitor* (*p21*) promoter regions and gene bodies. This exposes TP53 or tumor protein (p53) DNA binding sites and subsequent association with RNA polymerase II (Figure 1) [26,27]. These events enhance CPT1C and p21 expression, which play a role in cellular survival via activating autophagy and transporting long-chain fatty acids to the mitochondria, collectively enhancing β-oxidation and energy production [28–30]. Related to cellular survival, charged multivesicular body protein 1B (CHMP1b) is phosphorylated by AMPK, forms a "shell" around nucleosomes enriched with H3 phosphorylation and acetylation, and influences gene transcription [25]. This might play a role in the transition of active and inactive nucleosome regions and potentially heritable epigenetic marks [31].

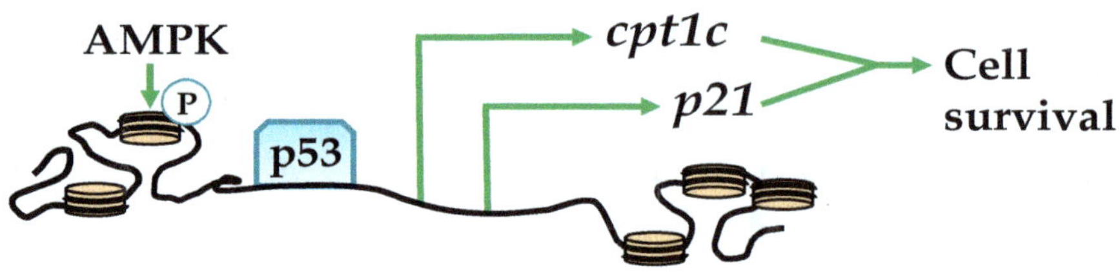

Figure 1. AMPK (AMP-activated protein kinase) promotes cell survival through histone phosphorylation. AMPK phosphorylates H2B to promote chromatin relaxation at tumor protein p53 (p53) recognized promoters and transcription of *carnitine palmitoyltransferase 1C* (*cpt1c*) and *cyclin-dependent kinase inhibitor* (*p21*) to enhance cell survival.

2.2. Histone Acetylation

Lysine acetylation of the N-terminal tails of histones disrupts the charge–charge interactions between DNA and histone tails, producing a euchromatin, a relaxed and active, chromatin state. This reaction changes histone–DNA associations, histone–histone associations between adjacent nucleosomes, and histone–regulatory protein interactions [32]. Histone acetylation occurs via transfer of an acetyl group from acetyl coenzyme-A (acetyl-CoA) to the to the ε-ammonium (NH_3^+) group of lysine catalyzed by histone acetyltransferases (HATs). HAT families are diverse and promote a spectrum of interactions and functions. Histone deacetylation restores the DNA-histone interaction promoting a heterochromatic, a condensed and silent chromatin state [33]. Antagonistically, histone deacetylation, catalyzed by histone deacetylases (HDACs), restores the DNA-histone interaction promoting a euchromatic chromatin state [33]. AMPK regulates the activity of both HATs and HDACs by influencing cofactor or substrate availability through direct phosphorylation.

Indirectly, 5-Aminoimidazole-4-carboxamide ribonucleotide (AICAR) activation of AMPK affects global HAT activity by increasing available acetyl-CoA levels through several mechanisms. AMPK phosphorylates acetyl-CoA carboxylase (ACC) to prevent the conversion of acetyl-CoA to malonyl-CoA increasing available acetyl donating groups for HATs (Figure 2A). AMPK also increases the formation of acetyl-CoA by phosphorylating acetyl-CoA synthetase short-chain family member 2 (ACSS2), causing its nuclear translocation for the conversion of acetate, the byproduct of HDAC histone deacetylation, to acetyl-CoA (Figure 2A). This mechanism has been shown to increase HAT acetylation of H3 at transcription factor EB (TFEB)-responsive promoters to activate genes important for autophagy and lysosomal function [34,35]. Yet, in general, increased acetyl-CoA levels in response to AICAR results in increased H3 acetylation at lysine 14 ($H3^{K14}$) and H4 acetylation at lysines 5,8,12,16 ($H4^{K5,8,12,16}$) [36]. AMPK activated by metformin regulates histone acetylation in a metabolic state-dependent manner. For example, metformin reduces H3 acetylation by decreasing bioavailability of mitochondrial acetyl-CoA in breast cancer, bringing acetylation status to normal in promoters of cancer-specific genes [37].

AMPK also regulates substrate availability of HDACs. The class III HDACs, sirtuins (SIRTs), couple deacetylation with nicotinamide adenine dinucleotide (NAD^+) hydrolysis to produce *O*-acetyl-ADP-ribose [38]. AMPK indirectly activates SIRT1, in part, by increasing the NAD^+/NADH ratio [38,39]. Once activated, SIRT1 plays a fundamental role in chromatin organization by interacting with and deacetylating a variety of transcription factors and coregulators [40]. However, in contrary to activating the SIRTs, AMPK indirectly inhibits class I and II HDACs by increasing β-hydroxybutyrate (βOHB) during fatty acid oxidation. βOHB acts similarly to the HDAC inhibitor butyrate, increasing global histone acetylation (Figure 2B) [41]. However, in addition to globally inhibiting class I and class II HDACs, AMPK activation also promotes HDACs 4, 5, and 7 hyperphosphorylation and translocation from the nucleus (Figure 2B) [42]. Both of these events increase global histone acetylation. However, despite these global effects, the regulation of gene expression is often promoter- and gene cluster-specific, suggesting the importance of direct phosphorylation events of AMPK on specific HATs and HDACs.

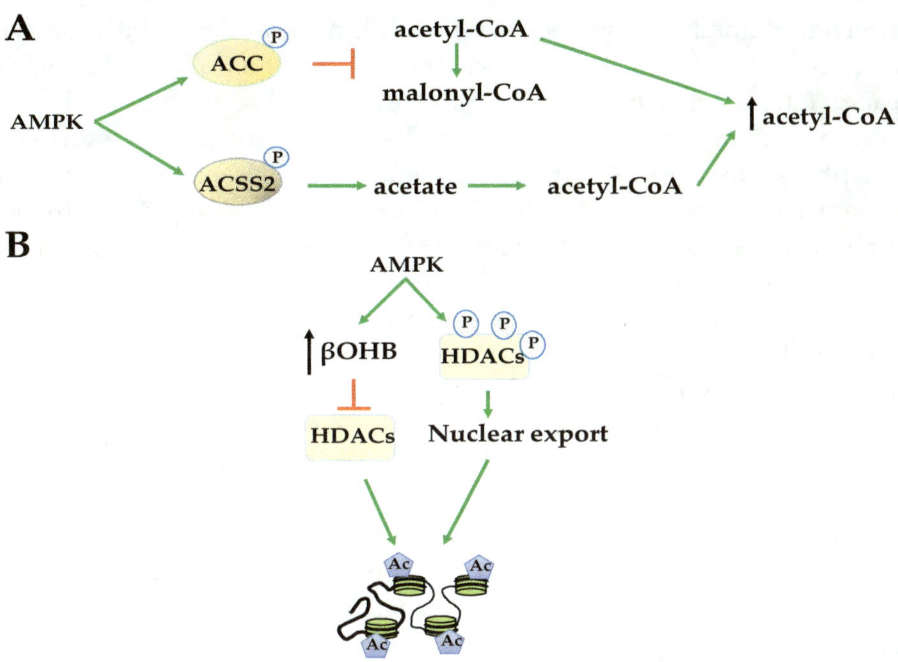

Figure 2. AMPK activation increases acetyl-CoA and promotes histone acetylation. (**A**) Through phosphorylation, AMPK inhibits acetyl-CoA carboxylase (ACC) while activating acetyl-CoA synthetase short-chain family member 2 (ACSS2) to increase acetyl-CoA availability for acetylation; (**B**) AMPK increases β-hydroxybutyrate (βOHB) to inhibit histone deacetylases (HDACs) and promotes HDAC nuclear export via hyperphosphorylation. Both of these events increase histone acetylation.

Investigations into epigenetically regulated networks identified HAT1 and retinoblastoma binding protein 7 (RBBP7) as direct targets of AMPK that dimerize following phosphorylation. This results in enhanced euchromatin structure at the promoters of *peroxisome proliferator–activated receptor gamma coactivator–1α* (*PGC-1α*), *transcription factor A* (*Tfam*), *nuclear respiratory factors 1* and *2* (*NRF1* and *NRF2*), and *uncoupling proteins 2* and *3* (*UCP2* and *UCP3*). The corresponding induction of these genes produces enhanced mitochondrial function (Figure 3A) [13]. In addition to HAT1 and RBBP7, AMPK phosphorylates HDAC5, promoting its dissociation from the promoters releasing its suppressive effects to increase expression of an array of metabolic genes, including glucose transporter type 4 (GLUT-4) (Figure 3B) [43].

Although less characterized, it is likely that AMPK phosphorylates an array of HATs including transcription regulator family member A (SIN3), CREB binding protein (CREBBP), elongator acetyltransferase complex subunits (ELP) 2, 3, and 4, and K(lysine) acetyltransferases (KAT) 2A, 2B, 6A, 6B, 7, and 8, and HDACs 1–5, 8–9, and 10, 11, as well as SIRTs 2, 3, 4, 5, 6, and 7; all of which contain AMPK consensus sequences and are opportune for future study [13].

2.3. Histone Methylation

Histone methylation occurs on lysine and arginine, and its effects on gene expression are dependent upon the site and degree of methylation. For example, methylation of $H3^{K4}$, $H3^{K36}$, and $H3^{K79}$ promotes an active euchromatin state; while methylation of $H3^{K9}$ and $H3^{K27}$ promotes a silent heterochromatic state [44–46]. Both lysine-specific and arginine-specific histone methyltransferases use S-adenosyl methionine (SAM) as cofactors and methyl donor catalyzing the transfer of one, two, or three methyl groups. The demethylases, however, serve a variety of functions and include two primary classes: flavin adenine dinucleotide (FAD)-dependent amine oxidase and Fe(II) and α-ketoglutarate-dependent hydroxylase. The regulation of demethylases via α-ketoglutarate, a tricarboxylic acid (TCA) cycle intermediate, suggests that AMPK may play a role in regulating histone methylation and demethylation because AMPK is activated by conditions where TCA cycle intermediates are depleted and, once activated, these intermediates are restored. For example, enhanced amino acid, fatty acid, or glycogen catabolism by AMPK increases α-ketoglutarate, which, in turn, may increase histone demethylase activity [47]. In addition to restoring α-ketoglutarate levels, AMPK phosphorylates and inhibits fumarase a TCA cycle enzyme that converts fumarate to malate. Subsequently, elevated fumarate levels inhibit lysine-specific demethylase 2A (KDM2A), restoring $H3^{K36}$ dimethylation ($H3^{K36me2}$) at promoters of genes mediating cell growth (Figure 4A) [48]. In addition to these indirect effects, AMPK activates the demethyltransferase lysine demethylase 5 (KDM5) and lysine-specific histone demethylase-1 (LSD1) that remove $H3^{K4}$ trimethylation ($H3^{K4me3}$) [49]. Although AMPK activates histone demethyltransferase activity, it inhibits several complexes that regulate methyltransferase activity including the methylation complex polycomb repressive complex 2 (PRC2) and the histone methyltransferase-containing COMPASS complex (complex proteins associated with Set1) [20,50]. While the inhibition of PRC2 occurs through the phosphorylation of histone methyl transferase enhancer of zeste homolog 2 (EZH2), which decreases $H2^{K27}$ monomethylation ($H2^{K27me1}$), inhibition of the COMPASS complex by AMPK decreases $H3^{K4me3}$, a marker of transcriptional activation (Figure 4B) [50–52]. Globally, inhibition of PRC2 results in the upregulation of tumor suppressor PRC2 target genes and suppression of tumor growth, while COMPASS complex inhibition is a protective mechanism that halts chromatin marking in the presence of nutrient deficiency to promote stress tolerance (Figure 4B) [20,50–52].

However, the orchestration between the methylation and demethylation activities of AMPK is promoter-specific. For example, AMPK recruits PRC2 and LSD1 to the promoter of *caudal type homeobox2* (*Cdx2*), increasing $H3^{K4me3}$ but not $H3^{K27me3}$ [53]. This increases CDX2 expression, which transactivates solute carrier family 5 member 8 (SLC5A8) to increase immune function and cell survival under several stress conditions (Figure 4C) [54,55].

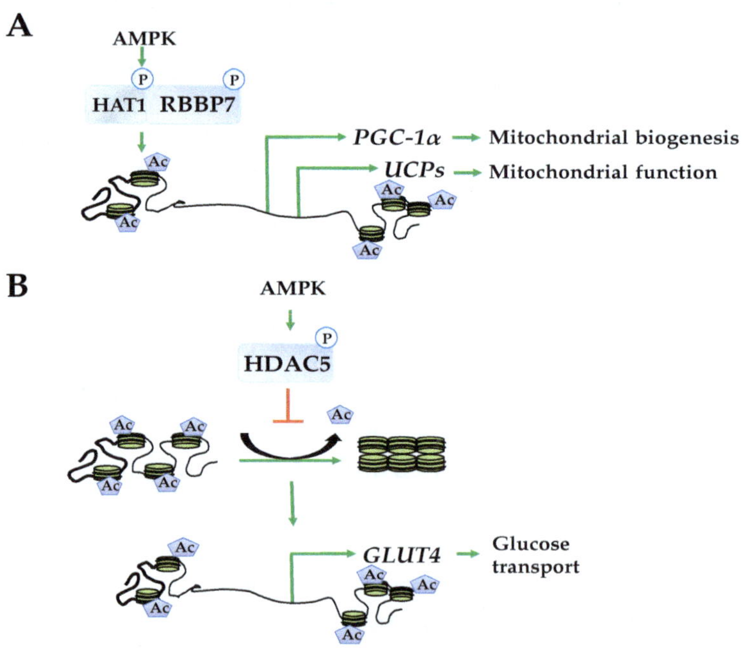

Figure 3. AMPK inhibits histone deacetylase activity to promote histone acetylation. (**A**) AMPK phosphorylates histone acetylase1 and retinoblastoma binding protein 7 (RBBP7) to form a complex that acetylates histones at promoters of *peroxisome proliferator-activated receptor gamma coactivator-1α* (*PGC-1α*), *uncoupling proteins* (*UCPs*) to increase mitochondrial biogenesis and function; (**B**) AMPK phosphorylates and inhibits HDAC5 enhancing acetylation at the *glucose transporter type 4* (*GLUT-4*) promoter increasing its transcription.

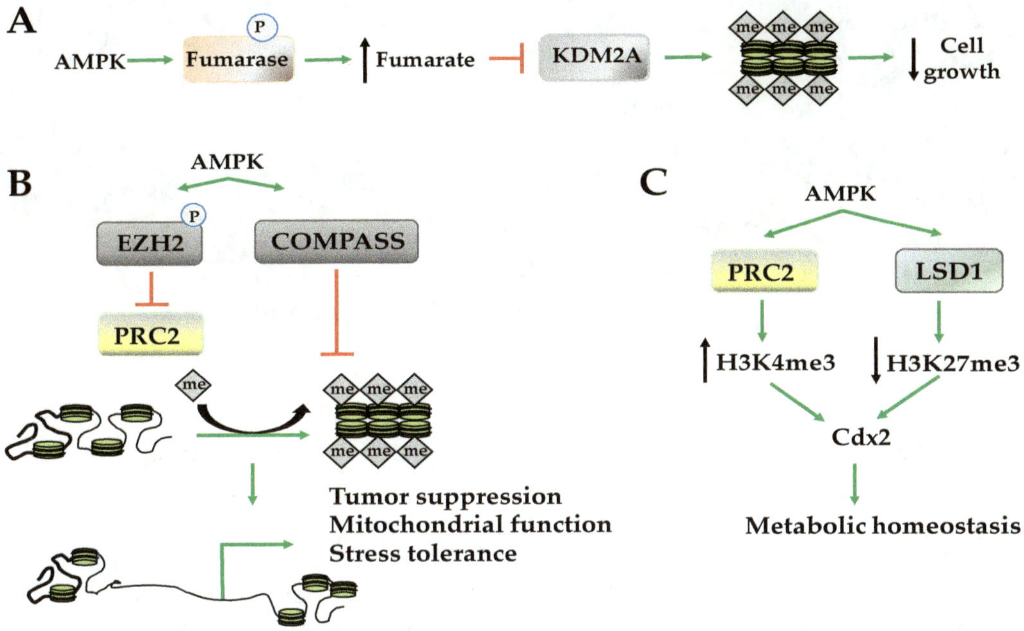

Figure 4. AMPK regulates histone methylation. (**A**) AMPK phosphorylates fumarase increase fumarate, which inhibits lysine-specific demethylase 2A (KDM2A), increasing histone methylation at cell growth; (**B**) AMPK decreases histone methylation through inhibition of polycomb repressive complex 2 (PRC2) and complex proteins associated with set1 (COMPASS complex); (**C**) AMPK modulates histone 3 methylations status to increase caudal type homeobox2 [cell differentiation]) (Cdx2) expression and metabolic homeostasis. LSD1: lysine-specific histone demethylase-1; EZH2: zeste homolog 2.

2.4. Histone O-GlcNAcylation

O-GlcNAcylation is the addition of an *O*-linked *N*-acetylglucosamine (*O*-GlcNAc) group to a serine or threonine by *O*-GlcNAc transferase (OGT). The donor substrate for *O*-GlcNAcylation, uridine diphosphate *N*-acetylglucosamine (UDP-GlcNAc), is a product of the hexamine biosynthetic pathway (HBP) in which glucose, fatty acid, amino acid, and ATP metabolism converge implicating its role as a nutrient and stress sensor [56]. *O*-GlcNAcase (OGA) removes *O*-GlcNAc through hydrolysis. The dynamics of histone *O*-GlcNAcylation regulate a variety of activities including other histone PTMs such as acetylation, methylation, and phosphorylation, and is highly sensitive to a spectrum of cellular stressors such as hypoxia, heat shock, and starvation (Figure 5A) [57]. AMPK phosphorylates *O*-GlcNAc transferase (OGT) promoting its dissociation from chromatin. This inhibits its *O*-GlcNAcylation of $H2B^{S112}$ in response to extracellular glucose through the hexosamine biosynthesis pathway (HBP) and promotes $H2B^{K120}$ mono-ubiquitination and transcriptional activation (Figure 5B) [56].

2.5. Histone Ribosylation

Histone ADP-ribosylation can occur as addition of mono- or poly-ADP-ribosylation units by mono-ADP ribosyltransferases (ARTs) or poly-(ADP-ribose) polymerases (PARPs), respectively. NAD^+ serves as the main source for histone ADP-ribosylation, which can occur on many amino acids restructuring chromatin into a euchromatin state [58,59]. AMPK phosphorylates and activates PARP1 [60,61]. Substrates of PARP1 include PARP1 itself and the tail of histones H1, H2A, H2B, H3, and H4, creating a poly(ADP-ribose)ylated mark at transcription start sites in transcriptionally necessary genes during metaphase [62]. Contrarily, when PARP1 is not active as a polymerase, it binds to DNA and promotes a heterochromatin state [59]. This is underscored by AMPK's phosphorylation of PARP1, causing its dissociation from the B-cell lymphoma protein 6 (Bcl-6) to increase its expression [61].

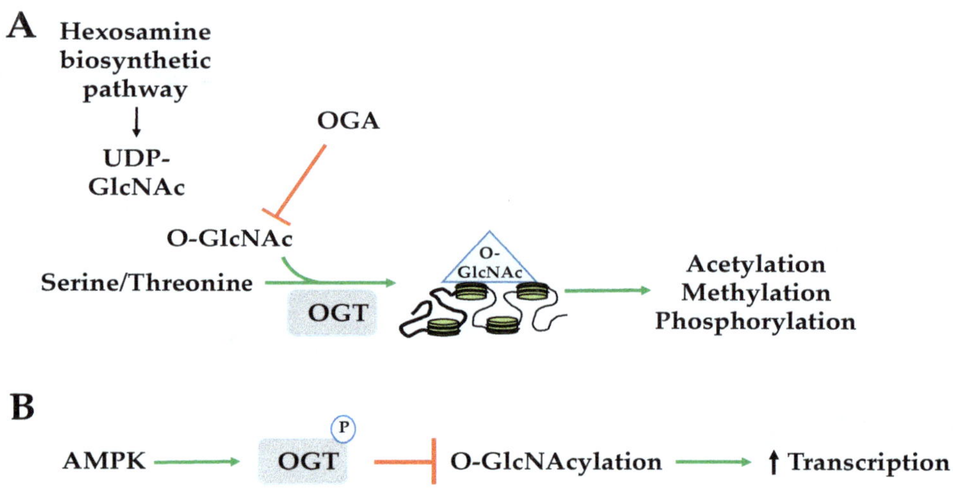

Figure 5. AMPK regulates histone *O*-GlcNAcylation and ADP-ribosylation. (**A**) *O*-GlcNAc transferase adds *O*-GlcNAc to a serine or threonine to produce *O*-GlcNAcylation. Nutrient and stress sensor stimulates hexosamine biosynthetic pathway (HBP) to create UDP-GlcNAc, which is a donor substrate for *O*-GlcNAcylation. Histone *O*-GlcNAcylation regulates acetylation, methylation, and phosphorylation; (**B**) AMPK phosphorylates *O*-GlcNAc transferase and inhibits *O*-GlcNAcylation to increase transcription. OGA: *O*-GlcNAcase; OGT: *O*-GlcNAc transferase.

3. DNA Modification

More abundantly enriched in promoter regions, cytosine-guanine dinucleotides (CpG) are susceptible to methylation at the 5′ cytosine position by a family of DNA methyltransferases (DNMTs). Each DNA methyltransferase member has a slightly different functional role in the global regulation of global DNA methylation. While DNMT 3a and 3b create the methyl "marks" that are carried through mitosis, DNMT1 maintains these methylation marks and regulates the dynamics of DNA modification and nucleosomal remodeling [63]. Following methylation, a number of functional changes occur. Due to its electron donating effects, methylation weakens Watson–Crick base pairing and recruits methyl-CpG binding domain proteins (MBDs), HDACs, and transcriptional repressors that collectively organize chromatin into a heterochromatic, a transcriptionally inactive conformation. These CpG methylations are highly dynamic- and stimulation-dependent. Removal of the CpG methylation requires oxidation of 5-methycytosine (5-mC) to 5-hydroxymethylcytosine (5-hmC), which is then converted to 5-formylcytosine (5-fC), and 5-carboxylcytosine (5-caC) followed by complete removal of the functional group by ten-eleven translocation hydroxylases (TETs) [64,65]. Although TETs are required to actively remove CpG methyl groups, the process occurs passively and semi-conservatively during DNA replication [66].

DNA Methylation

AMPK regulates global methylation by changing the substrates required for DNMT and TET activity. For example, SAM, the methyl donor for DNMT3a and DNMT3b, is converted to S-adenosylhomocysteine (SAH) upon CpG methylation [67]. Therefore, while SAM is required for DNMT3a and DNMT3b activity, SAH has an opposing inhibitory effect [67]. Supporting DNMT3a and DNMT3b activation, AMPK increases mitochondrial function and serine hydroxymethyltransferase 2 (SHMT2). Once activated, SHMT2 facilitates the one carbon transfer from folate to SAH, making SAM, and increasing the SAM/SAH ratio [67]. AMPK also increases the SAM/SAH ratio by transactivating let-7 micro RNA that subsequently degrades H19, relieving a direct inhibitory effect on S-adenosylhomocysteine hydrolase (SAHH). Subsequent activation of SAHH enables DNMT3b activation [68]. Taken together, these studies indicate that AMPK increases the SAM/SAH ratio that has an activating effect on DNMT3a and DNMT3b. These AMPK-dependent shifts in methylation also correlate to metabolic adaptive situations. For example, promoter methylation of *cytochrome C oxidase subunit 4I1* (*COX4I*) and *fatty acid binding protein 3* (*FABP3*) increases during exercise, while *peroxisome proliferator-activated receptor δ* (*PPARδ*) promoter methylation increases after fasting (Figure 6A) [69]. However, the overall effect AMPK has on DNA methylation is likely to be stimulation-, tissue-, promoter-, and DNMT isoform-specific. Paralleling its role as an epigenetic metabolic regulator in histone acetylation through HAT1 and RBBP7, AMPK also phosphorylates DNMT1^{S730}, inhibiting methyl CpG in *PGC-1α*, *Tfam*, *NRF1*, *NRF2*, *UCP2* and *UCP3* promoters [13]. This results in improved mitochondrial function, a requirement for cell survival [13,70].

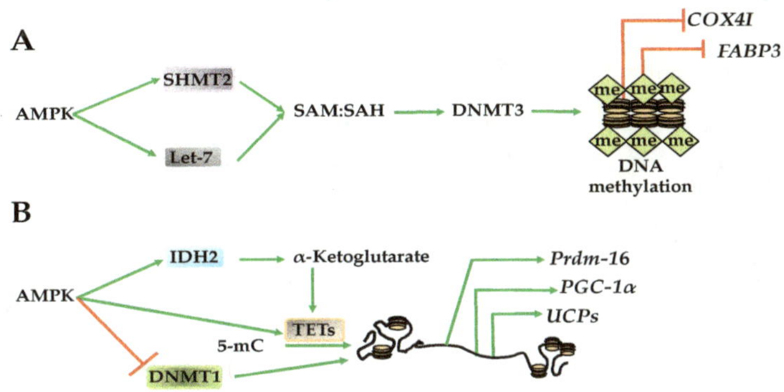

Figure 6. AMPK regulates DNA methylation. (**A**) AMPK activates serine hydroxymethyltransferase 2 (SHMT2) and Let-7 to increase SAM:SAH ratio providing substrate for DNA methyltransferase 3 (DNMT3) methylation of promoters such as *COX4I* and *FABP3*; (**B**) AMPK directly regulates isocitrase dehydrogenase 2 (IDH2) to yield α-ketoglutarates to promote ten-eleven translocation hydroxylases (TETs) activation. AMPK likely also directly regulates TETs catalyzing the conversion of 5-mc to unmethylated cytosine while inhibiting DNA methyltransferase 1 (DNMT1). These actions result in decreased promoter methylation of *PR domain containing 16* (*Prdm16*), *PGC-1α*, and the *uncoupler proteins*.

In addition to the DNMTs, AMPK also regulates TETs, which, in turn, govern loci-specific CpG methylation. The effect of AMPK on TETs occurs through its regulation of TCA cycle intermediates. For example, AMPK regulates isocitrase dehydrogenase 2 (IDH2) to increase the levels of α-ketoglutarate, an activator of TET1–TET4 resulting in CpG demethylation [47,71]. Activation of TET1–TET4 results in transactivation of the PR domain containing 16 (Prdm16) gene in progenitor cells that promotes brown adipogenesis, therefore supporting AMPK's role as an epigenetic regulator of metabolic function (Figure 6B) [47].

4. Approaches to Elucidating the AMPK-Modulated Epigenetic Landscape

Although biochemical studies have resolved a number of mechanistic insights into the role AMPK plays in the epigenetic regulation of chromatin structure, the study of AMPK on epigenetic function is still in its infancy. A number of predictive algorithms and studies have identified a potential role of AMPK on a number of epigenetic regulators. However, the role they may play under different physiological stimulations and their tissue specificity are still unknown. Because of the interconnected nature between regulatory cascades, the study of AMPK on epigenetic function in a disease-relevant capacity requires systems biological and bioinformatics approaches.

Scientific advances in the computational and Big Data arenas have resulted in novel experimental approaches, databases, and bioinformatics techniques that can be used in the exploration of AMPK epigenetic-regulated signaling pathways. Examples of tools used to explore chromatin characteristics include H3^{K27} acetylation H3^{K4} and monomethylation (H3^{K4me1}) immunoprecipitation sequencing (IP-Seq); assay for transposase-accessible chromatin using sequencing (ATAC-seq); assay for transposase-accessible chromatin using sequencing (ATAC-seq); and Hi-C [72]. The relationship between DNA, histone modifications, histone remodeling, followed by Hi-C structural mapping, provides an integrative understanding of how AMPK can influence chromatin architecture. The identification of AMPK-regulated networks can be explored using consensus sequence mapping and machine learning in computing environments such as R/Bioconductor computing [25,73–75]. The expression of effected genes and loci and their application of these loci to disease-relevant stimuli can be further explored via cross-referencing with expression profiles housed in the Gene Expression Omnibus (GEO) database, Sequence Read Archive (SRA), Single Nucleotide Polymorphism database (dbSNP), 3D Genome database C (3DGD), and CR2Cancer [76–79]. Integration of these datasets can provide a comprehensive picture of the influence AMPK has on epigenetic signaling cascades in addition to their genetic loci and disease-specific regulation (Figure 7).

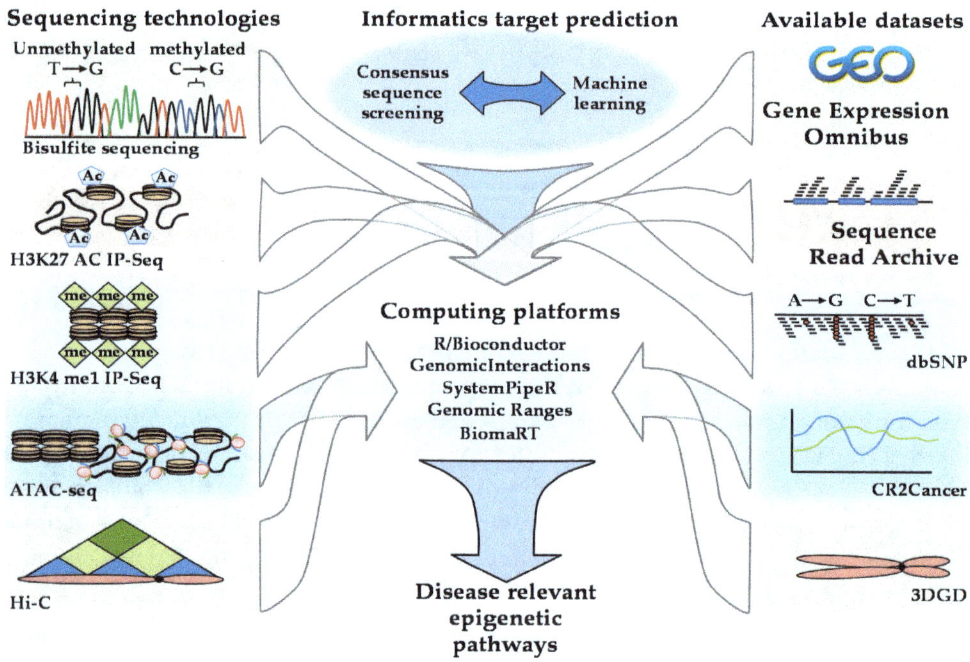

Figure 7. Methods for resolving disease-relevant AMPK-regulated epigenetic networks. Developed sequencing technologies are listed on the left-hand figure panels. Publicly available data repositories are listed on the right. Computing platforms R and relevant Bioconductor packages including Genomic interactions, SystemPipeR, Genomic Ranges, and BioMart can be used to integrate sequencing technologies, publicly available datasets, and AMPK target predictions as illustrated in the center panels.

5. Conclusions

AMPK's role as an epigenetic landscape modulator is underscored by its multifunctional kinase effects that regulate histones and epigenetic enzymes to mediate histone and DNA modifications. AMPK is crucial for cell survival and adaptation as evident by its activation upon nutrient depletion and stressors resulting in histone phosphorylation, acetylation, methylation, *O*-GlcNAcylation, and ribosylation. Additional, AMPK regulates DNA methylation mediated by influencing the SAM/SAH ratio, TET regulation and inhibition of DNMT1. Technological advances in data analysis continue to reveal the remarkable and diverse roles of AMPK in signaling pathways and epigenetic regulation, providing opportunities to exploit novel therapies relating to health and disease.

Acknowledgments: We would like to acknowledge David A. Johnson for his thoughtful review of the manuscript and comment to aid in clarity. We would also like to acknowledge John Shyy for his continued support and guidance.

Abbreviations

AMPK	AMP-activated protein kinase
CPT1C	carnitine palmitoyltransferase 1C
p21	cyclin-dependent kinase inhibitor
p53	TP53 or tumor protein
HAT	histone deacetylases
acetyl-CoA	acetyl coenzyme-A
HDAC	histone deacetylases
AICAR	5-Aminoimidazole-4-carboxamide ribonucleotide
ACC	acetyl-CoA carboxylase
ACSS2	acetyl-CoA synthetase short-chain family member 2
TFEB	transcription factor EB
SIRT	sirtuin
NAD+	nicotinamide adenine dinucleotide
βOHB	β-hydroxybutyrate
RBBP7	retinoblastoma binding protein 7
PGC-1α	peroxisome proliferator-activated receptor gamma coactivator-1α
Tfam	transcription factor A
NRF	nuclear respiratory factors
UCP	uncoupling proteins
SAM	S-adenosyl methionine
GLUT-4	glucose transporter type 4
KDM2A	lysine-specific demethylase 2A
KDM5	demethyltransferase lysine demethylase 5
LSD1	lysine-specific histone demethylase-1
PRC2	polycomb repressive complex 2
COMPASS	complex proteins associated with Set1
EZH2	enhancer of zeste homolog 2
Cdx2	caudal type homeobox2
SLC5A8	solute carrier family 5 member 8
O-GlcNAc	O-linked *N*-acetylglucosamine
OGT	*O*-GlcNAc transferase
UDP-GlcNAc	uridine diphosphate *N*-acetylglucosamine
HBP	hexosamine biosynthesis pathway
ART	mono-ADP ribosyltransferase
PTM	Post translational modification
PARP	poly-(ADP-ribose) polymerase
Bcl-6	B-cell Lymphoma protein 6
CpG	cytosine-guanine dinucleotides
DNMT	DNA methyltransferase
MBD	methyl-CpG binding domain proteins
5-mC	5-methycytosine
5-hmC	5-hydroxymethylcytosine
5-fC	5-formylcytosine
5-caC	5-carboxylcytosine
TETs	ten-eleven translocation hydroxylases
SAH	S-adenosylhomocysteine
SHMT2	serine hydroxymethyltransferase 2
SAHH	S-adenosylhomocysteine hydrolase
COX4I	cytochrome C oxidase subunit 4I1
FABP3	fatty acid binding protein 3
PPARδ	peroxisome proliferator-activated receptor δ
TCA	tricarboxylic acid
IDH2	isocitrase dehydrogenase 2

IP-Seq	immunoprecipitation sequencing
ATAC-seq	assay for transposase-accessible chromatin using sequencing
GEO	gene expression omnibus
SRA	sequence read archive
dbSNP	single nucleotide polymorphism database
3DGD	3D genome database

References

1. Trerotola, M.; Relli, V.; Simeone, P.; Alberti, S. Epigenetic inheritance and the missing heritability. *Hum. Genom.* **2015**, *28*, 9–17. [CrossRef] [PubMed]
2. Bunkar, N.; Pathak, N.; Lohiya, N.K.; Mishra, P.K. Epigenetics: A key paradigm in reproductive health. *Clin. Exp. Reprod. Med.* **2016**, *43*, 59–81. [CrossRef] [PubMed]
3. Boland, M.J.; Nazor, K.L.; Loring, J.F. Epigenetic regulation of pluripotency and differentiation. *Circ. Res.* **2014**, *115*, 311–324. [CrossRef] [PubMed]
4. Barrès, R.; Osler, M.E.; Yan, J.; Rune, A.; Fritz, T.; Caidahl, K.; Krook, A.; Zierath, J.R. Non-CpG methylation of the *PGC-1alpha* promoter through DNMT3B controls mitochondrial density. *Cell Metab.* **2009**, *10*, 189–198. [CrossRef] [PubMed]
5. Barrès, R.; Yan, J.; Egan, B.; Treebak, J.T.; Rasmussen, M.; Fritz, T.; Caidahl, K.; Krook, A.; O'Gorman, D.J.; Zierath, J.R. Acute exercise remodels promoter methylation in human skeletal muscle. *Cell Metab.* **2012**, *15*, 405–411. [CrossRef] [PubMed]
6. Nitert, M.D.; Dayeh, T.; Volkov, P.; Elgzyri, T.; Hall, E.; Nilsson, E.; Yang, B.T.; Lang, S.; Parikh, H.; Wessman, Y.; et al. Impact of an exercise intervention on DNA methylation in skeletal muscle from first-degree relatives of patients with type 2 diabetes. *Diabetes* **2012**, *61*, 3322–3332. [CrossRef] [PubMed]
7. Ferraro, E.; Giammarioli, A.M.; Chiandotto, S.; Spoletini, I.; Rosano, G. Exercise-induced skeletal muscle remodeling and metabolic adaptation: Redox signaling and role of autophagy. *Antioxid. Redox Signal.* **2014**, *21*, 154–176. [CrossRef] [PubMed]
8. Kim, J.; Yang, G.; Kim, Y.; Ha, J. AMPK activators: Mechanisms of action and physiological activities. *Exp. Mol. Med.* **2016**, *48*, e224. [CrossRef] [PubMed]
9. Rabinovitch, R.C.; Samborska, B.; Faubert, B.; Ma, E.H.; Gravel, S.P.; Andrzejewski, S.; Raissi, T.C.; Pause, A.; St-Pierre, J.; Jones, R.G. AMPK Maintains Cellular Metabolic Homeostasis through Regulation of Mitochondrial Reactive Oxygen Species. *Cell Rep.* **2017**, *21*, 1–9. [CrossRef] [PubMed]
10. Ross, F.A.; Jensen, T.E.; Hardie, D.G. Differential regulation by AMP and ADP of AMPK complexes containing different gamma subunit isoforms. *Biochem. J.* **2016**, *473*, 189–199. [CrossRef] [PubMed]
11. Vara-Ciruelos, D.; Dandapani, M.; Gray, A.; Egbani, E.O.; Evans, A.M.; Hardie, D.G. Genotoxic Damage Activates the AMPK-α1 Isoform in the Nucleus viaCa^{2+}/CaMKK2 Signaling to Enhance Tumor Cell Survival. *Mol. Cancer Res.* **2018**, *16*, 345–357. [CrossRef] [PubMed]
12. Langendorf, C.G.; Ngoei, K.R.; Scott, J.W.; Ling, N.X.; Issa, S.M.; Gorman, M.A.; Parker, M.W.; Sakamoto, K.; Oakhill, J.S.; Kemp, B.E. Structural basis of allosteric and synergistic activation of AMPK by furan-2-phosphonic derivative C2 binding. *Nat. Commun.* **2016**, *7*, 10912. [CrossRef] [PubMed]
13. Marin, T.L.; Gongol, B.; Zhang, F.; Martin, M.; Johnson, D.A.; Xiao, H.; Wang, Y.; Subramaniam, S.; Chien, S.; Shyy, J.Y. AMPK promotes mitochondrial biogenesis and function by phosphorylating the epigenetic factors DNMT1, RBBP7, and HAT1. *Sci. Signal.* **2017**, *10*, eaaf7478. [CrossRef] [PubMed]
14. Ratman, D.; Mylka, V.; Bougarne, N.; Pawlak, M.; Caron, S.; Hennuyer, N.; Paumelle, R.; De Cauwer, L.; Thommis, J.; Rider, M.H.; et al. Chromatin recruitment of activated AMPK drives fasting response genes co-controlled by GR and PPARα. *Nucleic Acids Res.* **2016**, *44*, 10539–10553. [CrossRef] [PubMed]
15. Satake, M.; Kawata, M.; McLysaght, A.; Makino, T. Evolution of vertebrate tissues driven by differential modes of gene duplication. *DNA Res.* **2012**, *19*, 305–316. [CrossRef] [PubMed]
16. Ross, F.A.; MacKintosh, C.; Hardie, D.G. AMP-activated protein kinase: A cellular energy sensor that comes in 12 flavours. *FEBS J.* **2016**, *283*, 2987–3001. [CrossRef] [PubMed]
17. Hardie, D.G. Keeping the home fires burning: AMP-activated protein kinase. *J. R. Soc. Interface* **2018**, *15*, 20170774. [CrossRef] [PubMed]

18. Laker, R.C.; Wlodek, M.E.; Connelly, J.J.; Yan, Z. Epigenetic origins of metabolic disease: The impact of the maternal condition to the offspring epigenome and later health consequences. *Food Sci. Hum. Wellness* **2013**, *2*, 1–11. [CrossRef]
19. Zhu, M.J.; Han, B.; Tong, J.; Ma, C.; Kimzey, J.M.; Underwood, K.R.; Xiao, Y.; Hess, B.W.; Ford, S.P.; Nathanielsz, P.W.; et al. AMP-activated protein kinase signaling pathways are down regulated and skeletal muscle development impaired in fetuses of obese, over-nourished sheep. *J. Physiol.* **2008**, *586*, 2651–2664. [CrossRef] [PubMed]
20. Demoinet, E.; Li, S.; Roy, R. AMPK blocks starvation-inducible transgenerational defects in *Caenorhabditis elegans*. *Proc. Natl. Acad. Sci. USA* **2017**, *114*, E2689–E2698. [CrossRef] [PubMed]
21. Bönisch, C.; Hake, S.B. Histone H2A variants in nucleosomes and chromatin: More or less stable? *Nucleic Acids Res.* **2012**, *40*, 10719–10741. [CrossRef] [PubMed]
22. Zlatanova, J.S.; Srebreva, L.N.; Banchev, T.B.; Tasheva, B.T.; Tsanev, R.G. Cytoplasmic pool of histone H1 in mammalian cells. *J. Cell Sci.* **1990**, *96*, 461–468. [PubMed]
23. Mariño-Ramírez, L.; Kann, M.G.; Shoemaker, B.A.; Landsman, D. Histone structure and nucleosome stability. *Expert Rev. Proteom.* **2005**, *2*, 719–729. [CrossRef] [PubMed]
24. Sawicka, A.; Seiser, C. Sensing core histone phosphorylation- a matter of perfect timing. *Biochim. Biophys. Acta* **2014**, *1839*, 711–718. [CrossRef] [PubMed]
25. Marin, T.L.; Gongol, B.; Martin, M.; King, S.J.; Smith, L.; Johnson, D.A.; Subramaniam, S.; Chien, S.; Shyy, J.Y. Identification of AMP-activated protein kinase targets by a consensus sequence search of the proteome. *BMC Syst. Biol.* **2015**, *9*, 13. [CrossRef] [PubMed]
26. Jones, R.G.; Plas, D.R.; Kubek, S.; Buzzai, M.; Mu, J.; Xu, Y.; Birnbaum, M.J.; Thompson, C.B. AMP-activated protein kinase induces a p53-dependent metabolic checkpoint. *Mol. Cell* **2005**, *18*, 283–293. [CrossRef] [PubMed]
27. Bungard, D.; Fuerth, B.J.; Zeng, P.Y.; Faubert, B.; Maas, N.L.; Viollet, B.; Carling, D.; Thompson, C.B.; Jones, R.G.; Berger, S.L. Signaling kinase AMPK activates stress-promoted transcription via histone H2B phosphorylation. *Science* **2010**, *329*, 1201–1205. [CrossRef] [PubMed]
28. Roa-Mansergas, X.; Fadó, R.; Atari, M.; Mir, J.F.; Muley, H.; Serra, D.; Casals, N. CPT1C promotes human mesenchymal stem cells survival under glucose deprivation through the modulation of autophagy. *Sci. Rep.* **2018**, *8*, 6997. [CrossRef] [PubMed]
29. Lee, I.H.; Kawai, Y.; Fergusson, M.M.; Rovira, I.I.; Bishop, A.J.; Motoyama, N.; Cao, L.; Finkel, T. Atg7 modulates p53 activity to regulate cell cycle and survival during metabolic stress. *Science* **2012**, *336*, 225–228. [CrossRef] [PubMed]
30. Itahana, Y.; Itahana, K. Emerging Roles of p53 Family Members in Glucose Metabolism. *Int. J. Mol. Sci.* **2018**, *19*, 776. [CrossRef] [PubMed]
31. Stauffer, D.R.; Howard, T.L.; Nyun, T.; Hollenberg, S.M. CHMP1 is a novel nuclear matrix protein affecting chromatin structure and cell-cycle progression. *J. Cell Sci.* **2001**, *114*, 2383–2393. [PubMed]
32. Javaid, N.; Choi, S. Acetylation- and Methylation-Related Epigenetic Proteins in the Context of Their Targets. *Genes* **2017**, *8*, 196. [CrossRef] [PubMed]
33. Marmorstein, R.; Zhou, M.M. Writers and readers of histone acetylation: Structure, mechanism, and inhibition. *Cold Spring Harb. Perspect. Biol.* **2014**, *6*, a018762. [CrossRef] [PubMed]
34. Bulusu, V.; Tumanov, S.; Michalopoulou, E.; van den Broek, N.J.; MacKay, G.; Nixon, C.; Dhayade, S.; Schug, Z.T.; Vande Voorde, J.; Blyth, K.; et al. Acetate recapturing by nuclear acetyl-CoA synthetase 2 prevents loss of histone acetylation during oxygen and serum limitation. *Cell Rep.* **2017**, *18*, 647–658. [CrossRef] [PubMed]
35. Li, X.; Yu, W.; Qian, X.; Xia, Y.; Zheng, Y.; Lee, J.H.; Li, W.; Lyu, J.; Rao, G.; Zhang, X.; et al. Nucleus-translocated ACSS2 promotes gene transcription for lysosomal biogenesis and autophagy. *Mol. Cell* **2017**, *66*, 684–697. [CrossRef] [PubMed]
36. Galdieri, L.; Gatla, H.; Vancurova, I.; Vancura, A. Activation of AMP-activated Protein Kinase by Metformin Induces Protein Acetylation in Prostate and Ovarian Cancer Cells. *J. Biol. Chem.* **2016**, *291*, 25154–25166. [CrossRef] [PubMed]
37. Cuyàs, E.; Fernández-Arroyo, S.; Joven, J.; Menendez, J.A. Metformin targets histone acetylation in cancer prone epithelial cells. *Cell Cycle.* **2016**, *15*, 3355–3361. [CrossRef] [PubMed]
38. Ruderman, N.B.; Xu, X.J.; Nelson, L.; Cacicedo, J.M.; Saha, A.K.; Lan, F.; Ido, Y. AMPK and SIRT1: A long-standing partnership? *Am. J. Physiol. Endocrinol. Metab.* **2010**, *298*, E751–E760. [CrossRef] [PubMed]

39. Salminen, A.; Kauppinen, A.; Kaarniranta, K. AMPK/Snf1 signaling regulates histone acetylation: Impact on gene expression and epigenetic functions. *Cell. Signal.* **2016**, *28*, 887–895. [CrossRef] [PubMed]
40. Zhang, T.; Kraus, W.L. SIRT1-dependent regulation of chromatin and transcription: Linking NAD(+) metabolism and signaling to the control of cellular functions. *Biochim. Biophys. Acta* **2010**, *1804*, 1666–1675. [CrossRef] [PubMed]
41. Shimazu, T.; Hirschey, M.D.; Newman, J.; He, W.; Shirakawa, K.; Le Moan, N.; Grueter, C.A.; Lim, H.; Saunders, L.R.; Stevens, R.D.; et al. Suppression of oxidative stress by β-hydroxybutyrate, an endogenous histone deacetylase inhibitor. *Science* **2013**, *339*, 211–214. [CrossRef] [PubMed]
42. Mihaylova, M.M.; Vasquez, D.S.; Ravnskjaer, K.; Denechaud, P.D.; Yu, R.T.; Alvarez, J.G.; Downes, M.; Evans, R.M.; Montminy, M.; Shaw, R.J. Class IIa histone deacetylases are hormone-activated regulators of FOXO and mammalian glucose homeostasis. *Cell* **2011**, *145*, 607–621. [CrossRef] [PubMed]
43. McGee, S.L.; van Denderen, B.J.; Howlett, K.F.; Mollica, J.; Schertzer, J.D.; Kemp, B.E.; Hargreaves, M. AMP-activated protein kinase regulates *GLUT4* transcription by phosphorylating histone deacetylase 5. *Diabetes* **2008**, *57*, 860–867. [CrossRef] [PubMed]
44. Ng, S.; Yue, W.; Oppermann, U.; Klose, R. Dynamic protein methylation in chromatin biology. *Cell. Mol. Life Sci.* **2009**, *66*, 407–422. [CrossRef] [PubMed]
45. Feitag, M. Histone Methylation by SET Domain Proteins in Fungi. *Annu. Rev. Microbiol.* **2017**, *71*, 413–439. [CrossRef] [PubMed]
46. Wei, S.; Li, C.; Yin, Z.; Wen, J.; Meng, H.; Xue, L.; Wang, J. Histone methylation in DNA repair and clinical practice: New findings during the past 5-years. *J. Cancer* **2018**, *9*, 2072–2081. [CrossRef] [PubMed]
47. Yang, Q.; Liang, X.; Sun, X.; Zhang, L.; Fu, X.; Rogers, C.; Berim, A.; Zhang, S.; Wang, S.; Wang, B.; et al. AMPK/α-ketoglutarate axis dynamically mediates DNA demethylation in the *Prdm16* promoter and brown adipogenesis. *Cell Metab.* **2016**, *24*, 542–554. [CrossRef] [PubMed]
48. Wang, T.; Yu, Q.; Li, J.; Hu, B.; Zhao, Q.; Ma, C.; Huang, W.; Zhuo, L.; Fang, H.; Liao, L.; et al. *O*-GlcNAcylation of fumarase maintains tumour growth under glucose deficiency. *Nat. Cell Biol.* **2017**, *19*, 833–843. [CrossRef] [PubMed]
49. Eissenberg, J.C.; Shilatifard, A. Histone H3 lysine 4 (H3K4) methylation in development and differentiation. *Dev. Biol.* **2010**, *339*, 240–249. [CrossRef] [PubMed]
50. Wan, L.; Xu, K.; Wei, Y.; Zhang, J.; Han, T.; Fry, C.; Zhang, Z.; Wang, Y.V.; Huang, L.; Yuan, M.; et al. Phosphorylation of EZH2 by AMPK Suppresses PRC2 Methyltransferase Activity and Oncogenic Function. *Mol. Cell* **2018**, *69*, 279–291. [CrossRef] [PubMed]
51. Tang, G.; Guo, J.; Zhu, Y.; Huang, Z.; Liu, T.; Cai, J.; Yu, L.; Wang, Z. Metformin inhibits ovarian cancer via decreasing H3K27 trimethylation. *Int. J. Oncol.* **2018**, *52*, 1899–1911. [CrossRef] [PubMed]
52. Aloia, L.; Di Stefano, B.; Di Croce, L. Polycomb complexes in stem cells and embryonic development. *Development* **2013**, *140*, 2525–2534. [CrossRef] [PubMed]
53. Sun, X.; Yang, Q.; Rogers, C.J.; Du, M.; Zhu, M.J. AMPK improves gut epithelial differentiation and barrier function via regulating Cdx2 expression. *Cell Death Differ.* **2017**, *24*, 819–831. [CrossRef] [PubMed]
54. Kakizaki, F.; Aoki, K.; Miyoshi, H.; Carrasco, N.; Aoki, M.; Taketo, M.M. CDX transcription factors positively regulate expression of solute carrier family 5, member 8 in the colonic epithelium. *Gastroenterology* **2010**, *138*, 627–635. [CrossRef] [PubMed]
55. Gurav, A.; Sivaprakasam, S.; Bhutia, Y.D.; Boettger, T.; Singh, N.; Ganapathy, V. Slc5a8, a Na^+-coupled high-affinity transporter for short-chain fatty acids, is a conditional tumour suppressor in colon that protects against colitis and colon cancer under low-fibre dietary conditions. *Biochem. J.* **2015**, *469*, 267–278. [CrossRef] [PubMed]
56. Yang, X.; Qian, K. Protein *O*-GlcNAcylation: Emerging mechanisms and functions. *Nat. Rev. Mol. Cell Biol.* **2017**, *18*, 452–465. [CrossRef] [PubMed]
57. Dehennaut, V.; Leprince, D.; Lefebvre, T. *O*-GlcNAcylation, an Epigenetic Mark. Focus on the Histone Code, TETFamily Proteins, and Polycomb Group Proteins. *Front. Endocrinol.* **2014**, *5*, 155. [CrossRef] [PubMed]
58. Poirier, G.G.; de Murcia, G.; Jongstra-Bilen, J.; Niedergang, C.; Mandel, P. Poly(ADP-ribosyl)ation of polynucleosomes causes relaxation of chromatin structure. *Proc. Natl. Acad. Sci. USA* **1982**, *79*, 3423–3427. [CrossRef] [PubMed]
59. Ciccarone, F.; Zampieri, M.; Caiafa, P. PARP1 orchestrates epigenetic events setting up chromatin domains. *Semin. Cell Dev. Biol.* **2017**, *63*, 123–134. [CrossRef] [PubMed]

60. Walker, J.W.; Jijon, H.B.; Madsen, K.L. AMP-activated protein kinase is a positive regulator of poly(ADP-ribose) polymerase. *Biochem. Biophys. Res. Commun.* **2006**, *342*, 336–341. [CrossRef] [PubMed]
61. Gongol, B.; Marin, T.; Peng, I.C.; Woo, B.; Martin, M.; King, S.; Sun, W.; Johnson, D.A.; Chien, S.; Shyy, J.Y. AMPKα2 exerts its anti-inflammatory effects through PARP-1 and Bcl-6. *Proc. Natl. Acad. Sci. USA* **2013**, *110*, 3161–3166. [CrossRef] [PubMed]
62. Lodhi, N.; Kossenkov, A.V.; Tulin, A.V. Bookmarking promoters in mitotic chromatin: Poly (ADP-ribose) polymerase-1 as an epigenetic mark. *Nucleic Acids Res.* **2014**, *42*, 7028–7038. [CrossRef] [PubMed]
63. Dunn, J.; Qiu, H.; Kim, S.; Jjingo, D.; Hoffman, R.; Kim, C.W.; Jang, I.; Son, D.J.; Kim, D.; Pan, C.; et al. Flow-dependent epigenetic DNA methylation regulates endothelial gene expression and atherosclerosis. *J. Clin. Investig.* **2014**, *124*, 3187–3199. [CrossRef] [PubMed]
64. Hill, P.W.; Amouroux, R.; Hajkova, P. DNA demethylation, Tet proteins and 5-hydroxymethylcytosine in epigenetic reprogramming: An emerging complex story. *Genomics* **2014**, *104*, 324–333. [CrossRef] [PubMed]
65. Ficz, G.; Branco, M.R.; Seisenberger, S.; Santos, F.; Krueger, F.; Hore, T.A.; Marques, C.J.; Andrews, S.; Reik, W. Dynamic regulation of 5-hydroxymethylcytosine in mouse ES cells and during differentiation. *Nature* **2011**, *473*, 398–402. [CrossRef] [PubMed]
66. Chen, Z.X.; Riggs, A.D. DNA methylation and demethylation in mammals. *J. Biol. Chem.* **2011**, *286*, 18347–18353. [CrossRef] [PubMed]
67. Cuyàs, E.; Fernández-Arroyo, S.; Verdura, S.; García, R.Á.; Stursa, J.; Werner, L.; Blanco-González, E.; Montes-Bayón, M.; Joven, J.; Viollet, B.; et al. Metformin regulates global DNA methylation via mitochondrial one-carbon metabolism. *Oncogene* **2018**, *37*, 963–970. [CrossRef] [PubMed]
68. Zhong, T.; Men, Y.; Lu, L.; Geng, T.; Zhou, J.; Mitsuhashi, A.; Shozu, M.; Maihle, N.J.; Carmichael, G.G.; Taylor, H.S.; et al. Metformin alters DNA methylation genome-wide via the H19/SAHH axis. *Oncogene* **2017**, *36*, 2345–2354. [CrossRef] [PubMed]
69. Lane, S.C.; Camera, D.M.; Lassiter, D.G.; Areta, J.L.; Bird, S.R.; Yeo, W.K.; Jeacocke, N.A.; Krook, A.; Zierath, J.R.; Burke, L.M.; et al. Effects of sleeping with reduced carbohydrate availability on acute training responses. *J. Appl. Physiol.* **2015**, *119*, 643–655. [CrossRef] [PubMed]
70. Snyder, C.M.; Chandel, N.S. Mitochondrial regulation of cell survival and death during low-oxygen conditions. *Antioxid. Redox Signal.* **2009**, *11*, 2673–2683. [CrossRef] [PubMed]
71. Wu, H.; Zhang, Y. Reversing DNA methylation: Mechanisms, genomics, and biological functions. *Cell* **2014**, *156*, 45–68. [CrossRef] [PubMed]
72. Jiang, S.; Mortazavi, A. Integrating ChIP-seq with other functional genomics data. *Brief. Funct. Genom.* **2018**, *17*, 104–115. [CrossRef] [PubMed]
73. Xue, L.; Tao, W.A. Current technologies to identify protein kinase substrates in high throughput. *Front. Biol.* **2013**, *8*, 216–227. [CrossRef] [PubMed]
74. Harmston, N.; Ing-Simmons, E.; Perry, M.; Barešić, A.; Lenhard, B. GenomicInteractions: An R/Bioconductor package for manipulating and investigating chromatin interaction data. *BMC Genom.* **2015**, *16*, 963. [CrossRef] [PubMed]
75. Backman, T.W.H.; Girke, T. systemPipeR: NGS workflow and report generation environment. *BMC Bioinform.* **2016**, *20*, 388. [CrossRef] [PubMed]
76. Deininger, P.; Morales, M.E.; White, T.B.; Baddoo, M.; Hedges, D.J.; Servant, G.; Srivastav, S.; Smither, M.E.; Concha, M.; DeHaro, D.L.; et al. A comprehensive approach to expressionof L1 loci. *Nucleic Acids Res.* **2017**, *45*, e31. [CrossRef] [PubMed]
77. Bhagwat, M. Searching NCBI's dbSNP database. *Curr. Protoc. Bioinform.* **2010**. [CrossRef]
78. Li, C.; Dong, X.; Fan, H.; Wang, C.; Ding, G.; Li, Y. The 3DGD: A database of genome 3D structure. *Bioinformatics* **2014**, *30*, 1640–1642. [CrossRef] [PubMed]
79. Ru, B.; Sun, J.; Tong, Y.; Wong, C.N.; Chandra, A.; Tang, A.T.S.; Chow, L.K.Y.; Wun, W.L.; Levitskaya, Z.; Zhang, J. CR2Cancer: A database for chromatin regulators in human cancer. *Nucleic Acids Res.* **2018**, *46*, D918–D924. [CrossRef] [PubMed]

6

Implication and Regulation of AMPK during Physiological and Pathological Myeloid Differentiation

Arnaud Jacquel [1,2,†], Frederic Luciano [1,2,†], Guillaume Robert [1,2,†] and Patrick Auberger [1,2,*,†]

1 Université Côte d'Azur, C3M Inserm U1065, 06204 Nice, France; jacquel@unice.fr (A.J.); fluciano@unice.fr (F.L.); robertg@unice.fr (G.R.)
2 Equipe Labellisée par la Fondation ARC, 94803 Villejuif, France
* Correspondence: auberger@unice.fr
† These authors contributed equally to this work.

Abstract: AMP-activated protein kinase (AMPK) is a heterotrimeric serine/threonine kinase consisting of the arrangement of various α β, and γ isoforms that are expressed differently depending on the tissue or the cell lineage. AMPK is one of the major sensors of energy status in mammalian cells and as such plays essential roles in the regulation of cellular homeostasis, metabolism, cell growth, differentiation, apoptosis, and autophagy. AMPK is activated by two upstream kinases, the tumor suppressor liver kinase B1 (LKB1) and the calcium/calmodulin-dependent protein kinase kinase 2 (CAMKK2) through phosphorylation of the kinase on Thr172, leading to its activation. In addition, AMPK inhibits the mTOR pathway through phosphorylation and activation of tuberous sclerosis protein 2 (TSC2) and causes direct activation of unc-51-like autophagy activating kinase 1 (ULK1) via phosphorylation of Ser555, thus promoting initiation of autophagy. Although it is well established that AMPK can control the differentiation of different cell lineages, including hematopoietic stem cells (HSCs), progenitors, and mature hematopoietic cells, the role of AMPK regarding myeloid cell differentiation is less documented. The differentiation of monocytes into macrophages triggered by colony stimulating factor 1 (CSF-1), a process during which both caspase activation (independently of apoptosis induction) and AMPK-dependent stimulation of autophagy are necessary, is one noticeable example of the involvement of AMPK in the physiological differentiation of myeloid cells. The present review focuses on the role of AMPK in the regulation of the physiological and pathological differentiation of myeloid cells. The mechanisms of autophagy induction by AMPK will also be addressed, as autophagy has been shown to be important for differentiation of hematopoietic cells. In addition, myeloid malignancies (myeloid leukemia or dysplasia) are characterized by profound defects in the establishment of proper differentiation programs. Reinduction of a normal differentiation process in myeloid malignancies has thus emerged as a valuable and promising therapeutic strategy. As AMPK seems to exert a key role in the differentiation of myeloid cells, notably through induction of autophagy, we will also discuss the potential to target this pathway as a pro-differentiating and anti-leukemic strategy in myeloid malignancies.

Keywords: AMPK; monocytes; macrophages; differentiation; autophagy; AML; MDS; CML; CMML

1. Introduction

AMPK is an essential regulator and sensor of cellular energy status in mammalian cells. This kinase integrates changes in the AMP/ATP and ADP/ATP ratios, fine-tuning the balance between

ATP consumption and synthesis [1,2]. AMPK acts to increase the rate of catabolic process and to decrease the rate of anabolic process to sustain intracellular energy homeostasis [3]. AMPK is activated by a wide range of stimuli, such as nutrient deprivation [4], cellular stresses [5], fasting or caloric restriction [6], caloric restriction mimetics [7–10], and nucleoside analogues such as (5-aminoimidazole-4-carboxamide-1-β-D-ribofuranoside; AICAR) [11], and has been evaluated in clinical trials for the treatment of metabolic diseases and cancers, including both hematopoietic malignancies and solid tumors [12–15]. In addition to its well-established role in maintaining energy homeostasis, AMPK is also extensively involved in promoting autophagy [16–18], longevity, and tumor suppression [19–22].

AMPK is a highly conserved heterotrimeric serine/threonine kinase that phosphorylates a plethora of cellular substrates implicated in numerous metabolic pathways through rapid or long-term adaptive responses to different metabolites [23]. The first thoroughly characterized AMPK substrate was acetyl-CoA-carboxylase, which catalyzes the carboxylation of acetyl-CoA to produce malonyl-CoA, highlighting an essential role for this kinase in the induction of fatty acid oxydation [24]. More generally, AMPK intervenes at the crossroad of different key cellular signaling pathways, monitoring cellular energy status, cell proliferation, differentiation and autophagy, notably through its negative control of the PI3K/AKT/mTOR pathway and its stimulatory effect on autophagy through direct phosphorylation of ULK1 (ATG1, a serine threonine kinase involved in the initiation of this catabolic process) [25–27]. AMPK negatively regulates energy-consuming cell proliferation/growth, whereas it induces an energy compensating program, such as autophagy. It is well established that AKT signaling favors glucose uptake and glycolysis, therefore increasing ATP production and reducing AMPK activation. To that end, AKT directly phosphorylates AMPK on ser487, hindering the activation of AMPK by LKB1 [28]. AKT signaling through mTORC1 stimulates ATP-consuming anabolic processes, whereas AMPK blocks anabolic metabolism to favor catabolic processes. In addition, AKT and AMPK exert opposing effects on mTORC1 signaling as well as nutrient and glycogen synthesis. In addition to regulating cell proliferation, AKT, through mTORC1, inhibits autophagy. Indeed, ULK1 activity is repressed by mTORC1 complex leading to autophagy inhibition. By contrast, phosphorylation of TSC2 by AMPK increases its activity, leading to mTORC1 inhibition and indirectly to autophagy induction. Finally, AMP directly phosphorylates ULK1 on ser555 to promote induction of autophagy. In conclusion, cross-talk between these two kinases is critical to switch from the catabolic to the anabolic states (and vice versa) and to balance autophagy [29].

Recent findings also indicate that AMPK can sense glucose availability independently of adenine nucleotide binding through the formation of a complex with axin at the lysosomal membrane [30,31]. Finally, in addition to its essential role in the regulation of cellular metabolism and autophagy, AMPK also plays a critical role in the regulation of anti-inflammatory responses [32,33] and cell growth and differentiation [4,16,34]. In this context, modulation of the AMPK pathway has emerged as a promising strategy in a wide range of human pathologies, including metabolic and inflammatory diseases and cancer.

There is increasing evidence that AMPK is also required for many types of cellular differentiation processes by providing through different mechanisms (increased metabolism and activation of autophagy, for instance) the fuel and nutrients (glucose, amino acids, lipids, and sugars) that sustain the modification of the transcription programs necessary for the completion of cellular differentiation. The present review emphasizes the role of AMPK in the differentiation of hematopoietic cells, with a special focus on the physiological and pathological differentiation of myeloid cells. The opportunity to use recently developed AMPK activators for the treatment of myeloid malignancies, including myelodysplastic syndrome (MDS), acute myeloid leukemia (AML), chronic myelogenous leukemia (CML), and chronic myelomonocytic leukemia (CMML) will also be addressed.

2. AMPK Structure and Activation

The structure and activation modalities of AMPK have been excellently reviewed recently [34,35]. Briefly, the AMPK catalytic subunits ($\alpha 1$ and $\alpha 2$) are encoded by two different genes called PRKAA1 and PRKAA2. The α subunits consist, from their N- to C-termini, of N- and C-lobes, which together form the kinase domain (α-KD) typical of serine/threonine kinases, an auto-inhibitory domain (α-AID), an α-linker domain, and finally a globular C-terminal domain (α-CTD). The scaffolding beta subunits ($\beta 1$ or $\beta 2$) are encoded by two different genes (PRKAB1 and PRKAB2). They display a myristoylated N-terminal region involved in the recruitment of AMPK to the membrane of mitochondria or phagophores, followed by a central carbohydrate-binding module domain (β-CBM) and a C-terminal subunit interaction domain (β-SID). The γ regulatory subunits ($\gamma 1$, $\gamma 2$, or $\gamma 3$) are encoded by three different genes: PRKAG1, PRKAG2, and PRKAG3.

The γ subunits are characterized by an N-terminal region of variable length and the presence of four tandem cystathionine-beta-synthase repeats (CBS1 to CBS4), that are involved in the binding of the regulatory nucleotides AMP, ADP, and ATP (Figure 1). The α, β, and γ AMPK subunits assemble into different heterotrimeric serine/threonine kinases depending on their tissue distribution and on the representativeness of the various isoforms (Figure 1). In myeloid cells, the $\alpha 1\beta 1\gamma 1$ complex is predominantly expressed. Of note, AMPK complexes composed of different gamma subunit isoforms ($\gamma 1$ to $\gamma 3$) exhibit exquisite variations in their sensitivity to increased AMP and ADP levels that enable subtle variations in the regulation of the kinase activity.

This suggests that AMPK could respond differently and selectively to changes in adenine nucleotide depending on the nature of the γ subunit present in the complex [36]. AMPK is activated by phosphorylation at Thr172 in the C-lobe domain, an event that is achieved by one of the three main upstream kinases, the tumor suppressor LKB1, CaMKK2, or TAK1 [37]. Other regulatory sites, such as Ser487 in the $\alpha 1$ subunit and Ser491 in the $\alpha 2$ subunit, that are phosphorylated by either protein kinase A, protein kinase B (AKT), or ribosomal S6 kinase are implicated in the negative regulation of AMPK and thus favor activation of the mTOR pathway. Once activated, AMPK phosphorylates a plethora of protein substrates that are involved in the regulation of cell energetic metabolism, autophagy, modulation of proliferation, regulation of inflammatory responses, and differentiation [38]. Thus far, more than 50 AMPK substrates have been identified in mammalian cells that are involved in energy homeostasis but also in other functions not directly linked to metabolism regulation [39]. However, the phosphorylome of AMPK is far from completely known, as recently attested by the identification of TET2 (Tet-eleven translocation 2) phosphorylation by AMPK linking diabetes to cancer [40].

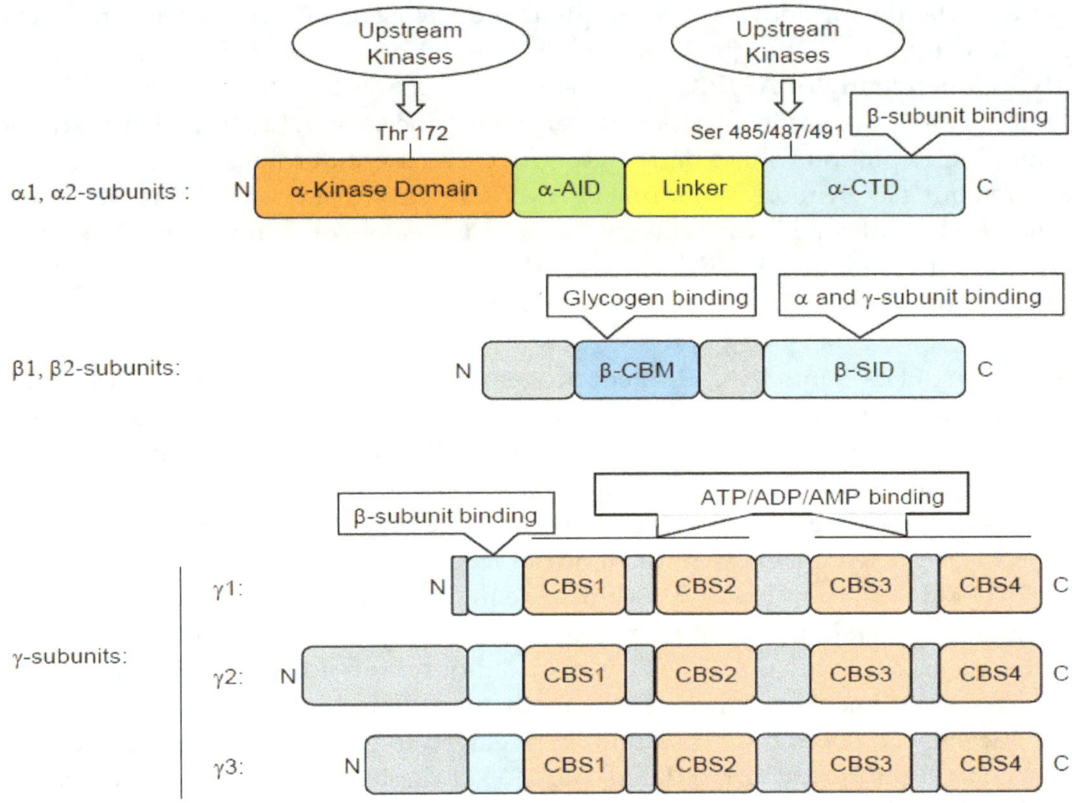

Figure 1. Structure of mammalian AMPK subunits. AMPK is a heterotrimeric protein consisting of 1 catalytic subunit (α subunit) and 2 regulatory subunits (β and γ subunits). The α subunit contains a kinase domain (α-KD), the activity of which relies on the phosphorylation of Thr172 by upstream AMPK kinases. The kinase domain is followed by an autoinhibitory domain (α-AID) that is joined to the COOH-terminal domains (α-CTD) by a less well conserved linker. The α-CTD domain binds to the C-terminal domain of the β subunit. The β-subunit contains two conserved regions: (1) a carbohydrate-binding module (β-CBM) that causes the mammalian complex to bind to glycogen particles and (2) a COOH- terminal subunit interaction domain (β-SID) that provides the bridge between the α- and γ-subunits. The γ-subunit contains variable NH2-terminal regions followed by a short sequence involved in binding to the β-subunit and by four tandem repeats of a cystathionine-β-synthase (CBS) motif that act in pairs to form the binding sites for adenine nucleotides (ATP, ADP and AMP).

3. AMPK and Differentiation of Hematopoietic Stem Cells

Hematopoietic stem cells (HSCs) are multipotent, self-renewing progenitor cells that can replenish all blood cell types in a process called "hematopoiesis." HSCs differentiate into two lineage-restricted, lymphoid and myelo-erythroid, oligopotent progenitor cells that will ultimately give rise to lymphocytes, granulocytes, macrophages, erythrocytes, and platelets. An alternative, "myeloid-based" model for blood lineage differentiation also implies a common myelo-lymphoid progenitor cell that generates progeny from both lineages. The mechanisms controlling HSC self-renewal and differentiation are influenced by a diverse set of cytokines, chemokines, receptors, and intracellular signaling molecules. Producing 500 billion cells a day from a limited number of HSCs is a highly energy-consuming process. How hematopoietic stem cells (HSCs) accommodate their energy requirement during growth and differentiation stages remains poorly understood. Recent studies in the literature have established that the LKB1 tumor suppressor, one of the AMPK upstream kinases, is critical for the maintenance of energy homeostasis in HSCs. In different studies, *Lkb1* invalidation in mice has been shown to cause loss of HSC quiescence, division, rapid depletion, and pancytopenia [41]. In addition, *Lkb1*-deficient bone marrow cells display drastic mitochondrial defects, defaults in lipid and nucleotide metabolism, and, as a consequence, decreased ATP consumption. Nevertheless, AMPK seems to exert only a marginal role in HSC depletion. While HSC exhaustion appears to occur largely independently of AMPK, the defects in mitochondrial function are also observed in *Ampk*-deficient mice [41], suggesting that AMPK can act in concert with LKB1 to regulate mitochondrial fitness but not HSC depletion. These data highlighted an essential role for LKB1 as an inhibitor of HSC proliferation but a likely more minor role of AMPK and mTOR in this process [42]. As mentioned previously, the LKB1/AMPK pathway is one of the main activators of autophagy [25,43]. It is therefore possible that the changes in mitochondrial mass and decreased ATP levels observed in the absence of LKB1 and AMPK could reflect alteration in the rates of autophagy in deficient mice.

Differentiation of HSCs into their different progeny, including myeloid precursors, requires high energy status levels. Recent studies have highlighted the crucial role of mitochondrial oxidative phosphorylation for the differentiation of HSCs. Indeed, specific invalidation of the PTEN-like mitochondrial phosphatase PTPMT-1 in HSCs resulted in an increase of the HSC pool and a block in differentiation [44]. Importantly, reintroduction of catalytically active PTPMT1, but not catalytically inactive PTPMT1 or PTPMP1 lacking mitochondrial localization signal, restores the differentiation capabilities of PTPMT1-deficient HSCs. This blockade in differentiation potential of HSCs is linked to altered mitochondrial metabolism in the absence of PTPMT1. While PTPMT1 seems to be essential for HSC commitment, depletion of PTPMT1 in myeloid or lymphoid progenitors failed to cause any defect in lineage-specific knock-out mice. It is likely that the altered mitochondrial metabolism caused by *PTPMT1* deficiency is sensed by AMPK and then relayed to the p53-p21/p57 pathway, as p21 and p57 expression was significantly upregulated in PTPMT1 depleted stem cells and early progenitors. AMPK on its own could exert its effect on mitochondria through phosphorylation and activation of PGC-1α and β, which are master transcriptional regulators of mitochondrial biogenesis [45].

4. AMPK and Physiological Monocyte Differentiation

Beyond their originally described role as conveyors of programmed cell death and inflammation, caspases are involved in a number of other cellular functions, the most prominent being cell differentiation, a conserved property across a plethora of cell types in divergent metazoan organisms (Figure 2). Caspases mediate DNA damage and morphological changes that are common to different cell fates. The specificity of their action may be controlled by a diversity of mechanisms, including protein–protein interactions, post-translational modifications, subcellular localization, and interaction with other fundamental cell processes such as autophagy and proteostasis. Macrophages, which are mainly derived from monocytes, are essential components of mammal tissue homeostasis. Some are seeded into tissues before birth, while others are continually replenished from blood monocytes. Monocytes are circulating blood leukocytes [46] that migrate into tissues where they differentiate into morphological and functionally heterogeneous cells, including macrophages, myeloid dendritic cells, and osteoclasts [47]. Their differentiation can be recapitulated ex vivo by incubation with cytokines, e.g., they differentiate into macrophages upon exposure to colony stimulating factor-1 (CSF-1) [48]. The biologic effects of CSF-1 are typically mediated by plasma membrane associated CSF-1R [49]. Downstream signaling pathways include PI3K-AKT and AMPK, which mediate caspase activation and autophagy, respectively [50,51]. We have depicted these two pathways in more detail. Monocyte differentiation triggered by CSF-1 receptor (CSF-1R) engagement is critically dependent on the oscillatory activation of the kinase AKT, leading within 2–3 days to the formation of a multi-molecular platform. This molecular complex includes the adaptor molecule Fas-associated death domain (FADD), the serine-threonine kinase receptor-interaction protein kinase1 (RIP1), the long and short isoforms of flice inhibitory protein (FLIP), and procaspase-8 [52,53]. In turn, active caspase-8 provokes a spatially restricted activation of caspase-3 and -7 that cleaves selected intracellular proteins to generate a resting macrophage phenotype. This proteolytic machinery is inhibited as soon as macrophages are activated with lipopolysaccharides, a condition that also impairs autophagy induction. Engaged CSF-1R also promotes autophagy [50] through increasing the expression of the purinergic receptor P2RY6 that activates the CAMKK2-AMPK-ULK1 pathway [51]. Caspase activation and autophagy induction are both required for proper generation of macrophages upon CSF-1 stimulation (Figure 3 provides a schematic view of signaling network involved). The molecular links between caspase activation and autophagy require further investigation. One possible hypothesis is that post-translational modifications of caspases and differentiation-specific cleavage sites in cellular protein targets may be instrumental in connecting caspases to autophagy in cells undergoing differentiation. Thus, the CAMKK2/AMPK/ULK1 axis appears essential for proper differentiation of monocytes into macrophages.

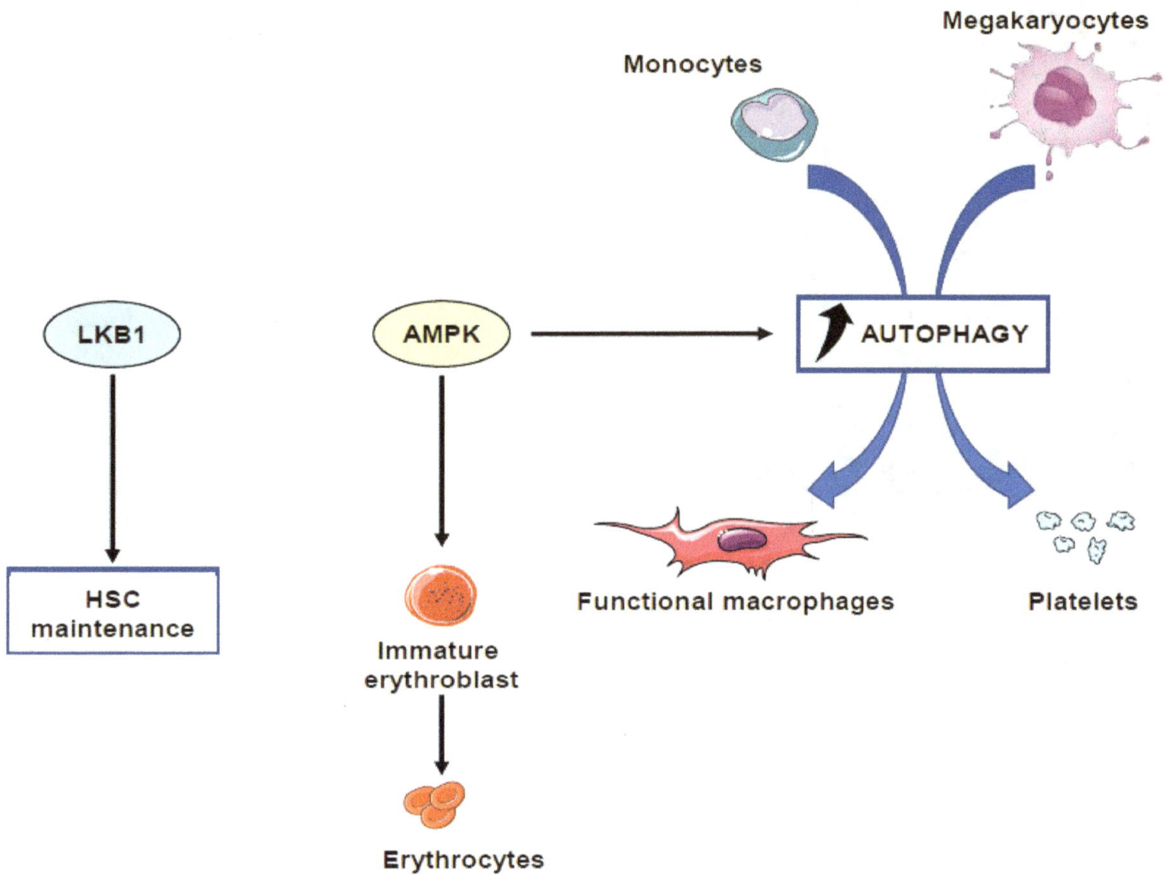

Figure 2. Role of LKB1 and AMPK in HSC maintenance and hematopoietic cell differentiation. There is genetic evidence that LKB1 is required for HSC maintenance since *Lkb1* depletion in mice results in loss of HSC quiescence, division and rapid depletion that contributes to pancytopenia. AMPK for its own is required for the early steps of erythroid differentiation, the production of functional macrophages from monocytes and the differentiation of HSCs into megakaryocytes and ultimately functional platelets.

Of note, in freshly isolated monocytes there is limited if any expression of AMPK subunit proteins. During CSF-1-mediated differentiation of monocytes into macrophages, there is a time-dependent increase in AMPKα1 expression and phosphorylation on Thr172 and a concomitant increase in ULK1 level and phosphorylation on Ser555 [51]. Increase in AMPKα1 and ULK1 expression at the protein but not the mRNA level, by a still unknown mechanism, does not rely on inhibition of the proteasome activity. This increase in AMPKα1 protein expression is further accompanied by elevated phosphorylation and activation of AMPK and ULK1 and correlates with the induction of autophagy and myeloid differentiation.

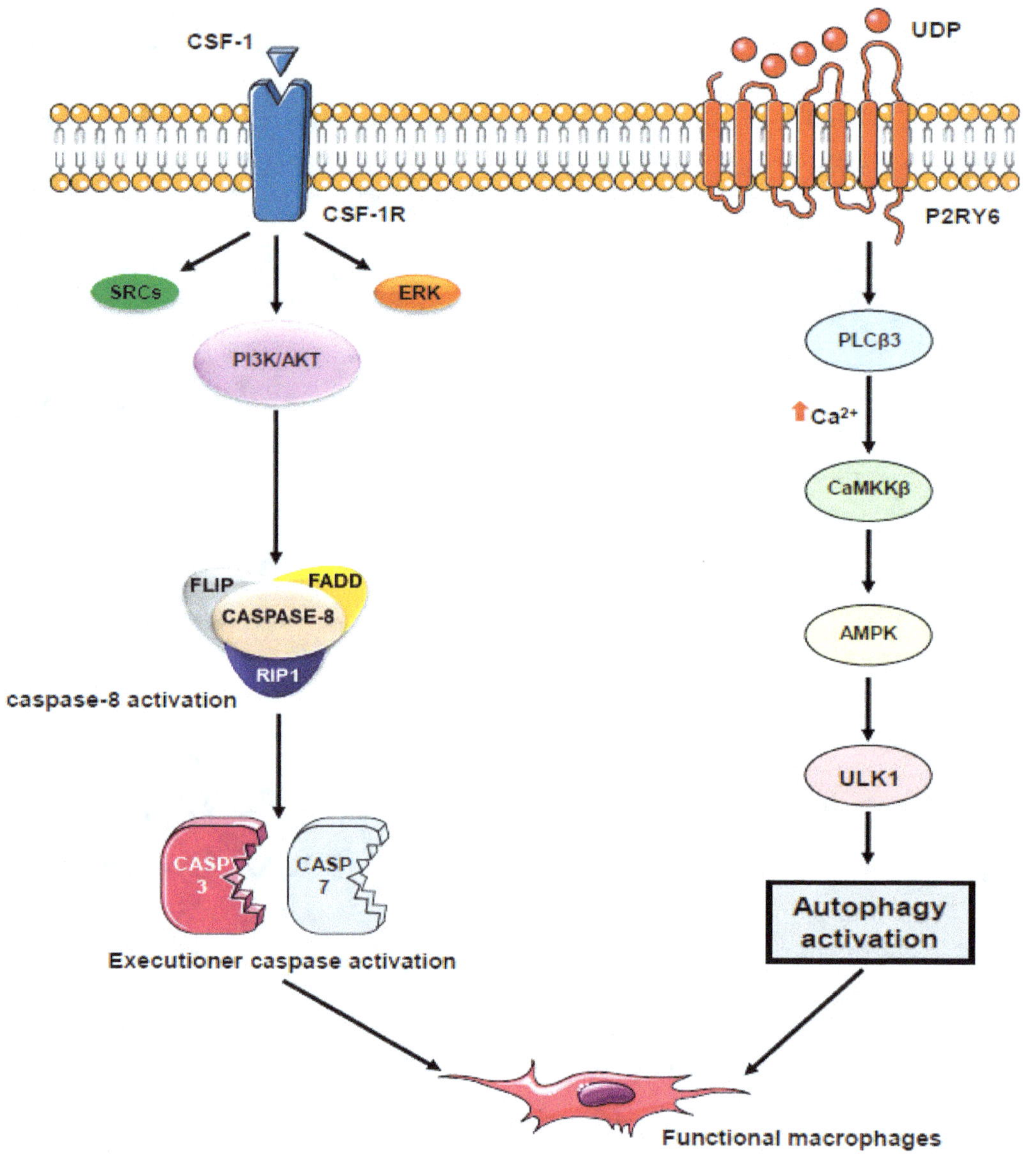

Figure 3. A schematic molecular view of the pathway involved in macrophagic differentiation. Engagement of the CSF-1 tyrosine kinase receptor by CSF-1 activates the PI3K/AKT pathway and induces caspase-8 activation within a FADD/RIP/FLIP multimolecular complex. In turn, active caspase-8 triggers a spatially restricted activation of caspase-3 and -7 that cleave selected intracellular proteins to generate a resting macrophage phenotype. Engaged CSF-1R also promotes autophagy through increasing the expression of the purinergic receptor P2RY6 that activates the CAMKK2-AMPK-ULK1 pathway. Caspase activation and autophagy induction are both required for proper generation of macrophages upon CSF-1 stimulation.

Other evidence for AMPK implication in myeloid cell differentiation comes from additional experiments performed in U937 and HEL acute myeloid leukemia (AML) cell lines, in which low doses of cytarabine were found to trigger the autophagy necessary for myeloid differentiation, as shown by increased expression of CD11b and morphologic features of differentiation. Inhibition of the mTOR pathway, likely independent of AMPK, was associated with cytarabine-mediated myeloid differentiation [54]. A last example of a role for AMPK in autophagy-mediated differentiation stems from in vivo experiments conducted with AMPKα1 mice. Indeed, AMPKα1 deficiency was shown to impair autophagy-mediated differentiation and decrease monocyte macrophage survival [55]. Finally, α1 and α2 AMPK isoforms appeared to display functional differences in the regulation of osteogenesis and osteoblast-associated induction of osteoclastogenesis [56].

5. The Role of AMPK in Hematological Cancers

AMPK has been reported to inhibit the growth of various hematological cancers. Indeed, in acute lymphoblastic leukemia (ALL) cell lines, AICAR triggered dose- and time-dependent inhibition of cell growth [57,58]. The pro-apoptotic effect of AMPK is mediated by activation of the p38 MAPK pathway, increased expression of cell-cycle inhibitors such as p27 and p53, and the downstream effects of the mTOR pathway. In B-cell chronic lymphocytic leukemia (B-CLL) cells, AMPK triggered apoptosis in a p53-independent manner [13]. In mantle cell lymphoma (MCL), AICAR-mediated stimulation of AMPK activity dampened phosphorylation of critical downstream effectors of mTOR signaling, such as 4E-BP1 and ribosomal protein S6, leading to cell growth inhibition [59]. In chronic myeloid leukemia (CML) and Philadelphia chromosome positive ALL, metformin and AICAR suppressed the mTOR activity and cell growth [58,60]. In myelodysplastic syndromes (MDS) and some acute myeloid leukemia (AML) cell lines, AICAR induced suppression of cell growth independently of apoptosis induction and AMPK activation [12]. Globally, indirect activators of AMPK induced suppression of cell growth and activation of cell death through both AMPK dependent and independent mechanisms in different hematological malignancies.

6. AMPK in the Regulation of Pathological Myeloid Differentiation

6.1. Myelodysplastic Syndromes

Myelodysplastic syndromes (MDS) constitute a set of myeloid progenitor cell diseases characterized by an ineffective production of mature and differentiated myeloid cells. This pathology strongly impacts the elderly, with an incidence of 20/100,000 people affected at 60 years and up to 50/100,000 at 80 years. This heterogeneous pathology is a multistep disease, the first stage being characterized by the appearance of abnormal cells in the bone marrow that are unable to differentiate and therefore to produce mature and functional cells in peripheral blood. At this stage, the disease is considered low-risk MDS and characterized by excessive apoptotic cell death of CD34-positive cells. In the next stage, CD34-positive progenitors accumulate mutations and hypermethylation of their DNA [61]. These genetic and epigenetic modifications render MDS cells less sensitive to apoptotic signals and contribute to abnormal cell growth. At this stage, the disease is defined as high-risk MDS and associated with an increase in medullar blast count, a higher probability for transformation to AML and a worse prognosis. Each stage of the disease is correlated with survival expectancy and probability for AML transformation by the Revised—International Prognostic Scoring System (IPSS-R). For classification of patients, IPSS-R takes into account the percentage of medullar blasts, the number of cytopenias and the cytogenetic status [62]. The recommended treatment in lower-risk MDS is the management of transfusion needs, whereas in high-risk MDS and AML, the only curative treatment remains bone marrow allograft, but few patients are eligible for this therapeutic option due to their advanced age. For non-eligible MDS and AML patients, the conventional chemotherapy consists of demethylating agents, including azacytidine or decitabine [63–65], or high-doses of cytarabine. Unfortunately, most patients fail to respond durably to these drugs, and there is currently no effective second-line option.

To adapt to the oxygen, nutrient, and glucose deprivation found during the different steps of tumor development, cancer cells modify their energetic needs through the fine-tuning of AMPK activity, a key metabolic sensor. As previously mentioned, AMPK is the master gene for the regulation of energy homeostasis. Its involvement in the regulation of apoptosis, autophagy and cell growth make this serine/threonine protein kinase a target of choice for the treatment of cancers in general and leukemia in particular [16,66,67]. Azacytidine-resistant MDS cell lines exhibited an altered response to apoptotic stimuli but displayed functional autophagy. Hence, triggering autophagy in resistant MDS cells induced autophagic cell death and bypassed the insensitivity to apoptosis [68,69]. Of note, it has recently been reported that the knockdown of *sf3b1*, a splicing factor mutated in 16% of MDS cases, deregulates the AMPK pathway [70]. This observation is of great interest since mutations of *sf3b1* are associated with good prognosis in MDS patients. Finally, cells issued from high-risk MDS patients exhibited a strong decrease of AMPK mRNA levels compared to those coming from low-risk MDS or healthy patients. This observation could explain the lower sensitivity of highly transformed MDS cells to variations in their energy environment.

6.2. Acute Myeloid Leukemia

The involvement of AMPK in the genesis and development of AML is well documented. Indeed, Saito et al. reported that leukemia initiating cells (LICs) responsible for the genesis of myeloid leukemia are protected from metabolic stress in the bone marrow through an AMPK-dependent mechanism [71]. This study clearly shows that dietary restriction induces metabolic cell survival. In contrast, depletion of AMPK expression in LIC reduces this cell population in the hypoxic bone marrow environment. Interestingly, human bone marrow stromal cells appear independent of the AMPK pathway, as knockout of AMPKα fails to sensitize these cells under dietary restriction or metformin treatment. Targeting AMPK depletes myeloid leukemia cells by disruption of glucose metabolism. In AML cell lines, exhibiting a high level of MAP kinase activation, AMPK activation was impaired. Indeed, Kawashima et al. demonstrated that glucose deprivation or metformin treatment could activate the AMPK pathway only in cells where ERK is weakly activated [72]. Moreover, in AML cells in which the MAP kinase pathway is overactivated, U0126, a MEK-specific inhibitor, was able to restore sustained AMPK activation under metformin treatment. The combination of an inhibitor of the MAPK pathway and an activator of the AMPK pathway leads to a significant decrease in cell growth and an increase in cell death in AML cells. This study suggests that both inhibition of an oncogenic pathway and activation of AMPK are necessary to trigger a strong anti-leukemic effect. The potent tumor suppressor role of the LKB1/AMPK pathway has been established in AML cells [73]. In this study, the authors showed that metformin activates the LKB1/AMPK pathway in AML and reduces tumor size in a mice xenograft model through the repression of mTOR-dependent oncogenic mRNA translation (c-MYC, CYCLIN-D, BCL-XL). It was recently demonstrated that co-activation of AMPK and mTORC1 could represent a good therapeutic strategy for AML [74]. Indeed, Sujobert et al. showed that GSK621, an AMPK direct activator, was highly cytotoxic for AML cells exhibiting a constitutive activation of mTORC1. As expected, this autophagy-dependent cell death induced by two concomitant signals was inhibited by rapamycin, an mTORC1 inhibitor.

Finally, although the strategy consisting of the use of AMPK agonists mimicking caloric restriction seems very promising, the development of an optimized and personalized therapy will need to block the predominant oncogenic pathway (overactivation due to mutations or translocation) and to activate AMPK in a concomitant manner. These complementary signals that will better impair cell growth and cell viability in AML will undoubtedly be a valuable option for patients in therapeutic failure.

6.3. Chronic Myelogenous Leukemia

Chronic myelogenous leukemia (CML) is a myeloproliferative syndrome linked to a hematopoietic stem cell disorder leading to increased production of granulocytes at all stages of differentiation. CML accounts for 15–20% of all cases of leukemia in adults [75] and is due to the t(9;22)(q34;q11) translocation, which encodes for the chimeric protein p210 BCR-ABL, a constitutively activated tyrosine kinase [76]. BCR-ABL expression leads to the engagement and activation of multiple pro-proliferative and anti-apoptotic cascades in transformed cells, including PI3K/AKT/mTOR and MAPK pathways [77–79]. Before the advent of targeted therapy, the gold standard for pharmacologic treatment of CML was α-interferon, but this treatment was associated with not-negligible toxicity and a median survival time of approximately five years [80]. In 2001, the identification of imatinib (Gleevec) as a small molecule ATP-pocket inhibitor of BCR-ABL dramatically re-defined the treatment of CML and had a major impact on the survival of patients with CML [81–83]. Imatinib mesylate along with second- (nilotinib, dasatinib, and bosutinib) and third-generation (ponatinib) tyrosine inhibitors (TKIs) have revolutionized the natural history of CML and have provided important treatment options for this leukemia that in the past was uniformly fatal [84]. Unfortunately, mutations rendering CML patients non-responsive to TKIs have been identified, including the threonine 315 to isoleucine (T315I) mutation and several others, which prevent binding of different TKIs to the active site of the ABL kinase, thereby avoiding its inhibition [85,86]. At present, more than 50 different mutation hotspots have been identified. BCR-ABL-independent mechanisms of resistance have also been reported to occur in tyrosine kinase inhibitor (TKI)-treated patients. In this regard, overexpression or hyperactivation of some members of the SRC family of kinases (LYN and HCK) have been described in cell lines and in some imatinib- and nilotinib-resistant patients [87–89]. Knowing that the PI3K/AKT/mTOR pathway is hyperactive in CML, indirect suppression of mTOR function by modulation of the AMPK pathway was proposed as an alternative therapeutic approach to overcome TKI resistances. Indeed, AMPK activation leads to mTOR inhibition through the phosphorylation and activation of the TSC1/2 complex [73,90] and/or a direct phosphorylation of the Raptor subunit on serines 722 and 792 [91], resulting in inactivation of the TORC1 complex. Most of the tested compounds that activate the AMPK pathway in the CML context are indirect activators. Among them, resveratrol, a naturally occurring substance found in grapes, triggers both apoptosis and autophagy in CML cells and is therefore able to overcome imatinib resistance [92,93]. Resveratrol-mediated autophagy is independent of BECLIN1 but mediated by a conjoint activation of AMPK and transcriptional upregulation of p62/SQSTM1 [93]. Interestingly, among the resistant cells that were sensitive to resveratrol, there were also cells expressing the T315I BCR-ABL mutant. Other studies have also pinpointed the anti-leukemic effects of AICAR or metformin on BCR-ABL transformed cells [12,60,94]. Unexpectedly, the effects of these compounds are often independent of AMPK activation. It has been established that the anti-leukemic effect of AICAR is dependent on protein kinase C-mediated autophagic cell death but independent of AMPK and apoptosis [12]. In conclusion, approaches to target cellular effectors of BCR-ABL, such as the PI3K/mTOR pathway, may provide alternative strategies to the use of TKIs to overcome resistance in refractory CML. The effect of direct AMPK activators on both TKI-sensitive and resistant CML cells has not been yet investigated.

6.4. Chronic Myelomonocytic Leukemia

Chronic myelomonocytic leukemia (CMML) is a paradigmatic chronic myeloid malignancy that associates features of myelodysplastic syndromes (MDS) and myeloproliferative neoplasms (MPN) [95]. CMML is a clonal disease of the hematopoietic stem cell characterized by a persistent monocytosis ($>1 \times 10^9$/L), due to an accumulation of classical monocytes [96], and the aleatory presence of immature dysplastic granulocytes (PolyMorphoNuclear-MDSC or PMN-MDSC) in the peripheral blood of CMML patients [97]. These PMN-MDSCs that belong to the same clone as the leukemic monocytes appear to have immunosuppressive properties resembling those of the myeloid-derived suppressor cells (MDSCs), widely described in solid tumors. Whether these immature granulocytes contribute to autoimmune manifestations or immune-escape and progression of CMML is a conundrum and remains to be investigated. In recent years, large numbers of gene mutations have been discovered in CMML, none of which are specific to this condition, as they can be encountered with different frequencies in other myeloid neoplasms. These mutated genes encode signaling proteins (NRAS, KRAS, CBL, JAK2, FLT3, and several members of the Notch pathway), epigenetic regulators (TET2, ASXL1, EZH2, IDH1, and IDH2), and splicing factors (SF3B1, SRSF2, and ZRSF2) [98]. Mutations in the transcription regulators RUNX1, NPM1, and TP53 have also been reported in CMML. However, the role of these mutations in leukemogenesis is still unclear. Allogeneic stem cell transplantation is the only potentially curative option in patients suffering CMML, but it is associated with significant morbidity and mortality. Consequently, CMML patients are often treated like MDS patients with supportive care and hypomethylating agents, such as 5-azacitidine and decitabine, with overall response rates of 30–40% and complete remission rates of 7–17%, although with no impact on mutational allele burdens [99–101]. To improve their efficacy, azacytidine and decitabine have been combined with another drug, typically in "pick the winner" clinical trials with DNA-damaging drugs, immune-modulating drugs and histone deacetylase inhibitors, but none of these strategies has proven effective, and several were even found to be toxic [102,103]. There is therefore a strong need for alternative strategies aiming at increasing CMML patient overall survival. In a recent study, we have shown that CMML is characterized by defects in monocyte to macrophage differentiation [97]. These differentiation defects can be partly attributed to the presence of PMN-MDSCs that secrete high levels of alpha-defensins HNP1-3, which antagonize the purinergic receptor P2RY6 and inhibit AMPK-mediated autophagy in CMML patients [51]. Interestingly, we demonstrated that the physiological P2RY6 ligand UDP and the specific P2RY6 agonist MRS2693 can restore normal monocyte differentiation through re-induction of AMPK-dependent autophagy in primary myeloid cells from some, but not all, CMML patients (Figure 4). These results highlight an essential role for P2RY6-mediated autophagy through AMPK activation during the differentiation of human monocytes and pave the way for future therapeutic interventions for CMML. The use of direct AMPK activator in the CMML context could therefore be a promising therapeutic strategy.

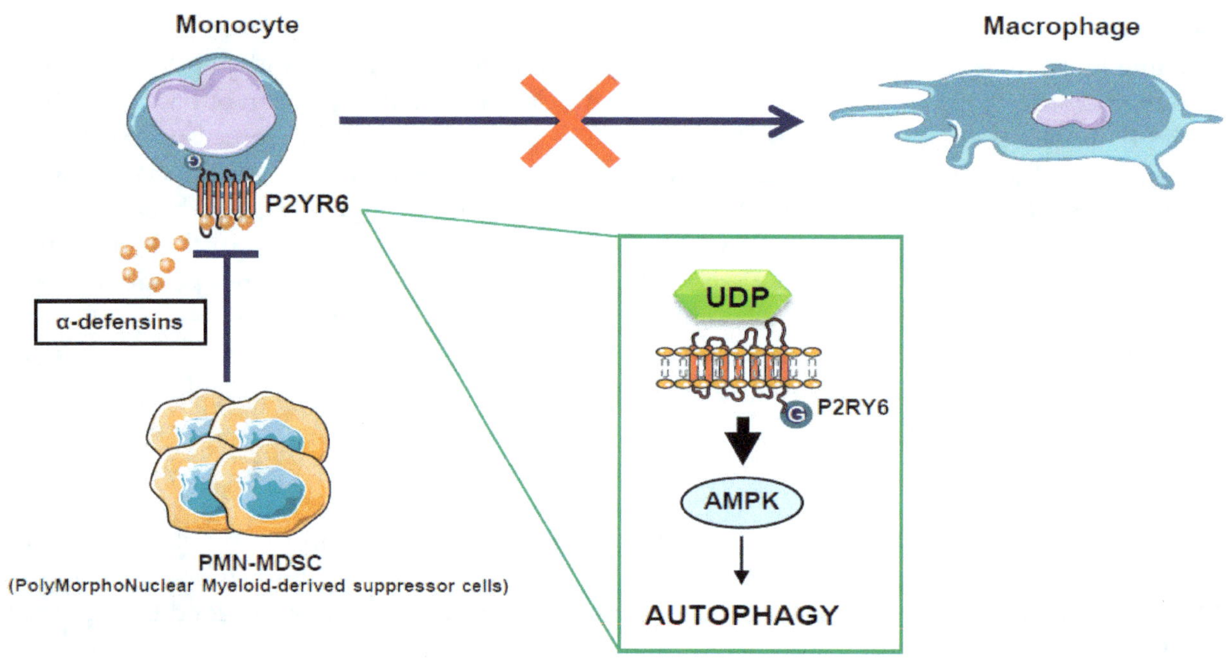

Figure 4. Mode of action for PMN-MDSC in the CMML context. In some CMML patients, PMN-MDSC secretes high levels of alpha-defensins, which antagonize the P2Y6 receptor, block autophagy activation, and inhibit the macrophagic differentiation of monocytes isolated from CMML patients. Alpha-defensin acts as a competitive inhibitor of UDP, the natural ligand of the P2Y6 receptor.

7. AMPK Modulators for the Treatment of Hematopoietic Malignancies

Due to its implication at the crossroad of cellular metabolism and proliferation, AMPK has emerged as an attractive and promising target for a great number of human pathological situations, including metabolic diseases and cancer. Pharmacological compounds, such as 5-aminoImidazole-4-carboxamide-1-β-D-ribofuranoside (AICAR), metformin, and natural occurring compounds such as resveratrol and spermidine, activate AMPK by direct or indirect mechanisms. It has been reported that metformin, AICAR, or resveratrol exert potent antitumoral effects in both solid tumors and hematopoietic malignancies either sensitive or resistant to their treatment of reference [12,60,93,104–106]. However, the anti-leukemic effects of some but not all of these molecules were linked to AMPK activation, making some natural AMPK activators promising molecules for the treatment of hematopoietic malignancies. Due to the potential therapeutic benefit of AMPK modulators in numerous diseases, direct and more specific modulators of AMPK have been developed that mainly target the β subunits, and less frequently the α and γ subunits (please refer to [35,107] for detailed reviews). These inhibitors included β1-selective compounds such as A-769662, PF-06409577, and GSK621. A-769662 has been evaluated in different settings and although efficient, appeared to exhibit off-target effects and cannot be considered per se as a specific AMPK activator. MT47-100 is an allosteric modulator of AMPK that activates the AMPK complexes containing the β1 isoform and inhibits those comprising the β2 subunits (Table 1). All these inhibitors have been tested in different settings, including cancers, with variable efficacy. Among the direct activators of AMPK, only GSK621, which selectively binds to the β1 subunit, has been studied in hematopoietic malignancies. Indeed, it was reported that GSK621 selectively kills AML cell lines and AML primary cells, sparing normal hematopoietic progenitors [74]. The lethality of GSK621 was abrogated by chemical inhibition or genetic ablation of mTORC1, suggesting a preferential effect in AML cells with over-activation of the mTORC1 pathway. Finally, the GSK621 cytotoxicity in AML cells was strictly dependent on the eIF2α/ATF4 signaling pathway activated through mTORC1. Although it is premature to speculate that activation of AMPK could represent a new strategic therapy for all myeloid malignancies, these promising findings strongly suggest that specific activators of AMPKα1, such as GSK621, may represent a therapeutic opportunity in AML and likely more globally in cancers in which mTORC1 is over-activated.

Table 1. AMPK activators include well-known pharmacological compounds, such as AICAR and metformin that act by increasing the AMP/ATP ratio, small molecules including A-769662, PF-06409577 and GSK621 that behave as AMPKβ1 subunit activator, and MT47-100, an AMPKβ1 activator and AMPKβ2 inhibitor. AMPK is also activated by a set of natural compounds including resveratrol and spermidine that indirectly increase the AMP/ATP ratio.

AMPK Activator	
Pharmacological Compounds	
AICAR Metformin	AMP:ATP ratio (up)
A-769662 PF-06409577 GSK621	AMPKβ1 subunit activators
MT47-100	AMPKβ1 subunit activator AMPKβ2 subunit inhibitor
Natural Compounds	
resveratrol spermidine	AMP:ATP ratio (up)

8. Conclusions and Outlook

In this review, we have discussed how AMPK is involved and regulated during myeloid lineage differentiation and how it can impact myeloid cell pathophysiology. As autophagy is known to play an important role in the process of hematopoietic cell differentiation [108], it was expected that AMPK would be a regulator of this process. However, the role of the AMPK pathway as an important actor of the metabolic modifications necessary for myeloid differentiation is just emerging. As an energy-consuming process, physiological myeloid differentiation requires adaptation of AMPK expression and activity in part to sustain autophagy that is necessary for this process. Hematopoietic malignancies are systematically characterized by profound defects in cell differentiation. Restoration of an effective differentiation process in myeloid malignancies has thus emerged as a pertinent therapeutic strategy. This notion is particularly well exemplified by the successful use of arsenic trioxide to promote acute promyelomonocytic leukemia redifferentiation, even leading to cure in a majority of patients. Alteration of AMPK expression and/or activity is found in hematological malignancies such as CMML, and activation of the AMPK pathway appears to restore normal differentiation in some CMML patients [51]. Beside myeloid differentiation, recent evidence in the literature indicates that AMPK could also play a key role during normal erythroid cell differentiation (Ladli, M et al., 2018, Haematologica, in press). While using small molecule AMPK activators to treat a panel of human metabolic and neurodegenerative diseases now appears achievable, further investigations should be carried out before such activators, direct or indirect, reach the clinic for the treatment of myeloid malignancies.

Acknowledgments: All authors participate equally to this work.

References

1. Lin, S.C.; Hardie, D.G. AMPK: Sensing Glucose as well as Cellular Energy Status. *Cell Metab.* **2018**, *27*, 299–313. [CrossRef] [PubMed]
2. Hardie, D.G. AMPK: Positive and negative regulation, and its role in whole-body energy homeostasis. *Curr. Opin. Cell Biol.* **2015**, *33*, 1–7. [CrossRef] [PubMed]
3. Hardie, D.G.; Schaffer, B.E.; Brunet, A. AMPK: An Energy-Sensing Pathway with Multiple Inputs and Outputs. *Trends Cell Biol.* **2016**, *26*, 190–201. [CrossRef] [PubMed]
4. Hardie, D.G. AMP-activated protein kinase: An energy sensor that regulates all aspects of cell function. *Genes Dev.* **2011**, *25*, 1895–1908. [CrossRef] [PubMed]
5. Steinberg, G.R.; Kemp, B.E. AMPK in Health and Disease. *Physiol. Rev.* **2009**, *89*, 1025–1078. [CrossRef] [PubMed]
6. Lopez-Lluch, G.; Navas, P. Calorie restriction as an intervention in ageing. *J. Physiol.* **2016**, *594*, 2043–2060. [CrossRef] [PubMed]
7. Pietrocola, F.; Castoldi, F.; Markaki, M.; Lachkar, S.; Chen, G.; Enot, D.P.; Durand, S.; Bossut, N.; Tong, M.; Malik, S.A.; et al. Aspirin Recapitulates Features of Caloric Restriction. *Cell Rep.* **2018**, *22*, 2395–2407. [CrossRef] [PubMed]
8. Pietrocola, F.; Pol, J.; Vacchelli, E.; Rao, S.; Enot, D.P.; Baracco, E.E.; Levesque, S.; Castoldi, F.; Jacquelot, N.; Yamazaki, T.; et al. Caloric Restriction Mimetics Enhance Anticancer Immunosurveillance. *Cancer Cell* **2016**, *30*, 147–160. [CrossRef] [PubMed]
9. Marino, G.; Pietrocola, F.; Madeo, F.; Kroemer, G. Caloric restriction mimetics: Natural/physiological pharmacological autophagy inducers. *Autophagy* **2014**, *10*, 1879–1882. [CrossRef] [PubMed]
10. Ben Sahra, I.; Le Marchand-Brustel, Y.; Tanti, J.F.; Bost, F. Metformin in cancer therapy: A new perspective for an old antidiabetic drug? *Mol. Cancer Ther.* **2010**, *9*, 1092–1099. [CrossRef] [PubMed]
11. Narkar, V.A.; Downes, M.; Yu, R.T.; Embler, E.; Wang, Y.X.; Banayo, E.; Mihaylova, M.M.; Nelson, M.C.; Zou, Y.; Juguilon, H.; et al. AMPK and PPARdelta agonists are exercise mimetics. *Cell* **2008**, *134*, 405–415. [CrossRef] [PubMed]
12. Robert, G.; Ben Sahra, I.; Puissant, A.; Colosetti, P.; Belhacene, N.; Gounon, P.; Hofman, P.; Bost, F.; Cassuto, J.P.; Auberger, P. Acadesine kills chronic myelogenous leukemia (CML) cells through PKC-dependent induction of autophagic cell death. *PLoS ONE* **2009**, *4*, e7889. [CrossRef] [PubMed]
13. Campas, C.; Lopez, J.M.; Santidrian, A.F.; Barragan, M.; Bellosillo, B.; Colomer, D.; Gil, J. Acadesine activates AMPK and induces apoptosis in B-cell chronic lymphocytic leukemia cells but not in T lymphocytes. *Blood* **2003**, *101*, 3674–3680. [CrossRef] [PubMed]
14. Guo, D.; Hildebrandt, I.J.; Prins, R.M.; Soto, H.; Mazzotta, M.M.; Dang, J.; Czernin, J.; Shyy, J.Y.; Watson, A.D.; Phelps, M.; et al. The AMPK agonist AICAR inhibits the growth of EGFRvIII-expressing glioblastomas by inhibiting lipogenesis. *Proc. Natl. Acad. Sci. USA* **2009**, *106*, 12932–12937. [CrossRef] [PubMed]
15. Su, R.Y.; Chao, Y.; Chen, T.Y.; Huang, D.Y.; Lin, W.W. 5-Aminoimidazole-4-carboxamide riboside sensitizes TRAIL- and TNFα-induced cytotoxicity in colon cancer cells through AMP-activated protein kinase signaling. *Mol. Cancer Ther.* **2007**, *6*, 1562–1571. [CrossRef] [PubMed]

16. Mihaylova, M.M.; Shaw, R.J. The AMPK signalling pathway coordinates cell growth, autophagy and metabolism. *Nat. Cell Biol.* **2011**, *13*, 1016–1023. [CrossRef] [PubMed]
17. Zhang, C.S.; Lin, S.C. AMPK Promotes Autophagy by Facilitating Mitochondrial Fission. *Cell Metab.* **2016**, *23*, 399–401. [CrossRef] [PubMed]
18. Herzig, S.; Shaw, R.J. AMPK: Guardian of metabolism and mitochondrial homeostasis. *Nat. Rev. Mol. Cell Biol.* **2018**, *19*, 121–135. [CrossRef] [PubMed]
19. Greer, E.L.; Dowlatshahi, D.; Banko, M.R.; Villen, J.; Hoang, K.; Blanchard, D.; Gygi, S.P.; Brunet, A. An AMPK-FOXO pathway mediates longevity induced by a novel method of dietary restriction in *C. elegans*. *Curr. Biol.* **2007**, *17*, 1646–1656. [CrossRef] [PubMed]
20. Mair, W.; Morantte, I.; Rodrigues, A.P.; Manning, G.; Montminy, M.; Shaw, R.J.; Dillin, A. Lifespan extension induced by AMPK and calcineurin is mediated by CRTC-1 and CREB. *Nature* **2011**, *470*, 404–408. [CrossRef] [PubMed]
21. Weir, H.J.; Yao, P.; Huynh, F.K.; Escoubas, C.C.; Goncalves, R.L.; Burkewitz, K.; Laboy, R.; Hirschey, M.D.; Mair, W.B. Dietary Restriction and AMPK Increase Lifespan via Mitochondrial Network and Peroxisome Remodeling. *Cell Metab.* **2017**, *26*, 884–896. [CrossRef] [PubMed]
22. Templeman, N.M.; Murphy, C.T. Regulation of reproduction and longevity by nutrient-sensing pathways. *J. Cell Biol.* **2018**, *217*, 93–106. [CrossRef] [PubMed]
23. Banko, M.R.; Allen, J.J.; Schaffer, B.E.; Wilker, E.W.; Tsou, P.; White, J.L.; Villen, J.; Wang, B.; Kim, S.R.; Sakamoto, K.; et al. Chemical genetic screen for AMPKalpha2 substrates uncovers a network of proteins involved in mitosis. *Mol. Cell* **2011**, *44*, 878–892. [CrossRef] [PubMed]
24. Winder, W.W.; Hardie, D.G. Inactivation of acetyl-CoA carboxylase and activation of AMP-activated protein kinase in muscle during exercise. *Am. J. Physiol.* **1996**, *270*, 299–304. [CrossRef] [PubMed]
25. Kim, J.; Kundu, M.; Viollet, B.; Guan, K.L. AMPK and mTOR regulate autophagy through direct phosphorylation of Ulk1. *Nat. Cell Biol.* **2011**, *13*, 132–141. [CrossRef] [PubMed]
26. Zhao, M.; Klionsky, D.J. AMPK-dependent phosphorylation of ULK1 induces autophagy. *Cell Metab.* **2011**, *13*, 119–120. [CrossRef] [PubMed]
27. Laker, R.C.; Drake, J.C.; Wilson, R.J.; Lira, V.A.; Lewellen, B.M.; Ryall, K.A.; Fisher, C.C.; Zhang, M.; Saucerman, J.J.; Goodyear, L.J.; et al. Ampk phosphorylation of Ulk1 is required for targeting of mitochondria to lysosomes in exercise-induced mitophagy. *Nat. Commun.* **2017**, *8*, 548. [CrossRef] [PubMed]
28. Hawley, S.A.; Ross, F.A.; Gowans, G.J.; Tibarewal, P.; Leslie, N.R.; Hardie, D.G. Phosphorylation by Akt within the ST loop of AMPK-alpha1 down-regulates its activation in tumour cells. *Biochem. J.* **2014**, *459*, 275–287. [CrossRef] [PubMed]
29. Manning, B.D.; Toker, A. AKT/PKB Signaling: Navigating the Network. *Cell* **2017**, *169*, 381–405. [CrossRef] [PubMed]
30. Zhang, Y.L.; Guo, H.; Zhang, C.S.; Lin, S.Y.; Yin, Z.; Peng, Y.; Luo, H.; Shi, Y.; Lian, G.; Zhang, C.; et al. AMP as a low-energy charge signal autonomously initiates assembly of AXIN-AMPK-LKB1 complex for AMPK activation. *Cell Metab.* **2013**, *18*, 546–555. [CrossRef] [PubMed]
31. Zhang, C.S.; Jiang, B.; Li, M.; Zhu, M.; Peng, Y.; Zhang, Y.L.; Wu, Y.Q.; Li, T.Y.; Liang, Y.; Lu, Z.; et al. The lysosomal v-ATPase-Ragulator complex is a common activator for AMPK and mTORC1, acting as a switch between catabolism and anabolism. *Cell Metab.* **2014**, *20*, 526–540. [CrossRef] [PubMed]
32. O'Neill, L.A.; Hardie, D.G. Metabolism of inflammation limited by AMPK and pseudo-starvation. *Nature* **2013**, *493*, 346–355. [CrossRef] [PubMed]
33. He, C.; Li, H.; Viollet, B.; Zou, M.H.; Xie, Z. AMPK Suppresses Vascular Inflammation In Vivo by Inhibiting Signal Transducer and Activator of Transcription-1. *Diabetes* **2015**, *64*, 4285–4297. [CrossRef] [PubMed]
34. Hardie, D.G.; Lin, S.C. AMP-activated protein kinase—Not just an energy sensor. *F1000Reseach* **2017**, *6*, 1724. [CrossRef] [PubMed]
35. Olivier, S.; Foretz, M.; Viollet, B. Promise and challenges for direct small molecule AMPK activators. *Biochem. Pharmacol.* **2018**, *153*, 147–158. [CrossRef] [PubMed]
36. Ross, F.A.; Jensen, T.E.; Hardie, D.G. Differential regulation by AMP and ADP of AMPK complexes containing different gamma subunit isoforms. *Biochem. J.* **2016**, *473*, 189–199. [CrossRef] [PubMed]
37. Neumann, D. Is TAK1 a Direct Upstream Kinase of AMPK? *Int. J. Mol. Sci.* **2018**, *19*, 2412. [CrossRef] [PubMed]

38. Carling, D.; Thornton, C.; Woods, A.; Sanders, M.J. AMP-activated protein kinase: New regulation, new roles? *Biochem. J.* **2012**, *445*, 11–27. [CrossRef] [PubMed]
39. Schaffer, B.E.; Levin, R.S.; Hertz, N.T.; Maures, T.J.; Schoof, M.L.; Hollstein, P.E.; Benayoun, B.A.; Banko, M.R.; Shaw, R.J.; Shokat, K.M.; et al. Identification of AMPK Phosphorylation Sites Reveals a Network of Proteins Involved in Cell Invasion and Facilitates Large-Scale Substrate Prediction. *Cell Metab.* **2015**, *22*, 907–921. [CrossRef] [PubMed]
40. Wu, D.; Hu, D.; Chen, H.; Shi, G.; Fetahu, I.S.; Wu, F.; Rabidou, K.; Fang, R.; Tan, L.; Xu, S.; et al. Glucose-regulated phosphorylation of TET2 by AMPK reveals a pathway linking diabetes to cancer. *Nature* **2018**, *559*, 637–641. [CrossRef] [PubMed]
41. Nakada, D.; Saunders, T.L.; Morrison, S.J. Lkb1 regulates cell cycle and energy metabolism in haematopoietic stem cells. *Nature* **2010**, *468*, 653–658. [CrossRef] [PubMed]
42. Gan, B.; Hu, J.; Jiang, S.; Liu, Y.; Sahin, E.; Zhuang, L.; Fletcher-Sananikone, E.; Colla, S.; Wang, Y.A.; Chin, L.; et al. Lkb1 regulates quiescence and metabolic homeostasis of haematopoietic stem cells. *Nature* **2010**, *468*, 701–704. [CrossRef] [PubMed]
43. Egan, D.; Kim, J.; Shaw, R.J.; Guan, K.L. The autophagy initiating kinase ULK1 is regulated via opposing phosphorylation by AMPK and mTOR. *Autophagy* **2011**, *7*, 643–644. [CrossRef] [PubMed]
44. Yu, W.M.; Liu, X.; Shen, J.; Jovanovic, O.; Pohl, E.E.; Gerson, S.L.; Finkel, T.; Broxmeyer, H.E.; Qu, C.K. Metabolic regulation by the mitochondrial phosphatase PTPMT1 is required for hematopoietic stem cell differentiation. *Cell Stem Cell* **2013**, *12*, 62–74. [CrossRef] [PubMed]
45. Jager, S.; Handschin, C.; St-Pierre, J.; Spiegelman, B.M. AMP-activated protein kinase (AMPK) action in skeletal muscle via direct phosphorylation of PGC-1alpha. *Proc. Natl. Acad. Sci. USA* **2007**, *104*, 12017–12022. [CrossRef] [PubMed]
46. Ginhoux, F.; Jung, S. Monocytes and macrophages: Developmental pathways and tissue homeostasis. *Nat. Rev. Immunol.* **2014**, *14*, 392–404. [CrossRef] [PubMed]
47. Jakubzick, C.V.; Randolph, G.J.; Henson, P.M. Monocyte differentiation and antigen-presenting functions. *Nat. Rev. Immunol.* **2017**, *17*, 349–362. [CrossRef] [PubMed]
48. Sordet, O.; Rebe, C.; Plenchette, S.; Zermati, Y.; Hermine, O.; Vainchenker, W.; Garrido, C.; Solary, E.; Dubrez-Daloz, L. Specific involvement of caspases in the differentiation of monocytes into macrophages. *Blood* **2002**, *100*, 4446–4453. [CrossRef] [PubMed]
49. Stanley, E.R.; Chitu, V. CSF-1 receptor signaling in myeloid cells. *Cold Spring Harb. Perspect. Biol.* **2014**, *6*. [CrossRef] [PubMed]
50. Jacquel, A.; Obba, S.; Boyer, L.; Dufies, M.; Robert, G.; Gounon, P.; Lemichez, E.; Luciano, F.; Solary, E.; Auberger, P. Autophagy is required for CSF-1-induced macrophagic differentiation and acquisition of phagocytic functions. *Blood* **2012**, *119*, 4527–4531. [CrossRef] [PubMed]
51. Obba, S.; Hizir, Z.; Boyer, L.; Selimoglu-Buet, D.; Pfeifer, A.; Michel, G.; Hamouda, M.A.; Goncalves, D.; Cerezo, M.; Marchetti, S.; et al. The PRKAA1/AMPKalpha1 pathway triggers autophagy during CSF1-induced human monocyte differentiation and is a potential target in CMML. *Autophagy* **2015**, *11*, 1114–1129. [CrossRef] [PubMed]
52. Rebe, C.; Cathelin, S.; Launay, S.; Filomenko, R.; Prevotat, L.; L'Ollivier, C.; Gyan, E.; Micheau, O.; Grant, S.; Dubart-Kupperschmitt, A.; et al. Caspase-8 prevents sustained activation of NF-kappaB in monocytes undergoing macrophagic differentiation. *Blood* **2007**, *109*, 1442–1450. [CrossRef] [PubMed]
53. Jacquel, A.; Benikhlef, N.; Paggetti, J.; Lalaoui, N.; Guery, L.; Dufour, E.K.; Ciudad, M.; Racoeur, C.; Micheau, O.; Delva, L.; et al. Colony-stimulating factor-1-induced oscillations in phosphatidylinositol-3 kinase/AKT are required for caspase activation in monocytes undergoing differentiation into macrophages. *Blood* **2009**, *114*, 3633–3641. [CrossRef] [PubMed]
54. Chen, L.; Guo, P.; Zhang, Y.; Li, X.; Jia, P.; Tong, J.; Li, J. Autophagy is an important event for low-dose cytarabine treatment in acute myeloid leukemia cells. *Leuk. Res.* **2017**, *60*, 44–52. [CrossRef] [PubMed]
55. Zhang, M.; Zhu, H.; Ding, Y.; Liu, Z.; Cai, Z.; Zou, M.H. AMP-activated protein kinase alpha1 promotes atherogenesis by increasing monocyte-to-macrophage differentiation. *J. Biol. Chem.* **2017**, *292*, 7888–7903. [CrossRef] [PubMed]
56. Wang, Y.G.; Han, X.G.; Yang, Y.; Qiao, H.; Dai, K.R.; Fan, Q.M.; Tang, T.T. Functional differences between AMPK alpha1 and alpha2 subunits in osteogenesis, osteoblast-associated induction of osteoclastogenesis, and adipogenesis. *Sci. Rep.* **2016**, *6*, 32771. [CrossRef] [PubMed]

57. Sengupta, T.K.; Leclerc, G.M.; Hsieh-Kinser, T.T.; Leclerc, G.J.; Singh, I.; Barredo, J.C. Cytotoxic effect of 5-aminoimidazole-4-carboxamide-1-beta-4-ribofuranoside (AICAR) on childhood acute lymphoblastic leukemia (ALL) cells: Implication for targeted therapy. *Mol. Cancer* **2007**, *6*, 46. [CrossRef] [PubMed]
58. Vakana, E.; Platanias, L.C. AMPK in BCR-ABL expressing leukemias. Regulatory effects and therapeutic implications. *Oncotarget* **2011**, *2*, 1322–1328. [PubMed]
59. Drakos, E.; Atsaves, V.; Li, J.; Leventaki, V.; Andreeff, M.; Medeiros, L.J.; Rassidakis, G.Z. Stabilization and activation of p53 downregulates mTOR signaling through AMPK in mantle cell lymphoma. *Leukemia* **2009**, *23*, 784–790. [CrossRef] [PubMed]
60. Vakana, E.; Altman, J.K.; Glaser, H.; Donato, N.J.; Platanias, L.C. Antileukemic effects of AMPK activators on BCR-ABL-expressing cells. *Blood* **2011**, *118*, 6399–6402. [CrossRef] [PubMed]
61. Shallis, R.M.; Ahmad, R.; Zeidan, A.M. The genetic and molecular pathogenesis of myelodysplastic syndromes. *Eur. J. Haematol.* **2018**, *101*, 260–271. [CrossRef] [PubMed]
62. Greenberg, P.L.; Tuechler, H.; Schanz, J.; Sanz, G.; Garcia-Manero, G.; Sole, F.; Bennett, J.M.; Bowen, D.; Fenaux, P.; Dreyfus, F.; et al. Revised international prognostic scoring system for myelodysplastic syndromes. *Blood* **2012**, *120*, 2454–2465. [CrossRef] [PubMed]
63. Fenaux, P.; Mufti, G.J.; Hellstrom-Lindberg, E.; Santini, V.; Finelli, C.; Giagounidis, A.; Schoch, R.; Gattermann, N.; Sanz, G.; List, A.; et al. Efficacy of azacitidine compared with that of conventional care regimens in the treatment of higher-risk myelodysplastic syndromes: A randomised, open-label, phase III study. *Lancet Oncol.* **2009**, *10*, 223–232. [CrossRef]
64. Kantarjian, H.; Oki, Y.; Garcia-Manero, G.; Huang, X.; O'Brien, S.; Cortes, J.; Faderl, S.; Bueso-Ramos, C.; Ravandi, F.; Estrov, Z.; et al. Results of a randomized study of 3 schedules of low-dose decitabine in higher-risk myelodysplastic syndrome and chronic myelomonocytic leukemia. *Blood* **2007**, *109*, 52–57. [CrossRef] [PubMed]
65. Steensma, D.P.; Baer, M.R.; Slack, J.L.; Buckstein, R.; Godley, L.A.; Garcia-Manero, G.; Albitar, M.; Larsen, J.S.; Arora, S.; Cullen, M.T.; et al. Multicenter study of decitabine administered daily for 5 days every 4 weeks to adults with myelodysplastic syndromes: The alternative dosing for outpatient treatment (ADOPT) trial. *J. Clin. Oncol.* **2009**, *27*, 3842–3848. [CrossRef] [PubMed]
66. Hardie, D.G.; Ross, F.A.; Hawley, S.A. AMPK: A nutrient and energy sensor that maintains energy homeostasis. *Nat. Rev. Mol. Cell Biol.* **2012**, *13*, 251–262. [CrossRef] [PubMed]
67. Laplante, M.; Sabatini, D.M. mTOR signaling in growth control and disease. *Cell* **2012**, *149*, 274–293. [CrossRef] [PubMed]
68. Cluzeau, T.; Robert, G.; Puissant, A.; Jean-Michel, K.; Cassuto, J.P.; Raynaud, S.; Auberger, P. Azacitidine-resistant SKM1 myeloid cells are defective for AZA-induced mitochondrial apoptosis and autophagy. *Cell Cycle* **2011**, *10*, 2339–2343. [CrossRef] [PubMed]
69. Cluzeau, T.; Robert, G.; Mounier, N.; Karsenti, J.M.; Dufies, M.; Puissant, A.; Jacquel, A.; Renneville, A.; Preudhomme, C.; Cassuto, J.P.; et al. BCL2L10 is a predictive factor for resistance to azacitidine in MDS and AML patients. *Oncotarget* **2012**, *3*, 490–501. [CrossRef] [PubMed]
70. Dolatshad, H.; Pellagatti, A.; Fernandez-Mercado, M.; Yip, B.H.; Malcovati, L.; Attwood, M.; Przychodzen, B.; Sahgal, N.; Kanapin, A.A.; Lockstone, H.; et al. Disruption of SF3B1 results in deregulated expression and splicing of key genes and pathways in myelodysplastic syndrome hematopoietic stem and progenitor cells. *Leukemia* **2015**, *29*, 1092–1103. [CrossRef] [PubMed]
71. Saito, Y.; Chapple, R.H.; Lin, A.; Kitano, A.; Nakada, D. AMPK Protects Leukemia-Initiating Cells in Myeloid Leukemias from Metabolic Stress in the Bone Marrow. *Cell Stem Cell* **2015**, *17*, 585–596. [CrossRef] [PubMed]
72. Kawashima, I.; Mitsumori, T.; Nozaki, Y.; Yamamoto, T.; Shobu-Sueki, Y.; Nakajima, K.; Kirito, K. Negative regulation of the LKB1/AMPK pathway by ERK in human acute myeloid leukemia cells. *Exp. Hematol.* **2015**, *43*, 524–533. [CrossRef] [PubMed]
73. Green, A.S.; Chapuis, N.; Maciel, T.T.; Willems, L.; Lambert, M.; Arnoult, C.; Boyer, O.; Bardet, V.; Park, S.; Foretz, M.; et al. The LKB1/AMPK signaling pathway has tumor suppressor activity in acute myeloid leukemia through the repression of mTOR-dependent oncogenic mRNA translation. *Blood* **2010**, *116*, 4262–4273. [CrossRef] [PubMed]
74. Sujobert, P.; Poulain, L.; Paubelle, E.; Zylbersztejn, F.; Grenier, A.; Lambert, M.; Townsend, E.C.; Brusq, J.M.; Nicodeme, E.; Decrooqc, J.; et al. Co-activation of AMPK and mTORC1 Induces Cytotoxicity in Acute Myeloid Leukemia. *Cell Rep.* **2015**, *11*, 1446–1457. [CrossRef] [PubMed]

75. Sawyers, C.L. Chronic myeloid leukemia. *N. Engl. J. Med.* **1999**, *340*, 1330–1340. [CrossRef] [PubMed]
76. Ren, R. Mechanisms of BCR-ABL in the pathogenesis of chronic myelogenous leukaemia. *Nat. Rev. Cancer* **2005**, *5*, 172–183. [CrossRef] [PubMed]
77. Steelman, L.S.; Pohnert, S.C.; Shelton, J.G.; Franklin, R.A.; Bertrand, F.E.; McCubrey, J.A. JAK/STAT, Raf/MEK/ERK, PI3K/Akt and BCR-ABL in cell cycle progression and leukemogenesis. *Leukemia* **2004**, *18*, 189–218. [CrossRef] [PubMed]
78. Kim, J.H.; Chu, S.C.; Gramlich, J.L.; Pride, Y.B.; Babendreier, E.; Chauhan, D.; Salgia, R.; Podar, K.; Griffin, J.D.; Sattler, M. Activation of the PI3K/mTOR pathway by BCR-ABL contributes to increased production of reactive oxygen species. *Blood* **2005**, *105*, 1717–1723. [CrossRef] [PubMed]
79. Redig, A.J.; Vakana, E.; Platanias, L.C. Regulation of mammalian target of rapamycin and mitogen activated protein kinase pathways by BCR-ABL. *Leuk. Lymphoma* **2011**, *52* (Suppl. 1), 45–53. [CrossRef] [PubMed]
80. Hehlmann, R.; Berger, U.; Pfirrmann, M.; Hochhaus, A.; Metzgeroth, G.; Maywald, O.; Hasford, J.; Reiter, A.; Hossfeld, D.K.; Kolb, H.J.; et al. Randomized comparison of interferon alpha and hydroxyurea with hydroxyurea monotherapy in chronic myeloid leukemia (CML-study II): Prolongation of survival by the combination of interferon alpha and hydroxyurea. *Leukemia* **2003**, *17*, 1529–1537. [CrossRef] [PubMed]
81. Druker, B.J.; Talpaz, M.; Resta, D.J.; Peng, B.; Buchdunger, E.; Ford, J.M.; Lydon, N.B.; Kantarjian, H.; Capdeville, R.; Ohno-Jones, S.; et al. Efficacy and safety of a specific inhibitor of the BCR-ABL tyrosine kinase in chronic myeloid leukemia. *N. Engl. J. Med.* **2001**, *344*, 1031–1037. [CrossRef] [PubMed]
82. Fang, G.; Kim, C.N.; Perkins, C.L.; Ramadevi, N.; Winton, E.; Wittmann, S.; Bhalla, K.N. CGP57148B (STI-571) induces differentiation and apoptosis and sensitizes Bcr-Abl-positive human leukemia cells to apoptosis due to antileukemic drugs. *Blood* **2000**, *96*, 2246–2253. [PubMed]
83. Dan, S.; Naito, M.; Tsuruo, T. Selective induction of apoptosis in Philadelphia chromosome-positive chronic myelogenous leukemia cells by an inhibitor of BCR—ABL tyrosine kinase, CGP 57148. *Cell Death Differ.* **1998**, *5*, 710–715. [CrossRef] [PubMed]
84. Pasic, I.; Lipton, J.H. Current approach to the treatment of chronic myeloid leukaemia. *Leuk. Res.* **2017**, *55*, 65–78. [CrossRef] [PubMed]
85. Patel, A.B.; O'Hare, T.; Deininger, M.W. Mechanisms of Resistance to ABL Kinase Inhibition in Chronic Myeloid Leukemia and the Development of Next Generation ABL Kinase Inhibitors. *Hematol. Oncol. Clin. N. Am.* **2017**, *31*, 589–612. [CrossRef] [PubMed]
86. Soverini, S.; Mancini, M.; Bavaro, L.; Cavo, M.; Martinelli, G. Chronic myeloid leukemia: The paradigm of targeting oncogenic tyrosine kinase signaling and counteracting resistance for successful cancer therapy. *Mol. Cancer* **2018**, *17*, 49. [CrossRef] [PubMed]
87. Donato, N.J.; Wu, J.Y.; Stapley, J.; Gallick, G.; Lin, H.; Arlinghaus, R.; Talpaz, M. BCR-ABL independence and LYN kinase overexpression in chronic myelogenous leukemia cells selected for resistance to STI571. *Blood* **2003**, *101*, 690–698. [CrossRef] [PubMed]
88. Ferri, C.; Bianchini, M.; Bengio, R.; Larripa, I. Expression of LYN and PTEN genes in chronic myeloid leukemia and their importance in therapeutic strategy. *Blood Cells Mol. Dis.* **2014**, *52*, 121–125. [CrossRef] [PubMed]
89. Grosso, S.; Puissant, A.; Dufies, M.; Colosetti, P.; Jacquel, A.; Lebrigand, K.; Barbry, P.; Deckert, M.; Cassuto, J.P.; Mari, B.; et al. Gene expression profiling of imatinib and PD166326-resistant CML cell lines identifies Fyn as a gene associated with resistance to BCR-ABL inhibitors. *Mol. Cancer Ther.* **2009**, *8*, 1924–1933. [CrossRef] [PubMed]
90. Inoki, K.; Ouyang, H.; Zhu, T.; Lindvall, C.; Wang, Y.; Zhang, X.; Yang, Q.; Bennett, C.; Harada, Y.; Stankunas, K.; et al. TSC2 integrates Wnt and energy signals via a coordinated phosphorylation by AMPK and GSK3 to regulate cell growth. *Cell* **2006**, *126*, 955–968. [CrossRef] [PubMed]
91. Gwinn, D.M.; Shackelford, D.B.; Egan, D.F.; Mihaylova, M.M.; Mery, A.; Vasquez, D.S.; Turk, B.E.; Shaw, R.J. AMPK phosphorylation of raptor mediates a metabolic checkpoint. *Mol. Cell* **2008**, *30*, 214–226. [CrossRef] [PubMed]
92. Puissant, A.; Grosso, S.; Jacquel, A.; Belhacene, N.; Colosetti, P.; Cassuto, J.P.; Auberger, P. Imatinib mesylate-resistant human chronic myelogenous leukemia cell lines exhibit high sensitivity to the phytoalexin resveratrol. *FASEB J.* **2008**, *22*, 1894–1904. [CrossRef] [PubMed]

93. Puissant, A.; Robert, G.; Fenouille, N.; Luciano, F.; Cassuto, J.P.; Raynaud, S.; Auberger, P. Resveratrol promotes autophagic cell death in chronic myelogenous leukemia cells via JNK-mediated p62/SQSTM1 expression and AMPK activation. *Cancer Res.* **2010**, *70*, 1042–1052. [CrossRef] [PubMed]
94. Shi, R.; Lin, J.; Gong, Y.; Yan, T.; Shi, F.; Yang, X.; Liu, X.; Naren, D. The antileukemia effect of metformin in the Philadelphia chromosome-positive leukemia cell line and patient primary leukemia cell. *Anticancer Drugs* **2015**, *26*, 913–922. [CrossRef] [PubMed]
95. Patnaik, M.M.; Tefferi, A. Chronic myelomonocytic leukemia: 2018 update on diagnosis, risk stratification and management. *Am. J. Hematol.* **2018**, *93*, 824–840. [CrossRef] [PubMed]
96. Selimoglu-Buet, D.; Badaoui, B.; Benayoun, E.; Toma, A.; Fenaux, P.; Quesnel, B.; Etienne, G.; Braun, T.; Abermil, N.; Morabito, M.; et al. Accumulation of classical monocytes defines a subgroup of MDS that frequently evolves into CMML. *Blood* **2017**, *130*, 832–835. [CrossRef] [PubMed]
97. Droin, N.; Jacquel, A.; Hendra, J.B.; Racoeur, C.; Truntzer, C.; Pecqueur, D.; Benikhlef, N.; Ciudad, M.; Guery, L.; Jooste, V.; et al. Alpha-defensins secreted by dysplastic granulocytes inhibit the differentiation of monocytes in chronic myelomonocytic leukemia. *Blood* **2010**, *115*, 78–88. [CrossRef] [PubMed]
98. Itzykson, R.; Duchmann, M.; Lucas, N.; Solary, E. CMML: Clinical and molecular aspects. *Int. J. Hematol.* **2017**, *105*, 711–719. [CrossRef] [PubMed]
99. Ades, L.; Sekeres, M.A.; Wolfromm, A.; Teichman, M.L.; Tiu, R.V.; Itzykson, R.; Maciejewski, J.P.; Dreyfus, F.; List, A.F.; Fenaux, P.; et al. Predictive factors of response and survival among chronic myelomonocytic leukemia patients treated with azacitidine. *Leuk. Res.* **2013**, *37*, 609–613. [CrossRef] [PubMed]
100. Fianchi, L.; Criscuolo, M.; Breccia, M.; Maurillo, L.; Salvi, F.; Musto, P.; Mansueto, G.; Gaidano, G.; Finelli, C.; Aloe-Spiriti, A.; et al. High rate of remissions in chronic myelomonocytic leukemia treated with 5-azacytidine: Results of an Italian retrospective study. *Leuk. Lymphoma* **2013**, *54*, 658–661. [CrossRef] [PubMed]
101. Wijermans, P.W.; Ruter, B.; Baer, M.R.; Slack, J.L.; Saba, H.I.; Lubbert, M. Efficacy of decitabine in the treatment of patients with chronic myelomonocytic leukemia (CMML). *Leuk. Res.* **2008**, *32*, 587–591. [CrossRef] [PubMed]
102. Garcia-Manero, G.; Sekeres, M.A.; Egyed, M.; Breccia, M.; Graux, C.; Cavenagh, J.D.; Salman, H.; Illes, A.; Fenaux, P.; DeAngelo, D.J.; et al. A phase 1b/2b multicenter study of oral panobinostat plus azacitidine in adults with MDS, CMML or AML with 30% blasts. *Leukemia* **2017**, *31*, 2799–2806. [CrossRef] [PubMed]
103. Kobayashi, Y.; Munakata, W.; Ogura, M.; Uchida, T.; Taniwaki, M.; Kobayashi, T.; Shimada, F.; Yonemura, M.; Matsuoka, F.; Tajima, T.; et al. Phase I study of panobinostat and 5-azacitidine in Japanese patients with myelodysplastic syndrome or chronic myelomonocytic leukemia. *Int. J. Hematol.* **2018**, *107*, 83–91. [CrossRef] [PubMed]
104. Puissant, A.; Robert, G.; Auberger, P. Targeting autophagy to fight hematopoietic malignancies. *Cell Cycle* **2010**, *9*, 3470–3478. [CrossRef] [PubMed]
105. Santidrian, A.F.; Gonzalez-Girones, D.M.; Iglesias-Serret, D.; Coll-Mulet, L.; Cosialls, A.M.; de Frias, M.; Campas, C.; Gonzalez-Barca, E.; Alonso, E.; Labi, V.; et al. AICAR induces apoptosis independently of AMPK and p53 through up-regulation of the BH3-only proteins BIM and NOXA in chronic lymphocytic leukemia cells. *Blood* **2010**, *116*, 3023–3032. [CrossRef] [PubMed]
106. Van Den Neste, E.; Cazin, B.; Janssens, A.; Gonzalez-Barca, E.; Terol, M.J.; Levy, V.; Perez de Oteyza, J.; Zachee, P.; Saunders, A.; de Frias, M.; et al. Acadesine for patients with relapsed/refractory chronic lymphocytic leukemia (CLL): A multicenter phase I/II study. *Cancer Chemother. Pharmacol.* **2013**, *71*, 581–591. [CrossRef] [PubMed]
107. Kim, J.; Yang, G.; Kim, Y.; Kim, J.; Ha, J. AMPK activators: Mechanisms of action and physiological activities. *Exp. Mol. Med.* **2016**, *48*, e224. [CrossRef] [PubMed]
108. Riffelmacher, T.; Simon, A.K. Mechanistic roles of autophagy in hematopoietic differentiation. *FEBS J.* **2017**, *284*, 1008–1020. [CrossRef] [PubMed]

7

AMPK Function in Mammalian Spermatozoa

David Martin-Hidalgo [1,2,†], Ana Hurtado de Llera [1,3,†], Violeta Calle-Guisado [1], Lauro Gonzalez-Fernandez [1], Luis Garcia-Marin [1] and M. Julia Bragado [1,*]

1 Research Group of Intracellular Signaling and Technology of Reproduction (SINTREP), Institute of Biotechnology in Agriculture and Livestock (INBIO G+C), University of Extremadura, 10003 Cáceres, Spain; davidmh@unex.es (D.M.-H.); anahl@unex.es (A.H.d.L.); violetacg@unex.es (V.C.-G.); lgonfer@unex.es (L.G.-F.); ljgarcia@unex.es (L.G.-M.)

2 Unit for Multidisciplinary Research in Biomedicine (UMIB), Laboratory of Cell Biology, Department of Microscopy, Institute of Biomedical Sciences Abel Salazar (ICBAS), University of Porto, 40050-313 Porto, Portugal

3 Hormones and Metabolism Research Group, Faculty of Health Sciences, University of Beira Interior, 6200-506 Covilhã, Portugal

* Correspondence: jbragado@unex.es

† These authors contributed equally to this work.

Abstract: AMP-activated protein kinase AMPK regulates cellular energy by controlling metabolism through the inhibition of anabolic pathways and the simultaneous stimulation of catabolic pathways. Given its central regulator role in cell metabolism, AMPK activity and its regulation have been the focus of relevant investigations, although only a few studies have focused on the AMPK function in the control of spermatozoa's ability to fertilize. This review summarizes the known cellular roles of AMPK that have been identified in mammalian spermatozoa. The involvement of AMPK activity is described in terms of the main physiological functions of mature spermatozoa, particularly in the regulation of suitable sperm motility adapted to the fluctuating extracellular medium, maintenance of the integrity of sperm membranes, and the mitochondrial membrane potential. In addition, the intracellular signaling pathways leading to AMPK activation in mammalian spermatozoa are reviewed. We also discuss the role of AMPK in assisted reproduction techniques, particularly during semen cryopreservation and preservation (at 17 °C). Finally, we reinforce the idea of AMPK as a key signaling kinase in spermatozoa that acts as an essential linker/bridge between metabolism energy and sperm's ability to fertilize.

Keywords: AMP-activated protein kinase (AMPK); spermatozoa; motility; mitochondria; membranes; signaling; stress; assisted reproduction techniques

---

1. Introduction

The AMP-activated protein kinase (AMPK) acts as a sensor molecule of cellular energy charge that maintains energy homeostasis at both the whole-body and cellular levels [1,2] by controlling key anabolic and catabolic pathways under different energy or stressful conditions. Thus, AMPK can be activated by physiological conditions, mediated by metabolic hormones, by different cell stress conditions, and also by pharmacological molecules [1,2]. This kinase is highly conserved in eukaryotic species and is ubiquitously expressed. Structurally, AMPK is a heterotrimeric protein including an alpha (α) catalytic subunit, a scaffolding beta (β), and regulatory gamma (γ) subunits, which are encoded by seven genes, allowing the formation of different heterotrimer combinations of α, β, and γ isoforms of AMPK (Figure 1).

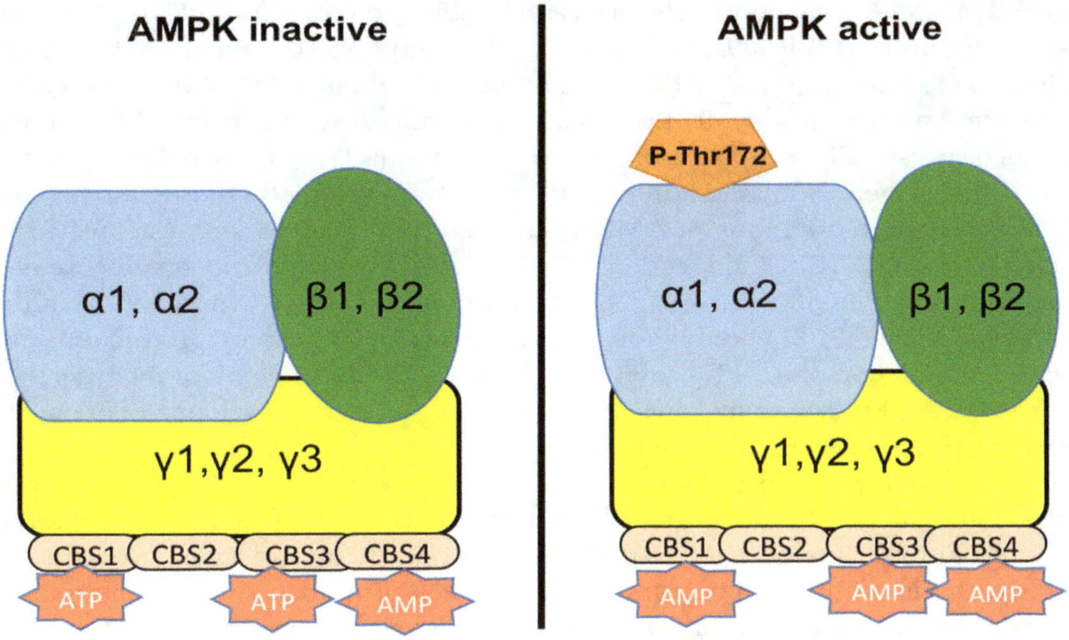

Figure 1. Structure of AMP-activated kinase (AMPK). AMPK is a heterotrimeric protein with 3 differents subunits: catalytic α (blue), scaffolding β (green) and regulatory γ (yelow), each one can be expressed as different isoforms, enabling distinct heterotrimeric combinations of AMPK. Regulatory subunit γ controls the activity of the α subunit through 4 tandem repeats of the motif CBS (cystathionine β-synthase) that bind adenosine nucleotides (AMP, ADP, ATP) with high affinity. AMPK remains inactive in presence of high levels of ATP, whereas during starvation or when ATP level decreases, CBS domains bind AMP, which favors the phosphorylation at Thr172 (catalytic α subunit) and subsequently leading to AMPK activation.

Given its central regulator role in metabolism and its pleiotropic action in many cellular processes, AMPK has received enormous interest as a therapeutic target, and therefore, there has been a great effort to develop pharmacological activators. In addition, AMPK has been the focus of relevant works, although the majority of them have been conducted in somatic cells. However, the ability to preserve energy homeostasis under fluctuating extracellular nutrients or conditions that physiologically occur through the female reproductive tract is a necessary characteristic of spermatozoa. As mentioned before, the direct molecular connector between the substrate's supply and energy requirement is AMPK. In 2012, two different research groups identified for the first time the presence of AMPK in mammalian spermatozoa [3,4]. Since then, interesting scientific efforts have been performed to elucidate the functions of AMPK in these male gametes in several species, including human, using different experimental approaches, including the use of α1AMPK knockout mice [3], and the pharmacological inhibitor and activators. To date, availability of AMPK inhibitors is very restricted and the most widely-used inhibitor of AMPK is the pyrazolopyrimidine derivate compound C or dorsomorphin (CC) that binds to the active site of AMPK and acts as an ATP-competitive inhibitor [5]. Very recently, a new pyrimidine derivative, SBI-0206965, has been discovered as a direct AMPK inhibitor with greater potency (40 fold) and lower kinase promiscuity than CC [6]. Regarding AMPK activators, great effort has been performed for the last few years in the development of direct small molecules and also in their mechanisms of action. The classical activator is the 5-aminoimidazol-4-carboxamide-1-B-D-ribofuranoside (AICAR), although its therapeutic use is not warranted in humans because of its elevated threshold for AMPK activation [7]. The first direct AMPK activator available to scientists was A769662, which activates AMPK both allosterically and by inhibiting AMPKα Thr172 dephosphorylation through binding at the ADaM site, and displays high specificity for complexes containing AMPK-β1 isoforms compared to AMPK-β2 [7]. An exhaustive review about other AMPK activators has been recently published [7] and includes cyclic benzimidazole derivatives (compound 991, PF-739, and MK-8722), the compound C2, the iminothiazolidione PT-1 and its optimized C24, the bi-quinoline JJO-1 and dihydroxyquinoline MT47-100, and recently indole- and indazole-acid-based AMPK synthetic activators, as well as two novel small activators, PXL770 and O304, which successfully completed phase I of clinical trials [7]. To date, the most clinically-used activator of AMPK is metformin and although its action was originally thought to be mediated by AMPK [8], likely through inhibition of AMP desaminase [9], it is currently established that many of metformin effects are AMPK-independent [10–16]. Thus, besides effects on glucose metabolism, metformin also directly inhibits complex I (NADH:ubiquinone oxidoreductase) of the mitochondrial electron transport chain [17,18] with obvious consequences in cell energy production. Overall, we believe that the modulation of AMPK activity would be an interesting tool for the future knowledge about exact roles of AMPK in mammalian spermatozoa, as well as for development of novel protocols for male fertility and/or sperm preservation in vitro.

In this review, we provide the current knowledge about AMPK function in spermatozoa under physiological and different extracellular conditions. AMPK activity is involved in the main physiological functions of mature spermatozoa, in particular in the maintenance of proper sperm motility, the integrity of sperm membranes, and the mitochondrial membrane potential, all of them adapted to the fluctuating medium of the female reproductive tract. Moreover, the intracellular signaling pathways leading to AMPK activation in mammalian spermatozoa are reviewed. We also discuss the role of AMPK in assisted reproduction techniques, particularly during semen cryopreservation and preservation at low temperature. Finally, we reinforce the idea of AMPK as a key signaling kinase in spermatozoa that acts as an essential connector between metabolism energy and sperm ability to fertilize.

2. Role of AMPK in Male Gonads and Spermatogenesis

Fertility can be affected by nutrition and energy metabolism; in this sense, AMPK plays an important role in the reproductive function, connecting, among other physiological regulators, the hypothalamus–pituitary–gonadal axis with energy balance [19]. Germ cells production in the testis occurs through spermatogenesis and terminates when the mature spermatozoa are released into the lumen of the seminiferous tubules. A proper energy balance is necessary for spermatozoa production in the testis, as well as for sperm function and quality after ejaculation. Nonetheless, only a few studies have focused on the role of AMPK in the control of male fertility.

In 2008, Towler et al. [20] indicated that kinases related to AMPK exert physiological functions in mammalian spermatozoa. Thereby, a shorter isoform of the tumor suppressor LKB1 (LKB1s) was shown to be mainly expressed in haploid spermatids [20] where it plays an essential role in spermiogenesis and in its fertilizing ability. *LKB1s* knockout mice have a high reduction in the number of mature spermatozoa in the epididymis, and the few spermatozoa produced are not motile and have abnormal head morphology, which causes sterile mice [20]. These data suggested that this variant of the LKB1 has a crucial role in spermiogenesis and fertility in mice. A new recent study has confirmed these results, where Kong et al. have shown that although LKB1s is the dominant isoform expressed in mice testis and plays a crucial role in spermiogenesis, the long isoform of the tumor suppressor $LKB1_L$ was also required [21]. This group suggests an indispensable role of $LKB1_L$ in spermatogonial stem cell maintenance and the cooperative regulations of both LKB1 isoforms in spermatids differentiation [21]. On the other hand, deletions of *TSSK1* and *TSSK2*, two members of testis specific TSSK family or Ser/Thr kinases, which belong to AMPK branch in the human kinome tree, cause infertility in chimera mice due to haploinsufficiency [22]. The above-mentioned studies point out that some AMPK-related kinases might play a key role in the spermatozoa production and function.

Male gonadal somatic cells include Sertoli (SCs) and Leydig cells (LCs). Proliferation and survival of somatic cells of the testis are crucial for fertility. In fact, testis size and sperm production are directly correlated to the total number of adult Sertoli cells. These cells play a central role in spermatogenesis by providing support and nutrition for spermatogenic cells; however, each SC nourishes a limited number of differentiating germ cells [23]. SCs maintain the blood–testis barrier, an essential feature of seminiferous tubules, which creates the proper environment for the spermatogenic process to be held [24]. It was demonstrated that maintenance of SCs cytoskeletal dynamics, polarity, and junctional communications are essential for fertility [25], and that disorders on their apical extensions, which direct migration of differentiating germ cells towards the lumen, leads to premature germ cell loss [26]. In 2012, Tanwar et al. showed the importance of the upstream AMPK kinase LKB1 signaling in rats SCs biology and spermatogenesis. They demonstrated that aberrant AMPK-mTOR signaling causes disruption of SCs polarity and spermatogenesis [27]. In the same year, Riera et al. demonstrated that the follicle-stimulating hormone (FSH) regulates SCs proliferation with the participation of the PI3K/Akt/mTORC1 pathway and that AMPK activation may be involved in the detention of proliferation by, at least in part, a decrease in mTORC1 signaling and an increase in cyclin-dependent kinase inhibitor (CDKI) expression [28]. Thus, well-coordinated AMPK-mTOR signaling is essential for SCs functions and survival in rats [27,28]. According to a recent study, heat treatment could reversibly perturb the expression of tight junction proteins in immature porcine SCs by inhibiting the AMPK signaling pathway [29]. The use of the AMPK activator AICAR can affect adhesion molecule expression and influences junction complex integrity in rat SCs, as it has been described in 2010 by Galardo et al. [30]. This group also suggested an important role of AMPK in modulating the nutritional function of SCs. They demonstrated that AMPK stimulation with AICAR increased lactate production and glucose transport [31]. On the other hand, the inactivation of α1AMPK in mice SCs reduces the expression of mitochondrial markers (cytochrome c and PGC1-α) and the ATP content, and conversely, increases the lactate production and lipid droplets [32]. Specific deletion of the α1AMPK gene in mice SCs resulted in a 25% reduction in male fertility associated with abnormal spermatozoa with a thin head. Testes showed no clear alterations of morphology or modification in the number of SCs in vivo, but a deregulation in energy metabolism in SCs was observed [32]. These authors suggested that deregulation of the energy-sensing machinery through disruption of α1AMPK in SCs leads to a reduction in the quality of germ cells and fertility [32]. These results supported those obtained by Tartarin et al. performed in mice, where the indirect AMPK activator metformin decreased testosterone production in vitro and reduced the testicular size and the population of SCs in vivo [3]. The involvement of AMPK activity has also been studied in SCs under different stimuli/conditions and from other species such as boar. Thus, Jiao et al., reported that 17β-estradiol inhibited immature boar SCs viability and cell cycle progression by activating the AMPK signaling pathway. Furthermore, they showed that AMPK activation by AICAR inhibited SCs viability, while the AMPK inhibitor compound C attenuated the effects of 17β-estradiol on SCs [33]. Besides these findings in mammalian testes, it has been reported that AMPK activity also exerts regulatory functions in avian gonads, as metformin treatment also reduced SCs proliferation and increased lactate secretion in chickens [34].

The precise composition of the extracellular environment of germinal cells in the seminiferous tubules remains unknown, but it is well established that SCs secrete molecules that are crucial for the spermatogenic process [35]. In this sense, SCs secrete proteins and cytokines, which participate in the control of spermatozoa movement from the testis to the epididymis, and in the control of the pH of the seminiferous fluid [36]. During spermatogenesis, lactate is the preferred substrate of spermatocytes and spermatids. Lactate production from glucose in SCs is an important point for the control of spermatogenesis [37]. Recently, it has been demonstrated that AMPK reduces heat-induced lactate secretion by decreasing the expression levels of the glucose transporter GLUT3, LDHA, and MCT1 [38]. This work also suggested that AMPK is a negative regulator of heat treatment-induced lactate secretion in cultured boar SCs [38].

It has been demonstrated that AMPK is involved in the regulation of spermatozoa quality through its action, not only on the proliferation of SCs, but also in another type of testicular somatic cells, Leydig cells (LCs), (Figure 2) which are responsible for producing testicular androgens, such as testosterone, for the paracrine regulation of spermatogenesis within the testis in adult life [39].

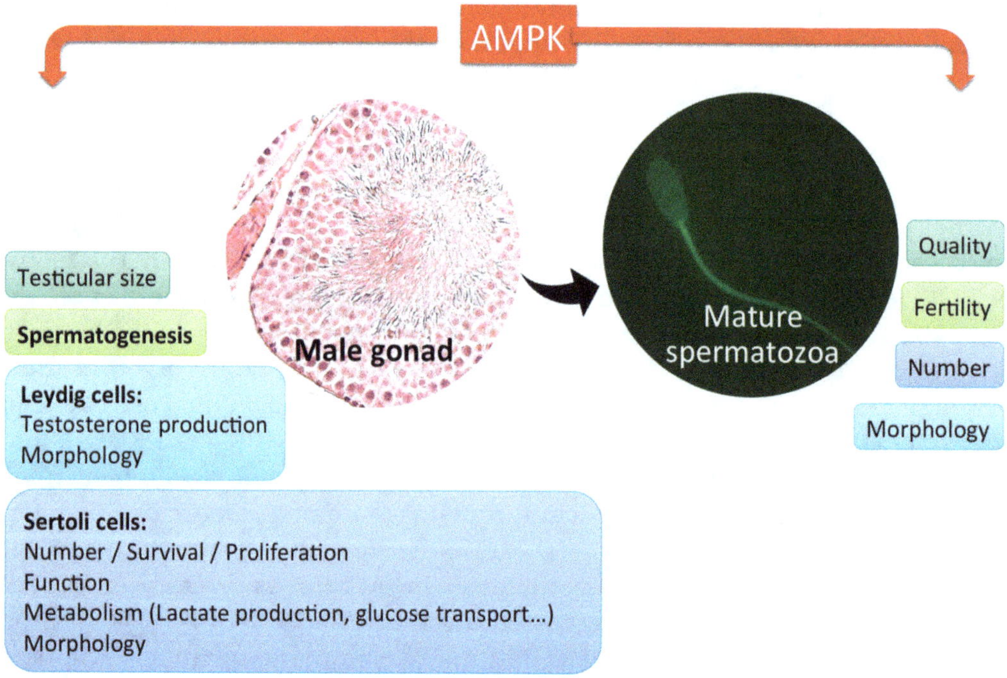

Figure 2. AMPK functions in male gonads and spermatozoa production. AMPK activity is involved in the control of the male gonad function, spermatozoa production and quality by its action also on the male gonadal somatic cells (Sertoli and Leydig cells).

Steroid hormones regulate essential physiological processes, thus, testosterone is also responsible for the differentiation and development of male reproductive organs, as well as for maintaining secondary sex characteristics and sexual function. Within the testis, the functions of the LCs are mainly controlled by the hypothalamus–pituitary axis. An *α1AMPK* knockout mice model has been studied by Tartarin et al. [3]. The male α1AMPK$^{-/-}$ mice have high levels of testosterone due to hyperactive LCs [3]. Indeed, the LCs of these animals have an increased volume, an altered endoplasmic reticulum area, a high intratesticular cholesterol concentration, and a greater expression of proteins involved in steroid production [3]. In agreement with this data, Abdou et al. identified AMPK as a molecular rheostat that actively represses steroid hormone biosynthesis to preserve cellular energy homeostasis and prevent excess steroid production in LCs [40]. Recently, the role of AMPK in steroidogenesis has also been investigated in vivo in ovine testis by Taibi et al. These authors report that active AMPK is expressed in this gonad, and more importantly, they demonstrated that AMPK activity is regulated by nutritional status [41]. However, to date, the expression patterns of the different AMPK isoforms in spermatozoa are unknown. One of the most recent studies revealed that resveratrol enhanced AMPK phosphorylation and may exert its cytoprotective role against oxidative injury by the activation of autophagy via AMPK/mTOR pathway in LCs [42]. Beside mammals, the effect of metformin has also been studied in avian species; thus in 6-week-old chickens, metformin induces a decrease in testosterone levels and reduces testis size [34].

To date, AMPK has also been studied in male gonads from other mammal species such as rats [28,43–48], mice [49–55], lambs [56], monkeys [48], and humans [57]. Taking into account all the aforementioned articles, it can be concluded that the AMPK pathway is involved in the control of the male gonad function, spermatozoa production, and quality by its action also on the male gonadal somatic cells and point out AMPK as a novel gatekeeper of steroidogenesis and a target for modulating steroid hormone production. Therefore, AMPK appears as a key signaling protein for the spermatozoa and male fertility control.

3. Localization of AMPK in Mammalian Spermatozoa

The presence of the AMPK protein in mature spermatozoa was demonstrated independently in 2012 in two mammalian species, boar [4] and mouse [3]. Later, AMPK protein localization has been confirmed in mouse cauda epididymal spermatozoa [58] and also in spermatozoa from stallions [59,60], rats [61], and very recently, goats [62] and humans [63,64]. Immunolocalization techniques reveal that AMPK localization in spermatozoa slightly varies between species. AMPK is localized at the entire acrosome and in the midpiece of flagellum in boar spermatozoa [65]. Interestingly, in this species, when AMPK becomes phosphorylated at Thr172 (active) is specifically restricted to the most apical part of the acrosome and to the subequatorial segment, remaining in the midpiece of the flagellum [65]. In human spermatozoa, AMPK is localized at the entire acrosome, the midpiece and along the tail of the flagellum [63,64], whereas its active form is found only at the most apical part of the acrosome and along the entire tail, with very slight amounts in the post-acrosomal region and in the midpiece of the flagellum [63]. In stallion spermatozoa, active AMPK is mainly detected in the subequatorial region in the head and in the principal piece of the tail [59]. AMPK localization has also been studied in spermatozoa from avian spermatozoa [66]. Thus, in chicken spermatozoa, AMPK is present in the acrosome, the intermediate part, and the whole flagellum, whereas active AMPK is mainly localized in the flagellum and the acrosome [66].

4. Functions of AMPK in Mammalian Spermatozoa

4.1. Role of AMPK in the Regulation of Spermatozoa Motility

Sperm metabolic plasticity, and particularly the capacity to switch substrates and also to respond to stimuli, renders a transcendent adaptation to sperm extracellular medium fluctuations, which might include exhausted energy, high-energy demand, and low oxygen availability. AMPK plays a central role in the regulation of energy metabolism homeostasis [1,2], and therefore, this kinase has recently emerged as a new signaling pathway that exerts a necessary control of sperm function. The specific and primordial function of spermatozoa fertilization is highly dependent on the proper sperm motility. Motility is a characteristic sperm functional process that is tightly regulated in response to the spermatozoa environment and to the subsequent changes in energy demands during its transit trough the female reproductive tract. AMPK involvement on sperm motility has been demonstrated by different experimental approaches that were independently performed in two mammalian species, boar [4] and mice [3]. Thus, transgenic mice lacking the catalytic subunit $\alpha 1$ gene (*α1AMPK* knockout) have a dramatic reduction in sperm motility and curvilinear velocity [3]. Similarly, a pharmacological approach in boar spermatozoa demonstrates that AMPK inhibition by CC significantly decreases the percentage of motile spermatozoa, the sperm curvilinear velocity and subsequently diminishes the percentage of rapid spermatozoa (with average velocity >80 μm/s), also affecting other motility parameters and coefficients evaluated specifically using a computer-assisted sperm analysis system [4]. Later, the important regulatory role of AMPK in spermatozoa motility was demonstrated for other mammalian species such as human [63,64], rat [61], stallion [60], and goat [62]. This clear role of AMPK in sperm motility suggests that this kinase might phosphorylate downstream substrates including proteins of the axoneme or in other related structures that are indispensable for sperm flagellar motility, as it has been previously demonstrated for the AMPK related kinase TSSK2 [67]. Thus, sperm TSKK2 phosphorylates the axoneme protein SPAG16L (sperm associated antigen 16) in vitro, a protein necessary for flagellar motility in mouse spermatozoa [68]. Surprisingly, by using a specific pharmacological AMPK activator, A769662 [69], it has been shown that an increase in AMPK activity above physiological levels also adversely influences sperm motility in two mammalian species, boar [70] and human [71]. Therefore, it has been proposed [70,72] that either up or down fluctuations of sperm AMPK activity (away from energy charge-regulated physiological levels) cause a negative role in mammalian sperm motility. A range of physiological levels of AMPK activity is essential to accomplish optimal sperm motility (Figure 3) that must adapt to the fluctuating extracellular conditions that spermatozoa can be physiologically exposed to [19,70,72].

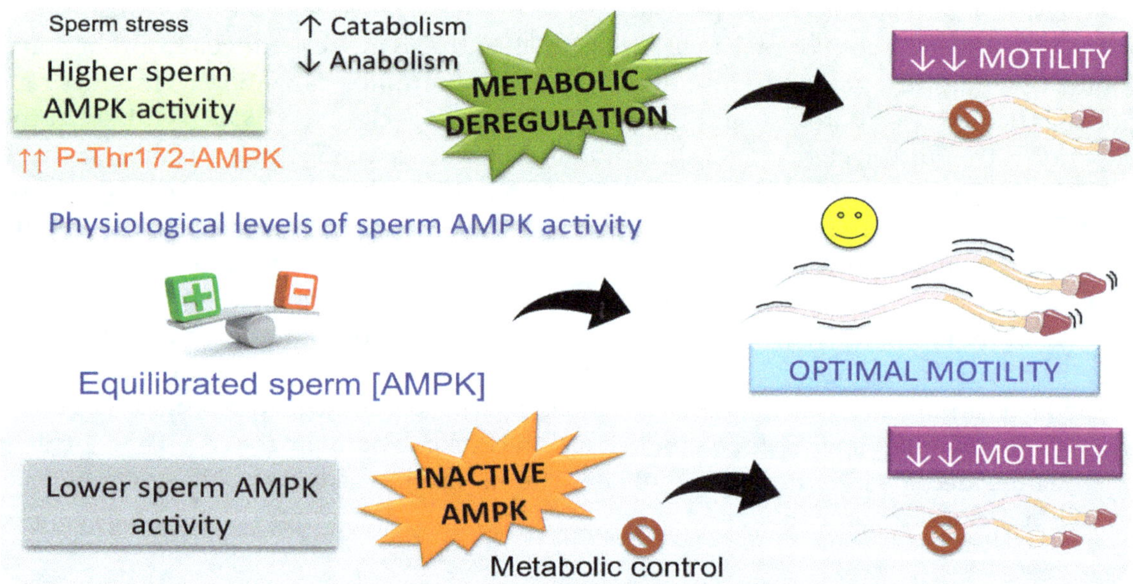

Figure 3. Different AMPK activity levels control spermatozoa motility. *In vivo*, during the spermatozoa transit trough the female reproductive tract, fluctuating levels of AMPK activity might occur in spermatozoa depending on the sperm energy demands. A physiological range of energy-charge regulated levels of AMPK activity is essential to accomplish optimal sperm motility that is adapted to the fluctuating extracellular conditions within the female reproductive tract. Up or down fluctuations of sperm AMPK activity cause a negative role in mammalian sperm motility.

In vivo, during the sperm transit trough the female reproductive tract, fluctuating levels of AMPK activity might occur in spermatozoa depending on the sperm energy demands. Whenever environment-depending demands of energy lead to a lower sperm AMPK activity that falls below the so-called physiological range, inactive AMPK is unable to adjust sperm metabolic pathways to control and maintain energy levels required for spermatozoa motility, which is negatively affected, as demonstrated in several species [3,4,62,63]. The proposal that a physiological range of AMPK activity is necessary for appropriate sperm motility is firmly supported by a study performed in transgenic mice lacking the catalytic subunit α1 gene [α1AMPK knockout (KO)] which presents a great reduction in sperm motility and curvilinear velocity [3]. Whenever environment-depending demands of energy lead to a sustained sperm AMPK activity raised above the physiological range (i.e., sperm stresses), the sustained AMPK activity would lead to a deregulation of sperm metabolism caused by a prolonged stimulation of ATP-generating catabolic pathways and by a sustained inhibition of ATP-consuming anabolic pathways (Figure 3). Thus, the deregulation of sperm metabolic pathways is not suitable to accomplish proper sperm motility under any extracellular conditions, as demonstrated in A769662-treated spermatozoa from boars [70] and humans [71]. In accordance with these findings, it has been demonstrated that spermatozoa exposure to stress stimuli, such as pollutants like hydrogen sulfide and/or ammonia, which lead to AMPK activation, causes a decline in boar spermatozoa motility [73]. In addition, this proposal is also supported by recent results obtained using a more physiological method of arresting the motility activation of rat caudal spermatozoa by isolating and lowering temperature (0–2 °C) without freezing [61]. AMPK activity in rat quiescent spermatozoa is higher than in motile spermatozoa [61], which supports the idea that AMPK is effectively regulating sperm motility by allowing these male gametes to adapt to extracellular-driven changes in sperm energy charge. Thus, the potent activation of AMPK in rat quiescent spermatozoa guarantees the most efficient pathways of energy metabolism and promotes their survival under a low energy-starved status with very limited energy resources [61]. Under these circumstances of an elevated AMPK activation, rat quiescent spermatozoa are not motile. When sperm substrates/stimuli become available during the initiation of sperm motility after ejaculation, AMPK activity levels diminish compared with quiescent spermatozoa [61], although they are still detectable and likely remain at the so-called physiological range of AMPK activity in rat spermatozoa. This female tract environment-driven shift in AMPK activity to a physiological level is necessary to adjust the proper sperm motility under these new requirements in energy demand [61].

However, very recently, one study found that other AMPK activators, such as AICAR and metformin, improve progressive motility in goat spermatozoa [62]. It is important to mention several arguments that might account for the different results found in this species: (a) The increase in phospho-AMPK levels obtained by AICAR and metformin in goat spermatozoa is very slight [62] compared with the high AMPK activity levels observed in boar [70] or rat [61] that lead to inhibition of sperm motility. Moreover, these two activators, AICAR and metformin, fail to activate AMPK in spermatozoa from other species such boar [70], where only A769662 was found to be effective activating AMPK. In addition, metformin also fails to effectively increase AMPK phosphorylation in human spermatozoa [71]. (b) It is reasonable to assume that besides great differences in the intensity of AMPK activation and the differences attributable to spermatozoa from distinct species, the specificity and effectiveness of each pharmacological activator should be kept in mind, also for further studies. This review tries to summarize the current knowledge about the role of AMPK activity in the regulation of sperm motility, and it seems clear that future research in this area is needed to elucidate the exact molecular mechanism by which AMPK exerts this regulation in sperm cells and also to discover AMPK protein substrates that belong to the intracellular signaling pathways controlling sperm motility.

4.2. Role of AMPK in the Regulation of Spermatozoa Mitochondrial Activity

As an energy regulator protein, AMPK has additionally been pointed out as important pathway that contributes to the maintenance of the mitochondrial membrane potential ($\Delta\Psi m$) in spermatozoa [3,65]. Different experimental approaches designed to inactivate sperm AMPK cause a decrease in sperm $\Delta\Psi m$ in different species: goat [62], boar [65], stallion [60], and α1AMPK knockout mouse [3]. Besides a fall (50%) in sperm $\Delta\Psi m$, α1AMPK knockout (KO) mice exhibited a lower number of mitochondria and had a significant lower consumption (60%) of basal oxygen [3]. Furthermore, a short up-activation of sperm AMPK using A769662 prevents the decrease in the number of Ca^{2+} and bicarbonate-stimulated spermatozoa presenting high $\Delta\Psi m$ in boar [70], whereas it has no significant effect in $\Delta\Psi m$ of human spermatozoa [71]. However, the indirect activator of AMPK resveratrol significantly increased $\Delta\Psi m$ in frozen–thawed human spermatozoa [64]. In mouse fresh spermatozoa, another indirect activator of AMPK metformin causes a concentration-dependent decrease of $\Delta\Psi m$ [74]. Thus, AMPK activity seems to be involved in the regulation of sperm $\Delta\Psi m$, although there are differences between species, the (direct or indirect) pharmacological compound used to activate AMPK and the cell status of spermatozoa (fresh or frozen–thawed). In any case, the effect of AMPK in sperm $\Delta\Psi m$, when demonstrated, is clearly dependent on sperm extracellular stimuli [70,72], pointing out AMPK as a metabolic checkpoint by integrating stimuli-induced signaling with sperm metabolism [70]. Therefore, it has been proposed that AMPK activity (at specific levels) is essential to maintain sperm mitochondrial membrane potential [70,72]. Thus, AMPK contributes to modulate the mitochondrial membrane potential according to extracellular stimuli or conditions-derived sperm energy, and subsequently promotes the proper sperm motility, at least in some species. The involvement of AMPK in the modulation of sperm $\Delta\Psi m$ is further supported by the localization of important amounts of active AMPK (Thr172 phosphorylated) at the midpiece of mammalian spermatozoa where mitochondria are located, as it has been demonstrated in boar [65], goats [62], stallions [60], and humans [63].

4.3. Role of AMPK in the Regulation of Spermatozoa Membranes

Mammalian spermatozoa need a tight modulation of energy also to maintain cellular structure, stability, and physiology of membranes during fluctuating extracellular conditions within the female reproductive tract that ultimately allow them oocyte fertilization [75,76]. An important factor that contributes to the correct function of spermatozoa is the degree of lipid organization of their plasma membrane. In fact, the energy charge-sensor kinase AMPK plays a role in the maintenance of sperm plasma membrane fluidity and lipid organization as it has been demonstrated in some species such as boar [65] and goat [62]. Thus, experiments using either an AMPK inhibitor (CC) in boar [65] and goat spermatozoa [62], or the activator A769662, demonstrated a significant increase in plasma membrane lipid disorganization [70]. Importantly, this effect of AMPK is clearly dependent of the stimuli present in the extracellular medium, as sperm plasma membrane disorganization is ≈3-fold higher when boar spermatozoa are incubated in the presence of capacitating stimuli (Ca^{2+}, bicarbonate, and serum albumin) than in a stimuli-free medium [65]. All together, these findings suggest that specific physiological levels of AMPK play a regulatory role in the preservation of the proper lipid organization and fluidity in sperm plasma membrane according to extracellular stimuli [65,77]. Furthermore, physiological levels of bicarbonate cause a quick collapse of the asymmetry of the sperm plasma membrane attributable to the scramblases activation, which translocate phospholipids such as phosphatidylethanolamine and phosphatidylserine (PS) outward of plasma membrane [78]. The PS externalization evidences plasma membrane scrambling which physiologically takes place in important sperm processes. Interestingly, AMPK inhibition in spermatozoa incubated in the presence of bicarbonate result in a significant inhibition of the outward exposure of PS in plasma membrane, suggesting that inhibition of AMPK, at least over a short time (4 h), could be causing a downstream inhibition of scramblases activity [65]. Moreover, an enhanced AMPK activity (for 24 h) induces a significant PS externalization in boar sperm plasma membrane [70]. These functional consequences of AMPK in boar sperm plasma membrane, which are lipid disorganization and the PS translocation, are presumably taking place at the plasma membrane surrounding the most apical acrosome region, where most of active phospho-Thr172 AMPK is restricted in this specie [65]. The involvement of AMPK in boar sperm membranes organization is supported by the finding that AMPK is downstream of the cAMP/PKA pathway [79], which is an essential regulatory pathway for the lipid architecture of the sperm plasma membrane [80,81].

Besides the regulatory role at the sperm plasma membrane, AMPK also regulates acrosome membrane integrity under capacitating conditions [65], which is additionally supported by the fact that high levels of active phospho-Thr172-AMPK are situated at this part of acrosome under physiological conditions [65]. By using different experimental approaches it has been demonstrated that any oscillation either up (A769662) or down (CC) of the AMPK activity outside of physiological levels lead to a loss of the outer acrosome membrane integrity [65,70]. In summary, physiological sperm AMPK activity is also essential to keep the integrity of acrosome membrane at a level suitable to the fluctuating extracellular conditions at which spermatozoa are physiologically exposed.

4.4. Role of AMPK in the Regulation of Spermatozoa Acrosome Reaction

Mammalian spermatozoa need to acquire the ability to reach the oocyte, penetrate the cumulus oophorus, and to bind to the zone pellucida of the oocyte, which subsequently triggers the acrosome reaction that leads to egg fertilization [76]. The acquisition of these spermatozoa functional competences occurs through important physiological and biochemical modifications that are collectively named sperm capacitation. Spermatozoa processes that necessarily must occur during capacitation are dependent on the energy levels of this gamete [82,83]. Therefore, it might be reasonable to assume that the energy charge-sensitive AMPK should be somehow involved in the acquisition of these sperm processes, although to date, few works have considered this issue. In fact, the activity of the energy charge-sensitive AMPK is regulated under capacitating conditions (including Ca^{2+}, bicarbonate, and serum albumin) in boar spermatozoa [65]. The inhibition of AMPK activity significantly reduces the integrity of the acrosomal membrane in a boar sperm-capacitating medium, whereas it has no effect in a stimuli-free medium [65]. These results indicate that AMPK is involved in the regulation of sperm events that take place at acrosome membrane during capacitation, at least in boar spermatozoa. However, although AMPK contributes to the maintenance of acrosome membrane integrity, is not likely involved in the sperm process of acrosome reaction, as the inhibition of AMPK in capacitated boar spermatozoa did not affect the acrosome reaction triggered by the calcium ionophore A23187 [65]. This lack of effect of AMPK activity in the acrosome reaction has been demonstrated also in fresh mouse spermatozoa using the indirect activator of AMPK metformin [74]. Moreover, AMPK up-activation does not affect the integrity of the sperm acrosome membrane. Thus, it has been demonstrated that the activator A769662 has no measurable effect in the integrity of the acrosome membrane in human [71] or boar spermatozoa, independently of stimuli present in the medium [79]. However, the effect of A769662 in boar seems to be dependent on the stimulation time as a sustained AMPK activity over 24 h causes a significant loss of outer acrosome membrane integrity [79]. In contrast to mammals, few works performed in avian sperm show that AMPK activity plays a role in the acrosome reaction [66,84]. These differences between mammals and avian spermatozoa highlight the importance of further research about the functional role of AMPK in the acrosome reaction. To date, it can be concluded that in parallel to modulate physiological changes (lipids organization) that occurs at the mammalian plasma membrane during the sperm capacitation, AMPK activity contributes to the maintenance of the outer acrosome membrane integrity, where a majority of phospho-Thr172-AMPK active is localized at physiological conditions [4].

5. Signaling Pathways Leading to AMPK Activation in Spermatozoa

The function of AMPK as a key regulatory molecule of the essential processes that contribute to the spermatozoa function of fertilization has been reviewed above. Therefore, an important issue in spermatozoa physiology is to elucidate the signaling pathways leading to AMPK activity (Figure 4). Several works using different experimental approaches demonstrate that soluble adenylate cyclase (sAC), cAMP [4], and cAMP-dependent protein kinase (PKA)-mediated pathway lie upstream of AMPK activity in boar spermatozoa [65,79].

The sperm cAMP/PKA pathway might promote AMPK activation through its upstream kinase LKB1, as it occurs in somatic cells [85], where it is reported that LKB1 can be directly phosphorylated at Ser431 by PKA in response to activation of adenylate cyclase by forskolin [86,87] or IBMX [86]. It has been demonstrated that the short splice variant of $LKB1_S$ is highly expressed in haploid spermatids in mice testis [20], where it is critically involved in spermiogenesis and fertility [88]. The knocking out $LKB1_S$ in male mice causes important reproductive consequences: sterility, a marked decrease in the number of mature spermatozoa, spermatozoa showing an abnormal head morphology, and are non-motile. An increase in cAMP concentration might also lead to AMPK activation trough cAMP degradation to AMP by phosphodiesterases, as occurs in somatic cells [89]. Thus, any sperm stimulus that increases intracellular cAMP levels could promote AMPK activation either (i) by direct (allosteric) activation of PKA, or (ii) by indirect activity of phosphodiesterases, which increase AMP levels that allosterically activate AMPK, or (iii) by both pathways.

The intracellular messenger Ca^{2+} is also an essential regulator of spermatozoa functional processes. Intracellular Ca^{2+} activates the specific sperm sAC and its downstream signaling through PKA, therefore potently activates AMPK in boar spermatozoa under physiological conditions [65,79]. However, spermatozoa from the same species under conditions considered high extracellular Ca^{2+} concentrations (>3 mM), sperm AMPK phosphorylation decreases [90]. The involvement of Ca^{2+} as an upstream regulator of AMPK activity has also been demonstrated in avian spermatozoa, as Ca^{2+} entry via store-operated Ca^{2+} channel (SOC) activates AMPK [84]. Moreover, Ca^{2+} can also lead to the activation of boar sperm AMPK through the activation of Ca^{2+}-calmodulin dependent kinase kinases II and CaMKKα/β as demonstrated in spermatozoa from boar [79] or chicken [91], which lie upstream of AMPK in somatic cells [92,93].

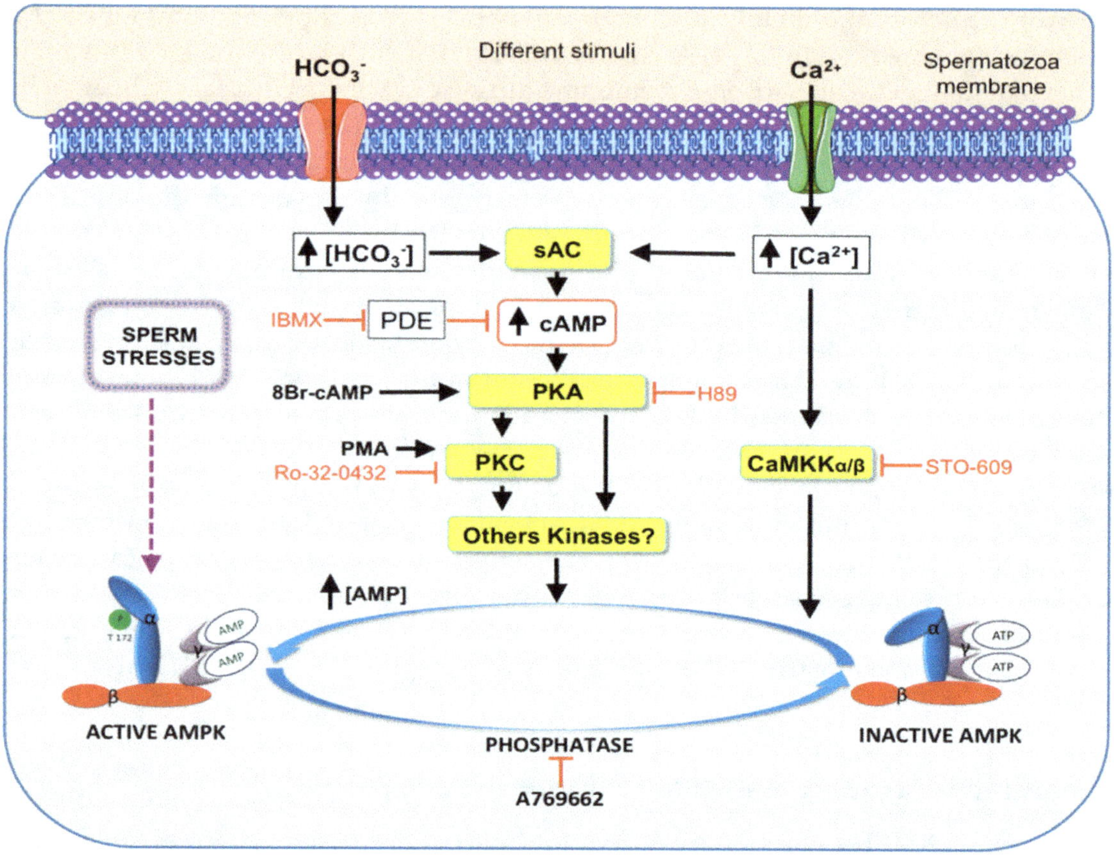

Figure 4. Signaling pathways underlying the regulation of AMPK activity in mammalian spermatozoa. This Figure summarizes the intracellular kinases and mechanisms that have been demonstrated to be involved in AMPK activation by Thr172 phosphorylation: soluble adenylyl cyclase (sAC), cAMP, protein kinase A (PKA), protein kinase C (PKC), intracellular Ca^{2+}, calcium and calmodulin kinase kinases α/β (CaMKK α/β) and different types of cell stresses detailed in the text (absence of Ca^{2+} by BAPTA-AM, hyperosmotic stress, inhibition of mitochondrial activity). Pharmacological inhibitors of different AMPK upstream kinases (H89, IBMX, STO-609 and Ro-0432) are also indicated with red lines.

Sperm AMPK phosphorylation is also stimulated by direct activation of PKC with phorbol 12-myristate 13-acetate (PMA), whereas the PKC inhibitor Ro-32-0432 inhibits HCO_3^- and Ca^{2+}-induced AMPK activation in boar spermatozoa [79]. These findings indicate that at least one isoform of PKC is upstream of sperm AMPK. Several isoforms of PKC have been identified in mammalian spermatozoa: PKCα and PKCβI in bovines [94], PKC-zeta in hamsters [95], and mouse spermatozoa [96]. It is therefore plausible that some of these PKC isoforms might exert in spermatozoa a similar function than in somatic cells by phosphorylating LKB1 and subsequently leading to AMPK activation. Thus, it has been demonstrated that PKC-zeta phosphorylates LKB1 at Ser 307 [97] and PKCζ phosphorylates LKB1 at Ser 399 [98]. An alternative explanation describing the pathway by which PKC is upstream of sperm AMPK activity is based on the fact that PKC activity lies downstream of PKA in the control pathway of boar sperm motility [99]. Previously, Harayama and Miyake (2006) demonstrated that cAMP/PKA signaling can induce the activation of calcium-sensitive PKCs, which are responsible for boar sperm hyperactivation [100]. Thus, it is proposed that another PKC isoform(s) besides PKC-zeta, which is not calcium sensitive, could likely be involved in AMPK activation, at least in response to an elevation of cAMP levels in boar spermatozoa [79].

In addition to the mentioned physiological mimicking conditions, AMPK becomes markedly activated in boar spermatozoa under different stimuli considered to be cell stress (Figure 1), such as inhibition of spermatozoa mitochondrial activity by blocking electron transport chain and sorbitol-induced hyperosmotic stress [79]. In somatic cells, cell stress-induced AMPK activation can be mediated by (i) an increase in AMP levels, and/or (ii) reactive oxygen species ROS generation that act as signaling molecules to activate AMPK [101] through LKB1 and CaMKKs pathways. Surprisingly, the incubation with the intracellular calcium chelator BAPTA-AM in a Ca^{2+}-free medium leads to a strong increase in AMPK activity [79], which may be mediated through an increase in nitric oxide NO· production. In this sense, de Lamirande et al. (2009) demonstrated in human sperm that BAPTA-AM promotes the production of a reactive oxygen specie, the nitric oxide NO· [102]. Accordingly, AMPK activation is also directly influenced by cellular redox status in somatic cells, as H_2O_2 activates AMPK through oxidative modification of cysteine residues in the AMPKα subunit [103]. An alternative or simultaneous explanation is that NO· produced by BAPTA-AM in boar spermatozoa might interacts with the cAMP pathway, as it occurs in human [102], leading to AMPK activity.

Moreover, other different types of stimuli that activate AMPK in mammalian spermatozoa have recently been described. Thus, pharmacological compounds, such as the anti-diabetic drug rosiglitazone, increase AMPK phosphorylation in stallions [60]. Also, some toxicant compounds that cause a marked reduction in sperm motility, such as the heavy metal cadmium, a major environmental toxicant, affected AMPK activity in mouse spermatozoa [104], or the air pollutants hydrogen sulfide and ammonia that activated AMPK activity in boar spermatozoa [73]. Additionally, the common natural mycotoxin ochratoxin A triggers AMPK activation to cause a clear decrease in boar sperm motility [105]. Based on all above-mentioned works, the exact molecular mechanisms involved in the signaling pathway(s) triggered by a variety of stimuli that lead to AMPK activity as well as the identity of AMPK downstream targets that ultimately control sperm function, undoubtedly deserve future investigations.

6. Role of AMPK during Assisted Reproduction Techniques: Semen Preservation

Assisted reproduction techniques (ART) include a wide range of technologies used to improve the chances to achieve pregnancy after the collection and handling of oocytes, sperm, and embryos in in vitro conditions. Nowadays, around 1.5 million human ART cycles are performed each year worldwide, with a reported 333,000 babies born. Data from human spermiograms show a decrease of sperm quality in the last decades [106–108], and as a consequence, an increase of the use of ART, such as intracytoplasmatic sperm injection (ICSI), sperm cryopreservation, in vitro fertilization (IVF), or the use of spermatozoa from testicle biopsies are the most commonly used ART. In the animal field, the use of ART is mostly aimed to quickly spread genetic material of selected animal to preserve gametes from endangered species, also to reduce disease transmission risk or to increase animal breading. This current and future reality leads to an increasing ART use and develop of new technical strategies to overcome fertility problems.

6.1. Why AMPK Protein Is Important in ART?

As mentioned before, sperm capacitation is a mandatory process to fertilize oocytes and the events related with sperm capacitation (i.e.,: hypermotility, acrosome reaction, protein phosphorylation, etc.) consume large amounts of energy (ATP). It was also stated that AMPK acts as an energy sensor and it is activated under physiological and/or stressful sperm conditions. Therefore, there is a growing interest in this kinase to improve ART procedures where gametes are handled and subjected to different types of stress such as mechanical (centrifugation, pipetting), changes in temperature, levels of CO_2 and O_2, medium composition, extracellular matrix (solids, plastics), or to light exposure [109] (Figure 5).

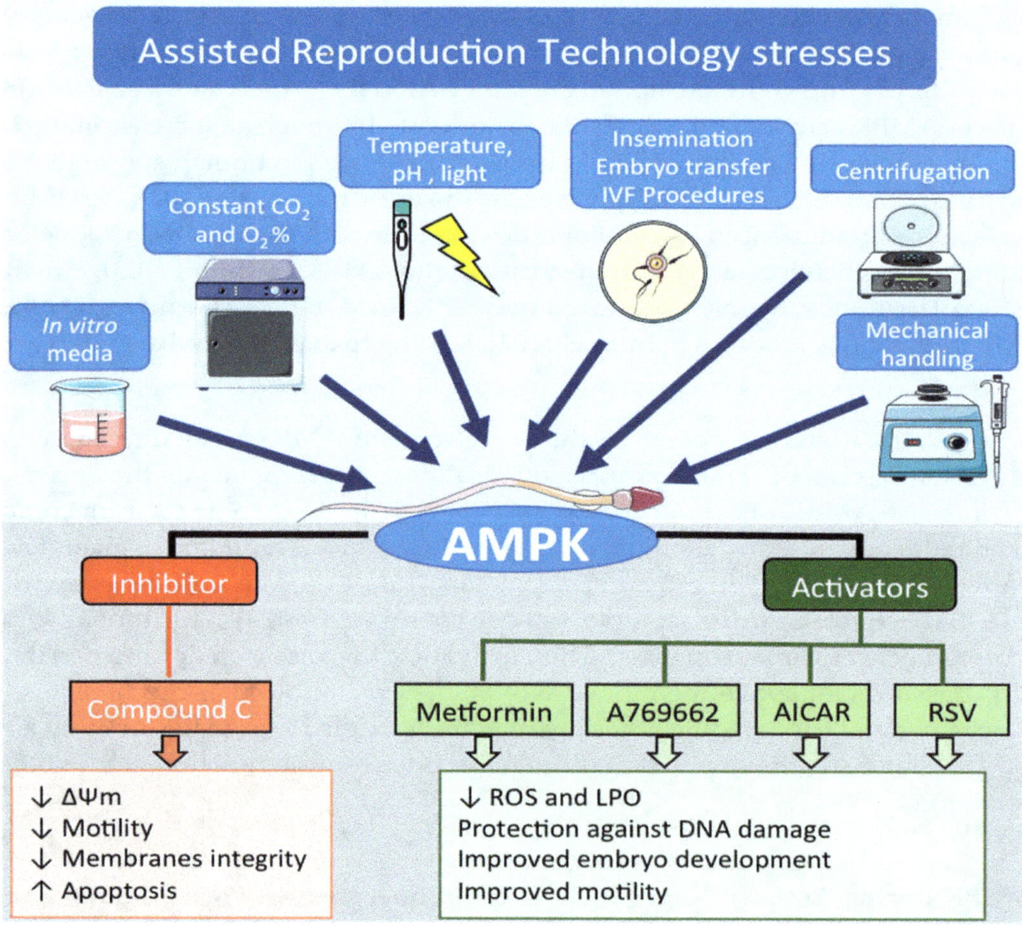

Figure 5. Effects of the modulation of AMPK activity (up or down) in spermatozoa during Assisted Reproduction Techniques (ART), including In Vitro Fertilization (IVF). During ART, spermatozoa are subjected to different stress conditions (non-physiological temperature and media, light exposure, mechanical handling, centrifugation or constant levels of CO_2 and O_2) that might modify AMPK activity. This figure summarizes the reported (beneficial or detrimental) actions in spermatozoa when extender media used in ART were supplemented with AMPK inhibitor (CC), AMPK activators (A769662 and AICAR) or indirect AMPK activators such as metformin and resveratrol (RSV).

6.2. AMPK as a Tool to Improve ART

The sector of animal farm is very interested in ART since livestock genetic background can be easily and successfully improved by the use of these techniques. Artificial Insemination (AI) was the first widely accepted ART for livestock. This technology allows genetically superior males to produce more offspring than it would be possible through conventional mating. More than 90% of western Europe's pig industry performs AI using boar seminal doses preserved at 17 °C as a routine technique [110]. The temperature decrease during this type of preservation is aimed to reduce sperm metabolism, which subsequently leads to lower rates of both acidification and reactive oxygen species (ROS) production in the storage extender, both derived from accumulation of CO_2 and lactic acid from oxidative phosphorylation and glycolysis respectively, obtaining the so called "quiescent spermatozoa". Based on the principle that AMPK acts as a stress and energy regulator, it was reported that AMPK becomes activated when quiescent rat spermatozoa are subjected to decreasing temperatures down to 4 °C [61]. Several works have studied AMPK with the aim to improve ART when spermatozoa are stored under non-physiological temperature during semen preservation techniques. A study about boar seminal doses conservation at a non-physiological temperature (17 °C), demonstrated that before semen preservation (day 0), sperm AMPK phosphorylation was undetectable but sperm AMPK activity increased during the following days of preservation, reaching a maximum phosphorylation at day 7 [77]. Therefore, during semen preservation at 17 °C, at least in boar spermatozoa, the activity of AMPK fluctuates according to extracellular conditions that include stress, and is tightly regulated during the conservation period. This AMPK activation could be associated with a decrease of intracellular ATP levels described during boar sperm conservation [111,112]. Effects of different pharmacological drugs that influence AMPK activity (anti-diabetic compounds, such as metformin and rosiglitazone, or the inhibitor CC) during semen preservation, have been studied in different species [60,77,113]. In boar spermatozoa, only the percentage of motile spermatozoa was improved after a short preservation period (<4 days) when compound C was added to the extender [77]. However, other beneficial effects in semen quality were not found with the addition of CC [77] or metformin [113] to the boar semen preservation media. Both pharmacological approaches (metformin and CC) used during boar sperm conservation for 10 days cause similar negative effects: a reduction of sperm motility and lower mitochondria membrane potential [77,113]. Therefore, as mentioned before, any alteration of AMPK activity away from physiological levels during semen preservation leads to a detrimental effect on the main sperm functions [72,77]. These results in boar have been supported by a recent work in equine spermatozoa where compound C had a detrimental effect by decreasing sperm motility [60]. However, incubation of equine spermatozoa at physiological temperature with rosiglitazone, another antidiabetic compound that activates AMPK [114], increases the percentages of motile and rapid spermatozoa [60]. These authors suggest that rosiglitazone either protects and enhances mitochondria metabolism or induces a shift in stallion spermatozoa to a glycolytic metabolism (glucose uptake is increased), subsequently decreasing ROS and increasing intracellular ATP levels [60]. These positive effects in sperm motility led to authors to test rosiglitazone effect on stallion seminal doses preserved at 24 °C and results showed an improvement of equine sperm motility, suggesting its addition to stallion extender during preservation [60].

Sperm cryopreservation is other ART widely used that allows preserved indefinitely spermatozoa. Nevertheless, this procedure is harmful to the cells causing mainly structural damages to sperm membrane, DNA fragmentation and metabolic changes that restrict its fertilization chances [115]. These deleterious effects during cryopreservation are caused by cold shock, intracellular ice formation, oxidative stress, hypertonic damage, and combinations of these and others stressful conditions [115]. Many studies have tried to improve male gamete cryopreservation by adding antioxidants to the semen extender to reduce oxidative stress [116]. In this regard, the natural compound resveratrol that activates AMPK in some somatic cell types, when added to cryopreservation media, had positive effects on human spermatozoa [64,117]. Thus, AMPK was activated and ROS levels were reduced in human spermatozoa cryopreserved in the presence of resveratrol [64]. Moreover, DNA damage on human cryopreserved spermatozoa was ameliorated when resveratrol was added before cryopreservation, likely mediated by AMPK [117]. However, when resveratrol was added to boar seminal doses preserved at 17 °C, no improvement was detected in any quality sperm parameter analyzed, and negative effects were reported in mitochondria membrane potential, intracellular ATP content and motility [118]. Thus, the possible beneficial effects of resveratrol during semen preservation seem to be dependent on the species studied or the preservation protocol, where different actions of resveratrol might preponderate depending of the temperature, i.e., at −196 °C, the scavenging capacity preponderates, while at 17 °C, resveratrol negative actions might be due to Ca^{2+} mobilization over time. Besides mammalian spermatozoa, the effects of the classical AMPK activator AICAR or the indirect activator metformin were studied in avian spermatozoa. Thus, chicken spermatozoa cryopreserved in the presence of AICAR or metformin had stimulated sperm anti-oxidative defenses by partially restoring superoxide dismutase, glutathione peroxidase, and glutathione reductase activities, which subsequently decreased ROS levels and lipid peroxidation, leading to an improved sperm cryopreserved quality [119]. Also, in order to decrease oxidative stress, metformin was added (5 to 5000 μM) before mouse spermatozoa cryopreservation [74]. Although only the higher concentration of metformin activated AMPK, any concentration tested doubled sperm survival, motility, and percentage of motile spermatozoa, as well as the fertilization rate and embryo development without any reduction in membrane lipid peroxidation [74]. These results observed in mice are surprising since it is assumed that mitochondria is the major cellular target of metformin that leads to an inhibition of mitochondrial respiration [120]. In fact, boar spermatozoa preserved at 17 °C in the presence of metformin effectively had lower mitochondria membrane potential and also inhibited sperm motility [113]. Interestingly, the quality of cryopreserved stallion spermatozoa was not improved when AMPK activity was modified by any drugs tested: CC, AICAR, or metformin [59]. No significant improvement was observed in stallion sperm survival, total motility, progressive motility or live sperm with a structurally intact acrosome [59].

In general, the effects of the manipulation (up or down) of AMPK activity during ART protocols seem contradictory and are basically dependent on the pharmacological agent, species, and/or the ART protocol studied. As an example, whereas resveratrol did not have any positive effect during boar spermatozoa preservation at 17 °C [118], positive effects were described in cryopreserved human spermatozoa [64,117]. While metformin did not have any positive effect during stallion spermatozoa cryopreservation [59], very positive outcomes were described for cryopreserved mouse spermatozoa [74]. Moreover, it should be kept in mind that the effects of the different AMPK activators in spermatozoa from the same species vary according to the ART used. For instance, AICAR or metformin do not exert any beneficial effect in stallion spermatozoa cryopreserved for a long period of time at −196 °C [59], whereas the indirect activator rosiglitazone has some positive actions when spermatozoa are preserved a room temperature [60]. Therefore, we cautiously suggest that experiments designed with pharmacological agents that act through modification of sperm AMPK activity should be performed in each particular animal species and also under each specific conditions of ART used in order to address stress-associated problems originated by ART protocols and that ultimately lead to a decrease in sperm quality.

Effects of the AMPK inhibitor CC supplementation to sperm conservation media during ART protocols are described in Table 1.

Table 1. Effects of the use of AMPK inhibitor C (compound C) in spermatozoa functions of several species during Assisted Reproduction Techniques (ART).

Specie	ART/Stress	Sperm Effects	Reference
Boar	Preserved (17 °C)	1. At short period (<4 days): Improves the % of motile sperm 2. At long period (≥4 days): Decreases sperm MMP Decreases acrosome membrane integrity Increases plasma membrane disorganization	[77]
Stallion	Cryopreserved (−196 °C)	No effect described	[59]
Chicken	Cryopreserved (−196 °C)	Decreases motility and antioxidant capacity	[119]
Stallion	Preserved (RT)	Reduces sperm motility	[60]
Human	Cryopreserved (−196 °C)	Deleterious effect in sperm motility and mitochondria Increases apoptotic-like spermatozoa	[64]

As expected, negative effects were found in spermatozoa when the AMPK pathway is inhibited by CC such as reduced mitochondrial membrane potential (by 50%) and basal oxygen consumption (by 60%) in *AMPKα1* knock out mice [3], and detrimental effects on important sperm functions in refrigerated boar spermatozoa [77], human cryopreserved spermatozoa [64], chicken cryopreserved spermatozoa [119], or room temperature preserved stallion spermatozoa [60]. These negative effects can be summarized as: (i) decrease of mitochondrial membrane potential; (ii) sperm motility inhibition; (iii) loss of acrosomal membrane integrity; (iv) increase of plasma membrane disorganization; and (v) increase of apoptotic-like spermatozoa (Table 1).

Nevertheless, although differences were observed between species and depending on the ART used, in general, beneficial effects were found when AMPK was shown to be over-activated by metformin in cryopreserved mouse spermatozoa [74], rosiglitazone in stallion spermatozoa preserved at room temperature [60], AICAR in cryopreserved chicken spermatozoa [119] and resveratrol in cryopreserved human spermatozoa [64,117]. The positive effects observed were: (i) improvement of sperm survival, motility, and velocity; (ii) improvement of sperm antioxidant defense and decrease of lipid peroxidation (LPO); and (iii) improvement of embryo quality (Table 2).

Table 2. Demonstrated effects of direct (AICAR) or indirect AMPK activators (resveratrol, metformin and rosiglitazone) in the spermatozoa functions of different species during Assisted Reproduction Techniques (ART). Cryopreservation was always at −196 °C. * It has not been showed that the indicated compound effectively activates AMPK in spermatozoa from the indicated specie.

Compound	Specie	ART/Stress	Sperm Effects	Ref.
Resveratrol *	Boar	Preserved (17 °C)	Decreases sperm motility, MMP and ATP	[118]
Metformin *	Stallion	Cryopreserved	No effect described	[59]
AICAR *	Stallion	Cryopreserved	No effect described	[59]
Metformin	Mouse	Cryopreserved	Increases sperm motility and viability Increases fertility rate and quality of embryos	[74]
AICAR	Chicken	Cryopreserved	Protects against ROS and lipid peroxidation Improves motility and % rapid spermatozoa	[119]
Metformin	Chicken	Cryopreserved	Protects against ROS and lipid peroxidation Improves motility and % rapid spermatozoa	[119]
Rosiglitazone	Stallion	Preserved (RT)	Improves % motile and % rapid spermatozoa Shifts to glycolytic metabolism, increases glucose uptake and reduces ROS	[60]
Resveratrol	Human	Cryopreserved	Reduces ROS and apoptosis-like spermatozoa	[64]
Resveratrol	Human	Cryopreserved	Protects against apoptotic-like spermatozoa and MMP damage	[117]
Metformin *	Boar	Preserved (17 °C)	Reduces sperm motility and MMP	[113]

However, this issue seems controversial, as negative effects have also been demonstrated, at least regarding metformin supplementation. Thus, sperm motility and mitochondrial membrane potential are negatively affected in boar spermatozoa preserved at 17 °C in the presence of metformin [113]. Above all, AMPK protein should be considered a suitable target to effectively improve assisted reproduction technologies' outcomes where spermatozoa are subjected to different types of stress.

7. AMPK in Spermatozoa: A Physiological Link between Fertility and Energy Metabolism

The central role of AMPK in energy metabolism is well known, although its involvement in fertility control has only recently been pointed out. AMPK is present in female and male gonads and seems to contribute to the different stages of maturation of germ cells and spermatozoa by modulating hormone production (steroidogenesis) and their interaction with nourishing gonadal somatic cells. In mature mammalian spermatozoa, the ability to preserve energy homeostasis under fluctuating extracellular nutrients, physiological stimuli, or even stress conditions that occur through the female reproductive tract is a necessary characteristic of successful spermatozoa. Moreover, the energy charge-sensitive kinase, AMPK, performs an important and essential regulator of the main physiological processes of spermatozoa and therefore represents a proper target to improve assisted reproduction technologies. The increasing scientific evidence about AMPK in the male fertility area points out AMPK as the physiological link between reproduction and energy metabolism.

Moreover, considerable progress has been performed recently in the molecular development of direct AMPK activators/inhibitors. A future scientific challenge will be to integrate some of these promising therapeutic compounds derived from pre-clinical animal studies into practical improvements in the field of assisted reproduction in animals and human.

Author Contributions: Conceptualization, D.M.-H., A.H.d.L., and M.J.B.; Writing-Original Draft Preparation, D.M.-H., A.H.d.L., and M.J.B.; Writing-Review & Editing, D.M.-H., A.H.d.L., V.C.-G., L.G.-F., and L.J.G.-M.; Supervision, M.J.B.; Project Administration, L.G.-M.; Funding Acquisition, L.G.-M., L.G.-F., and M.J.B.

Acknowledgments: D. Martin-Hidalgo and A. Hurtado de Llera are recipients of post-doctoral fellowships of Junta de Extremadura (Spain) and the Fondo Social Europeo (Refs. PO17020 and PO17022, respectively). V. Calle-Guisado is recipient of a PhD Fellowship (Ref. FPU14/03449) from the Ministry of Education, Culture and Sport, Spain. L. Gonzalez-Fernandez is supported by the Grant AGL2015-73249-JIN of Agencia Estatal de Investigación (AEI) of the Ministry of Economy, Industry and Competitiveness, Spain.

Abbreviations

AMPK	AMP-activated protein kinase
ART	Assisted reproduction techniques
ICSI	Intracytoplasmatic sperm injection
IVF	In Vitro Fertilization
SC	Sertoli cells
LC	Leydig cells
sAC	Soluble adenylate cyclase
cAMP	Cyclic adenosine mono phosphate
PKA	cAMP-dependent protein kinase
PKC	Protein kinase C
LKB1	Liver kinase B1
CaMKKα/β	Ca^{2+}-calmodulin dependent kinase kinases II
TSSK1/2	Testis specific serine/threonine kinases 1/2
AICAR	5-aminoimidazol-4-carboxamide-1-B-D-ribofuranoside
CC	Compound C or dorsomorphin
$\Delta\Psi m$	Mitochondrial membrane potential
PMA	Phorbol 12-myristate 13-acetate
ROS	Reactive oxygen species
RSV	Resveratrol

References

1. Carling, D. AMPK signalling in health and disease. *Curr. Opin. Cell Biol.* **2017**, *45*, 31–37. [CrossRef] [PubMed]
2. Hardie, D.G.; Schaffer, B.E.; Brunet, A. AMPK: An Energy-Sensing Pathway with Multiple Inputs and Outputs. *Trends Cell Biol.* **2016**, *26*, 190–201. [CrossRef] [PubMed]
3. Tartarin, P.; Guibert, E.; Toure, A.; Ouiste, C.; Leclerc, J.; Sanz, N.; Brière, S.; Dacheux, J.L.; Delaleu, B.; McNeilly, J.R.; et al. Inactivation of AMPKalpha1 induces asthenozoospermia and alters spermatozoa morphology. *Endocrinology* **2012**, *153*, 3468–3481. [CrossRef] [PubMed]
4. Hurtado de Llera, A.; Martin-Hidalgo, D.; Gil, M.C.; Garcia-Marin, L.J.; Bragado, M.J. AMP-Activated Kinase AMPK Is Expressed in Boar Spermatozoa and Regulates Motility. *PLoS ONE* **2012**, *7*, e38840. [CrossRef] [PubMed]
5. Handa, N.; Takagi, T.; Saijo, S.; Kishishita, S.; Takaya, D.; Toyama, M.; Terada, T.; Shirouzu, M.; Suzuki, A.; Lee, S.; et al. Structural basis for compound C inhibition of the human AMP-activated protein kinase alpha2 subunit kinase domain. *Acta Crystallogr. D Biol. Crystallogr.* **2011**, *67*, 480–487. [CrossRef] [PubMed]

6. Dite, T.A.; Langendorf, C.G.; Hoque, A.; Galic, S.; Rebello, R.J.; Ovens, A.J.; Lindqvist, L.M.; Ngoei, K.R.W.; Ling, N.X.Y.; Furic, L.; et al. AMP-activated protein kinase selectively inhibited by the type II inhibitor SBI-0206965. *J. Biol. Chem.* **2018**, *293*, 8874–8885. [CrossRef] [PubMed]
7. Olivier, S.; Foretz, M.; Viollet, B. Promise and challenges for direct small molecule AMPK activators. *Biochem. Pharmacol.* **2018**, *153*, 147–158. [CrossRef] [PubMed]
8. Zhou, G.; Myers, R.; Li, Y.; Chen, Y.; Shen, X.; Fenyk-Melody, J.; Wu, M.; Ventre, J.; Doebber, T.; Fujii, N.; et al. Role of AMP-activated protein kinase in mechanism of metformin action. *J. Clin. Investig.* **2001**, *108*, 1167–1174. [CrossRef] [PubMed]
9. Ouyang, J.; Parakhia, R.A.; Ochs, R.S. Metformin activates AMP kinase through inhibition of AMP deaminase. *J. Biol. Chem.* **2011**, *286*, 1–11. [CrossRef] [PubMed]
10. Ben-Sahra, I.; Regazzetti, C.; Robert, G.; Laurent, K.; Le Marchand-Brustel, Y.; Auberger, P.; Tanti, J.F.; Giorgetti-Peraldi, S.; Bost, F. Metformin, independent of AMPK, induces mTOR inhibition and cell-cycle arrest through REDD1. *Cancer Res.* **2011**, *71*, 4366–4372. [CrossRef] [PubMed]
11. Corominas-Faja, B.; Quirantes-Pine, R.; Oliveras-Ferraros, C.; Vazquez-Martin, A.; Cufí, S.; Martin-Castillo, B.; Micol, V.; Joven, J.; Segura-Carretero, A.; Menendez, J.A. Metabolomic fingerprint reveals that metformin impairs one-carbon metabolism in a manner similar to the antifolate class of chemotherapy drugs. *Aging* **2012**, *4*, 480–498. [CrossRef] [PubMed]
12. Kelly, B.; Tannahill, G.M.; Murphy, M.P.; O'Neill, L.A. Metformin Inhibits the Production of Reactive Oxygen Species from NADH: Ubiquinone Oxidoreductase to Limit Induction of Interleukin-1beta (IL-1beta) and Boosts Interleukin-10 (IL-10) in Lipopolysaccharide (LPS)-activated Macrophages. *J. Biol. Chem.* **2015**, *290*, 20348–20359. [CrossRef] [PubMed]
13. Miller, R.A.; Birnbaum, M.J. An energetic tale of AMPK-independent effects of metformin. *J. Clin. Investig.* **2010**, *120*, 2267–2270. [CrossRef] [PubMed]
14. Rosilio, C.; Ben-Sahra, I.; Bost, F.; Peyron, J.F. Metformin: A metabolic disruptor and anti-diabetic drug to target human leukemia. *Cancer Lett.* **2014**, *346*, 188–196. [CrossRef] [PubMed]
15. Saeedi, R.; Parsons, H.L.; Wambolt, R.B.; Paulson, K.; Sharma, V.; Dyck, J.R.; Brownsey, R.W.; Allard, M.F. Metabolic actions of metformin in the heart can occur by AMPK-independent mechanisms. *Am. J. Physiol. Heart Circ. Physiol.* **2008**, *294*, H2497–H2506. [CrossRef] [PubMed]
16. Scotland, S.; Saland, E.; Skuli, N.; de Toni, F.; Boutzen, H.; Micklow, E.; Sénégas, I.; Peyraud, R.; Peyriga, L.; Théodoro, F.; et al. Mitochondrial energetic and AKT status mediate metabolic effects and apoptosis of metformin in human leukemic cells. *Leukemia* **2013**, *27*, 2129–2138. [CrossRef] [PubMed]
17. Owen, M.R.; Doran, E.; Halestrap, A.P. Evidence that metformin exerts its anti-diabetic effects through inhibition of complex 1 of the mitochondrial respiratory chain. *Biochem. J.* **2000**, *348*, 607–614. [CrossRef] [PubMed]
18. El-Mir, M.Y.; Nogueira, V.; Fontaine, E.; Averet, N.; Rigoulet, M.; Leverve, X. Dimethylbiguanide inhibits cell respiration via an indirect effect targeted on the respiratory chain complex I. *J. Biol. Chem.* **2000**, *275*, 223–228. [CrossRef] [PubMed]
19. Bertoldo, M.J.; Faure, M.; Dupont, J.; Froment, P. AMPK: A master energy regulator for gonadal function. *Front. Neurosci.* **2015**, *9*, 235. [CrossRef] [PubMed]
20. Towler, M.C.; Fogarty, S.; Hawley, S.A.; Pan, D.A.; Martin, D.M.; Morrice, N.A.; McCarthy, A.; Galardo, M.N.; Meroni, S.B.; Cigorraga, S.B.; et al. A novel short splice variant of the tumour suppressor LKB1 is required for spermiogenesis. *Biochem. J.* **2008**, *416*, 1–14. [CrossRef] [PubMed]
21. Kong, F.; Wang, M.; Huang, X.; Yue, Q.; Wei, X.; Dou, X.; Peng, X.; Jia, Y.; Zheng, K.; Wu, T.; et al. Differential regulation of spermatogenic process by Lkb1 isoforms in mouse testis. *Cell Death Dis.* **2017**, *8*, e3121. [CrossRef] [PubMed]
22. Xu, B.; Hao, Z.; Jha, K.N.; Zhang, Z.; Urekar, C.; Digilio, L.; Pulido, S.; Strauss, J.F., 3rd; Flickinger, C.J.; Herr, J.C. Targeted deletion of Tssk1 and 2 causes male infertility due to haploinsufficiency. *Dev. Biol.* **2008**, *319*, 211–222. [CrossRef] [PubMed]
23. Russell, L.D.; Tallon-Doran, M.; Weber, J.E.; Wong, V.; Peterson, R.N. Three-dimensional reconstruction of a rat stage V Sertoli cell: III. A study of specific cellular relationships. *Am. J. Anat.* **1983**, *167*, 181–192. [CrossRef] [PubMed]

24. Crisostomo, L.; Alves, M.G.; Gorga, A.; Sousa, M.; Riera, M.F.; Galardo, M.N.; Meroni, S.B.; Oliveira, P.F. Molecular Mechanisms and Signaling Pathways Involved in the Nutritional Support of Spermatogenesis by Sertoli Cells. *Methods Mol. Biol.* **2018**, *1748*, 129–155. [CrossRef] [PubMed]
25. Dong, H.; Chen, Z.; Wang, C.; Xiong, Z.; Zhao, W.; Jia, C.; Lin, J.; Lin, Y.; Yuan, W.; Zhao, A.Z.; et al. Rictor Regulates Spermatogenesis by Controlling Sertoli Cell Cytoskeletal Organization and Cell Polarity in the Mouse Testis. *Endocrinology* **2015**, *156*, 4244–4256. [CrossRef] [PubMed]
26. Tanwar, P.S.; Zhang, L.; Teixeira, J.M. Adenomatous polyposis coli (APC) is essential for maintaining the integrity of the seminiferous epithelium. *Mol. Endocrinol.* **2011**, *25*, 1725–1739. [CrossRef] [PubMed]
27. Tanwar, P.S.; Kaneko-Tarui, T.; Zhang, L.; Teixeira, J.M. Altered LKB1/AMPK/TSC1/TSC2/mTOR signaling causes disruption of Sertoli cell polarity and spermatogenesis. *Hum. Mol. Genet.* **2012**, *21*, 4394–4405. [CrossRef] [PubMed]
28. Riera, M.F.; Regueira, M.; Galardo, M.N.; Pellizzari, E.H.; Meroni, S.B.; Cigorraga, S.B. Signal transduction pathways in FSH regulation of rat Sertoli cell proliferation. *Am. J. Physiol. Endocrinol. Metab.* **2012**, *302*, E914–E923. [CrossRef] [PubMed]
29. Yang, W.R.; Liao, T.T.; Bao, Z.Q.; Zhou, C.Q.; Luo, H.Y.; Lu, C.; Pan, M.H.; Wang, X.Z. Role of AMPK in the expression of tight junction proteins in heat-treated porcine Sertoli cells. *Theriogenology* **2018**, *121*, 42–52. [CrossRef] [PubMed]
30. Galardo, M.N.; Riera, M.F.; Pellizzari, E.H.; Sobarzo, C.; Scarcelli, R.; Denduchis, B.; Lustig, L.; Cigorraga, S.B.; Meroni, S.B. Adenosine regulates Sertoli cell function by activating AMPK. *Mol. Cell. Endocrinol.* **2010**, *330*, 49–58. [CrossRef] [PubMed]
31. Galardo, M.N.; Riera, M.F.; Pellizzari, E.H.; Cigorraga, S.B.; Meroni, S.B. The AMP-activated protein kinase activator, 5-aminoimidazole-4-carboxamide-1-b-D-ribonucleoside, regulates lactate production in rat Sertoli cells. *J. Mol. Endocrinol.* **2007**, *39*, 279–288. [CrossRef] [PubMed]
32. Bertoldo, M.J.; Guibert, E.; Faure, M.; Guillou, F.; Ramé, C.; Nadal-Desbarats, L.; Foretz, M.; Viollet, B.; Dupont, J.; Froment, P. Specific deletion of AMP-activated protein kinase (alpha1AMPK) in mouse Sertoli cells modifies germ cell quality. *Mol. Cell. Endocrinol.* **2016**, *423*, 96–112. [CrossRef] [PubMed]
33. Jiao, Z.J.; Yi, W.; Rong, Y.W.; Kee, J.D.; Zhong, W.X. MicroRNA-1285 Regulates 17beta-Estradiol-Inhibited Immature Boar Sertoli Cell Proliferation via Adenosine Monophosphate-Activated Protein Kinase Activation. *Endocrinology* **2015**, *156*, 4059–4070. [CrossRef] [PubMed]
34. Faure, M.; Guibert, E.; Alves, S.; Pain, B.; Ramé, C.; Dupont, J.; Brillard, J.P.; Froment, P. The insulin sensitiser metformin regulates chicken Sertoli and germ cell populations. *Reproduction* **2016**, *151*, 527–538. [CrossRef] [PubMed]
35. Paillamanque, J.; Sanchez-Tusie, A.; Carmona, E.M.; Trevino, C.L.; Sandoval, C.; Nualart, F.; Osses, N.; Reyes, J.G. Arachidonic acid triggers [Ca^{2+}]i increases in rat round spermatids by a likely GPR activation, ERK signalling and ER/acidic compartments Ca^{2+} release. *PLoS ONE* **2017**, *12*, e0172128. [CrossRef] [PubMed]
36. Oliveira, P.F.; Sousa, M.; Barros, A.; Moura, T.; Rebelo da Costa, A. Intracellular pH regulation in human Sertoli cells: Role of membrane transporters. *Reproduction* **2009**, *137*, 353–359. [CrossRef] [PubMed]
37. Rato, L.; Alves, M.G.; Socorro, S.; Duarte, A.I.; Cavaco, J.E.; Oliveira, P.F. Metabolic regulation is important for spermatogenesis. *Nat. Rev. Urol.* **2012**, *9*, 330–338. [CrossRef] [PubMed]
38. Yu, C.L.; Guan, J.Y.; Ding, J.; Huang, S.; Lian, Y.; Luo, H.Y.; Wang, X.Z. AMP-activated protein kinase negatively regulates heat treatment-induced lactate secretion in cultured boar sertoli cells. *Theriogenology* **2018**, *121*, 35–41. [CrossRef] [PubMed]
39. Teerds, K.J.; Huhtaniemi, I.T. Morphological and functional maturation of Leydig cells: From rodent models to primates. *Hum. Reprod. Update* **2015**, *21*, 310–328. [CrossRef] [PubMed]
40. Abdou, H.S.; Bergeron, F.; Tremblay, J.J. A cell-autonomous molecular cascade initiated by AMP-activated protein kinase represses steroidogenesis. *Mol. Cell. Biol.* **2014**, *34*, 4257–4271. [CrossRef] [PubMed]
41. Taibi, N.; Dupont, J.; Bouguermouh, Z.; Froment, P.; Ramé, C.; Anane, A.; Amirat, Z.; Khammar, F. Expression of adenosine 5′-monophosphate-Activated protein kinase (AMPK) in ovine testis (*Ovis aries*): In vivo regulation by nutritional state. *Anim. Reprod. Sci.* **2017**, *178*, 9–22. [CrossRef] [PubMed]
42. Liu, S.; Sun, Y.; Li, Z. Resveratrol protects Leydig cells from nicotine-induced oxidative damage through enhanced autophagy. *Clin. Exp. Pharmacol. Physiol.* **2018**, *45*, 573–580. [CrossRef] [PubMed]

43. Chen, X.; Dong, Y.; Tian, E.; Xie, L.; Wang, G.; Li, X.; Chen, X.; Chen, Y.; Lv, Y.; Ni, C.; et al. 4-Bromodiphenyl ether delays pubertal Leydig cell development in rats. *Chemosphere* **2018**, *211*, 986–997. [CrossRef] [PubMed]
44. Cheung, P.C.; Salt, I.P.; Davies, S.P.; Hardie, D.G.; Carling, D. Characterization of AMP-activated protein kinase gamma-subunit isoforms and their role in AMP binding. *Biochem. J.* **2000**, *346*, 659–669. [CrossRef] [PubMed]
45. Duan, P.; Hu, C.; Quan, C.; Yu, T.; Zhou, W.; Yuan, M.; Shi, Y.; Yang, K. 4-Nonylphenol induces apoptosis, autophagy and necrosis in Sertoli cells: Involvement of ROS-mediated AMPK/AKT-mTOR and JNK pathways. *Toxicology* **2016**, *341–343*, 28–40. [CrossRef] [PubMed]
46. Duan, P.; Hu, C.; Quan, C.; Yu, T.; Huang, W.; Chen, W.; Tang, S.; Shi, Y.; Martin, F.L.; Yang, K. 4-Nonylphenol induces autophagy and attenuates mTOR-p70S6K/4EBP1 signaling by modulating AMPK activation in Sertoli cells. *Toxicol. Lett.* **2017**, *267*, 21–31. [CrossRef] [PubMed]
47. Guo, Z.; Yan, X.; Wang, L.; Wu, J.; Jing, X.; Liu, J. Effect of Telmisartan or Insulin on the Expression of Adiponectin and its Receptors in the Testis of Streptozotocin-Induced Diabetic Rats. *Horm. Metab. Res.* **2016**, *48*, 404–412. [CrossRef] [PubMed]
48. Hallows, K.R.; Alzamora, R.; Li, H.; Gong, F.; Smolak, C.; Neumann, D.; Pastor-Soler, N.M. AMP-activated protein kinase inhibits alkaline pH- and PKA-induced apical vacuolar H+-ATPase accumulation in epididymal clear cells. *Am. J. Physiol. Cell Physiol.* **2009**, *296*, C672–C681. [CrossRef] [PubMed]
49. Ahn, S.W.; Nedumaran, B.; Xie, Y.; Kim, D.K.; Kim, Y.D.; Choi, H.S. Bisphenol A bis(2,3-dihydroxypropyl) ether (BADGE.2H2O) induces orphan nuclear receptor Nur77 gene expression and increases steroidogenesis in mouse testicular Leydig cells. *Mol. Cells* **2008**, *26*, 74–80. [PubMed]
50. Cheng, Y.; Chen, G.; Wang, L.; Kong, J.; Pan, J.; Xi, Y.; Shen, F.; Huang, Z. Triptolide-induced mitochondrial damage dysregulates fatty acid metabolism in mouse sertoli cells. *Toxicol. Lett.* **2018**, *292*, 136–150. [CrossRef] [PubMed]
51. Li, R.; Luo, X.; Zhu, Y.; Zhao, L.; Li, L.; Peng, Q.; Ma, M.; Gao, Y. ATM signals to AMPK to promote autophagy and positively regulate DNA damage in response to cadmium-induced ROS in mouse spermatocytes. *Environ. Pollut.* **2017**, *231*, 1560–1568. [CrossRef] [PubMed]
52. Li, W.; Fu, J.; Zhang, S.; Zhao, J.; Xie, N.; Cai, G. The proteasome inhibitor bortezomib induces testicular toxicity by upregulation of oxidative stress, AMP-activated protein kinase (AMPK) activation and deregulation of germ cell development in adult murine testis. *Toxicol. Appl. Pharmacol.* **2015**, *285*, 98–109. [CrossRef] [PubMed]
53. Mancilla, H.; Maldonado, R.; Cereceda, K.; Villarroel-Espindola, F.; Montes de Oca, M.; Angulo, C.; Castro, M.A.; Slebe, J.C.; Vera, J.C.; Lavandero, S.; et al. Glutathione Depletion Induces Spermatogonial Cell Autophagy. *J. Cell. Biochem.* **2015**, *116*, 2283–2292. [CrossRef] [PubMed]
54. Wu, L.; Xu, B.; Fan, W.; Zhu, X.; Wang, G.; Zhang, A. Adiponectin protects Leydig cells against proinflammatory cytokines by suppressing the nuclear factor-kappaB signaling pathway. *FEBS J.* **2013**, *280*, 3920–3927. [CrossRef] [PubMed]
55. Zhang, J.; Zhu, Y.; Shi, Y.; Han, Y.; Liang, C.; Feng, Z.; Zheng, H.; Eng, M.; Wang, J. Fluoride-Induced Autophagy via the Regulation of Phosphorylation of Mammalian Targets of Rapamycin in Mice Leydig Cells. *J. Agric. Food Chem.* **2017**, *65*, 8966–8976. [CrossRef] [PubMed]
56. Pang, J.; Li, F.; Feng, X.; Yang, H.; Han, L.; Fan, Y.; Nie, H.; Wang, Z.; Wang, F.; Zhang, Y. Influences of different dietary energy level on sheep testicular development associated with AMPK/ULK1/autophagy pathway. *Theriogenology* **2018**, *108*, 362–370. [CrossRef] [PubMed]
57. Ham, S.; Brown, K.A.; Simpson, E.R.; Meachem, S.J. Immunolocalisation of aromatase regulators liver kinase B1, phosphorylated AMP-activated protein kinase and cAMP response element-binding protein-regulated transcription co-activators in the human testis. *Reprod. Fertil. Dev.* **2016**. [CrossRef] [PubMed]
58. Vadnais, M.L.; Kirkwood, R.N.; Tempelman, R.J.; Sprecher, D.J.; Chou, K. Effect of cooling and seminal plasma on the capacitation status of fresh boar sperm as determined using chlortetracycline assay. *Anim. Reprod. Sci.* **2005**, *87*, 121–132. [CrossRef] [PubMed]
59. Cordova, A.; Strobel, P.; Vallejo, A.; Valenzuela, P.; Ulloa, O.; Burgos, R.A.; Menarim, B.; Rodríguez-Gil, J.E.; Ratto, M.; Ramírez-Reveco, A. Use of hypometabolic TRIS extenders and high cooling rate refrigeration for cryopreservation of stallion sperm: Presence and sensitivity of 5′ AMP-activated protein kinase (AMPK). *Cryobiology* **2014**, *69*, 473–481. [CrossRef] [PubMed]

60. Swegen, A.; Lambourne, S.R.; Aitken, R.J.; Gibb, Z. Rosiglitazone Improves Stallion Sperm Motility, ATP Content, and Mitochondrial Function. *Biol. Reprod.* **2016**, *95*, 107. [CrossRef] [PubMed]
61. Kumar, L.; Yadav, S.K.; Kushwaha, B.; Pandey, A.; Sharma, V.; Verma, V.; Maikhuri, J.P.; Rajender, S.; Sharma, V.L.; Gupta, G. Energy Utilization for Survival and Fertilization-Parsimonious Quiescent Sperm Turn Extravagant on Motility Activation in Rat. *Biol. Reprod.* **2016**, *94*, 96. [CrossRef] [PubMed]
62. Zhu, Z.; Li, R.; Ma, G.; Bai, W.; Fan, X.; Lv, Y.; Luo, J.; Zeng, W. 5′-AMP-Activated Protein Kinase Regulates Goat Sperm Functions via Energy Metabolism in Vitro. *Cell. Physiol. Biochem.* **2018**, *47*, 2420–2431. [CrossRef] [PubMed]
63. Calle-Guisado, V.; Hurtado de Llera, A.; Martin-Hidalgo, D.; Mijares, J.; Gil, M.C.; Alvarez, I.S.; Bragado, M.J.; Garcia-Marin, L.J. AMP-activated kinase in human spermatozoa: Identification, intracellular localization, and key function in the regulation of sperm motility. *Asian J. Androl.* **2016**, *185*, 848.
64. Shabani, N.M.; Amidi, F.; Sedighi Gilani, M.A.; Aleyasin, A.; Bakhshalizadeh, S.; Naji, M.; Nekoonam, S. Protective features of resveratrol on human spermatozoa cryopreservation may be mediated through 5′ AMP-activated protein kinase activation. *Andrology* **2017**, *5*, 313–326. [CrossRef] [PubMed]
65. Hurtado de Llera, A.; Martin-Hidalgo, D.; Rodriguez-Gil, J.E.; Gil, M.C.; Garcia-Marin, L.J.; Bragado, M.J. AMP-activated kinase, AMPK, is involved in the maintenance of plasma membrane organization in boar spermatozoa. *Biochim. Biophys. Acta* **2013**, *1828*, 2143–2151. [CrossRef] [PubMed]
66. Nguyen, T.M.; Alves, S.; Grasseau, I.; Metayer-Coustard, S.; Praud, C.; Froment, P.; Blesbois, E. Central role of 5′-AMP-activated protein kinase in chicken sperm functions. *Biol. Reprod.* **2014**, *91*, 121. [CrossRef] [PubMed]
67. Xu, B.; Hao, Z.; Jha, K.N.; Digilio, L.; Urekar, C.; Kim, Y.H.; Pulido, S.; Flickinger, C.J.; Herr, J.C. Validation of a testis specific serine/threonine kinase [TSSK] family and the substrate of TSSK1 & 2, TSKS, as contraceptive targets. *Soc. Reprod. Fertil. Suppl.* **2007**, *63*, 87–101. [PubMed]
68. Zhang, Z.; Kostetskii, I.; Tang, W.; Haig-Ladewig, L.; Sapiro, R.; Wei, Z.; Patel, A.M.; Bennett, J.; Gerton, G.L.; Moss, S.B.; et al. Deficiency of SPAG16L causes male infertility associated with impaired sperm motility. *Biol. Reprod.* **2006**, *74*, 751–759. [CrossRef] [PubMed]
69. Cool, B.; Zinker, B.; Chiou, W.; Kifle, L.; Cao, N.; Perham, M.; Dickinson, R.; Adler, A.; Gagne, G.; Iyengar, R.; et al. Identification and characterization of a small molecule AMPK activator that treats key components of type 2 diabetes and the metabolic syndrome. *Cell Metab.* **2006**, *3*, 403–416. [CrossRef] [PubMed]
70. Hurtado de Llera, A.; Martin-Hidalgo, D.; Gil, M.C.; Garcia-Marin, L.J.; Bragado, M.J. AMPK up-activation reduces motility and regulates other functions of boar spermatozoa. *Mol. Hum. Reprod.* **2015**, *21*, 31–45. [CrossRef] [PubMed]
71. Calle-Guisado, V.; de Hurtado, L.A.; Gonzalez-Fernandez, L.; Bragado, M.J.; Garcia-Marin, L.J. Human sperm motility is downregulated by the AMPK activator A769662. *Andrology* **2017**, *5*, 1131–1140. [CrossRef] [PubMed]
72. De Hurtado, L.A.; Martin-Hidalgo, D.; Gil, M.C.; Garcia-Marin, L.J.; Bragado, M.J. New insights into transduction pathways that regulate boar sperm function. *Theriogenology* **2016**, *85*, 12–20. [CrossRef] [PubMed]
73. Zhao, Y.; Zhang, W.D.; Liu, X.Q.; Zhang, P.F.; Hao, Y.N.; Li, L.; Chen, L.; Shen, W.; Tang, X.F.; Min, L.J.; et al. Hydrogen Sulfide and/or Ammonia Reduces Spermatozoa Motility through AMPK/AKT Related Pathways. *Sci. Rep.* **2016**, *6*, 37884. [CrossRef] [PubMed]
74. Bertoldo, M.J.; Guibert, E.; Tartarin, P.; Guillory, V.; Ramé, C.; Nadal-Desbarats, L.; Foretz, M.; Viollet, B.; Dupont, J.; Froment, P. Effect of metformin on the fertilizing ability of mouse spermatozoa. *Cryobiology* **2014**, *68*, 262–268. [CrossRef] [PubMed]
75. Harayama, H. Roles of intracellular cyclic AMP signal transduction in the capacitation and subsequent hyperactivation of mouse and boar spermatozoa. *J. Reprod. Dev.* **2013**, *59*, 421–430. [CrossRef] [PubMed]
76. Yanagimachi, R. Mammalian fertilization. In *The Physiology of Reproduction*; Knobil, E., Neil, J.D., Eds.; Raven Press: New York, NY, USA, 1994; pp. 189–317.
77. Martin-Hidalgo, D.; Hurtado de Llera, A.; Yeste, M.; Gil, M.C.; Bragado, M.J.; Garcia-Marin, L.J. Adenosine monophosphate-activated kinase, AMPK, is involved in the maintenance of the quality of extended boar semen during long-term storage. *Theriogenology* **2013**. [CrossRef] [PubMed]

78. Gadella, B.M.; Harrison, R.A. The capacitating agent bicarbonate induces protein kinase A-dependent changes in phospholipid transbilayer behavior in the sperm plasma membrane. *Development* **2000**, *127*, 2407–2420. [PubMed]
79. Hurtado de Llera, A.; Martin-Hidalgo, D.; Gil, M.C.; Garcia-Marin, L.J.; Bragado, M.J. The calcium/CaMKKalpha/beta and the cAMP/PKA pathways are essential upstream regulators of AMPK activity in boar spermatozoa. *Biol. Reprod.* **2014**, *90*, 29. [CrossRef] [PubMed]
80. Gadella, B.M.; van Gestel, R.A. Bicarbonate and its role in mammalian sperm function. *Anim. Reprod. Sci.* **2004**, *82–83*, 307–319. [CrossRef] [PubMed]
81. Harrison, R.A. Rapid PKA-catalysed phosphorylation of boar sperm proteins induced by the capacitating agent bicarbonate. *Mol. Reprod. Dev.* **2004**, *67*, 337–352. [CrossRef] [PubMed]
82. Garrett, L.J.; Revell, S.G.; Leese, H.J. Adenosine triphosphate production by bovine spermatozoa and its relationship to semen fertilizing ability. *J. Androl.* **2008**, *29*, 449–458. [CrossRef] [PubMed]
83. Miki, K. Energy metabolism and sperm function. *Soc. Reprod. Fertil. Suppl.* **2007**, *65*, 309–325. [PubMed]
84. Nguyen, T.M.; Duittoz, A.; Praud, C.; Combarnous, Y.; Blesbois, E. Calcium channels in chicken sperm regulate motility and the acrosome reaction. *FEBS J.* **2016**, *283*, 1902–1920. [CrossRef] [PubMed]
85. Woods, A.; Johnstone, S.R.; Dickerson, K.; Leiper, F.C.; Fryer, L.G.; Neumann, D.; Schlattner, U.; Wallimann, T.; Carlson, M.; Carling, D. LKB1 is the upstream kinase in the AMP-activated protein kinase cascade. *Curr. Biol.* **2003**, *13*, 2004–2008. [CrossRef] [PubMed]
86. Collins, S.P.; Reoma, J.L.; Gamm, D.M.; Uhler, M.D. LKB1, a novel serine/threonine protein kinase and potential tumour suppressor, is phosphorylated by cAMP-dependent protein kinase (PKA) and prenylated in vivo. *Biochem. J.* **2000**, *345*, 673–680. [CrossRef] [PubMed]
87. Sapkota, G.P.; Kieloch, A.; Lizcano, J.M.; Lain, S.; Arthur, J.S.; Williams, M.R.; Morrice, N.; Deak, M.; Alessi, D.R. Phosphorylation of the protein kinase mutated in Peutz-Jeghers cancer syndrome, LKB1/STK11, at Ser431 by p90(RSK) and cAMP-dependent protein kinase, but not its farnesylation at Cys(433), is essential for LKB1 to suppress cell vrowth. *J. Biol. Chem.* **2001**, *276*, 19469–19482. [CrossRef] [PubMed]
88. Denison, F.C.; Smith, L.B.; Muckett, P.J.; O'Hara, L.; Carling, D.; Woods, A. LKB1 is an essential regulator of spermatozoa release during spermiation in the mammalian testis. *PLoS ONE* **2011**, *6*, e28306. [CrossRef] [PubMed]
89. Omar, B.; Zmuda-Trzebiatowska, E.; Manganiello, V.; Goransson, O.; Degerman, E. Regulation of AMP-activated protein kinase by cAMP in adipocytes: Roles for phosphodiesterases, protein kinase B, protein kinase A, Epac and lipolysis. *Cell Signal.* **2009**, *21*, 760–766. [CrossRef] [PubMed]
90. Li, X.; Wang, L.; Li, Y.; Zhao, N.; Zhen, L.; Fu, J.; Yang, Q. Calcium regulates motility and protein phosphorylation by changing cAMP and ATP concentrations in boar sperm in vitro. *Anim. Reprod. Sci.* **2016**, *172*, 39–51. [CrossRef] [PubMed]
91. Nguyen, T.M.; Combarnous, Y.; Praud, C.; Duittoz, A.; Blesbois, E. Ca^{2+}/Calmodulin-Dependent Protein Kinase Kinases (CaMKKs) Effects on AMP-Activated Protein Kinase (AMPK) Regulation of Chicken Sperm Functions. *PLoS ONE* **2016**, *11*, e0147559. [CrossRef] [PubMed]
92. Hawley, S.A.; Pan, D.A.; Mustard, K.J.; Ross, L.; Bain, J.; Edelman, A.M.; Frenguelli, B.G.; Hardie, D.G. Calmodulin-dependent protein kinase kinase-beta is an alternative upstream kinase for AMP-activated protein kinase. *Cell Metab.* **2005**, *2*, 9–19. [CrossRef] [PubMed]
93. Woods, A.; Dickerson, K.; Heath, R.; Hong, S.P.; Momcilovic, M.; Johnstone, S.R.; Carlson, M.; Carling, D. Ca^{2+}/calmodulin-dependent protein kinase kinase-beta acts upstream of AMP-activated protein kinase in mammalian cells. *Cell Metab.* **2005**, *2*, 21–33. [CrossRef] [PubMed]
94. Breitbart, H.; Naor, Z. Protein kinases in mammalian sperm capacitation and the acrosome reaction. *Rev. Reprod.* **1999**, *4*, 151–159. [CrossRef] [PubMed]
95. NagDas, S.K.; Winfrey, V.P.; Olson, G.E. Identification of ras and its downstream signaling elements and their potential role in hamster sperm motility. *Biol. Reprod.* **2002**, *67*, 1058–1066. [CrossRef] [PubMed]
96. Jungnickel, M.K.; Sutton, K.A.; Wang, Y.; Florman, H.M. Phosphoinositide-dependent pathways in mouse sperm are regulated by egg ZP3 and drive the acrosome reaction. *Dev. Biol.* **2007**, *304*, 116–126. [CrossRef] [PubMed]
97. Xie, Z.; Dong, Y.; Zhang, J.; Scholz, R.; Neumann, D.; Zou, M.H. Identification of the serine 307 of LKB1 as a novel phosphorylation site essential for its nucleocytoplasmic transport and endothelial cell angiogenesis. *Mol. Cell. Biol.* **2009**, *29*, 3582–3596. [CrossRef] [PubMed]

98. Zhu, H.; Moriasi, C.M.; Zhang, M.; Zhao, Y.; Zou, M.H. Phosphorylation of Serine 399 in LKB1 Protein Short Form by Protein Kinase Czeta Is Required for Its Nucleocytoplasmic Transport and Consequent AMP-activated Protein Kinase (AMPK) Activation. *J. Biol. Chem.* **2013**, *288*, 16495–16505. [CrossRef] [PubMed]
99. Bragado, M.J.; Aparicio, I.M.; Gil, M.C.; Garcia-Marin, L.J. Protein kinases A and C and phosphatidylinositol 3 kinase regulate glycogen synthase kinase-3A serine 21 phosphorylation in boar spermatozoa. *J. Cell. Biochem.* **2010**, *109*, 65–73. [CrossRef] [PubMed]
100. Harayama, H.; Miyake, M. A cyclic adenosine 3′,5′-monophosphate-dependent protein kinase C activation is involved in the hyperactivation of boar spermatozoa. *Mol. Reprod. Dev.* **2006**, *73*, 1169–1178. [CrossRef] [PubMed]
101. Emerling, B.M.; Weinberg, F.; Snyder, C.; Burgess, Z.; Mutlu, G.M.; Viollet, B.; Budinger, G.R.; Chandel, N.S. Hypoxic activation of AMPK is dependent on mitochondrial ROS but independent of an increase in AMP/ATP ratio. *Free Radic. Biol. Med.* **2009**, *46*, 1386–1391. [CrossRef] [PubMed]
102. De Lamirande, E.; Lamothe, G.; Villemure, M. Control of superoxide and nitric oxide formation during human sperm capacitation. *Free Radic. Biol. Med.* **2009**, *46*, 1420–1427. [CrossRef] [PubMed]
103. Zmijewski, J.W.; Banerjee, S.; Bae, H.; Friggeri, A.; Lazarowski, E.R.; Abraham, E. Exposure to hydrogen peroxide induces oxidation and activation of AMP-activated protein kinase. *J. Biol. Chem.* **2010**, *285*, 33154–33164. [CrossRef] [PubMed]
104. Wang, L.; Li, Y.; Fu, J.; Zhen, L.; Zhao, N.; Yang, Q.; Li, S.; Li, X. Cadmium inhibits mouse sperm motility through inducing tyrosine phosphorylation in a specific subset of proteins. *Reprod. Toxicol.* **2016**, *63*, 96–106. [CrossRef] [PubMed]
105. Zhang, T.Y.; Wu, R.Y.; Zhao, Y.; Xu, C.S.; Zhang, W.D.; Ge, W.; Liu, J.; Sun, Z.Y.; Zou, S.H.; Shen, W. Ochratoxin A exposure decreased sperm motility via the AMPK and PTEN signaling pathways. *Toxicol. Appl. Pharmacol.* **2018**, *340*, 49–57. [CrossRef] [PubMed]
106. Carlsen, E.; Giwercman, A.; Keiding, N.; Skakkebaek, N.E. Evidence for decreasing quality of semen during past 50 years. *BMJ* **1992**, *305*, 609–613. [CrossRef] [PubMed]
107. Levine, H.; Jorgensen, N.; Martino-Andrade, A.; Mendiola, J.; Weksler-Derri, D.; Mindlis, I.; Pinotti, R.; Swan, S.H. Temporal trends in sperm count: A systematic review and meta-regression analysis. *Hum. Reprod. Update* **2017**, *23*, 646–659. [CrossRef] [PubMed]
108. Swan, S.H.; Elkin, E.P.; Fenster, L. The question of declining sperm density revisited: An analysis of 101 studies published 1934-1996. *Environ. Health Perspect.* **2000**, *108*, 961–966. [CrossRef] [PubMed]
109. Agarwal, A.; Durairajanayagam, D.; du Plessis, S.S. Utility of antioxidants during assisted reproductive techniques: An evidence based review. *Reprod. Biol. Endocrinol.* **2014**, *12*, 112. [CrossRef] [PubMed]
110. Knox, R.V. Artificial insemination in pigs today. *Theriogenology* **2016**, *85*, 83–93. [CrossRef] [PubMed]
111. Gogol, P.; Szczesniak-Fabianczyk, B.; Wierzchos-Hilczer, A. The photon emission, ATP level and motility of boar spermatozoa during liquid storage. *Reprod. Biol.* **2009**, *9*, 39–49. [CrossRef]
112. Nguyen, Q.T.; Wallner, U.; Schmicke, M.; Waberski, D.; Henning, H. Energy metabolic state in hypothermically stored boar spermatozoa using a revised protocol for efficient ATP extraction. *Biol. Open* **2016**, *5*, 1743–1751. [CrossRef] [PubMed]
113. Hurtado de Llera, A.; Martin-Hidalgo, D.; Garcia-Marin, L.J.; Bragado, M.J. Metformin blocks mitochondrial membrane potential and inhibits sperm motility in fresh and refrigerated boar spermatozoa. *Reprod. Domest. Anim.* **2018**, *53*, 733–741. [CrossRef] [PubMed]
114. Fryer, L.G.; Parbu-Patel, A.; Carling, D. The Anti-diabetic drugs rosiglitazone and metformin stimulate AMP-activated protein kinase through distinct signaling pathways. *J. Biol. Chem.* **2002**, *277*, 25226–25232. [CrossRef] [PubMed]
115. Yeste, M. Sperm cryopreservation update: Cryodamage, markers, and factors affecting the sperm freezability in pigs. *Theriogenology* **2016**, *85*, 47–64. [CrossRef] [PubMed]
116. Amidi, F.; Pazhohan, A.; Shabani, N.M.; Khodarahmian, M.; Nekoonam, S. The role of antioxidants in sperm freezing: A review. *Cell Tissue Bank.* **2016**, *17*, 745–756. [CrossRef] [PubMed]
117. Shabani, N.M.; Nekoonam, S.; Naji, M.; Bakhshalizadeh, S.; Amidi, F. Cryoprotective effect of resveratrol on DNA damage and crucial human sperm messenger RNAs, possibly through 5′ AMP-activated protein kinase activation. *Cell Tissue Bank.* **2018**, *19*, 87–95. [CrossRef] [PubMed]

118. Martin-Hidalgo, D.; Hurtado de Llera, A.; Henning, H.; Wallner, U.; Waberski, D.; Bragado, M.J.; Gil, M.C.; Garcia-Marin, L.J. The Effect of Resveratrol on the Quality of Extended Boar Semen during Storage at 17 °C. *J. Agric. Sci.* **2013**, *5*, 231–242.
119. Nguyen, T.M.; Seigneurin, F.; Froment, P.; Combarnous, Y.; Blesbois, E. The 5′-AMP-Activated Protein Kinase (AMPK) Is Involved in the Augmentation of Antioxidant Defenses in Cryopreserved Chicken Sperm. *PLoS ONE* **2015**, *10*, e0134420. [CrossRef] [PubMed]
120. Bridges, H.R.; Jones, A.J.; Pollak, M.N.; Hirst, J. Effects of metformin and other biguanides on oxidative phosphorylation in mitochondria. *Biochem. J.* **2014**, *462*, 475–487. [CrossRef] [PubMed]

8

Is TAK1 a Direct Upstream Kinase of AMPK?

Dietbert Neumann

Department of Pathology, CARIM School for Cardiovascular Diseases, Faculty of Health, Medicine and Life Sciences, Maastricht University, 6200 MD Maastricht, The Netherlands; d.neumann@maastrichtuniversity.nl

Abstract: Alongside Liver kinase B1 (LKB1) and Ca^{2+}/Calmodulin-dependent protein kinase kinase 2 (CaMKK2), Transforming growth factor-β (TGF-β)-activated kinase 1 (TAK1) has been suggested as a direct upstream kinase of AMP-activated protein kinase (AMPK). Several subsequent studies have reported on the TAK1-AMPK relationship, but the interpretation of the respective data has led to conflicting views. Therefore, to date the acceptance of TAK1 as a genuine AMPK kinase is lagging behind. This review provides with argumentation, whether or not TAK1 functions as a direct upstream kinase of AMPK. Several specific open questions that may have precluded the consensus are discussed based on available data. In brief, TAK1 can function as direct AMPK upstream kinase in specific contexts and in response to a subset of TAK1 activating stimuli. Further research is needed to define the intricate signals that are conditional for TAK1 to phosphorylate and activate AMPKα at T172.

Keywords: TAK1; AMPK; phosphorylation; AMPK kinase

---

1. About AMPK and TAK1

This review addresses questions that are relevant for experts in the field already familiar with AMPK and TAK1. To begin with, I will not discuss whether AMPK and TAK1 are a disparate couple or ideal affiliates, but rather provide entry points for further reading, in case readers are in need of information about AMPK or TAK1. I will not go into any detail with AMPK, because this review is part of the Special Issue on AMPK. Moreover, multiple authors (repeatedly) reviewed AMPK. In addition, there is a growing base of reviews focusing on different aspects of AMPK, such as the functions of AMPK in various tissues, or (patho-)physiological contexts (e.g., [1–4]). For AMPK novices, Hardie provides an excellent overview (e.g., [5–8]). In a nutshell, AMPK is an energy-sensing kinase that functions to maintain cellular and whole body energy balance [9]. AMPK is part of a protein kinase cascade [10]. T172 phosphorylation of the AMPKα subunit activates the kinase, which is dependent on upstream kinases, called AMPK kinases, identified as LKB1 and CaMKK2 [11–14].

TAK1 has been proposed as an alternative third AMPK kinase, which has received varying appreciation. This is the topic of this review. TAK1 is a serine/threonine protein kinase of the mitogen-activated protein kinase kinase kinase (MAP3K) family, playing a crucial role in regulating cell survival, differentiation, apoptosis, and inflammatory responses [15,16]. It forms complexes by binding to its accessory subunits, the TAK1-binding proteins (TAB1, TAB2, TAB3). TAK1 is activated by interleukin-1 (IL-1) and TGF-β receptors, tumour necrosis factor (TNF)-α, Toll-like receptors (TLR), CD40, and the B cell receptor. TAK1 is also involved in activating several intracellular kinases, p38 mitogen-activated protein kinase (p38MAPK), c-Jun N-terminal kinase (JNK), and IκB kinase complex (IKK). Therefore, TAK1 has been described as a regulator of nuclear factor κ-light-chain-enhancer of activated B cells (NF-κB) and MAPKs in proinflammatory signalling. More recently, this picture has been significantly amended, with the roles of TAK1 in tissue homeostasis (reviewed in [17]), as also further discussed below.

2. The Origin of the Debate

In 2006, after the discovery of LKB1 and CaMKK2 as upstream kinases of AMPK, TAK1 was identified as the third kinase capable of activating AMPK [18]. However, different from LKB1 and CaMKK2, TAK1 to date remains a disputed AMPK activating kinase. The reactions of the scientific community range from complete ignorance, to questioning TAK1 as an AMPK kinase, to acceptance without question. In this review, I will provide an overview on TAK1, with respect to its (putative) role as (direct) upstream activating kinase of AMPK.

In yeast, three alternative upstream kinases (Sak1, Tos3, and Elm1) have been described to activate the AMPK ortholog Snf1; knockout of all three kinases replicates the Snf1 knockout phenotype [19,20]. In search for alternative AMPK activating kinases in mammalian cells and by applying a screening approach in yeast, TAK1 was identified as in vitro upstream kinase of AMPK [18]. The in vivo relevance remained unknown, since the authors based their conclusion solely on cell-free and cell-based approaches. Notably, the study included evidence for TAK1 action on AMPK in LKB1-deficient HeLa cells. In the same year, cardiac-specific dominant-negative TAK1 mice were reported to show Wolff-Parkinson-White (WPW)-like phenotype [21], i.e., consistent with the idea that AMPK loss-of-function mutations in AMPKγ2 underlie WPW [22]. In the same study, TAK1 knockout embryos were shown to exhibit defective AMPK signalling. Due to observed midgestation embryonic lethality, the authors subsequently went on to acutely knock out floxed TAK1 alleles in cells using virally delivered Cre. The obtained results again generally supported a role of TAK1 upstream of AMPK, but the authors concluded that LKB1 could have been the intermediate of TAK1 action. The reason for this reservation was that acute loss of TAK1 interfered with the kinase activity of adenovirally delivered LKB1 complex, i.e., consisting of LKB1, mouse protein 25 (MO25) and STE20-related kinase adapter protein (STRAD). It should be noted that LKB1 is considered to be constitutively active upon complex formation with MO25 and STRAD [23,24]. Therefore, the mechanism of LKB1 inhibition, as observed by Xie et al., remains elusive. Accordingly, from these two early publications some discrepancy on the role of TAK1 upstream of AMPK primarily evolved around LKB1, and whether or not it mediates TAK1 effects on AMPK [18,21]. On the other hand, both reports agree on TAK1 as an important regulator of AMPK. In subsequent work on upstream kinases of AMPK, almost all studies have dealt with LKB1 and CaMKK2, which are firmly confirmed without any question. In contrast, the role of TAK1 as direct or indirect AMPK kinase remained obscure.

Until today only few studies further addressed TAK1-AMPK signalling, of which the majority applied chemical tools (such as kinase inhibitors) that are prone to misinterpretation because of possible off-target effects. Moreover, a few reports also indicate signalling of AMPK to TAK1, i.e., turning AMPK into a possible activating kinase of TAK1 [25,26]. In this review, I am focusing on studies using genetic tools and offering clues on the exact role of TAK1 upstream of AMPK, but will also try to integrate controversial findings. As a guide to the reader, I am asking specific remaining open questions that are subsequently either partly or wholly answered, based on scientific evidence.

3. Is TAK1 Capable of Directly Phosphorylating AMPKα at T172 in Cell Free Assays?

In the original paper, Momcilovic et al. used a purified GST-fusion of the isolated Snf1 kinase domain that was directly incubated with an artificial construct of TAK1-TAB1 fusion protein purified from insect cells. Snf1 kinase domain was phosphorylated by a TAK1-TAB1 fusion protein at T210; the site equivalent to T172 in AMPK [18]. In my lab a bacterial co-expression strategy for TAK1 with TAB1 or with TAB2 was developed [27]. TAK1-TAB1 (but not TAK1-TAB2) was active upon co-expression in bacteria, strongly suggesting that the formation of TAK1-TAB1 complexes is sufficient for kinase activation. In contrast, recombinant AMPK heterotrimers after purification from bacteria are not phosphorylated in the α-subunit at T172 [28], but received this modification in presence of either LKB1-MO25-STRAD or TAK1-TAB1 complexes [23,27]. Therefore, mammalian AMPK heterotrimeric complexes can be directly activated by human TAK1-TAB1 complexes, in a process not requiring but resembling LKB1 complex, as shown in cell-free systems.

4. Is TAK1 Activating Cellular AMPK in Absence of LKB1?

This question has been already been addressed in the original work of Momcilovic et al., by using LKB1-deficient HeLa cells transfected with TAK1 and TAB1 plasmids, i.e., revealing that AMPK is activated by the wild-type but not the kinase defective TAK1-TAB1 complex [18]. In my lab, we obtained similar results in HeLa cells using wild type and mutant AMPK [27]. This approach rules out that TAB1 scaffolding is sufficient for AMPK activation. This is an important detail, because TAB1 has been shown to activate p38MAPK employing an unusual autophosphorylation mechanism [29]. In fact, even before TAK1 was suggested as a new upstream kinase, TAB1 was shown to co-immunoprecipitate with AMPK in cardiomyocytes [30]. This latter study also suggested that the association of AMPK with TAB1 did not require prior AMPK activation. Therefore, it seems unlikely that LKB1 action is needed for AMPK-TAB1 interaction, and TAB1 should be able to recruit TAK1 independent of LKB1. Further direct evidence for TAK1-AMPK signalling came from an unexpected pathway: in TNF-related apoptosis-inducing ligand (TRAIL)-treated epithelial cells AMPK was activated by TAK1 [31]. In this study, LKB1 and CaMKK2 were knocked down without affecting the ability of TRAIL to activate AMPK. In contrast, siRNA against TAK1, along with over-expression of kinase-defective TAK1 efficiently interfered with AMPK activation. These data establish TAK1-dependency, as well as LKB1-independency, at least in this particular setting.

5. Is Stimulation of TAK1 Sufficient for Activation of AMPK?

Many researchers doubt whether TAK1 can be considered a genuine upstream kinase, if TAK1 activation is seen in situations where AMPK is not activated. Indeed, Herrero-Martin et al. also observed that TNF-treatment activates TAK1 (as seen by IκB phosphorylation), but did not activate AMPK [31]. In addition, TAK1 activation originally was seen as an intracellular mediator of pro-inflammatory signals (such as TNF-α), giving rise to the development of TAK1 inhibitors for possible treatment of inflammatory disorders [32]. This prevalent view compounded scepticism about TAK1 as a genuine AMPK kinase, since reported AMPK effects are summarized to be the inverse, i.e., anti-inflammatory [3]. Therefore, the role of TAK1 as an upstream kinase of AMPK may be relevant only in certain physiological situations, or in response to specific signals. It should also be noted that LKB1 and CaMKK2 do not share the same input, and may well be active in situations where AMPK is not. In particular, LKB1, a constitutively active kinase upon complex formation, more efficiently phosphorylates AMPK in response to a drop in cellular energy level (through allosteric regulation of AMPK by AMP and ADP). CaMKK2 may be dependent on extracellular input (operating downstream of G-protein coupled receptors), but it is not clear whether transient Ca^{2+} waves, such as those occurring in contracting myocytes, are sufficient to activate AMPK in this cell type. However, both signals, LKB1 and CaMKK2 may also act synergistically [33]. Thus, AMPK phosphorylation at T172 increases through different pathways, downstream of various signals that can be intra- or extracellular. If TAK1 activation per se is insufficient for AMPK activation, TAK1 may still activate AMPK conditionally in response to specific upstream signals.

6. What Is the Cellular Condition Where TAK1 Acts as an Upstream Kinase of AMPK?

In several recent studies, a possible role of TAK1 as upstream mediator of AMPK activation was verified by applying genetic knockdown strategies [34–39]. Although not verifying the role of TAK1 as a direct AMPK kinase, this approach puts TAK1 as an upstream AMPK activating signal into various cellular contexts. Moreover, TRAIL is an example of a distinct extracellular signal that activates TAK1-AMPK signalling [31]. Thus, the question may be asked, whether we can recognise a pattern of cellular challenges or signals where TAK1 is acting as activating AMPK kinase.

In recent literature, TAK1 is interpreted as a regulator of cell death and survival [17], which is well in accordance with the known functions of death ligands, such as TNFα and TRAIL [40]. Notably, TRAIL-induced TAK1-AMPK signalling was shown to induce cytoprotective autophagy in untransformed cells [31], whereas TRAIL induces apoptosis in several cancer cell types. Autophagy is a survival mechanism, which can be elicited by various sublethal stresses as a response to fluctuating external conditions, ranging from extracellular signals, to a change in pH, temperature or oxygen tension [41]. Some of these stresses are not predicted to directly affect cellular energy levels. The known role of AMPK in the control of autophagy in response to nutrient starvation is commonly linked to LKB1 signalling, whereas TRAIL elicits autophagy via TAK1-AMPK [31]. Of note, independent of the upstream signalling pathway, autophagy is an important survival mechanism, providing the cell with building blocks and metabolites. This integrates well with one of AMPKs more general roles; limiting cell proliferation and growth, as well as energy expense, in times of nutrient scarcity, while also enhancing the cell's ability to survive stresses, such as hypoxia and glucose deprivation [42].

As already indicated, TAK1 was shown to activate AMPK in response to various stimuli and different cell types. Receptor activator of NF-κB ligand (RANKL) activated AMPK in osteoclast precursors, and siRNA-mediated TAK1 knockdown blocked RANKL-induced activation of AMPK [34]. RANKL is a member of the TNF superfamily, supporting the idea that a subset of TAK1 activating signals could physiologically activate AMPK. In endothelial cells, Vascular endothelial growth factor (VEGF) stimulated TAK1 and AMPK, whereas TAK1 downregulation by shRNA also inhibited VEGF–stimulated phosphorylation of several kinases, including AMPK [36].

Belinostat promoted reactive oxygen species (ROS) production in PANC-1 cells and increased the ROS induced TAK1/AMPK association resulting in AMPK activation. Anti-oxidants, as well as TAK1 shRNA knockdown, suppressed Belinostat-induced AMPK activation and PANC-1 cell apoptosis [35].

Fasted mice deficient of TAK1 in hepatocytes exhibited severe hepatosteatosis with increased mTORC1 activity, and suppression of autophagy compared with their WT counterparts [43], suggesting reduced AMPK function in these livers. TAK1-deficient hepatocytes exhibited autophagy and suppressed AMPK activity in response to starvation or metformin treatment; however, ectopic activation of AMPK restored autophagy in these cells. These data indicate that TAK1 regulates hepatic lipid metabolism and tumorigenesis via the AMPK/mTORC1 axis [43]. Therefore, it was proposed that TAK1-mediated autophagy in the liver plays a role in preventing excessive lipid accumulation induced by starvation and fat overload [44]. Knockdown of TAK1 decreased the AMPK phosphorylation induced by overexpression of a dominant-negative form of p38α [38], which the authors interpreted as a negative feedback loop. Recent data suggested that TAK1 could be the upstream kinase for AMPK activation by *Helicobacter pylori*, since partial depletion of TAK1 by shRNAs not only inhibited AMPK activation, but also suppressed survival of *H. pylori*-infected gastric epithelial cells [37]. Activation of TAK1 was also found to restrict *Salmonella typhimurium* growth by inducing AMPK activation and autophagy [45]. In this study, TAK1 siRNA led to the inhibition of *S. typhimurium*-induced phosphorylation of AMPK T172, ULK1 S317, and ACC S79. The authors concluded that TAK1 activation leads to AMPK activation, which activates ULK1 by phosphorylating ULK1 S317 and suppressing mTOR activity and ULK1 S757 phosphorylation.

In conclusion, published data indicate TAK1-dependent AMPK activation could be required for induction of autophagy, as a possible survival mechanism in response to acute and specific life-threatening challenges. TAK1-induced autophagy may thus occur in the absence of an energy challenge, such as those elicited through extracellular factors (e.g., TRAIL), or bacterial infections (e.g., *H. pylori*, *S. typhimurium*), and oxidative stress (e.g., Belinostat).

Further conditions promoting TAK1-dependent AMPK activation are likely to be identified.

7. Does AMPK Have a Role in Activating TAK1?

AMPK has been reported to activate TAK1 and mediate pro-inflammatory effects in THP-1 cells [25]. In this study, it was shown that pro-inflammatory signals activated TAK1 signalling, which was then inhibited by AMPKα knockdown. Taking into account the ability of AMPK to bind TAB1 [21], and considering the role of TAB1 in activating TAK1, the interpretation of AMPK as upstream kinase of TAK1 could consequently be challenged. For, example, could the lack of AMPK reduce the availability of TAB1 for subsequent activation of TAK1? Notably, binding of TAB1 to TAK1 in a sequence of molecular events, activates TAK1 by autophosphorylation of T184/T187 [27], and does not require any upstream kinase. Interestingly, the authors of the same study confirmed AMPK-TAK1 interaction in their model, which required both the AMPKα autoinhibitory-domain, and the TAB1-binding domain of TAK1 [25]. The possible AMPK-TAB1 complex formation, and putative requirement of TAB1 as a mediator of AMPK-TAK1 binding in THP-1 cells was not investigated.

In another recent study, AMPKα1 was suggested to participate in renal TAK1 activation and TAK1-dependent signalling induced by angiotensin-II [26]. Angiotensin-II increased the phosphorylation of TAK1 (S412) in renal tissue of AMPKα1+/+ mice but not AMPKα1−/− mice. Notably, S412 is targeted by PKA [46]. Furthermore, the authors also observe that angiotensin-II upregulates the AMPKα1 isoform in renal tissue, and increased TAK1-target gene mRNA and renal protein expression in AMPKα1+/+ mice, but less-so in AMPKα1−/− mice [47]. Using the same argumentation as above, if AMPKα indeed acts as a scaffold for TAB1, one could predict that TAK1 activity is downregulated in AMPKα knockouts.

Therefore, AMPK may be involved in TAK1 activation, but not necessarily as an upstream kinase. Importantly, to date, there is only circumstantial evidence for AMPK to activate TAK1, whereas biochemical proof is available and functional data is accumulating to support TAK1 as a genuine direct AMPK activating kinase.

8. Conclusions

About 12 years after the original publication reporting TAK1 as a 'candidate' AMPK kinase [1], as argued above, the collective data rather confirms the suggested authentic role. Thus, I propose to accept TAK1, in addition to LKB1 and CaMKK2, as the third genuine upstream kinase of AMPK (Figure 1).

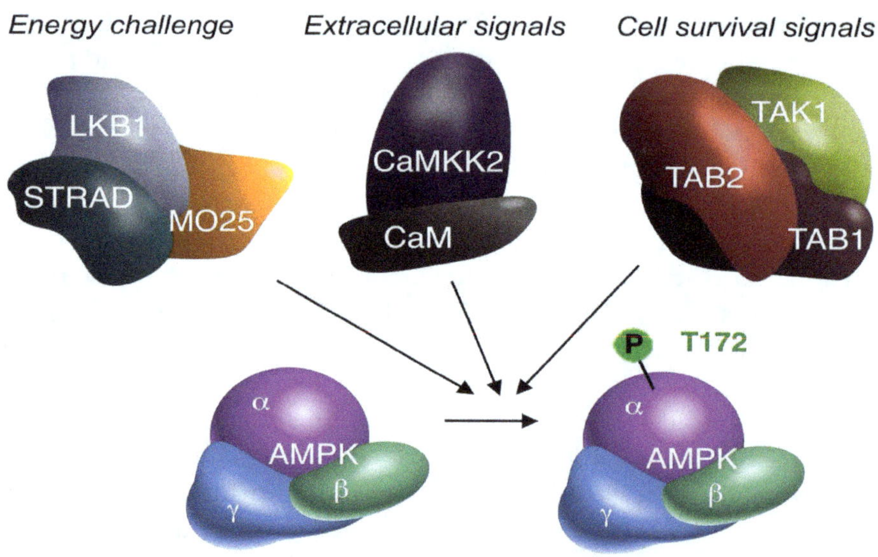

Figure 1. The three alternative AMPK kinases. Biochemical (cell-free), cell biological (in vitro) and animal (in vivo) experimentation suggest that TAK1 can activate AMPKα by phosphorylation of the critical T172 residue. Summative evidence therefore supports TAK1 as an additional AMPK upstream kinase, besides LKB1 and CaMKK2. AMPK may receive (simultaneous) activation from all three upstream kinases. The original signal leading to AMPK activation may differ per upstream kinase, as suggested above. All four kinases are depicted with their accessory subunits, as functional protein complexes. The requirement of TAB1/TAB2/TAB3 for AMPK activation has not been fully elucidated. However, to date TAB1 and/or TAB2 are the most likely candidates. TAB1 may also bind to AMPK independent of TAK1 [30]. MO25: mouse protein 25; STRAD: STE20-related kinase adapter protein; CaM: Calmodulin.

Acknowledgments: I thank Erik Biessen and his team members in the Department of Pathology at Maastricht University for support and the kind atmosphere during the preparation of this manuscript. I also thank the anonymous reviewers for careful reading and Olivia Waring for language editing.

Abbreviations

AMPK	AMP-activated protein kinase
CaMKK2	Ca^{2+}/Calmodulin-dependent protein kinase kinase 2
LKB1	Liver kinase B1
T172	Threonine 172 residue (of AMPKα)
TAB1	TAK1 binding protein 1
TAB2	TAK1 binding protein 2
TAB3	TAK1 binding protein 3
TAK1	Transforming growth factor β-activated protein kinase
TNF-α	Tumour necrosis factor α
TRAIL	Tumor necrosis factor related apoptosis inducing ligand

References

1. Hardie, D.G. Molecular pathways: Is AMPK a friend or a foe in cancer? *Clin. Cancer Res.* **2015**, *21*, 3836–3840. [CrossRef] [PubMed]
2. Lopez, M.; Nogueiras, R.; Tena-Sempere, M.; Dieguez, C. Hypothalamic AMPK: A canonical regulator of whole-body energy balance. *Nat. Rev. Endocrinol.* **2016**, *12*, 421–432. [CrossRef] [PubMed]
3. Day, E.A.; Ford, R.J.; Steinberg, G.R. AMPK as a therapeutic target for treating metabolic diseases. *Trends Endocrinol. Metab.* **2017**, *28*, 545–560. [CrossRef] [PubMed]
4. Salt, I.P.; Hardie, D.G. AMP-activated protein kinase: An ubiquitous signaling pathway with key roles in the cardiovascular system. *Circ. Res.* **2017**, *120*, 1825–1841. [CrossRef] [PubMed]
5. Hardie, D.G.; Ross, F.A.; Hawley, S.A. AMPK: A nutrient and energy sensor that maintains energy homeostasis. *Nat. Rev. Mol. Cell Biol.* **2012**, *13*, 251–262. [CrossRef] [PubMed]
6. Lin, S.C.; Hardie, D.G. AMPK: Sensing glucose as well as cellular energy status. *Cell Metab.* **2018**, *27*, 299–313. [CrossRef] [PubMed]
7. Hardie, D.G.; Schaffer, B.E.; Brunet, A. AMPK: An energy-sensing pathway with multiple inputs and outputs. *Trends Cell Biol.* **2016**, *26*, 190–201. [CrossRef] [PubMed]
8. Hardie, D.G. Keeping the home fires burning: AMP-activated protein kinase. *J. R. Soc. Interface* **2018**, *15*. [CrossRef] [PubMed]
9. Hardie, D.G.; Ashford, M.L. AMPK: Regulating energy balance at the cellular and whole body levels. *Physiology* **2014**, *29*, 99–107. [CrossRef] [PubMed]
10. Carling, D.; Mayer, F.V.; Sanders, M.J.; Gamblin, S.J. AMP-activated protein kinase: Nature's energy sensor. *Nat. Chem. Biol.* **2011**, *7*, 512–518. [CrossRef] [PubMed]
11. Hawley, S.A.; Boudeau, J.; Reid, J.L.; Mustard, K.J.; Udd, L.; Makela, T.P.; Alessi, D.R.; Hardie, D.G. Complexes between the LKB1 tumor suppressor, STRADα/β and MO25α/β are upstream kinases in the AMP-activated protein kinase cascade. *J. Biol.* **2003**, *2*, 28. [CrossRef] [PubMed]
12. Woods, A.; Johnstone, S.R.; Dickerson, K.; Leiper, F.C.; Fryer, L.G.; Neumann, D.; Schlattner, U.; Wallimann, T.; Carlson, M.; Carling, D. LKB1 is the upstream kinase in the AMP-activated protein kinase cascade. *Curr. Biol.* **2003**, *13*, 2004–2008. [CrossRef] [PubMed]
13. Hawley, S.A.; Pan, D.A.; Mustard, K.J.; Ross, L.; Bain, J.; Edelman, A.M.; Frenguelli, B.G.; Hardie, D.G. Calmodulin-dependent protein kinase kinase-β is an alternative upstream kinase for AMP-activated protein kinase. *Cell Metab.* **2005**, *2*, 9–19. [CrossRef] [PubMed]
14. Woods, A.; Dickerson, K.; Heath, R.; Hong, S.P.; Momcilovic, M.; Johnstone, S.R.; Carlson, M.; Carling, D. Ca^{2+}/calmodulin-dependent protein kinase kinase-β acts upstream of AMP-activated protein kinase in mammalian cells. *Cell Metab.* **2005**, *2*, 21–33. [CrossRef] [PubMed]
15. Ajibade, A.A.; Wang, H.Y.; Wang, R.F. Cell type-specific function of TAK1 in innate immune signaling. *Trends Immunol.* **2013** *34*, 307–316. [CrossRef] [PubMed]

16. Dai, L.; Aye Thu, C.; Liu, X.Y.; Xi, J.; Cheung, P.C. TAK1, more than just innate immunity. *IUBMB Life* **2012**, *64*, 825–834. [CrossRef] [PubMed]
17. Mihaly, S.R.; Ninomiya-Tsuji, J.; Morioka, S. TAK1 control of cell death. *Cell Death Differ.* **2014**, *21*, 1667–1676. [CrossRef] [PubMed]
18. Momcilovic, M.; Hong, S.P.; Carlson, M. Mammalian TAK1 activates SNF1 protein kinase in yeast and phosphorylates AMP-activated protein kinase in vitro. *J. Biol. Chem.* **2006**, *281*, 25336–25343. [CrossRef] [PubMed]
19. Hong, S.P.; Leiper, F.C.; Woods, A.; Carling, D.; Carlson, M. Activation of yeast SNF1 and mammalian AMP-activated protein kinase by upstream kinases. *Proc. Natl. Acad. Sci. USA* **2003**, *100*, 8839–8843. [CrossRef] [PubMed]
20. Sutherland, C.M.; Hawley, S.A.; McCartney, R.R.; Leech, A.; Stark, M.J.; Schmidt, M.C.; Hardie, D.G. Elm1p is one of three upstream kinases for the *Saccharomyces cerevisiae* SNF1 complex. *Curr. Biol.* **2003**, *13*, 1299–1305. [CrossRef]
21. Xie, M.; Zhang, D.; Dyck, J.R.; Li, Y.; Zhang, H.; Morishima, M.; Mann, D.L.; Taffet, G.E.; Baldini, A.; Khoury, D.S.; et al. A pivotal role for endogenous TGF-β-activated kinase-1 in the LKB1/AMP-activated protein kinase energy-sensor pathway. *Proc. Natl. Acad. Sci. USA* **2006**, *103*, 17378–17383. [CrossRef] [PubMed]
22. Grahame Hardie, D. AMP-activated protein kinase: A key regulator of energy balance with many roles in human disease. *J. Intern. Med.* **2014**, *276*, 543–559. [CrossRef] [PubMed]
23. Neumann, D.; Suter, M.; Tuerk, R.; Riek, U.; Wallimann, T. Co-expression of LKB1, MO25α and STRADα in bacteria yield the functional and active heterotrimeric complex. *Mol. Biotechnol.* **2007**, *36*, 220–231. [CrossRef] [PubMed]
24. Alessi, D.R.; Sakamoto, K.; Bayascas, J.R. LKB1-dependent signaling pathways. *Annu. Rev. Biochem.* **2006**, *75*, 137–163. [CrossRef] [PubMed]
25. Kim, S.Y.; Jeong, S.; Jung, E.; Baik, K.H.; Chang, M.H.; Kim, S.A.; Shim, J.H.; Chun, E.; Lee, K.Y. AMP-activated protein kinase-α1 as an activating kinase of TGF-β-activated kinase 1 has a key role in inflammatory signals. *Cell Death Dis.* **2012**, *3*, e357. [CrossRef] [PubMed]
26. Mia, S.; Castor, T.; Musculus, K.; Voelkl, J.; Alesutan, I.; Lang, F. Role of AMP-activated protein kinase α1 in angiotensin-II-induced renal TGFβ-activated kinase 1 activation. *Biochem. Biophys. Res. Commun.* **2016**, *476*, 267–272. [CrossRef] [PubMed]
27. Scholz, R.; Sidler, C.L.; Thali, R.F.; Winssinger, N.; Cheung, P.C.; Neumann, D. Autoactivation of transforming growth factor β-activated kinase 1 is a sequential bimolecular process. *J. Biol. Chem.* **2010**, *285*, 25753–25766. [CrossRef] [PubMed]
28. Neumann, D.; Woods, A.; Carling, D.; Wallimann, T.; Schlattner, U. Mammalian AMP-activated protein kinase: Functional, heterotrimeric complexes by co-expression of subunits in *Escherichia coli*. *Protein Expr. Purif.* **2003**, *30*, 230–237. [CrossRef]
29. Tanno, M.; Bassi, R.; Gorog, D.A.; Saurin, A.T.; Jiang, J.; Heads, R.J.; Martin, J.L.; Davis, R.J.; Flavell, R.A.; Marber, M.S. Diverse mechanisms of myocardial p38 mitogen-activated protein kinase activation: Evidence for MKK-independent activation by a TAB1-associated mechanism contributing to injury during myocardial ischemia. *Circ. Res.* **2003**, *93*, 254–261. [CrossRef] [PubMed]
30. Li, J.; Miller, E.J.; Ninomiya-Tsuji, J.; Russell, R.R., 3rd; Young, L.H. AMP-activated protein kinase activates p38 mitogen-activated protein kinase by increasing recruitment of p38 MAPK to TAB1 in the ischemic heart. *Circ. Res.* **2005**, *97*, 872–879. [CrossRef] [PubMed]
31. Herrero-Martin, G.; Hoyer-Hansen, M.; Garcia-Garcia, C.; Fumarola, C.; Farkas, T.; Lopez-Rivas, A.; Jaattela, M. TAK1 activates AMPK-dependent cytoprotective autophagy in trail-treated epithelial cells. *EMBO J.* **2009**, *28*, 677–685. [CrossRef] [PubMed]
32. Sakurai, H. Targeting of TAK1 in inflammatory disorders and cancer. *Trends Pharmacol. Sci.* **2012**, *33*, 522–530. [CrossRef] [PubMed]
33. Hardie, D.G. AMPK–sensing energy while talking to other signaling pathways. *Cell Metab.* **2014**, *20*, 939–952. [CrossRef] [PubMed]
34. Lee, Y.S.; Kim, Y.S.; Lee, S.Y.; Kim, G.H.; Kim, B.J.; Lee, S.H.; Lee, K.U.; Kim, G.S.; Kim, S.W.; Koh, J.M. AMP kinase acts as a negative regulator of RANKL in the differentiation of osteoclasts. *Bone* **2010**, *47*, 926–937. [CrossRef] [PubMed]

35. Wang, B.; Wang, X.B.; Chen, L.Y.; Huang, L.; Dong, R.Z. Belinostat-induced apoptosis and growth inhibition in pancreatic cancer cells involve activation of TAK1-AMPK signaling axis. *Biochem. Biophys. Res. Commun.* **2013**, *437*, 1–6. [CrossRef] [PubMed]
36. Zippel, N.; Malik, R.A.; Fromel, T.; Popp, R.; Bess, E.; Strilic, B.; Wettschureck, N.; Fleming, I.; Fisslthaler, B. Transforming Growth Factor-β–Activated Kinase 1 Regulates Angiogenesis via AMP-Activated Protein Kinase-α1 and Redox Balance in Endothelial Cells. *Arterioscler. Thromb. Vasc. Biol.* **2013**, *33*, 2792–2799. [CrossRef] [PubMed]
37. Lv, G.; Zhu, H.; Zhou, F.; Lin, Z.; Lin, G.; Li, C. Amp-activated protein kinase activation protects gastric epithelial cells from helicobacter pylori-induced apoptosis. *Biochem. Biophys. Res. Commun.* **2014**, *453*, 13–18. [CrossRef] [PubMed]
38. Jing, Y.; Liu, W.; Cao, H.; Zhang, D.; Yao, X.; Zhang, S.; Xia, H.; Li, D.; Wang, Y.C.; Yan, J.; et al. Hepatic p38α regulates gluconeogenesis by suppressing AMPK. *J. Hepatol.* **2015**, *62*, 1319–1327. [CrossRef] [PubMed]
39. Xu, X.; Sun, J.; Song, R.; Doscas, M.E.; Williamson, A.J.; Zhou, J.; Sun, J.; Jiao, X.; Liu, X.; Li, Y. Inhibition of p70 S6 kinase (S6k1) activity by a77 1726, the active metabolite of leflunomide, induces autophagy through tak1-mediated AMPK and JNK activation. *Oncotarget* **2017**, *8*, 30438–30454. [CrossRef] [PubMed]
40. Flusberg, D.A.; Sorger, P.K. Surviving apoptosis: Life-death signaling in single cells. *Trends Cell Biol.* **2015**, *25*, 446–458. [CrossRef] [PubMed]
41. Kroemer, G.; Marino, G.; Levine, B. Autophagy and the integrated stress response. *Mol. Cell* **2010**, *40*, 280–293. [CrossRef] [PubMed]
42. Zadra, G.; Batista, J.L.; Loda, M. Dissecting the dual role of AMPK in cancer: From experimental to human studies. *Mol. Cancer Res.* **2015**, *13*, 1059–1072. [CrossRef] [PubMed]
43. Inokuchi-Shimizu, S.; Park, E.J.; Roh, Y.S.; Yang, L.; Zhang, B.; Song, J.; Liang, S.; Pimienta, M.; Taniguchi, K.; Wu, X.; et al. TAK1-mediated autophagy and fatty acid oxidation prevent hepatosteatosis and tumorigenesis. *J. Clin. Investig.* **2014**, *124*, 3566–3578. [CrossRef] [PubMed]
44. Seki, E. TAK1-dependent autophagy: A suppressor of fatty liver disease and hepatic oncogenesis. *Mol. Cell. Oncol.* **2014**, *1*, e968507. [CrossRef] [PubMed]
45. Liu, W.; Jiang, Y.; Sun, J.; Geng, S.; Pan, Z.; Prinz, R.A.; Wang, C.; Sun, J.; Jiao, X.; Xu, X. Activation of TGF-β-activated kinase 1 (TAK1) restricts *Salmonella* Typhimurium growth by inducing AMPK activation and autophagy. *Cell Death Dis.* **2018**, *9*, 570. [CrossRef] [PubMed]
46. Kobayashi, Y.; Mizoguchi, T.; Take, I.; Kurihara, S.; Udagawa, N.; Takahashi, N. Prostaglandin E2 enhances osteoclastic differentiation of precursor cells through protein kinase A-dependent phosphorylation of TAK1. *J. Biol. Chem.* **2005**, *280*, 11395–11403. [CrossRef] [PubMed]
47. Mia, S.; Federico, G.; Feger, M.; Pakladok, T.; Meissner, A.; Voelkl, J.; Groene, H.J.; Alesutan, I.; Lang, F. Impact of AMP-activated protein kinase α1 deficiency on tissue injury following unilateral ureteral obstruction. *PLoS ONE* **2015**, *10*, e0135235. [CrossRef] [PubMed]

9

AMP-Activated Protein Kinase as a Reprogramming Strategy for Hypertension and Kidney Disease of Developmental Origin

You-Lin Tain [1,2] and Chien-Ning Hsu [3,*]

1 Departments of Pediatrics, Kaohsiung Chang Gung Memorial Hospital and Chang Gung University College of Medicine, Kaohsiung 833, Taiwan; tainyl@hotmail.com
2 Institute for Translational Research in Biomedicine, Kaohsiung Chang Gung Memorial Hospital and Chang Gung University College of Medicine, Kaohsiung 833, Taiwan
3 Department of Pharmacy, Kaohsiung Chang Gung Memorial Hospital, Kaohsiung 833, Taiwan
* Correspondence: chien_ning_hsu@hotmail.com

Abstract: Suboptimal early-life conditions affect the developing kidney, resulting in long-term programming effects, namely renal programming. Adverse renal programming increases the risk for developing hypertension and kidney disease in adulthood. Conversely, reprogramming is a strategy aimed at reversing the programming processes in early life. AMP-activated protein kinase (AMPK) plays a key role in normal renal physiology and the pathogenesis of hypertension and kidney disease. This review discusses the regulation of AMPK in the kidney and provides hypothetical mechanisms linking AMPK to renal programming. This will be followed by studies targeting AMPK activators like metformin, resveratrol, thiazolidinediones, and polyphenols as reprogramming strategies to prevent hypertension and kidney disease. Further studies that broaden our understanding of AMPK isoform- and tissue-specific effects on renal programming are needed to ultimately develop reprogramming strategies. Despite the fact that animal models have provided interesting results with regard to reprogramming strategies targeting AMPK signaling to protect against hypertension and kidney disease with developmental origins, these results await further clinical translation.

Keywords: AMP-activated protein kinase; developmental origins of health and disease (DOHaD); hypertension; kidney disease; nutrient-sensing signals; oxidative stress; renin-angiotensin system

1. Introduction

Hypertension and kidney disease have a significant impact on morbidity and mortality worldwide. Hypertension and kidney disease can be the cause and consequence of one another. Importantly, both disorders can originate in early life. Kidneys play a key role in blood pressure (BP) regulation. The developing kidney is highly vulnerable to environmental effects in fetal and infantile life, leading to long-term programming effects on the morphology and functioning of the kidney [1,2]. Adverse renal programming increases the risk for developing hypertension and kidney disease in adulthood [3]. This notion has become a globally recognized concept as developmental origins of health and disease (DOHaD) [4]. Conversely, the DOHaD concept also allows reprogramming [5], a strategy aimed at reversing the initial programming processes prior to the onset of hypertension and kidney disease, in order to shift therapeutic interventions from adulthood to early life. A growing body of evidence suggests that AMP-activated protein kinase (AMPK) plays a decisive role in the normal renal physiology and pathogenesis of hypertension and kidney disease [6,7]. Based on the two aspects of the DOHaD concept, this review will first present the evidence for the link between AMPK signaling and programming mechanisms that may lead to hypertension and kidney disease of developmental origin, with a focus on the kidney. This will be followed by potential pharmacological interventions targeting AMPK signaling that may serve as reprogramming strategies to halt the growing epidemic of hypertension and kidney disease.

2. AMP-Activated Protein Kinase in the Renal System

2.1. The Structure and Function of AMP-Activated Protein Kinase

AMPK is a phylogenetically conserved, ubiquitously expressed serine/threonine protein kinase. AMPK is a heterotrimer, composed of an α (α1, α2) catalytic subunit, a regulatory and structurally crucial β (β1, β2) subunit, and a regulatory γ (γ1, γ2, γ3) subunit [8–10]. These isoforms are encoded by distinct genes and differentially expressed, and have unique tissue-specific expression profiles, creating the potential to generate a diverse collection of 12 $\alpha\beta\gamma$ heterotrimer combinations. AMPK has a diverse range of biological functions, including cellular energy homeostasis, glucose metabolism, lipid metabolism, protein synthesis, redox regulation, mitochondria biogenesis, autophagy, ion transport, tumor suppression, anti-inflammation, and nitric oxide (NO) synthesis [8–11]. The structure and function of these different isoforms have been reviewed in detail previously [8–10], and it is not within the scope of the current review to thoroughly outline these further. However, for the purposes of the discussion below, it is important to note that activation of AMPK leads to the biological functions that are linked to renal pathophysiology.

2.2. Regulation of AMP-Activated Protein Kinase AMP-Activated Protein Kinase in the Kidney

AMPK is strongly expressed in the kidney, where it is involved in diverse physiological and pathologic processes, including sensing cellular energy status, sodium and ion transport, podocyte function, BP control, the epithelial-to-mesenchymal transition, and NO production [7–11]. In the rat kidney, α1 and β1 subunits are predominant [7,11]. Except for the muscle-specific γ3 isoform, both γ1 and γ2 subunits are similarly expressed in the kidney. However, little is known regarding the differences in AMPK subunit expression between different cell types within the kidney.

The activity of AMPK is mainly regulated by the AMP and adenosine triphosphate (ATP) ratio. AMPK is activated both allosterically and by post-translational modifications. The most well-defined mechanisms of AMPK activation are phosphorylation at αThr^{172} by upstream AMPK kinases and by AMP or adenosine diphosphate (ADP) binding to the γ subunit. So far, at least three kinases and three phosphatases have been identified as upstream AMPK-activating kinases, including liver kinase B1 (LKB1), TGFβ-activated kinase 1 (TAK1), Ca^{2+}-/calmodulin-dependent protein kinase β (CaMKKβ), protein phosphatase 2A (PP2A), protein phosphatase 2C (PP2C), and Mg^{2+}-/Mn^{2+}-dependent protein phosphatase 1E (PPM1E) [12]. Additionally, AMPK can also be regulated by intracellular calcium and oxidant signaling, as well as extracellular signaling like hormones and cytokines [13]. Furthermore, AMPK is the target of a growing number of pharmacological activators [14].

AMPK has transcriptional effects on numerous enzymes that mediate cellular energy metabolism. AMPK can induce mitochondrial biogenesis by activating the peroxisome proliferator-activated receptor-γ (PPARγ) coactivator-1α (PGC-1α), either directly or through the silent information regulator transcript 1 (SIRT1) [14,15]. Additionally, AMPK and SIRT1 can mediate phosphorylation and deacetylation of PGC-1α, respectively [15], to regulate the expression of PPAR target genes. As reviewed elsewhere, several PPAR target genes contribute to renal programming and hypertension of developmental origins, such as *Sod2*, *Nrf2*, *Sirt7*, *Ren*, *Nos2*, *Nos3*, and *Sgk1* [16]. Another important downstream effect of AMPK is the inhibition of the mammalian target of rapamycin (mTOR). Both AMPK and mTOR can oppositely regulate unc-51-like kinase 1/2 (ULK1/2) activity by phosphorylation to mediate autophagy, a cellular catabolic process in which key organelles are transported to lysosomes for degradation. In addition to activating ULK1/2, AMPK can promote autophagy through SIRT1 by de-acetylating several autophagy-related proteins [17]. Given that autophagy is involved in the pathogenesis of many kidney diseases, and that AMPK regulates autophagic protection against kidney injury, AMPK is becoming a potential target for kidney disease therapy [18]. Additionally, AMPK has been shown to exert anti-inflammatory, antioxidant, and anti-apoptosis effects. Moreover, AMPK regulates many sodium and ion transport proteins in the renal tubular cells, including the epithelial Na^+ channel (ENaC) [19], the Na^+–K^+–$2Cl^-$ cotransporter (NKCC2) [20], Na^+/K^+-ATPase (NaKATPase) [21], the vacuolar H^+-ATPase (V-ATPase) [22], and others [7,11,23]. The activation of AMPK and its biochemical pathways are illustrated in Figure 1.

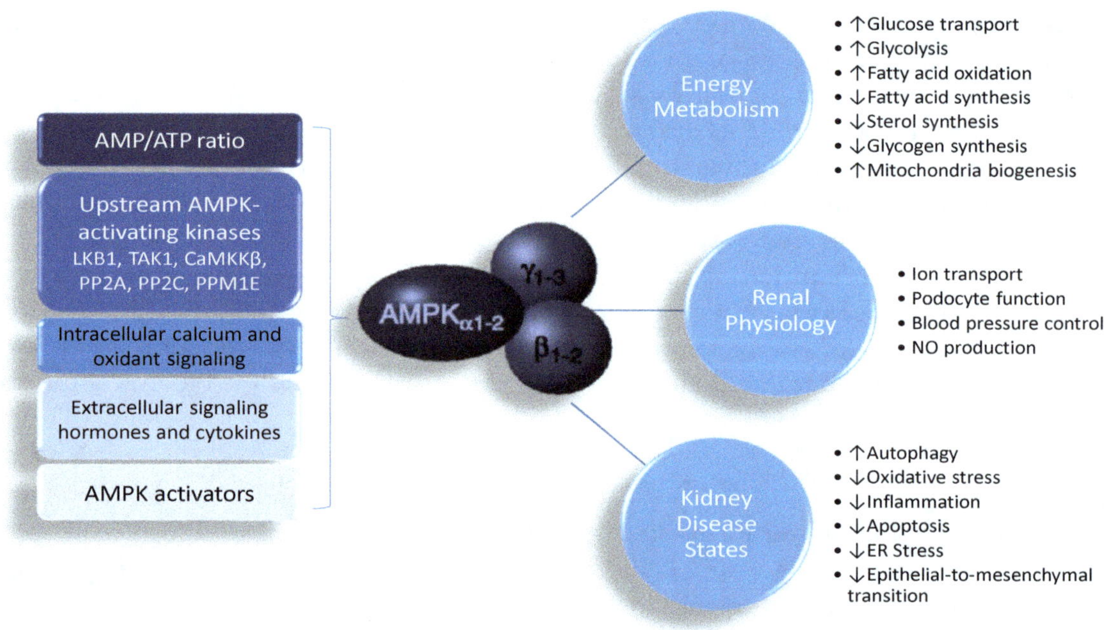

Figure 1. Schematic representation of AMP-activated protein kinase (AMPK) actions and its biochemical functions in the kidney. ↑ = increased. ↓ = decreased.

2.3. AMP-Activated Protein Kinase in Hypertension and Kidney Disease

Emerging evidence suggests that dysfunction in the AMPK signaling pathway is involved with the development of various cardiovascular diseases, including hypertension [6]. Despite several AMPK activators having been assessed in a number of human studies, interventions necessary to provide a reprogramming strategy and prove causation remain undeveloped. It is for this reason that much of our knowledge of potential mechanisms of renal programming, the impacts of AMPK in renal programming, and reprogramming strategies targeting AMPK signaling come from studies using animal models.

In a genetic hypertension model of a spontaneously hypertensive rat (SHR), AMPK activation was reduced in the aorta of the SHR, while 5-aminoimidazole-4carboxamide riboside (AICAR), a direct AMPK activator, lowered BP [24]. Our previous report showed that metformin, a known AMPK activator, blocks the development of hypertension in SHRs, which is associated with increased renal NO production [25]. Like metformin, activation of AMPK by perinatal resveratrol supplementation has been shown to mitigate the development of hypertension in adult SHRs [26]. Additionally, we recently found that maternal plus post-weaning high-fat diets induced hypertension and reduced renal cortical protein levels of phosphorylated AMPK2α in offspring kidneys, which was prevented by resveratrol therapy [27]. Similarly, AMPK activator metformin was reported to protect adult offspring against the developmental programming of hypertension induced by a maternal plus post-weaning high-fat diet [28]. However, the reprogramming effects of the AMPK activator have not been fully assessed in other developmental programming models for hypertension. Therefore, further investigation is needed to reveal the precise role of AMPK in hypertension of developmental origins, especially the reprogramming effects of AMPK activation.

Apart from its role in the development of hypertension, as mentioned above, there have been many studies on the effects of AMPK in kidney diseases, notably diabetic nephropathy, autosomal dominant polycystic kidney disease, subtotal nephrectomy, lupus nephritis, and renal fibrosis [7]. Although AMPK signals have been studied in established kidney diseases, so far there remains a lack of data on the role of AMPK in renal programming and kidney disease of developmental origin.

3. Common Mechanisms Link AMP-Activated Protein Kinase to Renal Programming

Despite the fact that several organ systems can be programmed in response to adverse environmental exposures in early life, renal programming is considered to be decisive in the development of hypertension, as well as kidney disease [3,5,29,30]. Thus far, a number of proposed mechanisms, including dysregulation of the renin–angiotensin system (RAS), impaired sodium transporters, a nutrient-sensing signal, and oxidative stress have been linked to renal programming [3,5,29,30]. Each mechanism related to AMPK signaling will be discussed in turn.

3.1. Renin–Angiotensin System

The RAS is a central regulator of BP and renal function. The RAS consists of two opposing axes: the angiotensin converting enzyme (ACE)-angiotensin (Ang) II type 1 receptor (AT1R) classical axis, mediated primarily by Ang II; and the ACE2-angiotensin-(1–7)-Mas receptor axis, mediated mainly by Angiotensin-(1–7) [31]. In contrast to ACE, ACE2 appears to adjust the angiotensin II type 2 receptor (AT2R) and the angiotensin (1–7) receptor Mas in a way that opposes the development of hypertension [31]. Over-activation of the classical RAS leads to hypertension and kidney disease [31]. Notably, this hormone-signaling pathway controls kidney development [32]. Both RAS axes and the above-mentioned RAS components have been linked to fetal programming [33–36]. There is a biphasic response with reduced classical RAS expression at birth that increases with age. Early-life renal programming might activate the classical RAS, leading to hypertension and kidney disease development in later life. Conversely, early blockade of the classical RAS has been shown to prevent the development of hypertension and kidney disease [37–39]. Decreased renal AMPK expression has been found in uninephretomized rats with the activation of the RAS [40], which was prevented by the blockage of the RAS. AMPKα2 knockout mice expressed high ACE levels, resulting in vasoconstriction [41]. Conversely, prenatal metformin therapy has been shown to restore the maternal high-fructose plus post-weaning high-fat diet-induced increases of RAS components *Ren*, *Atp6ap2*, *Agt*, *Ace*, and *Agtr1a* in the kidney cortex, resulting in protection from hypertension [28]. Given that resveratrol, an indirect AMPK activator, was reported to exert its protective effects in association with increased expression of the AT2R and Mas receptors [42], further studies are required to determine whether AMPK has a role in activation of the ACE2-angiotensin-(1–7)-Mas receptor axis and AT2R to prevent hypertension, and whether this contributes to the reprogramming effects of AMPK activators. Nevertheless, the detailed mechanisms underlying the modulation of RAS by AMPK and its contributions to protection from programmed hypertension kidney disease still await for further study in different models of developmental programming.

3.2. Sodium Transporters

Hypertension and kidney disease of developmental origin have been associated with enhanced sodium reabsorption, attributed to the increased expression of sodium transporters [1,5,29,30]. Several adverse environmental impacts during early life leading to predisposition toward impaired sodium transporters have been reported, including maternal low-protein diet, maternal high-fat diet, maternal exposure to continuous light, and prenatal glucocorticoid exposure [1,5,29]. Several sodium transporters have been identified in the programming processes, including Na^+/Cl^- cotransporter (NCC), type 3 sodium hydrogen exchanger (NHE3), NKCC2, and NaKATPase. We previously showed that a maternal high-fructose diet plus a postnatal high-salt diet increased renal cortical protein levels of NKCC2, NHE3, and NCC in a two-hit model of programmed hypertension [43]. Notably, AMPK regulates several sodium transporters, such as NKCC2 and NaKATPase, which may account for its beneficial effects on hypertension and kidney disease. However, there are as yet no studies examining the role of AMPK in sodium transporters in the kidneys from animal models of programmed hypertension and kidney disease.

3.3. Nutrient-Sensing Signals

Nutrient-sensing signals regulate fetal metabolism and development in response to maternal nutritional input. AMPK is a well-known nutrient-sensing signal [44]. In addition to AMPK, known nutrient-sensing signals exist in the kidney, including SIRT, PPARs, PGC-1α, and mTOR [44]. The interplay between AMPK and other nutrient-sensing signals, driven by early-life input, can regulate PPARs and their target genes, thereby promoting programmed hypertension and kidney disease [16,45]. Our previous work demonstrated that resveratrol, an AMPK activator, prevents the development of hypertension programmed by maternal plus post-weaning high-fructose diets, via regulation of nutrient-sensing signals [46], supporting the notion that nutrient-sensing signals might be a common mechanism underlying the pathogenesis of hypertension and kidney disease of developmental origin. Since AMPK is a crucial hub for the nutrient-sensing signals network, further studies are required to determine whether AMPK has a role in the regulation of renal programming, and whether AMPK activators can serve as reprogramming strategies to prevent the developmental programming of hypertension and kidney disease.

3.4. Oxidative Stress

Oxidative stress is an oxidative shift characterized by an imbalance between oxidants (e.g., reactive oxygen species (ROS)) and antioxidants, in favor of oxidants. The developing fetus is highly vulnerable to oxidative stress damage, due to its low antioxidant capacity [47]. As reviewed elsewhere [29,45], numerous pre- and peri-natal inputs have been linked to renal programming attributed to oxidative stress, including imbalanced maternal nutrition, maternal diabetes, preeclampsia, prenatal hypoxia, maternal nicotine exposure, maternal inflammation, prenatal glucocorticoid exposure, and a high-fat maternal diet. Conversely, some reprogramming interventions have targeted antioxidants in order to reduce oxidative stress, and accordingly, prevent hypertension and kidney disease of developmental origin [30]. As nutrient-sensing is interconnected with redox regulation, AMPK has a key role in regulating antioxidant defense during oxidative stress. AMPK has been reported to upregulate several antioxidant genes, such as superoxide dismutase (SOD), uncoupling protein 2 (UCP2), and nuclear factor erythroid-2-related factor (NRF2) [48]. Additionally, AMPK activation was shown to suppress nicotinamide adenine dinucleotide phosphate (NADPH) oxidase, a primary source of ROS [49]. Furthermore, AMPK promotes autophagy. Since mitochondria are another major source of ROS within cells, activation of mitochondrial autophagy driven by AMPK is also beneficial for reducing oxidative stress. Thus, these findings suggest that the interplay between AMPK and oxidative stress contributes to programmed hypertension and kidney disease.

All of these observations demonstrate a close link between AMPK and other hypothetical mechanisms involved in renal programming. Nevertheless, there remains no definite conclusion that AMPK plays a central role on mediating other mechanisms leading to hypertension and developmental kidney disease.

4. Reprogramming Strategy Targeting AMP-Activated Protein Kinase Signaling

Reprogramming strategies to counterbalance the programming processes that have been employed range from nutritional intervention and lifestyle modification to pharmacological therapy [1,5,50]. Currently, a variety of therapeutic regimens have been reported to either activate or inhibit AMPK activity and its downstream signaling pathway. Since AMPK inhibition, either by AMPK silencing or AMPK inhibitors (e.g., compound C), contributes to hypertension [51,52], treatment modalities for AMPK activation have become more attractive reprogramming strategies. Both indirect and direct AMPK activators have been studied in established hypertension and kidney disease [6,13,18,53]. Modulators that cause AMP or calcium accumulation without a direct interaction with AMPK are classified as "indirect AMPK activators". Several kinds of indirect AMPK activators have been studied in relation to cardiovascular and kidney disease [13], including metformin, resveratrol, thiazolidinediones (TZDs), polyphenols, berberine, ginsenoside, α-lipoic acid, quercetin, and so on. On the other hand, direct AMPK activators induce conformational changes in the AMPK complex through direct interaction with a specific subunit of AMPK. While some are potent pan-activators (e.g., AICAR) for all 12 heterotrimetric AMPK complexes, others show isoform-specific activations for the α1 (e.g., compound-13), β1 (e.g., PF-06409577 and PF-249), β1/β2 (e.g., GSK621), or γ1 isoforms (e.g., PT-1) [13,48]. However, present knowledge of AMPK activators in the kidney is significantly less advanced than that for other organs, such as the liver, muscles, and heart. So far, only one report has shown that selective AMPKβ1 activators PF-06409577 and PF-249 protect against kidney damage in a rat model of diabetic nephropathy [54]. In the current review, we will primarily be limited to pharmacological therapies aimed at AMPK signaling as a reprogramming strategy to prevent hypertension and kidney disease of developmental origin. Of note, pharmacotherapies will be narrowly restricted to those beginning prior to the onset of hypertension and kidney disease. SHRs, for example, reveal a rise in BP starting from six weeks of age, and a steep increase between 6 and 24 weeks. We therefore restrict our discussion mainly to therapies starting before six weeks of age, using the SHR model. These AMPK activation modalities are listed in Table 1 [25–28,46,55–62]. Because the field of DOHaD research is beginning to emerge, this list is by no means complete and is expected to grow rapidly.

Table 1. Reprogramming strategy targeted on AMPK signaling in animal models of programmed hypertension and kidney disease.

Animal Models	Gender/Species	Age at Evaluation	Dose and Period of Treatment	Reprogramming Effects	Ref.
		Metformin			
SHR [1]	Male SHR	12 weeks	Metformin (500 mg/kg/day) between 4 to 12 weeks of age	Prevented hypertension	[25]
Maternal high-fructose plus post-weaning high-fat diet	Male SD [2] rats	12 weeks	Metformin (500 mg/kg/day) for 3 weeks during pregnancy	Attenuated hypertension;	[28]
		Resveratrol and other polyphenols			
SHR	Male SHR	11 weeks	Resveratrol (50 mg/L) in drinking water between 3–11 weeks of age	Attenuated hypertension	[55]
SHR	Male and female SHR	12 weeks	Resveratrol (4g/kg of diet) between gestational day 0.5 and postnatal day 21	Attenuated hypertension	[26]
SHR	Male SHR	13 weeks	Resveratrol (50 mg/L) in drinking water between 3–13 weeks of age	Attenuated hypertension	[56]
Prenatal hypoxia and postnatal high-fat diet	Male SD rats	12 weeks	Resveratrol (4g/kg of diet) between 3–12 weeks of age	Prevented hypertension	[57]
Maternal plus post-weaning high-fructose diets	Male SD rats	12 weeks	Resveratrol (50 mg/L) in drinking water from weaning to three months of age	Prevented hypertension	[46]
Maternal plus post-weaning high-fat diets	Male SD rats	16 weeks	0.5% resveratrol in drinking water between 2 and 4 months of age	Prevented hypertension	[27]
SHR	Male SHR	7 weeks	Magnolol (100 mg/kg/day) between 4 to 7 weeks of age	Attenuated hypertension	[58]
SHR	Male SHR	20 weeks	Berberine (100 mg/kg/day) between 3 to 20 weeks of age	Attenuated hypertension and kidney damage	[59]

Table 1. *Cont.*

Animal Models	Gender/Species	Age at Evaluation	Dose and Period of Treatment	Reprogramming Effects	Ref.
High-salt stroke-prone SHR	Male stroke-prone SHR	16 weeks	Genistein (0.06% wt/wt diet) between 7 to 16 weeks of age	Attenuated hypertension and kidney damage	[60]
			Thiazolidinediones		
SHR	Male SHR	7 weeks	Pioglitazone (10 mg/kg/day) between 5 to 7 weeks of age	Attenuated hypertension	[61]
SHR	Male SHR	13 weeks	Rosiglitazone (150 mg/kg/day) between 5 to 13 weeks of age	Attenuated hypertension	[62]

[1] SHR: Spontaneously hypertensive rat; [2] SD rats: Sprague–Dawley rats.

4.1. *Metformin*

Metformin, the most commonly prescribed first-line antidiabetic drug in the world, exerts its beneficial effects primarily by AMPK activation. Despite a growing body of evidence indicating the protective effects of metformin in established cardiovascular and kidney diseases [63,64], only a few studies have been conducted to explore its reprogramming effects on programmed hypertension and kidney disease. Early metformin treatment in the pre-hypertensive stage blocks the development of hypertension in SHRs [25]. Additionally, maternal metformin therapy protects adult offspring against the developmental programming of hypertension induced by a maternal plus post-weaning high-fat diet [28]. However, a concern raised by these studies is that AMPK-independent effects on metformin were also reported. A better understanding of the AMPK-dependent and -independent mechanisms responsible for the protective effects of metformin on programmed hypertension and kidney disease is therefore warranted.

4.2. *Resveratrol and Other Polyphenols*

Polyphenols are a large group of phytochemicals found in plant-based food. Resveratrol is a naturally occurring polyphenol phytoalexin. It has been considered to have cardiovascular protective effects, including against hypertension [65]. Mechanisms of activation of AMPK by resveratrol appear to elevate AMP levels and inhibit mitochondrial ATP production. Early resveratrol therapy mitigates the development of hypertension in adult SHRs of both sexes [26,55,56]. Using the prenatal hypoxia and postnatal high-fat diet rat model, post-weaning resveratrol treatment protects adult offspring against programmed hypertension [57]. Our previous report showed that a maternal plus post-weaning high-fat diet induced hypertension and reduced protein levels of phosphorylated AMPK2α in the offspring kidney cortex, which resveratrol therapy prevented [27]. Additionally, early post-weaning resveratrol therapy was reported to prevent the development of hypertension of adult offspring exposed to maternal and post-weaning high-fructose consumption [46].

Interestingly, increased AMP levels and ATP depletion leading to uric acid production have been demonstrated as key mediators in the pathogenesis of fructose-induced metabolic syndrome and hypertension [66]. Our previous study showed that the mechanisms underlying the development of hypertension in offspring exposed to maternal high-fructose consumption are different from those in adult rats fed with a high-fructose diet [67]. This may explain why resveratrol increases AMP levels to activate AMPK, resulting in a beneficial effect on the offspring's BP. However, the reprogramming effect of maternal resveratrol on kidney disease has not been fully assessed in developmental programming models. Of note is that resveratrol has multifaceted biological functions; however, to what extent does its reprogramming effect on hypertension and kidney disease can be attributed to AMPK activation deserves further elucidation.

In addition to resveratrol, several polyphenols are capable of activating AMPK, including quercetin, genistein, epigallocatechin gallate, anthocyanin, magnolol, berberine, and so on. Early treatment with magnolol, berberine, or genistein offered protective effects against programmed hypertension in adult SHRs [58–60], whereas quercetin did not [68]. While many polyphenols are antioxidants and exert beneficial effects on oxidative stress-related disorders [69], evidence for their reprogramming effects on hypertension and kidney disease as AMPK activators is equivocal. A better understanding of the differential mechanisms of various polyphenols in the prevention and treatment of programmed hypertension and kidney disease is therefore warranted.

4.3. Thiazolidinediones

Thiazolidinediones (TZDs) are a class of insulin-sensitizing drugs, including pioglitazone, rosiglitazone, and troglitazone. TZDs exert their effects mainly by activating PPARγ. They are also known to act in part through AMPK activation. TZDs that activate AMPK are associated with the accumulation of AMP. Treatment with pioglitazone and rosiglitazone prior to the onset of hypertension can be protective in SHRs, by attenuating the development of hypertension [61,62]. However, another study showed that perinatal pioglitazone treatment fails to confer antihypertensive or renoprotective effects in adult fawn-hooded hypertensive rats [70]. Thus, further examination is required to understand the protective effects of TZDs in programmed hypertension and kidney disease, which are exerted mainly via AMPK or PPARγ signaling pathway.

4.4. Others

Despite progress made in recent years in discovering direct AMPK activators [13,53], little is known regarding their reprogramming effects on hypertension and kidney disease of developmental origins. The first direct AMPK activator, AICAR, has been reported to lower BP in adult SHRs [24]. However, the reprogramming effect of AICAR on programmed hypertension and kidney disease has not been examined yet in developmental programming models. Additionally, oxidative modification of the AMPKα subunit appears to be a major mechanism by which AMPK is activated under conditions of oxidative stress. Therefore, any modulators that induce intracellular ROS generation might serve as AMPK activators. Furthermore, it is of great importance to understand the interplay between the AMPK signaling and other mechanisms underlying renal programming; the application of reprogramming strategies targeting the above-mentioned mechanisms is also feasible for early intervention.

5. Conclusions and Future Perspectives

Hypertension and kidney disease in adult life can be programmed by early-life input. This concept opens a new window for preventing or delaying the onset of hypertension and kidney disease via reprogramming. Studies in short-lived animals, with controlled interventions across their lifespan, have provided interesting results from reprogramming interventions to prevent programmed hypertension and kidney disease via targeting AMPK signaling.

Regardless of recent advances in pharmacotherapies for hypertension and kidney disease, only a few studies have targeted their potential for reprogramming. In the current review, the beneficial effects of these treatments are all coming from indirect AMPK activators that are known to act in both an AMPK-dependent and -independent manner. There remains a lack of data regarding AMPK isoforms, specific knockdown models, or direct AMPK activators for renal programming and developmental programming of hypertension. Using modern genomic techniques [71,72], the identification of nephron segment-specific pathways regulated by AMPK isoforms will be an emerging area of interest. Additionally, another question raised from the current review is that the follow-up periods after the cessation of treatment in most cited reprogramming studies was relatively short. Of note is that the reprogramming effects of some perinatal interventions seem to persist beyond six months of age in female, but not male, SHRs [73,74]. Since sex differences appear in AMPK signaling [75], we must determine the long-term effects of AMPK activators in different programming models, and whether there is a sex-dependent response. Furthermore, a major concern in the translation from animal model to human use of AMPK activators is the activators' still-unknown adverse effects. Because pharmacological activation of AMPK is required to reach a specific target tissue, off-target effects may counter the therapeutic effects we are aiming for. For example, one possible effect of AMPK being off-target is the stimulation of the hypothalamus to increase food intake, despite the fact that our target organ is the kidney.

The evidence of the reprogramming effects of AMPK is just the beginning of the field. It is worth noting that instead of fully elucidating the potential mechanisms, these studies pointed out several key mechanisms linking AMPK to renal programming. It is clear that better understanding of the isoform- and tissue-specific effects of AMPK for programmed hypertension and kidney disease are required before a reprogramming strategy targeting AMPK could be translated from animal studies to human trials.

Author Contributions: C.-N.H.: contributed to concept generation, data interpretation, drafting of the manuscript, critical revision of the manuscript, and approval of the article; Y.-L.T.: contributed to concept generation, data interpretation, critical revision of the manuscript, and approval of the article.

Acknowledgments: This work was supported by the Grants CMRPG8F0023, CMRPG8G0672, and CMRPG8H0081 from Chang Gung Memorial Hospital, Kaohsiung, Taiwan.

Abbreviations

ACE	Angiotensin converting enzyme
AICAR	5-aminoimidazole-4carboxamide riboside
AMPK	AMP-activated protein kinase
AT1R	Angiotensin II type 1 receptor
CaMKKβ	Ca^{2+}-/calmodulin-dependent protein kinase β
DOHaD	Developmental origins of health and disease
LKB1	Liver kinase B1
mTOR	Mammalian target of rapamycin
NCC	Na^{+}/Cl^{-} cotransporter
NHE3	Type 3 sodium hydrogen exchanger
NKCC2	Na-K-2Cl cotransporter
PGC-1α	Peroxisome proliferator-activated receptor γ coactivator-1α
PPAR	Peroxisome proliferator-activated receptor
PPM1E	Mg^{2+}-/Mn^{2+}-dependent protein phosphatase 1E
PP2A	Protein phosphatase 2A
PP2C	protein phosphatase 2C
RAS	Renin-angiotensin system
SD	Sprague–Dawley
SHR	Spontaneously hypertensive rat
SIRT	Silent information regulator transcript
TAK1	TGFβ-activated kinase 1
Ulk1	Unc-51-like kinase 1

References

1. Chong, E.; Yosypiv, I.V. Developmental programming of hypertension and kidney disease. *Int. J. Nephrol.* **2012**, *2012*, 15. [CrossRef] [PubMed]
2. Luyckx, V.A.; Bertram, J.F.; Brenner, B.M.; Fall, C.; Hoy, W.E.; Ozanne, S.E.; Vikse, B.E. Effect of fetal and child health on kidney development and long-term risk of hypertension and kidney disease. *Lancet* **2013**, *382*, 273–283. [CrossRef]

3. Kett, M.M.; Denton, K.M. Renal programming: Cause for concern? *Am. J. Physiol. Regul. Integr. Comp. Physiol.* **2011**, *300*, R791–R803. [CrossRef] [PubMed]
4. Haugen, A.C.; Schug, T.T.; Collman, G.; Heindel, J.J. Evolution of DOHaD: The impact of environmental health sciences. *J. Dev. Orig. Health Dis.* **2015**, *6*, 55–64. [CrossRef] [PubMed]
5. Tain, Y.L.; Joles, J.A. Reprogramming: A preventive strategy in hypertension focusing on the kidney. *Int. J. Mol. Sci.* **2015**, *17*, 23. [CrossRef] [PubMed]
6. Xu, Q.; Si, L.Y. Protective effects of AMP-activated protein kinase in the cardiovascular system. *J. Cell Mol. Med.* **2010**, *14*, 2604–2613. [CrossRef] [PubMed]
7. Rajani, R.; Pastor-Soler, N.M.; Hallows, K.R. Role of AMP-activated protein kinase in kidney tubular transport, metabolism, and disease. *Curr. Opin. Nephrol. Hypertens.* **2017**, *26*, 375–383. [CrossRef] [PubMed]
8. Hardie, D.G.; Ross, F.A.; Hawley, S.A. AMPK: A nutrient and energy sensor that maintains energy homeostasis. *Nat. Rev. Mol. Cell Biol.* **2012**, *13*, 251–262. [CrossRef] [PubMed]
9. Grahame Hardie, D. AMP-activated protein kinase: A key regulator of energy balance with many roles in human disease. *J. Intern. Med.* **2014**, *276*, 543–559. [CrossRef] [PubMed]
10. Moreira, D.; Silvestre, R.; Cordeiro-da-Silva, A.; Estaquier, J.; Foretz, M.; Viollet, B. AMP-activated Protein Kinase as a Target For Pathogens: Friends Or Foes? *Curr. Drug Targets* **2016**, *17*, 942–953. [CrossRef] [PubMed]
11. Fraser, S.; Mount, P.; Hill, R.; Levidiotis, V.; Katsis, F.; Stapleton, D.; Kemp, B.E.; Power, D.A. Regulation of the energy sensor AMP-activated protein kinase in the kidney by dietary salt intake and osmolality. *Am. J. Physiol. Renal Physiol.* **2005**, *288*, F578–F586. [CrossRef] [PubMed]
12. Jeon, S.M. Regulation and function of AMPK in physiology and diseases. *Exp. Mol. Med.* **2016**, *48*, e245. [CrossRef] [PubMed]
13. Kim, J.; Yang, G.; Kim, Y.; Kim, J.; Ha, J. AMPK activators: Mechanisms of action and physiological activities. *Exp. Mol. Med.* **2016**, *48*, e224. [CrossRef] [PubMed]
14. Sugden, M.C.; Caton, P.W.; Holness, M.J. PPAR control: It's SIRTainly as easy as PGC. *J. Endocrinol.* **2010**, *204*, 93–104. [CrossRef] [PubMed]
15. Finck, B.N.; Kelly, D.P. Peroxisome proliferator-activated receptor gamma coactivator-1 (PGC-1) regulatory cascade in cardiac physiology and disease. *Circulation* **2007**, *115*, 2540–2548. [CrossRef] [PubMed]
16. Tain, Y.L.; Hsu, C.N.; Chan, J.Y. PPARs Link Early Life Nutritional insults to later programmed hypertension and metabolic syndrome. *Int. J. Mol. Sci.* **2015**, *17*, 20. [CrossRef] [PubMed]
17. Dutta, D.; Calvani, R.; Bernabei, R.; Leeuwenburgh, C.; Marzetti, E. Contribution of impaired mitochondrial autophagy to cardiac aging: Mechanisms and therapeutic opportunities. *Circ. Res.* **2012**, *110*, 1125–1138. [CrossRef] [PubMed]
18. Allouch, S.; Munusamy, S. AMP-activated Protein Kinase as a Drug Target in Chronic Kidney Disease. *Curr. Drug Targets* **2018**, *19*, 709–720. [CrossRef] [PubMed]
19. Carattino, M.D.; Edinger, R.S.; Grieser, H.J.; Wise, R.; Neumann, D.; Schlattner, U.; Johnson, J.P.; Kleyman, T.R.; Hallows, K.R. Epithelial sodium channel inhibition by AMP-activated protein kinase in oocytes and polarized renal epithelial cells. *J. Biol. Chem.* **2005**, *280*, 17608–17616. [CrossRef] [PubMed]
20. Fraser, S.A.; Gimenez, I.; Cook, N.; Jennings, I.; Katerelos, M.; Katsis, F.; Levidiotis, V.; Kemp, B.E.; Power, D.A. Regulation of the renal-specific Na^+–K^+–$2Cl^-$ co-transporter NKCC2 by AMP-activated protein kinase (AMPK). *Biochem. J.* **2007**, *405*, 85–93. [CrossRef] [PubMed]
21. Vadasz, I.; Dada, L.A.; Briva, A.; Trejo, H.E.; Welch, L.C.; Chen, J.; Toth, P.T.; Lecuona, E.; Witters, L.A.; Schumacker, P.T.; et al. AMP-activated protein kinase regulates CO_2-induced alveolar epithelial dysfunction in rats and human cells by promoting Na,K-ATPase endocytosis. *J. Clin. Invest.* **2008**, *118*, 752–762. [CrossRef] [PubMed]
22. Hallows, K.R.; Alzamora, R.; Li, H.; Gong, F.; Smolak, C.; Neumann, D.; Pastor-Soler, N.M. AMP-activated protein kinase inhibits alkaline pH and PKA-induced apical vacuolar H^+-ATPase accumulation in epididymal clear cells. *Am. J. Physiol. Cell Physiol.* **2009**, *296*, C672–C681. [CrossRef] [PubMed]
23. Pastor-Soler, N.M.; Hallows, K.R. AMP-activated protein kinase regulation of kidney tubular transport. *Curr. Opin. Nephrol. Hypertens.* **2012**, *21*, 523–533. [CrossRef] [PubMed]
24. Ford, R.J.; Teschke, S.R.; Reid, E.B.; Durham, K.K.; Kroetsch, J.T.; Rush, J.W. AMP-activated protein kinase activator AICAR acutely lowers blood pressure and relaxes isolated resistance arteries of hypertensive rats. *J. Hypertens.* **2012**, *30*, 725–733. [CrossRef] [PubMed]

25. Tsai, C.M.; Kuo, H.C.; Hsu, C.N.; Huang, L.T.; Tain, Y.L. Metformin reduces asymmetric dimethylarginine and prevents hypertension in spontaneously hypertensive rats. *Transl. Res.* **2014**, *164*, 452–459. [CrossRef] [PubMed]
26. Care, A.S.; Sung, M.M.; Panahi, S.; Gragasin, F.S.; Dyck, J.R.; Davidge, S.T.; Bourque, S.L. Perinatal Resveratrol Supplementation to Spontaneously Hypertensive Rat Dams Mitigates the Development of Hypertension in Adult Offspring. *Hypertension* **2016**, *67*, 1038–1044. [CrossRef] [PubMed]
27. Tain, Y.L.; Lin, Y.J.; Sheen, J.M.; Lin, I.C.; Yu, H.R.; Huang, L.T.; Hsu, C.N. Resveratrol prevents the combined maternal plus postweaning high-fat-diets-induced hypertension in male offspring. *J. Nutr. Biochem.* **2017**, *48*, 120–127. [CrossRef] [PubMed]
28. Tain, Y.L.; Wu, K.L.H.; Lee, W.C.; Leu, S.; Chan, J.Y.H. Prenatal Metformin Therapy Attenuates Hypertension of Developmental Origin in Male Adult Offspring Exposed to Maternal High-Fructose and Post-Weaning High-Fat Diets. *Int. J. Mol. Sci.* **2018**, *19*, 1066. [CrossRef] [PubMed]
29. Tain, Y.L.; Hsu, C.N. Developmental Origins of Chronic Kidney Disease: Should We Focus on Early Life? *Int. J. Mol. Sci.* **2017**, *18*, 381. [CrossRef] [PubMed]
30. Tain, Y.L.; Chan, S.H.H.; Chan, J.Y.H. Biochemical basis for pharmacological intervention as a reprogramming strategy against hypertension and kidney disease of developmental origin. *Biochem. Pharmacol.* **2018**, *153*, 82–90. [CrossRef] [PubMed]
31. Te Riet, L.; van Esch, J.H.; Roks, A.J.; van den Meiracker, A.H.; Danser, A.H. Hypertension: Renin-angiotensin-aldosterone system alterations. *Circ. Res.* **2015**, *116*, 960–975. [CrossRef] [PubMed]
32. Yosypiv, I.V. Renin-angiotensin system in ureteric bud branching morphogenesis: Insights into the mechanisms. *Pediatr. Nephrol.* **2011**, *26*, 1499–1512. [CrossRef] [PubMed]
33. Bogdarina, I.; Welham, S.; King, P.J.; Burns, S.P.; Clark, A.J. Epigenetic modification of the renin-angiotensin system in the fetal programming of hypertension. *Circ. Res.* **2007**, *100*, 520–526. [CrossRef] [PubMed]
34. Chappell, M.C.; Marshall, A.C.; Alzayadneh, E.M.; Shaltout, H.A.; Diz, D.I. Update on the Angiotensin converting enzyme 2-Angiotensin (1-7)-MAS receptor axis: Fetal programing, sex differences, and intracellular pathways. *Front. Endocrinol.* **2014**, *4*, 201. [CrossRef] [PubMed]
35. Siragy, H.M. The angiotensin II type 2 receptor and the kidney. *J. Renin Angiotensin Aldosterone Syst.* **2010**, *11*, 33–36. [CrossRef] [PubMed]
36. Ali, Q.; Dhande, I.; Samuel, P.; Hussain, T. Angiotensin type 2 receptor null mice express reduced levels of renal angiotensin II type 2 receptor/angiotensin (1-7)/Mas receptor and exhibit greater high-fat diet-induced kidney injury. *J. Renin Angiotensin Aldosterone Syst.* **2016**, *17*. [CrossRef] [PubMed]
37. Sherman, R.C.; Langley-Evans, S.C. Early administration of angiotensin-converting enzyme inhibitor captopril, prevents the development of hypertension programmed by intrauterine exposure to a maternal low-protein diet in the rat. *Clin. Sci.* **1998**, *94*, 373–381. [CrossRef] [PubMed]
38. Hsu, C.N.; Lee, C.T.; Huang, L.T.; Tain, Y.L. Aliskiren in early postnatal life prevents hypertension and reduces asymmetric dimethylarginine in offspring exposed to maternal caloric restriction. *J. Renin Angiotensin Aldosterone Syst.* **2015**, *16*, 506–513. [CrossRef] [PubMed]
39. Hsu, C.N.; Wu, K.L.; Lee, W.C.; Leu, S.; Chan, J.Y.; Tain, Y.L. Aliskiren administration during early postnatal life sex-specifically alleviates hypertension programmed by maternal high fructose consumption. *Front. Physiol.* **2016**, *7*, 299. [CrossRef] [PubMed]
40. Yang, K.K.; Sui, Y.; Zhou, H.R.; Shen, J.; Tan, N.; Huang, Y.M.; Li, S.S.; Pan, Y.H.; Zhang, X.X.; Zhao, H.L. Cross-talk between AMP-activated protein kinase and renin-angiotensin system in uninephrectomised rats. *J. Renin Angiotensin Aldosterone Syst.* **2016**, *17*. [CrossRef] [PubMed]
41. Kohlstedt, K.; Trouvain, C.; Boettger, T.; Shi, L.; Fisslthaler, B.; Fleming, I. AMP-activated protein kinase regulates endothelial cell angiotensin-converting enzyme expression via p53 and the post-transcriptional regulation of microRNA-143/145. *Circ. Res.* **2013**, *112*, 1150–1158. [CrossRef] [PubMed]
42. Kim, E.N.; Kim, M.Y.; Lim, J.H.; Kim, Y.; Shin, S.J.; Park, C.W.; Kim, Y.S.; Chang, Y.S.; Yoon, H.E.; Choi, B.S. The protective effect of resveratrol on vascular aging by modulation of the renin-angiotensin system. *Atherosclerosis* **2018**, *270*, 123–131. [CrossRef] [PubMed]
43. Tain, Y.L.; Lee, W.C.; Leu, S.; Wu, K.; Chan, J. High salt exacerbates programmed hypertension in maternal fructose-fed male offspring. *Nutr. Metab. Cardiovasc. Dis.* **2015**, *25*, 1146–1151. [CrossRef] [PubMed]
44. Efeyan, A.; Comb, W.C.; Sabatini, D.M. Nutrient-sensing mechanisms and pathways. *Nature* **2015**, *517*, 302–310. [CrossRef] [PubMed]

45. Tain, Y.L.; Hsu, C.N. Interplay between oxidative stress and nutrient sensing signaling in the developmental origins of cardiovascular disease. *Int. J. Mol. Sci.* **2017**, *18*, 841. [CrossRef] [PubMed]
46. Tain, Y.L.; Lee, W.C.; Wu, K.; Leu, S.; Chan, J.Y.H. Resveratrol prevents the development of hypertension programmed by maternal plus post-weaning high-fructose consumption through modulation of oxidative stress, nutrient-sensing signals, and gut microbiota. *Mol. Nutr. Food Res.* **2018**, in press. [CrossRef] [PubMed]
47. Dennery, P.A. Oxidative stress in development: Nature or nurture? *Free Radic. Biol. Med.* **2010**, *49*, 1147–1151. [CrossRef] [PubMed]
48. Trewin, A.J.; Berry, B.J.; Wojtovich, A.P. Exercise and Mitochondrial Dynamics: Keeping in Shape with ROS and AMPK. *Antioxidants* **2018**, *7*, 7. [CrossRef] [PubMed]
49. Song, P.; Zou, M.H. Regulation of NAD(P)H oxidases by AMPK in cardiovascular systems. *Free Radic. Biol. Med.* **2012**, *52*, 1607–1619. [CrossRef] [PubMed]
50. Nüsken, E.; Dötsch, J.; Weber, L.T.; Nüsken, K.D. Developmental Programming of Renal Function and Re-Programming Approaches. *Front. Pediatr.* **2018**, *6*, 36. [CrossRef] [PubMed]
51. Cao, X.; Luo, T.; Luo, X.; Tang, Z. Resveratrol prevents Ang II-induced hypertension via AMPK activation and RhoA/ROCK suppression in mice. *Hypertens. Res.* **2014**, *37*, 803–810. [CrossRef] [PubMed]
52. Ford, R.J.; Rush, J.W. Endothelium-dependent vasorelaxation to the AMPK activator AICAR is enhanced in aorta from hypertensive rats and is NO and EDCF dependent. *Am. J. Physiol. Heart Circ. Physiol.* **2011**, *300*, H64–H75. [CrossRef] [PubMed]
53. Olivier, S.; Foretz, M.; Viollet, B. Promise and challenges for direct small molecule AMPK activators. *Biochem. Pharmacol.* **2018**, *153*, 147–158. [CrossRef] [PubMed]
54. Salatto, C.T.; Miller, R.A.; Cameron, K.O.; Cokorinos, E.; Reyes, A.; Ward, J.; Calabrese, M.F.; Kurumbail, R.G.; Rajamohan, F.; Kalgutkar, A.S.; et al. Selective Activation of AMPK β1-Containing Isoforms Improves Kidney Function in a Rat Model of Diabetic Nephropathy. *J. Pharmacol. Exp. Ther.* **2017**, *361*, 303–311. [CrossRef] [PubMed]
55. Javkhedkar, A.A.; Banday, A.A. Antioxidant resveratrol restores renal sodium transport regulation in SHR. *Physiol. Rep.* **2015**, *3*, e12618. [CrossRef] [PubMed]
56. Bhatt, S.R.; Lokhandwala, M.F.; Banday, A.A. Resveratrol prevents endothelial nitric oxide synthase uncoupling and attenuates development of hypertension in spontaneously hypertensive rats. *Eur. J. Pharmacol.* **2011**, *667*, 258–264. [CrossRef] [PubMed]
57. Rueda-Clausen, C.F.; Morton, J.S.; Dolinsky, V.W.; Dyck, J.R.; Davidge, S.T. Synergistic effects of prenatal hypoxia and postnatal high-fat diet in the development of cardiovascular pathology in young rats. *Am. J. Physiol. Regul. Integr. Comp. Physiol.* **2012**, *303*, R418–R426. [CrossRef] [PubMed]
58. Liang, X.; Xing, W.; He, J.; Fu, F.; Zhang, W.; Su, F.; Liu, F.; Ji, L.; Gao, F.; Su, H.; et al. Magnolol administration in normotensive young spontaneously hypertensive rats postpones the development of hypertension: Role of increased PPARγ, reduced TRB3 and resultant alleviative vascular insulin resistance. *PLoS ONE* **2015**, *10*, e0120366. [CrossRef] [PubMed]
59. Guo, Z.; Sun, H.; Zhang, H.; Zhang, Y. Anti-hypertensive and renoprotective effects of berberine in spontaneously hypertensive rats. *Clin. Exp. Hypertens.* **2015**, *37*, 332–339. [CrossRef] [PubMed]
60. Cho, T.M.; Peng, N.; Clark, J.T.; Novak, L.; Roysommuti, S.; Prasain, J.; Wyss, J.M. Genistein attenuates the hypertensive effects of dietary NaCl in hypertensive male rats. *Endocrinology* **2007**, *148*, 5396–5402. [CrossRef] [PubMed]
61. Dovinová, I.; Barancik, M.; Majzunova, M.; Zorad, S.; Gajdosechová, L.; Gresová, L.; Cacanyiova, S.; Kristek, F.; Balis, P.; Chan, J.Y. Effects of PPARγ agonist pioglitazone on redox-sensitive cellular signaling in young spontaneously hypertensive rats. *PPAR Res.* **2013**, *2013*, 11. [CrossRef] [PubMed]
62. Wu, L.; Wang, R.; de Champlain, J.; Wilson, T.W. Beneficial and deleterious effects of rosiglitazone on hypertension development in spontaneously hypertensive rats. *Am. J. Hypertens.* **2004**, *17*, 749–756. [CrossRef] [PubMed]
63. Wang, Y.W.; He, S.J.; Feng, X.; Cheng, J.; Luo, Y.T.; Tian, L.; Huang, Q. Metformin: A review of its potential indications. *Drug Des. Dev. Ther.* **2017**, *11*, 2421–2429. [CrossRef] [PubMed]
64. De Broe, M.E.; Kajbaf, F.; Lalau, J.D. Renoprotective Effects of Metformin. *Nephron* **2018**, *138*, 261–274. [CrossRef] [PubMed]
65. Li, H.; Xia, N.; Förstermann, U. Cardiovascular effects and molecular targets of resveratrol. *Nitric Oxide* **2012**, *26*, 102–110. [CrossRef] [PubMed]

66. Johnson, R.J.; Sanchez-Lozada, L.G.; Nakagawa, T. The effect of fructose on renal biology and disease. *J. Am. Soc. Nephrol.* **2010**, *21*, 2036–2039. [CrossRef] [PubMed]
67. Tain, Y.L.; Leu, S.; Wu, K.L.; Lee, W.C.; Chan, J.Y. Melatonin prevents maternal fructose intake-induced programmed hypertension in the offspring: Roles of nitric oxide and arachidonic acid metabolites. *J. Pineal Res.* **2014**, *57*, 80–89. [CrossRef] [PubMed]
68. Carlstrom, J.; Symons, J.D.; Wu, T.C.; Bruno, R.S.; Litwin, S.E.; Jalili, T. A quercetin supplemented diet does not prevent cardiovascular complications in spontaneously hypertensive rats. *J. Nutr.* **2007**, *137*, 628–633. [CrossRef] [PubMed]
69. Goszcz, K.; Duthie, G.G.; Stewart, D.; Leslie, S.J.; Megson, I.L. Bioactive polyphenols and cardiovascular disease: Chemical antagonists, pharmacological agents or xenobiotics that drive an adaptive response? *Br. J. Pharmacol.* **2017**, *174*, 1209–1225. [CrossRef] [PubMed]
70. Koeners, M.P.; Wesseling, S.; Sánchez, M.; Braam, B.; Joles, J.A. Perinatal Inhibition of NF-κB has long-term antihypertensive and renoprotective effects in fawn-hooded hypertensive rats. *Am. J. Hypertens.* **2016**, *29*, 123–131. [CrossRef] [PubMed]
71. Lee, J.W.; Chou, C.L.; Knepper, M.A. Deep Sequencing in Microdissected Renal Tubules Identifies Nephron Segment-Specific Transcriptomes. *J. Am. Soc. Nephrol.* **2015**, *26*, 2669–2677. [CrossRef] [PubMed]
72. Gonzalez-Vicente, A.; Hopfer, U.; Garvin, J.L. Developing Tools for Analysis of Renal Genomic Data: An Invitation to Participate. *J. Am. Soc. Nephrol.* **2017**, *28*, 3438–3440. [CrossRef] [PubMed]
73. Koeners, M.P.; van Faassen, E.E.; Wesseling, S.; de Sain-van der Velden, M.; Koomans, H.A.; Braam, B.; Joles, J.A. Maternal supplementation with citrulline increases renal nitric oxide in young spontaneously hypertensive rats and has long-term antihypertensive effects. *Hypertension* **2007**, *50*, 1077–1084. [CrossRef] [PubMed]
74. Koeners, M.P.; Braam, B.; Joles, J.A. Perinatal inhibition of NF-kappaB has long-term antihypertensive effects in spontaneously hypertensive rats. *J. Hypertens.* **2011**, *29*, 1160–1166. [CrossRef] [PubMed]
75. Mukai, Y.; Ozaki, H.; Serita, Y.; Sato, S. Maternal fructose intake during pregnancy modulates hepatic and hypothalamic AMP-activated protein kinase signalling in a sex-specific manner in offspring. *Clin. Exp. Pharmacol. Physiol.* **2014**, *41*, 331–337. [CrossRef] [PubMed]

Co-Expression Network Analysis of AMPK and Autophagy Gene Products during Adipocyte Differentiation

Mahmoud Ahmed [1], Jin Seok Hwang [1], Trang Huyen Lai [1], Sahib Zada [1], Huynh Quoc Nguyen [1], Trang Min Pham [1], Miyong Yun [2] and Deok Ryong Kim [1,*]

1 Department of Biochemistry and Convergence Medical Sciences and Institute of Health Sciences, Gyeongsang National University School of Medicine, Jinju 527-27, Korea; ma7moud_sha3ban@hotmail.com (M.A.); cloud8104@naver.com (J.S.H.); tranghuyen20493@gmail.com (T.H.L.); s.zada.qau@gmail.com (S.Z.); nqh2412@gmail.com (H.Q.N.); phamminhtrang010895@gmail.com (T.M.P.)

2 Department of Bioindustry and Bioresource Engineering, College of Life Sciences, Sejong University, Seoul 05006, Korea; myyun91@gmail.com

* Correspondence: drkim@gnu.ac.kr

Abstract: Autophagy is involved in the development and differentiation of many cell types. It is essential for the pre-adipocytes to respond to the differentiation stimuli and may contribute to reorganizing the intracellulum to adapt the morphological and metabolic demands. Although AMPK, an energy sensor, has been associated with autophagy in several cellular processes, how it connects to autophagy during the adipocyte differentiation remains to be investigated. Here, we studied the interaction between AMPK and autophagy gene products at the mRNA level during adipocyte differentiation using public-access datasets. We used the weighted-gene co-expression analysis to detect and validate multiple interconnected modules of co-expressed genes in a dataset of MDI-induced 3T3-L1 pre-adipocytes. These modules were found to be highly correlated with the differentiation course of the adipocytes. Several novel interactions between AMPK and autophagy gene products were identified. Together, it is possible that AMPK-autophagy interaction is temporally and locally modulated in response to the differentiation stimuli.

Keywords: AMPK; autophagy; co-expression; microarrays; 3T3-L1; adipocyte; differentiation

1. Introduction

Autophagy is essential for the white adipocyte differentiation. The knockdown of *Atg5* and/or *Atg7* gene in the 3T3-L1 pre-adipocyte prevents its maturation upon the chemical induction [1,2]. Secondary to that, the pre-adipocyte fails to accumulate triglycerides and to form the fat droplets, which is a characteristic of mature white adipocytes [3]. This is yet to be reconciled with another observation in cells known to contain large quantities of lipids (e.g., hepatocytes) where autophagy takes part in lipid degradation [4]. AMP-activated protein kinase (AMPK), which can be activated at the low level of energy such as starvation, stimulates autophagy through the inhibition of the mTOR activity [5,6] and/or the direct phosphorylation of ULK1 [7,8]. Autophagy and AMPK regulate several aspects of the lipid metabolism and the cell response to changing energy levels. Therefore, the AMPK-autophagy interaction could be consequential in the context of adipocyte differentiation.

3T3-L1 pre-adipocyte is a mouse fibroblast with the potential to differentiate into a mature adipocyte when treated with the MDI differentiation induction medium (160 nM insulin, 250 nM dexamethasone, and 0.5 mM 1-methyl-3-isobutylxanthine) [9]. Upon induction, the pre-adipocyte undergoes multiple metabolic and morphological changes to reach maturation. Evidently, several of these changes can be observed at the transcription level of multiple adipogenic and lipogenic markers [10,11].

The aim of this work is to identify the potential AMPK-autophagy connections, in the broad sense of the pathways, that are both novel and consequential in the adipocyte differentiation, mainly using the weighted-gene co-expression network analysis (WGCNA) [12]. This approach focuses on identifying co-expressed pairs of genes across the differentiation stages, and enables the direct use of similar datasets to test and validate the findings even though they might be performed on different platforms.

In this study, we applied the WGCNA approach to a microarrays dataset of MDI-induced adipocyte at eight different time points corresponding to three differentiation stages. We identified two networks/modules among autophagy and AMPK gene products that were correlated with the differentiation course. By analyzing these modules in one of the datasets, we were able to specify several potential novel AMPK-autophagy interactions and connect these to candidate functions through the known annotations. Finally, we checked these findings in three independent datasets of similar design and found the networks/modules to be well preserved.

2. Results

2.1. Preparing Data and Annotations

First, we retrieved several microarray datasets from the Gene expression omnibus (GEO) [13]. We sought arrays of MDI-induced 3T3-L1 pre-adipocytes at three or more time points. After excluding the ones with varying designs and limited annotations, four datasets were included in the analysis (Table 1); one dataset (GSE34150) was chosen for the main analysis, and the rest were reserved for testing and validation. In GSE34150, the total RNA from 24 samples of MDI-induced pre-adipocytes were collected at eight different time points corresponding to three adipocyte differentiation stages (0 day, undifferentiated; two and four days, differentiating; 6–18 days, maturating). The initial quality assessment included checking the distribution of the intensities from all probes at the log scale, hierarchal clustering and multi-dimensional scaling analysis (MDS). Groups of samples from different stage of differentiation showed similar distributions, appropriate clustering and separation across the two different dimensions of MDS. Furthermore, to ensure the reliability of the analysis, we examined the expression of a number of differentiation and lipogenesis markers (Appendix A). The *Pparg*, *Cebpa* and *Lpl* genes, essential factors for adipocyte differentiation, were highly expressed in differentiating and maturing cells compared to the undifferentiated cells. Expression of most lipogenic genes (*Pparg, Cebpa, Lpl, Scd1, Scd2, Dgat1, Dgat2* and *Fasn*) was correlated with the development of 3T3-L1 pre-adipocytes into mature adipocytes.

Table 1. MDI-induced 3T3-L1 microarrays' datasets.

Series ID	Platform ID	Samples	Included	(Contact, Year)	Reference
GSE15018	GPL6845	54	18	(Chin, 2009)	[14]
GSE20696	GPL1261	8	8	(Mikkelsen, 2010)	[15]
GSE34150	GPL6885	24	24	(Irmler, 2011)	[16]
GSE69313	GPL6246	48	12	(Renbin, 2015)	[17]

The gene ontology (GO) terms: AMP-activated protein kinase activity (AMPK) and autophagy were used to identify 14 and 167 genes of known functions in the corresponding biological processes (BP), respectively. A total of 181 genes was used in the downstream analysis to limit the input to WGCNA, over-representation and defining novel interactions between and among AMPK and autophagy pathways. GO was also used to define the terms involving these interacting gene products and link them to known molecular functions (MF) and cellular components (CC). Appendix A contains a detailed discussion for the data inclusion criteria, quality assessment and obtaining the GO annotations.

2.2. Detecting Co-Expression Modules of AMPK and Autophagy Genes in Differentiating Adipocytes

Constructing co-expression networks is a multi-step process. First, the Pearson's correlation coefficient was calculated between each pair of the genes of interest (n = 181) across all samples (n = 24). Second, these correlations were raised to the power 5 to obtain an adjacency matrix of all possible pairs. Third, the adjacency matrix was used to calculate the Topological Overlap Matrix (TOM) as a reliable similarity measure. Finally, TOM similarity between pairs of genes were used to calculate the weight of their connection in a network of all possible pairs and a distance (1 - TOM) to cluster the pairs into highly interconnected modules/colors (Figure 1). Appendix B provides a detailed discussion of the previous steps and the rationale for the different choices that were made in this analysis.

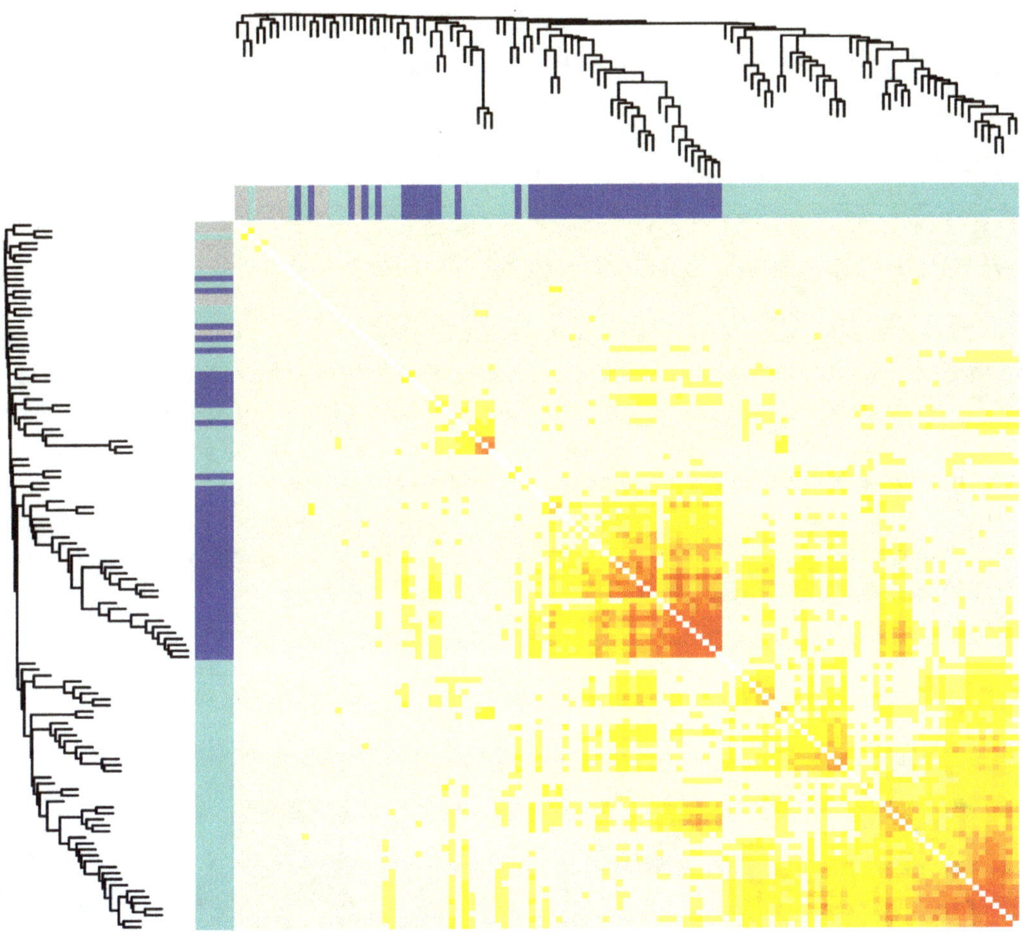

Figure 1. Clustering of AMPK and autophagy genes by their pairwise distances. Pairwise topological overlap matrix (TOM) similarities of AMPK and autophagy genes (n = 181) were calculated from their expression values in the GSE34150 dataset. Distances between each pair of genes were derived as 1 - TOM and shown as color values (small, red or large, yellow). A hierarchal tree and colored segments of the clusters were shown on the top and side.

Among all possible pairwise correlations between the 181 genes of interest, two groups/modules of highly co-expressed gene products were formed (blue, 42; turquoise, 66), and the rest were unassigned (gray, 10). Genes that code for the subunits of the AMPK complex fell into different modules; four AMPK genes, *Prkaa2, Prkab2, Prkag2* and *Prkag3*, in the blue module along with 38 of the genes involved in autophagy; and two AMPK genes, *Prkab2 and Prkag1*, in the turquoise module together with 63 of the autophagy genes (Table 2). In the following sections, we describe the significance and the interactions of the individual members of these modules.

Table 2. AMPK and autophagy genes in different modules/colors.

Module/Color	AMPK	Autophagy
blue	*Prkaa2, Prkab1, Prkag2, Prkag3, Smok3b*	*Acbd5, Atg5, Atm, Bmf, Bok, Casp1, Cln3, Dapk1, Dhrsx, Fbxl2, Fbxo7, Fis1, Hif1a, Lep, Map1lc3a, Mcl1, Mid2, Optn, Pik3c2a, Pink1, Prkaa2, Rab39b, Rab8a, Rragc, Sh3bp4, Sh3glb1, Sqstm1, Tbc1d5, Tnfaip3, Tpcn1, Tpcn2, Trim8, Trp53inp2, Vcp, Wdr45, Yod1, Zc3h12a, Zfyve1*
turquoise	*4921509C19Rik, Prkab2, Prkag1*	*Ager, Akt1, Bcl2, Becn1, Capn10, Cdkn2a, D17Wsu92e, Dap, Dcn, Depdc5, Ei24, Eif4g1, Eif4g2, Foxo1, Fundc1, Fundc2, Hmgb1, Hspa8, Htr2b, Ifng, Lamp2, Lars, Lmx1b, Lrrk2, Map1lc3b, Map2k1, Mapt, Mt3, Nbr1, Nlrp6, Pik3c3, Pik3r2, Pik3r4, Pim2, Plaa, Plekhf1, Plk2, Pycard, Rasip1, Rnf5, Rraga, Rragb, Sirt2, Smcr8, Smurf1, Stk11, Tcirg1, Tmem74, Trim21, Trp53inp1, Tsc2, Ubqln1, Ulk1, Usp10, Usp13, Usp30, Usp33, Vps4a, Vps4b, Wdr6, Wipi1, Wipi2, Xbp1*

2.3. Correlating the Detected Modules to the Stage of Differentiation

To establish the biological significance of these modules, we used the expression values of their individual members to calculate a representative summary—the first principal component (PC)—for each module. Then, we calculated the Pearson's correlation coefficient for the first PC with the stage of differentiation (undifferentiated, differentiating or maturating) of all 24 samples. Both modules showed a reasonable correlation with sample stages (>0.8 for the blue and >0.3 for the turquoise module) (Figure 2A). In other words, the expression values of the members of the blue module, and to a less extent the turquoise, capture a lot of the observed differences between the cells as they progress from a differentiation stage to the next.

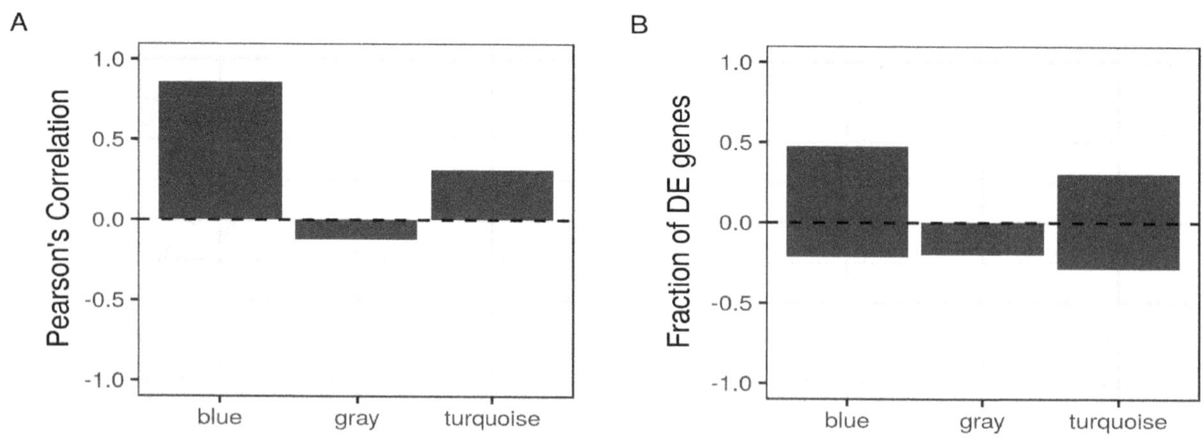

Figure 2. Correlations and over-representation of the detected modules in differentiation stages. The expression values of the members of the detected modules in the GSE34150 dataset (42, blue; 10, gray; and 66, turquoise) were used to calculate two representative summary statistics. (**A**) the first principal component (PC) across samples were correlated to the sample stages using Pearson's correlation (bars); (**B**) the fraction of differentially expressed (DE) genes across differentiation stages (bars).

2.4. *Testing the Over-Representation of the Modules over the Differentiation Course*

Again, we considered the expression values of the individual members of each module to calculate the fraction of the differentially expressed genes (DE) across differentiation stages. Both blue and turquoise modules had a significant fraction of their member genes (>0.5) either up or downregulated at the differentiating or maturating stage compared to the control undifferentiated cell stage (Figure 2B). These fractions were significantly higher than the expected fractions of DE genes in randomly selected modules of the corresponding sizes. The calculated p-values were adjusted for multiple testing using the False Discovery Rate (FDR). Adjusted p-values less than 0.1 were considered significant.

2.5. *Visualizing Modules and Identifying Novel AMPK-Autophagy Interactions*

To visually explore the detected modules, we treated each of their members as a node in a network graph. Nodes were divided into two networks based on the module to which they belong. Each pair of nodes was connected by an edge that has a weight calculated from the TOM similarity measure between the corresponding pair of genes. Edges with weights less than a minimum threshold (0.1) were excluded to obtain a less condense network (Figure 3). Evidently, some nodes did not share edges that passes this threshold and were not included in the network graph. In addition, nodes were labeled with the corresponding official gene symbol and colored as AMPK or autophagy genes; and the edges were colored by the novelty of the connection. The latter was determined mainly based on previous reports in the STRING database (textmining evidence), which is evidence extracted from abstracts of scientific literature.

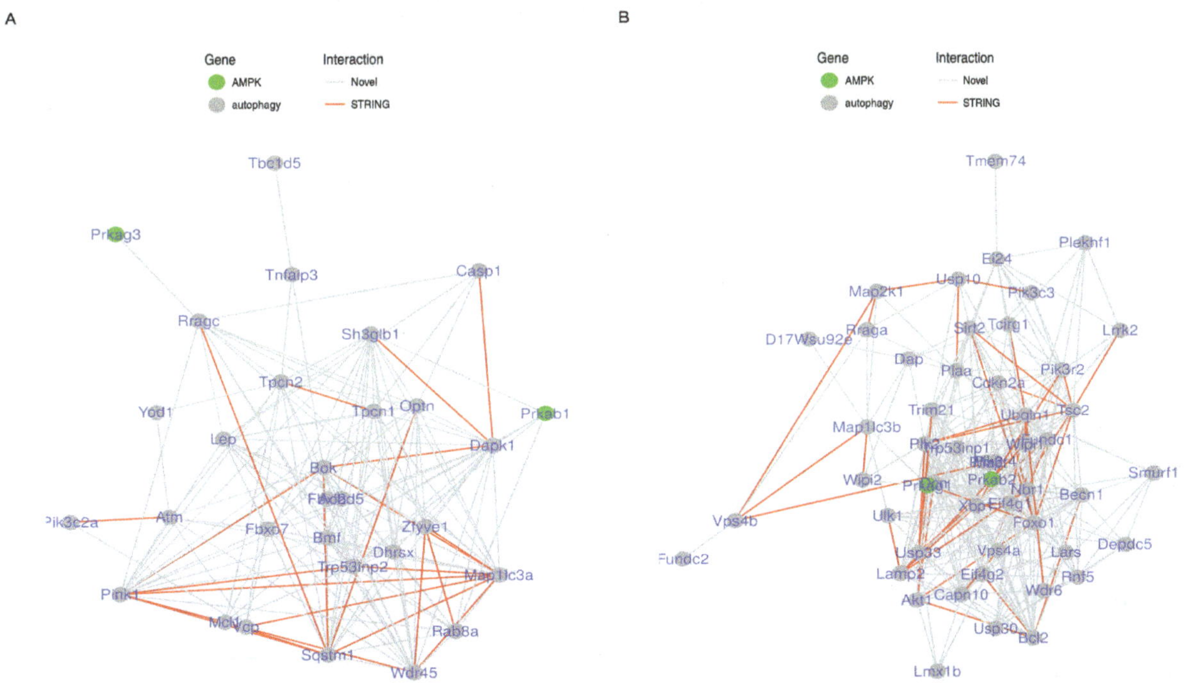

Figure 3. Network representation of the AMPK and autophagy modules. Members of the blue (**A**) and turquoise (**B**) modules are shown as a nodes. Each pair of nodes is connected by an edge if the corresponding pairwise topological overlap matrix (TOM) similarity/weight is above the threshold 0.1. Nodes are colored by gene category (AMPK, green or autophagy, gray). Edges are colored by type of interaction (STRING, red or Novel, gray).

By representing the modules in graphs, we were able to calculate different statistics to identify influential genes/nodes and important interactions/edges. Considering various centrality measures, we ranked the genes in each module by their influence on the modules (Table 3). *Trp53inp2, Map1lc3a, Wadr45, Pink1* and *Dapk1* genes were the most influential nodes in the graph of the blue module with a hub score more than 0.95, while *Foxo1, Dcn* and *Xbp1* genes had the highest scores in the turquoise module. Edges between AMPK and autophagy nodes that were not previously reported in the STRING database (text-mining evidence) were considered novel potential interactions (Table 4). The protein kinase AMP-activated non-catalytic subunit beta 1 (Prkab1) showed a potential interaction with several autophagy gene products including Bcl2-modifying factor (Bmf), death associated protein kinase 1 (Dapk1), Ras-associated protein Rab8a (Rab8a), SH3 domain, GRB2-like, endophilin B1 (Sh3glb1) and transformation related protein 53 inducible nuclear protein 2 (Trp53inp2) as part of the blue module. Similarly, the gamma subunit 1 (Prkag1) in the turquoise module revealed a novel binding ability to some well-known autophagy-related gene products such as Becn1, Fundc1, Lamp2 or Map1lc3b and also showed a novel interaction with some other gene products including calpain-like cysteine protease (Capn10), cyclin-dependent kinase inhibitor 2a (Cdkn2a) for p16INK4a and p14ARF, and ubiquitin-associated proteins (Trp53inp1, Nbr1, Usp33). In addition, there were some novel interactions between AMPK and autophagy gene products across the two modules, indicated as "inbetween" in Table 4.

Table 3. Top five hubs in the different module networks.

Module/Color	Gene	Degree	Betweenness	Closeness	Hub Score
blue	*Trp53inp2*	19	14.67	0.16	1
	Map1lc3a	19	22.71	0.16	0.99
	Wdr45	18	11.23	0.16	0.98
	Pink1	18	12.91	0.16	0.97
	Dapk1	18	14.48	0.16	0.96
turquoise	*Foxo1*	28	70.89	0.28	1
	Dcn	25	36.23	0.28	0.96
	Xbp1	24	29.59	0.28	0.93
	Plk2	24	60.28	0.27	0.9
	Eif4g1	24	41.01	0.27	0.9

Table 4. Summary of reported and novel AMPK-autophagy interactions.

Module/Color	AMPK	Autophagy
blue	*Prkab1*	*Bmf, Dapk1, Rab8a, Sh3glb1, Trp53inp2*
	Prkag3	*Rragc* [4]
inbetween	*Prkab1*	*Tsc2* [2,4], *Ubqln1, Wipi1*
	Prkab2	*Zc3h12a*
	Prkag3	*Usp33*
turquoise	*Prkag1*	*Akt1* [1,3,4], *Bcl2, Becn1, Capn10, Cdkn2a, Dcn* [3,4], *Eif4g1, Foxo1* [2], *Fundc1, Lamp2, Lars, Map1lc3b, Nbr1, Plk2, Sirt2* [3,4], *Trim21, Trp53inp1, Usp33, Vps4a, Wipi2* [3,4], *Xbp1*

[1] Coexpression in the same or in other species (transferred by homology). [2] Database gathered from curated databases. [3] Experiments gathered from other protein–protein interaction databases. [4] Textmining extracted from the abstracts of scientific literature.

2.6. *Testing for Molecular Functions and Cellular Components Enrichment by the Detected Modules*

We used a list-based enrichment to specify the contributions of the modules to the differentiation process. The mouse gene ontology Molecular Function (MF) and Cellular Components (CC) terms were submitted to an enrichment analysis by the gene members of the detected modules (42 for blue and 66 for turquoise). The significant terms (FDR < 0.1) are shown in Figure 4 stratified by the category (MF/CC) and the module (blue/turquoise). As expected, the two modules share a number of MF terms, namely; ubiquitin-like protein binding, ubiquitinyl hydrolase activity and phospholipid binding. At the same time, several terms had significance by mutually exclusive enrichment by the modules. This includes the nucleoside binding and the ubiquitin-like protein transferase activity by the blue module; and a few protein kinase terms by the turquoise module. Similarly, two of the CC terms; extrinsic component of membrane and outer membrane were enriched by both modules, while others related to only one of the two modules.

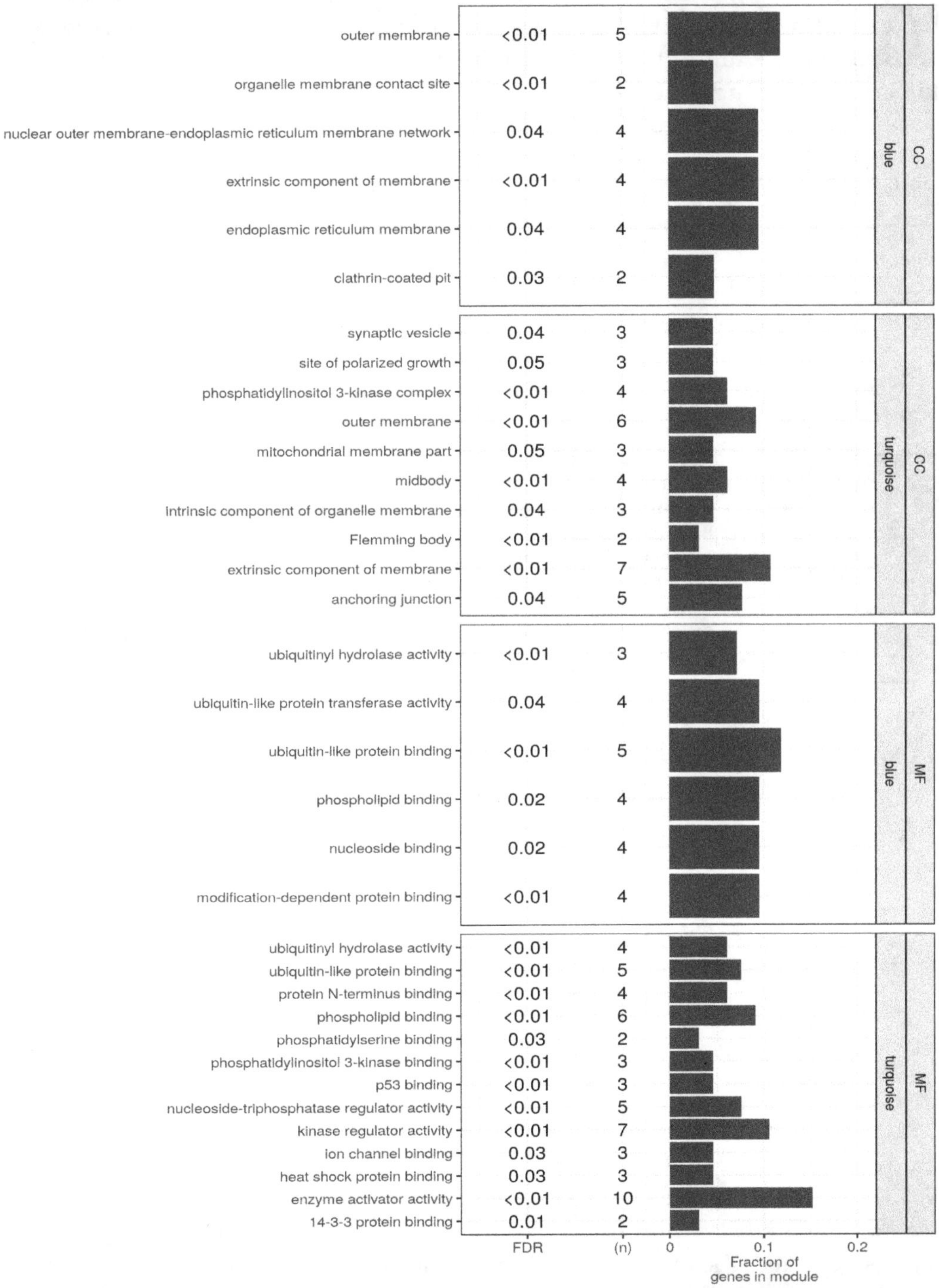

Figure 4. Enrichment of the gene ontology terms by the detected modules. The list of genes in the two detected modules (42, blue and 66, turquoise) were used to test for gene ontology terms enrichment. All terms in the Molecular Function (MF) and the Cellular component (CC) categories of the gene ontology were considered. Only significant terms at a false discovery rate (FDR) less than 0.1 are shown. For each term, the count (n) and the fractions of hits (bars) in the module are shown.

Building on these two pieces of the analysis, the suggested novel AMPK interactions and the gene ontology enrichment by the members of the modules, we set out to specify the kind of functions that the AMPK-autophagy interactions were likely to be involved in. Table 5 shows how AMPK is functionally connected to autophagy gene products through several gene ontology terms such as membrane components, and regulation of kinases, enzymatic and ubiquitin activity. For examples, Prkab1 interacts with Sh3glb1, a Bax-interacting protein at the mitochondrial outer membrane and also with Trp53Inp2 for the uniquitin-like activity. In addition, Prkag1 associates with many autophagy gene products such as Usp33 for the uniquitin activity at the membrane anchoring junction, Cdkn2a for the regulation of kinase activity, and other cellular components for their molecular functions.

Table 5. AMPK and autophagy interactions by gene ontology term.

Module/Color	Ontology	AMPK	Term	Autophgy
blue	CC	*Prkab1*	outer membrane	*Sh3glb1*
	MF		nucleoside binding	*Dapk1, Rab8a*
			ubiquitin-like protein binding	*Trp53inp2*
		Prkag3	nucleoside binding	*Rragc*
turquoise	CC	*Prkag1*	anchoring junction	*Usp33*
			extrinsic component of membrane	*Becn1, Wipi2*
			Flemming body	*Vps4a*
			intrinsic component of organelle membrane	*Fundc1, Lamp2*
			midbody	*Sirt2, Vps4a*
			mitochondrial membrane part	*Fundc1*
			outer membrane	*Bcl2, Capn10, Fundc1*
			phosphatidylinositol 3-kinase complex	*Becn1*
	MF		14-3-3 protein binding	*Akt1*
			enzyme activator activity	*Lars*
			kinase regulator activity	*Cdkn2a, Dcn*
			nucleoside-triphosphatase regulator activity	*Lars*
			p53 binding	*Cdkn2a*
			phosphatidylinositol 3-kinase binding	*Becn1, Xbp1*
			phospholipid binding	*Akt1, Wipi2*
			protein N-terminus binding	*Cdkn2a, Dcn*
			ubiquitin-like protein binding	*Nbr1, Sirt2*
			ubiquitinyl hydrolase activity	*Usp33*

2.7. Preservation of AMPK-Autophagy Networks across Independent Datasets

Finally, we validated these findings in three independent datasets of similar MDI-induced 3T3-L1 cells at different time points or differentiation stages. Three GEO microarray datasets (GSE15018, GSE20696 and GSE69313) were used to perform this step of the analysis (Table 1). The average log expression of the 181 genes of interest from the three datasets were first compared to these in the main dataset (Figure 5). As expected, the averages are highly correlated between the datasets (>0.74), a pre-requisite for the following module preservation analysis. A moderate to high preservation of the modules was observed in the three independent datasets (Figure 6). Generally, modules with a Z summary values between 5 and 10 are considered moderately preserved and these above 10 are considered highly preserved. In fact, the two modules: blue and turquoise, showed a summary statistics in that first category with at least 6 and 7, respectively, indicating that the interaction modules of the main dataset are well preserved in other independent datasets.

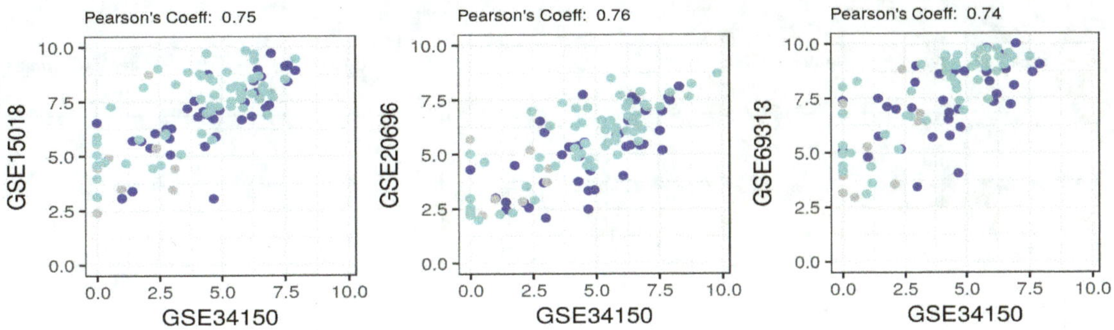

Figure 5. Average expression of AMPK and autophagy in multiple MDI-induced 3T3-L1 microarrays datasets. The log average expression values of AMPK and autophagy genes (n = 181) in the MDI-induced 3T3-L1 datasets (GSE15018, GSE20696 and GSE69313) are compared to the corresponding averages in the main dataset (GSE34150). Individual values are shown as colored points by their assigned modules. The Pearson's correlation coefficient of the corresponding values is shown on top.

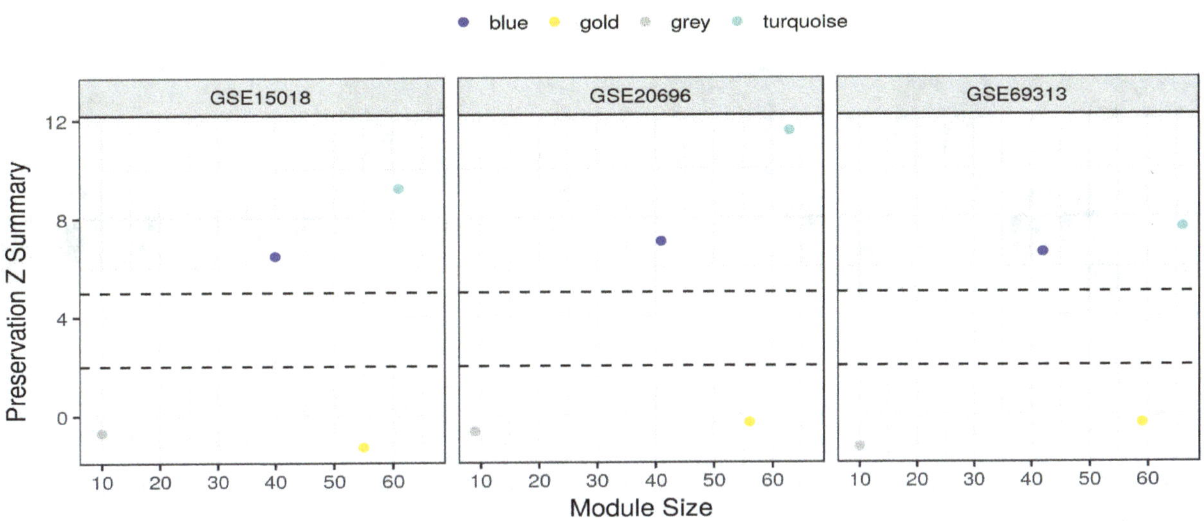

Figure 6. Module preservation Z summary across multiple MDI-induced 3T3-L1 microarrays datasets. The GSE34150 dataset was used to detect the highly co-expressed modules among AMPK and autophagy genes (42, blue; 66, turquoise; 10, gray, unassigned; and 55, gold , randomly assigned). The detected modules were used as a reference to calculate several preservation statistics in three independent datasets of similar design (GSE15018, GSE20696 and GSE69313). Z summary statistics and sizes of four modules are shown as colored points.

2.8. Validation of Selected Gene Products Correlations with Prkab1 and Prkag1

We selected several autophagy gene products that are highly correlated with the AMPK subunits for experimental validation using RT-qPCR (Figure 7A,B). The relative mRNA level of each group of genes were used to calculate the Pearson's correlation coefficients with two AMPK subunits: Prkab1 and Prkag1. Although the resulting coefficient may vary, these calculated earlier due to the different sensitivities between microarrays and RT-qPCR, the directions of the correlation were the same as ones that we observed in the dataset (Figure 7C,D). In agreement with the suggested potential interaction of Prkab1 with Wipi1, Rab8a and Trp53inp2, strong correlations were validated. Prkag1 showed strong to moderate correlations with Becn1, Sirt2 and Trim21 as previously predicted by WGCNA.

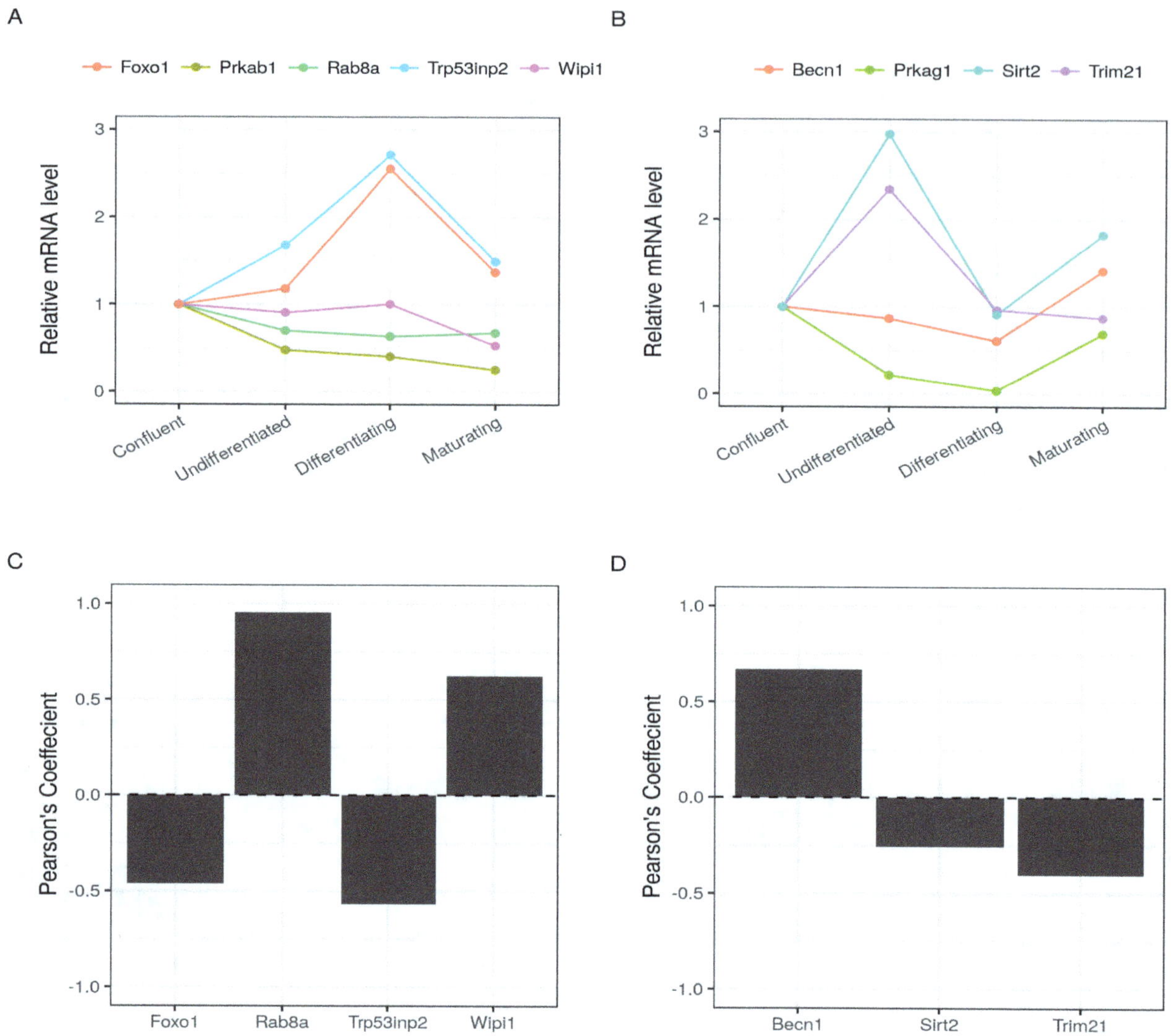

Figure 7. Validation of selected gene products expression and co-expression with AMPK subunits. Three independent samples of MDI-induced 3T3-L1 cells at four different time points corresponding to confluent, undifferentiated, differentiating and maturating stages were used to check the mRNA level of several gene products. (**A**,**B**) the $\Delta\Delta C_t$ values of five and four gene products, respectively, normalized by 18S and relative to the confluent cell stage are shown as points; (**C**,**D**) the Pearson's coefficient of four and three gene products with Prkab1 and Prkag1, respectively, are shown as bars.

3. Discussion

The adipocyte differentiation is a well regulated complex process. On one hand, this complexity allows for flexibility in response to different stimuli. For example, the over-expression of LC3 in 3T3-L1 pre-adipocytes produced a downstream activation of key regulators of adipogenesis and resulted in a differentiation pattern similar to that of the MDI induction [18]. On the other hand, this process likely involves a wide range of changes in transcription, translation and protein modification. In a previous study from our laboratory, we suggested that many autophagy genes were functionally associated with adipocyte differentiation using the RNA-Seq expression data [19]. We also showed that the mRNA level of key autophagy genes is specifically regulated at different time points, and clusters of these genes respond to the differentiation stimulus in a time-dependent manner. In particular, the subsets of organelle specific autophagy (e.g., mitophagy, reticulophagy, etc.) are highly regulated, suggesting a role in reorganizing the interacellulum and removing parts of the cell to adapt the morphological and metabolic changes of the mature adipocyte. Here, we explore the connection between autophagy and AMPK, which was established in conditions such as starvation, as it applies to the pre-adipocytes response to differentiation stimuli.

One aspect of this connection can be deduced from the observed positive correlation of the detected modules with the differentiation course (Figure 2A). In addition, the blue and the turquoise modules scores a significant (p-value < 0.001) protein–protein interaction (PPI) enrichment with an average clustering co-efficient of 0.4 and 0.5, respectively (STRING web interface). Although a detailed molecular link would be less clearer, AMPK gene products were evenly split among these modules, and had multiple edges with highly influential nodes (hubs) of well studied autophagy genes (Tables 3 and 4). Thus, AMPK gene products are part of biologically connected autophagy modules, which seem to be fairly consequential in the process of adipocyte differentiation.

The AMPK complex is formed of one catalytic subunit (α) and two non-catalytic regulatory subunits (β and γ), each has more than one isoform encoded by a separate gene [20]. The different subunits contribute to the stability and activity of the complex, whereas the combinations of the different isoforms give rise to complexes that behave differently and/or are specific to certain tissues [21,22]. Probes corresponding to the genes that code the different isoforms of the subunits were consistently expressed at different levels. Moreover, they showed varying correlations with the cell differentiation stage (data not shown). We considered the subunits and the isoforms of the AMPK complex individually. *Prkab1* and *Prkag1* were expressed at higher levels, and they, therefore, are the main AMPK side of the reported interaction with the autophagy pathway (Tables 4 and 5).

The adipocyte differentiation is characterized by events of increased lipogenesis and intracellular remodeling. AMPK is known for inhibiting the former and stimulating the latter. The absence of a reliable signal from the catalytic subunits of AMPK in our analysis may only enable a partial view. Nevertheless, we observe enrichment of ubiquitin activity and certain cellular component terms, probably akin to a form of localization, by autophagy genes co-expressed with the regulatory (β and γ) subunits of AMPK (Figure 4). One of these terms, ubiquitin-like protein binding, includes two gene products, *Trp53inp2* and *Nbr1*. Both are known to help the formation of autophagosome and the selective removal of ubiquitinated proteins through binding to LC3 [23–25]. Few other terms related to the organelle membranes appear interesting. Perhaps, the localization of AMPK at certain intracellular locations mediates the autophagy selectivity, as suggested before [26]. This is consistent with the description of two emerging mechanisms of AMPK regulation, namely by ubiquitination and sub-cellular distribution [27].

Together, it is possible that AMPK-autophagy connection is determined by the energy supply and demand of the differentiating cells. It is more likely that the two pathways interact dexterously with some temporospatial agility. For example, AMPK activates autophagy early in the differentiation course in response to the differentiation stimulus. In a later stage, autophagy might remove ubiquitinated AMPK to allow the accelerated fat accumulation. Finally, the localization of AMPK to certain intracellular organelles could guide their recycling or removal by selective autophagy. The implications of these features do not escape us, both AMPK and autophagy are involved in disorders such as obesity and diabetes [28,29]. The manipulation of one pathway could affect the outcomes controlled by the other. In addition, the timely intervention during adipocyte differentiation could preferentially favor certain consequences of the pathways' interplay.

Nassiri and colleagues provided a system view of the adipogenesis and ranked the involved biological processes to identify the coordinated activity among them using the NASFinder method of publicly available omics data [30]. According to their analyses, the translational machinery, mitochondrial associated pathways, PPAR signaling, insulin and leptin signaling, and some membrane associated complexes are coordinately upregulated after adipocyte induction. Although they might be associated with AMPK or autophagy at the broad concept, they do not show any specific coordination between AMPK pathway and autophagy process. Here, we used the widely known WGCNA method to detect the conserved genetic networks of AMPK-autophagy gene products that might contribute to the process of differentiation [31,32]. Typically, it needs to input the list of differentially expressed genes among three or more experimental conditions [12]. In this study, we limited the analysis to the probes that mapped uniquely to AMPK and autophagy genes as defined in their gene ontology terms. The downside of limiting the analysis to a predefined set of genes is that the prospective findings would be limited to the available annotation, potential loss of signals from probes that map to genes not in the predefined gene set and the inclusion of probes that map to genes that are not actively changing among the conditions. On the other hand, this approach allows for simplifying the analysis steps and the interpretation of the results. The detected networks are more likely to have biologically meaningful consequences since they are formed of nodes that are known for certain functions in their pathways/gene sets. In addition, this allows for including genes that are highly correlated even though they don't show the highest degree of differentiation among conditions. Certainly, some of these genes are involved in the biology of adipocyte differentiation either by maintaining essential cellular processes or they show subtle changes that wouldn't be typically picked by the differential expression approach.

4. Materials and Methods

4.1. Data and Annotation Sources

4.1.1. Gene Ontology

The Gene Ontology (GO) terms AMP-activated protein kinase (AMPK) (GO:0004679) and autophagy (GO:0006914) were used to identify the gene products (14 and 167, respectively) with known functions in the corresponding biological processes [33]. Similarly, GO was used to identify the molecular function (MF) and cellular component (CC) terms containing these gene products. GO was accessed through the GO.db and the mouse organism package org.Mm.eg.db [34,35].

4.1.2. Microarrays Expression Data

To identify the relevant datasets, we queried the NCBI Gene Expression Omnibus (GEO) metadata by GEOmetadb [36]. The term '3T3-L1' was used to search the titles of all entries, the query results were then searched manually and datasets of similar induction time-course design were included. The expression and the annotation data were then obtained using a GEOquery [37]. Table 1 summaries the four datasets that were used in this analysis. GSE34150 consists of 24 samples of MDI-induced 3T3-L1 pre-adipocytes at eight different time points corresponding to three differentiation stages (0 day, undifferentiated; two and four days, differentiating; 6–18 days, maturating).

4.1.3. Protein–Protein Interactions

The STRING database was used to query all possible AMPK-autophagy protein–protein interactions that are reported with different evidence types [38]. The HUGO symbols of 181 genes were mapped to the ENSEMBL IDs before querying the database. STRINGdb was used to do the mapping, construct the query and obtain the results. The interactions were matched against the edges of the co-expression networks of the detected modules to label the edges with the type of evidence when they were previously reported.

4.2. Weighted-Gene Co-Expression Network Analysis

The package WGCNA was used to apply most of the necessary steps for weighted-gene co-expression network analysis on the GSE34150 dataset as described in the original publications [39]. Briefly, a co-expression measure (Pearson's correlation coefficient) was calculated between each pair of genes. The coefficients were raised to the power of 5 to form an adjacency matrix. The adjacency matrix was then used to calculate the topological overlap similarity matrix (TOM). To detect modules and assign genes to them, a dissimilarity matrix is obtained (1 − TOM) and used as distances between genes. A hierarchical clustering was then performed and a gene tree is built. Upon cutting the tree at a certain height, genes nearby are assigned to modules, referred to as colors (names are arbitrarily assigned). The detected modules were then used to find the correlation with the phenotype and the preservation in independent datasets. To correlate the modules to the sample phenotypes or to each other, an eigengen or the principal components (PC) were calculated from the expression of their respective members and used as a representative summary. Finally, a module preservation analysis was performed by calculating various summary statistics on the detected modules in the test datasets [40].

4.3. Network Visualization and Analysis

The igraph package was used to visualize and analyze the detected modules [41]. The genes of interest were treated as nodes in a network graph and were connected by an edge if its weight—calculated from the TOM similarity between each pair of genes—passed a minimum threshold. Several graph statistics were used to determine the importance/centrality of genes and their interactions.

4.4. Gene Modules Over-Representation

The limma package was used to test for the over-representation of the detected modules in the GSE34150 dataset [42]. An index of the modules as gene sets, the expression data and comparison matrix based on the differentiation stage were used as input. A gene set is considered *over-represented* when it has a significantly higher fraction of differentially expressed genes than a randomly selected module of the same size. The clusterProfiler package was used to apply a similar list-based enrichment of GO terms by the detected modules [43]. Tests were adjusted for multiple testing using the False Discovery Rate (FDR) and a cutoff (0.1) was applied.

4.5. Cell Culture and RT-qPCR

3T3-L1 pre-adipocytes were cultured and induced for differentiation using MDI protocol as described before [19]. Total RNA was collected at four different time points corresponding the the major differentiation stages of the adipocytes (−2 day, full confluence; 0 day, undifferentiated; 10 h differentiating; and −6 day, maturating). The list of the primers that were used in the reaction are provided in (Appendix A). The C_t values from the RT-qPCR reaction were normalized by a reference gene 18S and calibrated by the confluent samples ($\Delta\Delta C_t$) using the pcr R package [44].

4.6. Software Environment and Reproducibility

The data were obtained, processed and analyzed in an R environment and using multiple Bioconductor packages [45,46]. The full analysis was done and reproduced in an isolated environment based on docker (bioconductor/release_base2) [47]. The scripts for reproducing the analysis, figures and tables are available at https://github.com/MahShaaban/aacna. The instructions for reproducing the analysis are described in Appendix C.

5. Conclusions

In summary, we used the WGCNA to investigate the interactions of AMPK and autophagy gene products in the context of adipocyte differentiation. Two co-expression networks were found to be highly correlated with the time course of differentiation. We were able to validate the case of these networks in other independent datasets of similar experimental designs. These networks appear to be consequential in the response of the pre-adipocyte to the differentiation stimulus. Finally, we present several novel potential interactions between AMPK and autophagy gene products and link them to potential functions and cellular sites.

Author Contributions: M.A. conceived, performed and reported the analysis; J.S.H. performed the RT-qPCR experiment; H.Q.N., S.Z., T.H.L., T.M.P. performed the cell culture and the differentiation induction assay; M.Y. consulted on the design and the execution of the study; D.R.K. supervised the study and contributed to writing the manuscript

Acknowledgments: This study was supported by the Basic
Science Research Program through the National Research Foundation of Korea (NRF) funded by the Ministry of Education Science and Technology (2018R1D1A1B07043715) and by the Ministry of Science, ICT and Future Planning (NRF-2015R1A5A2008833).

Abbreviations

AMPK	AMP-Activated Protein Kinase
CC	Cellular Components
FDR	False Discovery Rate
GEO	Gene Expression Omnibus
GO	Gene Ontology
MDS	Multi-Dimensional Scaling
mTOR	Mechanistic Target Of Rapamycin Kinase
MF	Molecular Function
PC	Principal Component
PPI	Protein–Protein Interaction
RT-qPCR	Real-Time Quantitative Polymerase Chain Reaction
TOM	Topological Overlap Matrix
ULK1	Unc-51 Like Autophagy Activating Kinase 1
WGCNA	Weighted Gene Co-expression Network Analysis

Appendix A. Datasets and Annotations

This appendix contains details of the datasets and the gene annotation used in the study.

Appendix A.1. Time Point of Samples in the Microarrays Datasets

We queried the metadata of the Gene Expression omnibus (GEO) for microarrays datasets of MDI-induced 3T3-L1 pre-adipocytes at different time points that covers the various differentiation stages. The datasets were then manually checked for containing sufficient phenotype and annotation data. A few datasets were generated using custom microarrays chips of a few thousand probes and were excluded for not containing a sufficient number of the probes of interest. In Table A1, we listed four datasets included in this study and the time points (in hours) of their samples.

Table A1. Sample time points in the different datasets.

Time Point (hours)	GSE15018	GSE20696	GSE34150	GSE69313
−48		2		
0		2	3	3
1	1			
2	1			
3	1			
4	1			
5	1			
6	1			3
7	1			

Table A1. *Cont.*

Time Point (hours)	GSE15018	GSE20696	GSE34150	GSE69313
8	1			
9	1			
10	1			
11	1			
12	1			
13	1			
14	1			
15	1			
16	1			
17	1			
18	1			
24				3
48		2	3	
72				3
96			3	
144			3	
168		2		
192			3	
240			3	
336			3	
432			3	

Appendix A.2. Comparing the Average Expression in Four Datasets

To ensure that the different datasets are exhibiting comparable probe expression, we compared the average expression of all common probes in three datasets (GSE15018, GSE20696 and GSE69313) with the main (GSE34150) dataset (Figure A1).

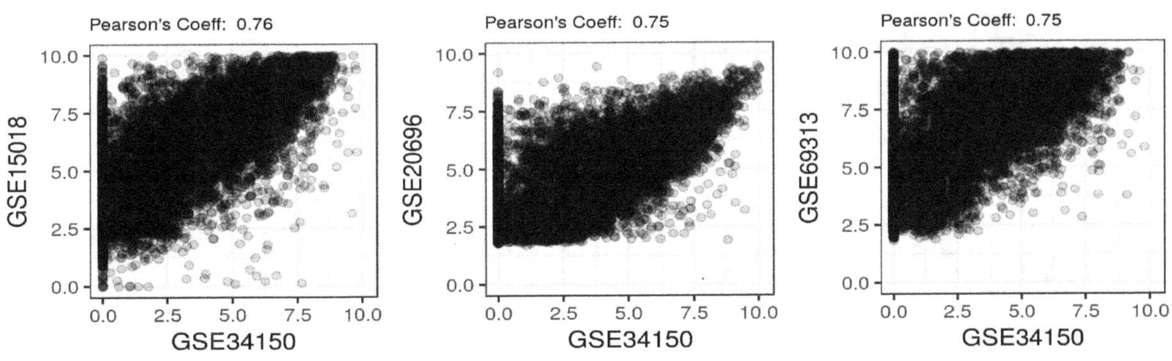

Figure A1. Average expression of all probes in multiple MDI-induced 3T3-L1 microarrays datasets. The log average expression values of all probes in the MDI-induced 3T3-L1 datasets (GSE15018, GSE20696 and GSE69313) are compared to the corresponding averages in the main dataset (GSE34150). Individual values are shown as points. The Pearson's correlation coefficient of the corresponding values is shown on top.

Appendix A.3. Data Quality Assessment

The accession number (GSE34150) was used to obtain the expression matrix and the metadata of the dataset. Several quality assessment measures were applied to ensure the suitability of the data for the downstream analysis (Figure A2).

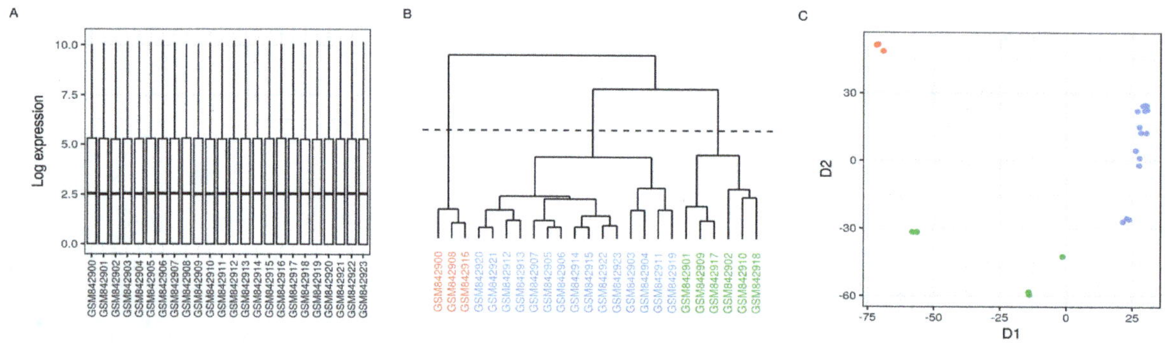

Figure A2. Quality assessment and exploration of the main microarrays dataset. Twenty-four samples of MDI-treated 3T3-L1 cells of the microarray series (GSE34150) were obtained from GEO along with the corresponding annotation data (GPL6885); (**A**) the distribution of the log expression of (n = 25,697) probes from all samples as box plots; (**B**) hierarchical clustering based on the euclidean distances of all samples; (**C**) multi-dimensional scaling (MDS) of all samples. Colors represent the cell stage/time point (green, undifferentiated; dark green, differentiating and red, maturating).

Appendix A.4. Confirming Differentiation and Lipogenesis

Several adipogenic and lipogenic of markers transcriptional changes are expected during the course of the 3T3-L1 cell differentiation. Figure A3 show the log expression level of some of these markers at different stages in the GSE34150 dataset.

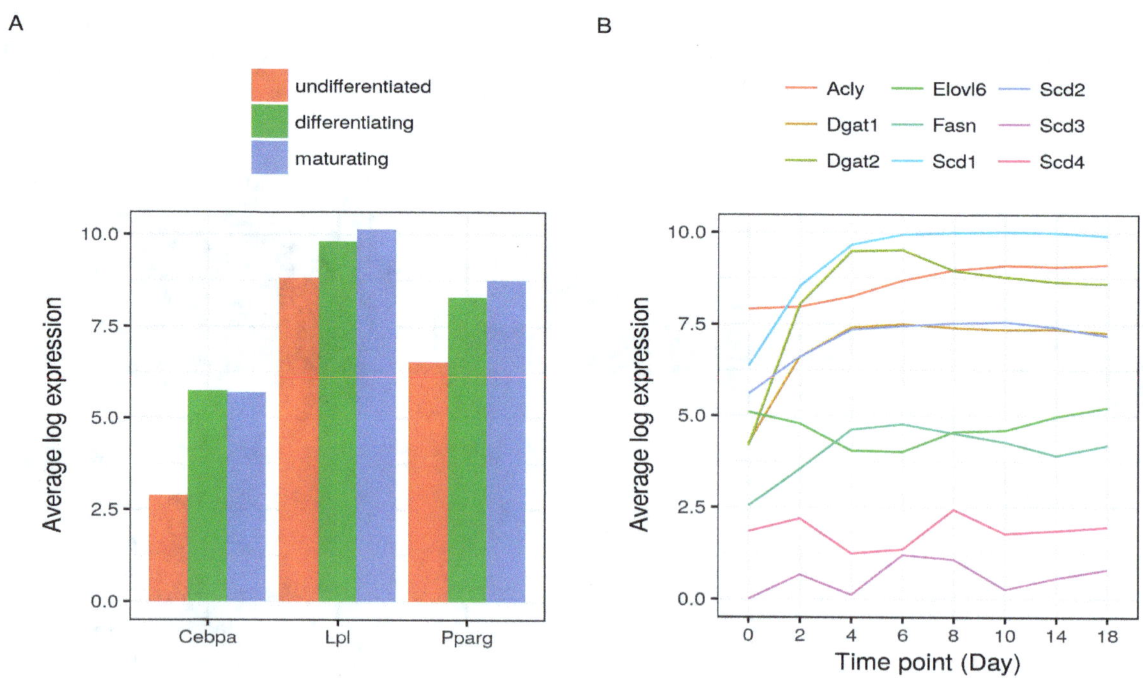

Figure A3. Differentiation and lipogensis markers in differentiating adipocytes. Average log expression values from 24 samples of MDI-induced 3T3-L1 cells (GSE34150) at 3 differentiation stages and 8 time points (0 day, undifferentiated (red); 2 and 4 days, differentiating (green); 6–18 days, maturating (blue)) from (**A**) differentiation markers and (**B**) lipogenesis markers are shown as bars and lines, respectively. *Cebpa*, CCAAT/enhancer binding protein (C/EBP), alpha; *Lpl*, lipoprotein lipase; *Pparg*, peroxisome proliferator activated receptor gamma; *Acly*, ATP Citrate Lyase; *Dgat*, Diacylglycerol *O*-Acyltransferase; *Elov6*, Fatty Acid Elongase 6; *Fasn*, Fatty Acid Synthase; *Scd*, Stearoyl-CoA Desaturase.

Appendix A.5. RT-qPCR Primer Sequences

The following nine primers were used to validate the expression and correlations of the corresponding genes products with the subunits of AMPK during the time course of 3T3-L1 differentiation (Table A2).

Table A2. RT-qPCR primer sequences.

Name	Forward (5′ to 3′)	Reverse (3′ to 5′)
18S	ACCGCAGCTAGGAATAATGGA	GCCTCAGTTCCGAAAACCA
Becn1	CAGGAACTCACAGCTCCATTAC	CCATCCTGGCGAGTTTCAATA
Prkab1	GAGATCAAGGCTCCAGAGAAAG	GTTGAAGGACCCAGACAAGTAG
Prkag1	GAACTGGAGGAGCACAAGATAG	GGGAGCCTGTGGATCTTATTT
Rab8a	GCTCGATGGCAAGAGGATTA	CTGTAGTAGGCTGTCGTGATTG
Sirt2	CATAGCCTCTAACCACCATAGC	GTAGCCTGTTGTCTGGGAATAA
Trim21	GATAGCCCAGAATACCAAGAAGAG	GCCCATCTTCCTCACAGAATAG
Trp53inp2	GGTGAAGCGCTGGAACAT	CACAACTACCTCAGCGCAGC
Wipi1	GTGTGTCTAGACGACGAGAATG	GACTTCTGAGGTAGGCTTCTTG

Appendix A.6. Gene Ontology Annotation

To define the sets of genes involved in the AMPK and autophagy pathways, we turned to the gene ontology (GO) annotations. GO identifies the AMP-dependent protein kinase activity (GO:0004679) as catalysis of the reaction: ATP + a protein = ADP + a phosphoprotein. In total, 14 genes were identified to be involved in the pathway and its regulation and are referred to as AMPK genes in the manuscript. Similarly, GO defines the term autophagy (GO:0006914) as the catabolic process in which the cells digest parts of their own cytoplasm. It contains 10 children/subcategory terms and 167 genes. The gene symbols of AMPK and autophagy gene sets are listed in Table A3.

Table A3. Gene members of the AMPK and autopahgy gene ontology terms.

Category	Term	Genes
AMPK	AMP-activated protein kinase activity	*Prkab1, Prkag1, Smok1, Smok2a, Prkaa1, Prkaa2, Prkab2, Prkag2, Smok2b, Prkag3, 4921509C19Rik, Smok3a, Smok3b, Smok3c*
autophagy	autophagy of mitochondrion	*Atg5, Rb1cc1, Cdkn2a, Capn10, Park2, Wipi1, Wdr45, Becn1, Fis1, Atg4b, Map1lc3a, Wdr45b, Cisd2, Fundc2, Map1lc3b, Atg12, Atg3, Pink1, Fbxo7, Fundc1, Atg7, Wipi2, Atg2b, Usp30, Atg9b, Ambra1, Atg4d, Atg4c, Atg9a, Atg2a, Atg4a*
	autophagy of nucleus	*Atg5, Wipi1, Wdr45, Becn1, Atg4b, Wdr45b, Atg12, Atg3, Wipi2, Trappc8, Atg2b, Becn2, Atg4d, Atg4c, Atg2a, Atg4a*
	autophagy of peroxisome	*Rb1cc1, Acbd5, Pik3r4, Trappc8, Pik3c3*
	chaperone-mediated autophagy	*Hspa8, Lamp2*
	late endosomal microautophagy	*Hspa8, Vps4b, Vps4a*
	macroautophagy	*Atg5, Cln3, Ei24, Nbr1, Sqstm1, Pik3c2a, Plaa, Ulk1, Tcirg1, Ubqln1, Becn1, Ubxn6, Map1lc3a, Map1lc3b, Atg3, Trp53inp2, Tbc1d5, Atg7, Pik3r4, Zfyve1, D17Wsu92e, Pik3c3, Yod1, Tmem74, Pik3c2b, Vcp, Atg14*
	negative regulation of autophagy	*Akt1, Bcl2, Eif4g2, Htr2b, Il3, Lep, Lepr, Mcl1, Mt3, Ptpn22, Tnfaip3, Rnf5, Mtor, Sirt2, Rraga, Washc1, Rasip1, Zkscan3, Dapl1, Wdr6, Tbc1d14, Lars, Bmf, Eif4g1, Dap, Kdm4a, Herc1, Rubcn*
	positive regulation of autophagy	*Ager, Dcn, Hif1a, Ifng, Pim2, Prkd1, Plk2, Trim21, Stk11, Tfeb, Tsc2, Xbp1, Mid2, Map2k1, Irgm2, Mefv, Sh3glb1, Nprl2, Becn1, Tmem59, Foxo1, Trp53inp1, Lrrk2, Mtdh, Trp53inp2, Dapk1, Optn, Plekhf1, Atg7, Svip, Uvrag, Tlr9, Trim8, Sh3bp4, Prkaa1, Ticam1, Prkaa2, Flcn, Ambra1, Zc3h12a, Dhrsx, Tpcn1, Rnf152, Trim65, Atg14*
	protein targeting to vacuole involved in autophagy	*Smurf1*
	regulation of autophagy	*Atm, Bcl2, Casp1, Hmgb1, Lmx1b, Rab8a, Mapt, Pik3r2, Pip4k2a, Usp10, Xbp1, Park2, Bok, Rragd, Rragc, Iigp1, Trp53inp1, Lrrk2, Pycard, Cisd2, Dram2, Soga3, Kat8, Rab39b, Rraga, Mtcl1, Dram1, Mfsd8, Fbxl2, Usp13, 3110043O21Rik, Rptor, Chmp4b, Nlrp6, Pip4k2b, Pip4k2c, Usp33, Wdr41, Tpcn2, Smcr8, Lamp3, Rragb, Wdr24, Depdc5, Soga1, Fbxw7as1*

Appendix B. Rational of Analysis Directive

This appendix contains a brief discussion of some of the decision that were made at different steps of the analysis and the rationals behind them. Specifically, we describe the steps of constructing the co-expression networks and the centrality measures that were applied to them.

Appendix B.1. Choosing the Network Power Threshold

A critical choice in constructing the co-expression networks is setting the soft threshold (power) to which the adjacency matrix is raised. A Scale Free Topology (SFT) measure is calculated by multiplying the a slope and a fitted R square and a mean connectivity of the networks at different power values (Figure A4). A power value of 5 was chosen to satisfy the SFT the most.

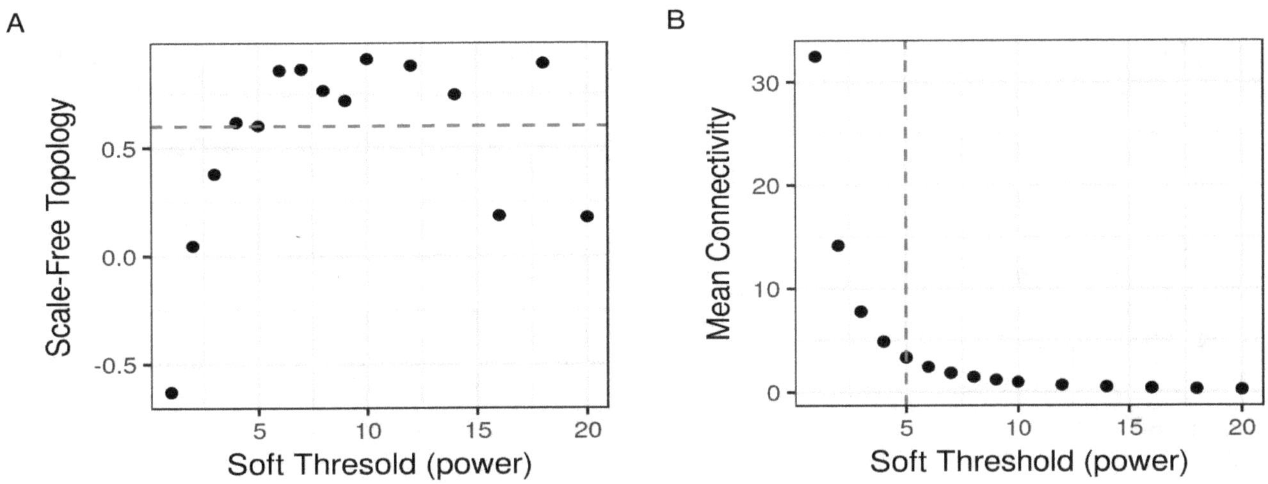

Figure A4. Scale free topology for multiple power values. Expression data from 24 samples (GSE34150) MDI-induced 3T3-L1 at different time points were used to calculate an $(n \times n)$ similarity matrix ($n = 181$ genes), which were used to obtain the weighed networks by raising them to multiple power values. For each value, a scale free topology index was calculated. (**A**) the fit indices, the slopes multiplied by the R squared R^2 values, are shown for each power value as points; (**B**) the mean connectivity, average edges shared by a node, for the resultant network at each power value are shown as points. Red lines represent the choice of power that satisfy both high R squared R^2 values and high connectivity.

Appendix B.2. Steps of Constructing the Weighed Co-Expression Networks

In this section, we discuss the different steps for calculating the similarity measures that was used in constructing the networks as well as comparing to the intermediary forms in representing the notation of co-expression. Particularly, the issue of penalizing the low correlation values. Three main steps are necessary:

- The absolute values of either Pearson's (default) or Spearman's coefficient can be used to provide an initial similarity measure s_{ij} between each pair of nodes (ij) as in: $s_{ij} = \mid cor(ij) \mid$.
- The similarity matrix is then transformed to and adjacency matrix by elevating it to a selected power β as in: $a_{ij} = s_{ij}^{\beta}$.
- This matrix is then used to calculate the connectivity/weight of each pair of nodes as follows:

$$w_{ij} = \frac{l_{ij} + a_{ij}}{min\{k_i + k_j\} + 1 - a_{ij}},$$

where $l_{ij} = \sum_u a_{iu} a_{uj}$ and $k_i = \sum_u a_{iu}$. These weights (also not shown) are finally used to calculate a dissimilarity measure for clustering and detecting the gene modules as in: $d_{ij}^{w} = 1 - w_{ij}$.

Figure A5A shows the cumulative distribution functions of the final TOM similarity measure and the two intermediaries; Pearson's correlation and the adjacency (after being raised to the power of 5).

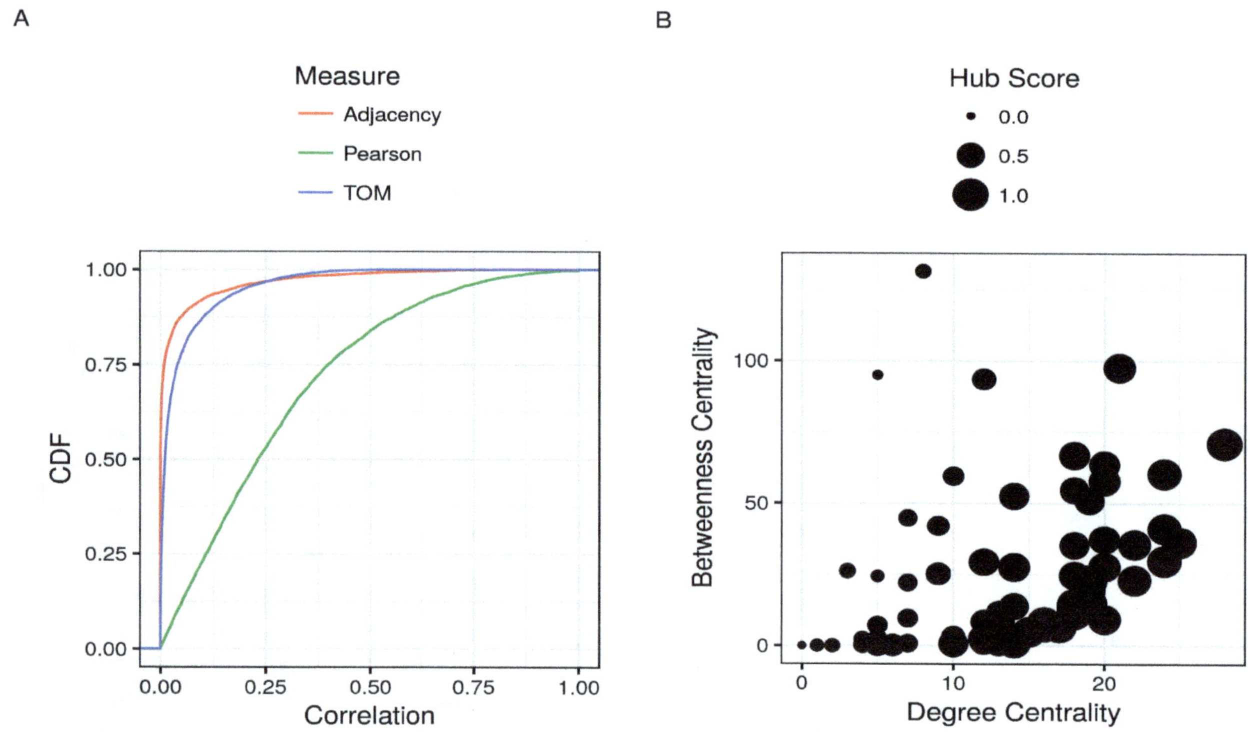

Figure A5. Gene similarity and node centrality measures. (**A**) the cumulative distribution function (CDF) of three correlation/similarity measures of AMPK and autophagy genes ($n = 181$) are shown as colored lines (green, Pearson's correlation coefficients; red, adjacency; and blue, TOM); (**B**) three centrality measures for all nodes in the two detected modules are as points. The degree centrality on the x-axis, the betweenness centrality on the y-axis and the hub score as the point size.

Appendix B.3. Node Centrality Measures

To determine the importance of each node/gene in the networks, we relied on multiple measures of centrality or node influence. These measures are calculated as follows:

- Degree Centrality: the number of edges/connections shared by a node. The Degree of a node v is given by:

$$Degree(v) = \sum_j a_{v,j},$$

where $a_{v,j}$ is the adjacency matrix of the network or the sum of the corresponding row.
- Betweenness Centrality: The number of the occasion of a node falls on the shortest between two other nodes. The Betweenness of a node v is given by:

$$Betweenness(v) = \sum_{s \neq v \neq t \in V} \frac{\sigma_{st}(v)}{\sigma_{st}}$$

or the fraction of the shortest paths between each pair of nodes (s, t) that passes through v.
- Hub Score/Eigenvector: a measure of the influence of a given node in a network. The Eigenvector of a node v is given by:

$$Eigenvector(v) = \frac{1}{\lambda} \sum_{t \in M(v)} x_t = \frac{1}{\lambda} \sum_{t \in G} a_{vt} x_t$$

or the sum of the scores x_i of its neighboring nodes $M(v)$ in the network G.

Figure A5B shows the different centrality measures and the correlations between them. Specifically, we found that the degree centrality is highly correlated with the hub scores (Pearson's coefficient about 0.73) and less so with the betweenness centrality (Pearson's coefficient about 0.44).

Appendix B.4. Network Preservation

The authors of the WGCNA method suggests using composite preservation summaries to evaluate the evidence of the preservation of the detected modules in the test dataset/s as opposed to using individual statistics as they measure different aspect of the preservation. In the main text, we showed the $Z_{summary}$ for the preservation of the two detected modules in three test datasets. Here, we briefly expand on what this statistics composed of and show additional statistics supporting the preservation of the modules.

The $Z_{summary}$ statistics is given by:

$$Z_{summary} = \frac{Z_{connectivity} + Z_{density}}{2}.$$

Each of the two Z summaries are further composed of several statistics, the details of which are provided in the references.

As opposed to the $Z_{summary}$ as evidence for module preservation, the median rank of the preservation is more robust to the module sizes. However, it is more informative in comparing preservation of modules relative to each other, as it is based on the ranks of the observed preservation statistics of the module. The lower the median rank of the module, the more preservation it exhibits in the test set. Figure A6 shows the median ranks/relative preservation of the modules with their sizes.

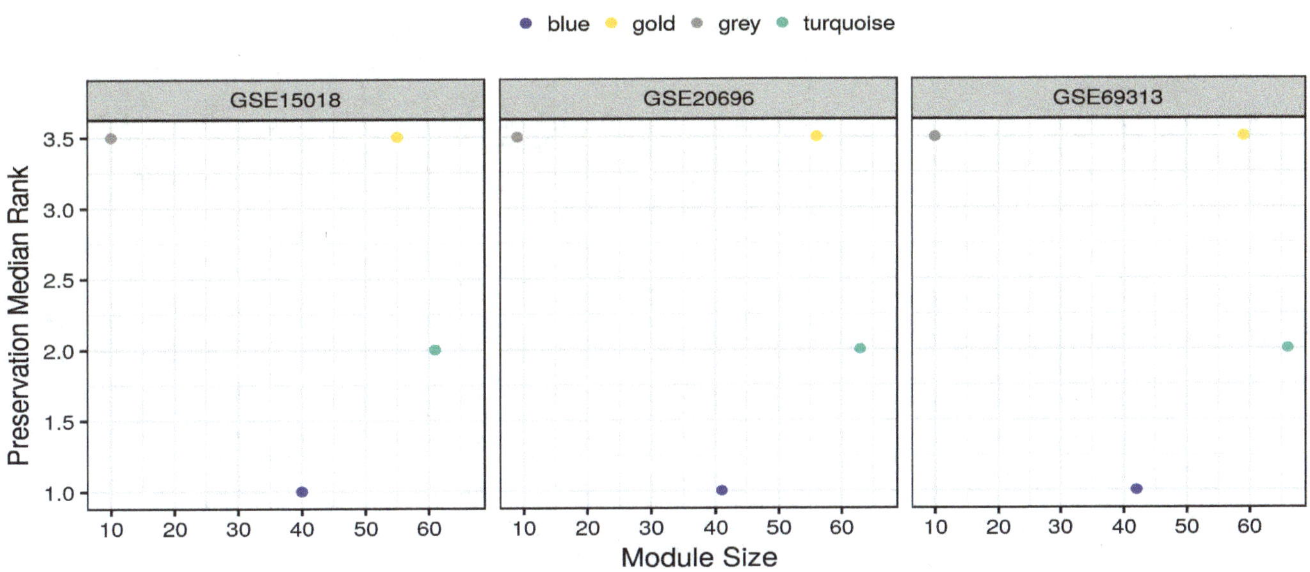

Figure A6. Module preservation ranks across multiple MDI-induced 3T3-L1 microarrays datasets. The GSE34150 dataset was used to detect the highly co-expressed modules among AMPK and autophagy genes (42, blue; 66, turquoise; 10, gray, unassigned; and 55, gold, randomly assigned). The detected modules were used as a reference to calculate several preservation statistics in three independent datasets of similar design (GSE15018, GSE20696 and GSE69313). The median ranks of the preservation statistics and the sizes of four modules are shown as colored points.

Appendix C. A Note on Reproducing the Analysis

This is a detailed description of the details required to reproduce the analysis from the source code. We first introduce the way to obtain the source code, the software environment and the commands to run the analysis script and generate the figures and tables that appear in this manuscript. In Figure A7, we show the workflow of the study listing the steps, datasets and software packages that were used in the analysis.

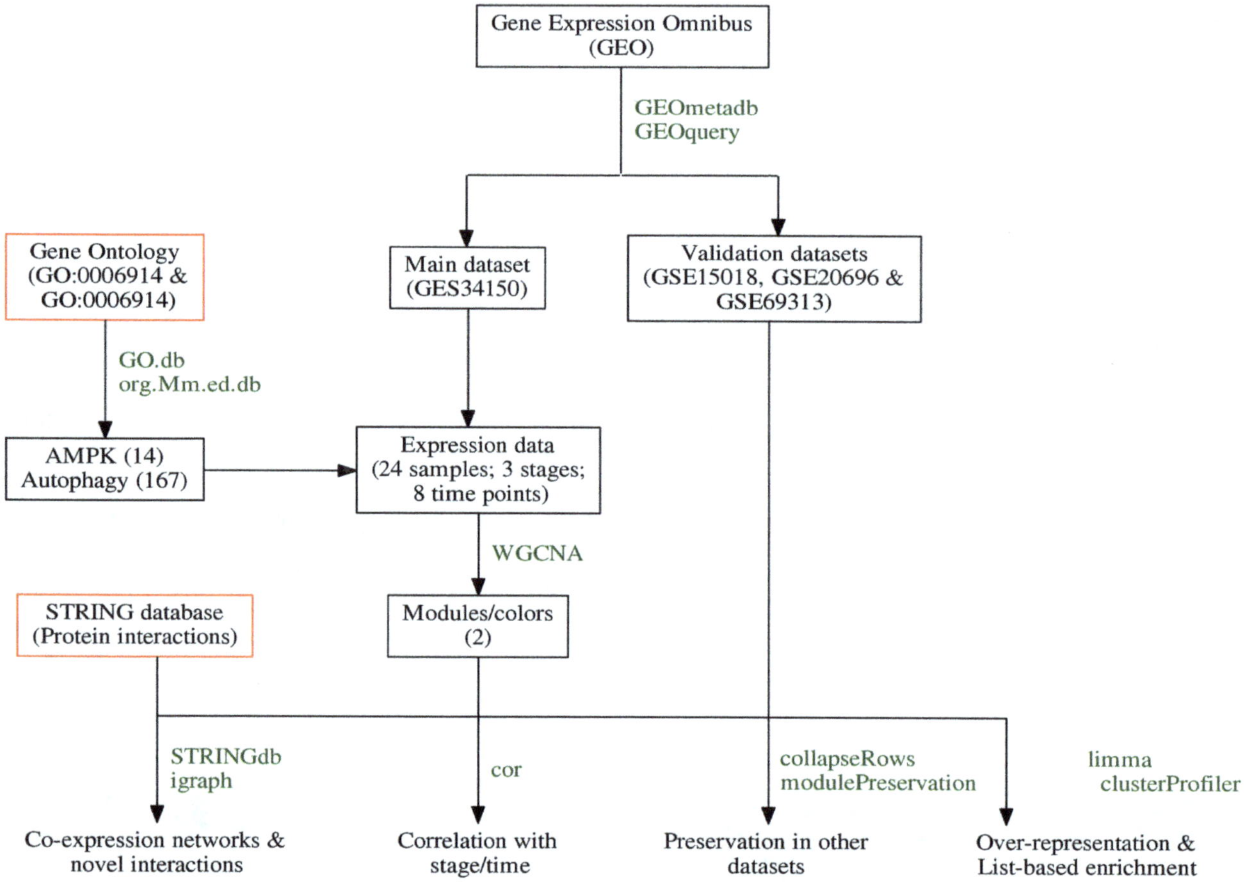

Figure A7. Workflow of the study.

Appendix C.1. Setting up the Docker Environment

The analysis was run on a docker image based on the the latest bioconductor/release_base2. Other R packages were added to the image and were made available as an image that can be obtained and launched on any local machine running docker:

```
$ docker pull mahshaaban/analysis_containers:bioc_wgcna,
$ docker run -it mahshaaban/analysis_containers:bioc_wgcna bash.
```

Appendix C.2. Obtaining the Source Code

The source code is hosted publicly on a repository on github in a form of research compendium. This includes the functions used throughout the analysis as an R package, the scripts to run the analysis and finally the scripts to reproduce the figures and tables in this manuscript. From within the container, git can be used to cloned the source code. The cloned repository contains a sub-folder called 'analysis/scripts', which can be used to reproduce the analysis from scratch:

- 01.analysis.R This script loads the required libraries, download the data and run all the steps of the analysis described in the manuscript,
- figures/A sub-folder with a separate file for each graph in the manuscript,
- tables/A sub-folder with a sepearte file for each table in the manuscript.

The following code clones the repository containing the source code:

```
$ git clone http://github.com/MahShaaban/aacna.
```

Appendix C.3. Running the Analysis

The analysis scripts is organized to be ran using a single 'make' command. This will first load the necessary functions and run the main analysis and save the data in an R object 'wgcna.rda'. This will be used to generate the figures and graphs. In addition, a log file is generated in the sub-folder 'log/' for each script that can be used for troubleshooting.

To do that, the 'make' command should be invoked from withing the 'analysis/' sub-folder.

```
$ cd aacna/analysis/
$ make
```

Appendix C.4. Details of the R Environment

The version of R that was used to perform this analysis is the 3.4.2 (28 September 2017) on x86_64-pc-linux-gnu. The 'DESCRIPTION' file in the main repository contains further details about the dependencies and the license of this work.

References

1. Zhang, Y.; Goldman, S.; Baerga, R.; Zhao, Y.; Komatsu, M.; Jin, S. Adipose-specific deletion of autophagy-related gene 7 (*atg7*) in mice reveals a role in adipogenesis. *Proc. Natl. Acad. Sci. USA* **2009**, *106*, 19860–19865. [CrossRef] [PubMed]
2. Baerga, R.; Zhang, Y.; Chen, P.H.; Goldman, S.; Jin, S. Targeted deletion of autophagy-related 5 *(atg5)* impairs adipogenesis in a cellular model and in mice. *Autophagy* **2009**, *5*, 1118–1130. [CrossRef] [PubMed]
3. Singh, R.; Xiang, Y.; Wang, Y.; Baikati, K.; Cuervo, A.M.; Luu, Y.K.; Tang, Y.; Pessin, J.E.; Schwartz, G.J.; Czaja, M.J. Autophagy regulates adipose mass and differentiation in mice. *J. Clin. Investig.* **2009**, *119*, 3329–3339. [CrossRef] [PubMed]
4. Singh, R.; Kaushik, S.; Wang, Y.; Xiang, Y.; Novak, I.; Komatsu, M.; Tanaka, K.; Cuervo, A.M.; Czaja, M.J. Autophagy regulates lipid metabolism. *Nature* **2009**, *458*, 1131–1135. [CrossRef] [PubMed]
5. Gwinn, D.M.; Shackelford, D.B.; Egan, D.F.; Mihaylova, M.M.; Mery, A.; Vasquez, D.S.; Turk, B.E.; Shaw, R.J. AMPK Phosphorylation of raptor mediates a metabolic checkpoint. *Mol. Cell* **2008**, *30*, 214–226. [CrossRef] [PubMed]
6. Inoki, K.; Zhu, T.; Guan, K.L. TSC2 Mediates Cellular Energy Response to Control Cell Growth and Survival. *Cell* **2003**, *115*, 577–590. [CrossRef]
7. Kim, J.; Kundu, M.; Viollet, B.; Guan, K.L. AMPK and mTOR regulate autophagy through direct phosphorylation of Ulk1. *Nat. Cell Biol.* **2011**, *13*, 132–141. [CrossRef] [PubMed]
8. Egan, D.F.; Shackelford, D.B.; Mihaylova, M.M.; Gelino, S.; Kohnz, R.A.; Mair, W.; Vasquez, D.S.; Joshi, A.; Gwinn, D.M.; Taylor, R.; et al. Phosphorylation of ULK1 (hATG1) by AMP-activated protein kinase connects energy sensing to mitophagy. *Science* **2011**, *331*, 456–461. [CrossRef] [PubMed]
9. Green, H.; Kehinde, O. An established preadipose cell line and its differentiation in culture II. Factors affecting the adipose conversion. *Cell* **1975**, *5*, 19–27. [CrossRef]
10. Ntambi, J.M.; Young-Cheul, K. Adipocyte differentiation and gene expression. *J. Nutr.* **2000**, *130*, 3122S–3126S. [CrossRef] [PubMed]
11. Roberts, R.; Hodson, L.; Dennis, A.L.; Neville, M.J.; Humphreys, S.M.; Harnden, K.E.; Micklem, K.J.; Frayn, K.N. Markers of de novo lipogenesis in adipose tissue: Associations with small adipocytes and insulin sensitivity in humans. *Diabetologia* **2009**, *52*, 882–890. [CrossRef] [PubMed]
12. Zhang, B.; Horvath, S. A General Framework for Weighted Gene Co-Expression Network Analysis. *Stat. Appl. Genet. Mol. Biol.* **2005**, *4*, Article17. [CrossRef] [PubMed]
13. Edgar, R.; Domrachev, M.; Lash, A.E. Gene Expression Omnibus: NCBI gene expression and hybridization array data repository. *Nucleic Acids Res.* **2002**, *30*, 207–210. [CrossRef] [PubMed]
14. Chin K. *Dataset: A Time Course Analysis of the Effects of Prieurianin in the Mouse Preadipocytes 3T3-L1 Cells*; The University of Toledo College of Medicine: Toledo, OH, USA, 2010.
15. Mikkelsen, T.S.; Xu, Z.; Zhang, X.; Wang, L.; Gimble, J.M.; Lander, E.S.; Rosen, E.D. Comparative epigenomic analysis of murine and human adipogenesis. *Cell* **2010**, *143*, 156–169. [CrossRef] [PubMed]

16. Horsch, M.; Beckers, J.; Adamski, J.; Halama, A. *Dataset: Genome-Wide Expression Profiling Analysis of a Time Course of Differentiating Adipocytes*; Helmholtz Zentrum Munchen GmbH: Neuherberg, Germany, 2015.
17. Zhang, M.; Zhang, Y.; Ma, J.; Guo, F.; Cao, Q.; Zhang, Y.; Zhou, B.; Chai, J.; Zhao, W.; Zhao, R. *Dataset: Effect of siRNA Knock-Down of FTO on 3T3-L1 Cell Differentiation*; China Academy of Space Technology: Beijing, China, 2015.
18. Hahm, J.R.; Ahmed, M.; Kim, D.R. RKIP phosphorylation–dependent ERK1 activation stimulates adipogenic lipid accumulation in 3T3-L1 preadipocytes overexpressing LC3. *Biochem. Biophys. Res. Commun.* **2016**, *478*, 12–17. [CrossRef] [PubMed]
19. Ahmed, M.; Quoc Nguyen, H.; Seok Hwang, J.; Zada, S.; Huyen Lai, T.; Soo Kang, S.; Ryong Kim, D. Systematic characterization of autophagy-related genes during the adipocyte differentiation using public-access data. *Oncotarget* **2018**, *9*, 15526–15541. [CrossRef] [PubMed]
20. Davies, S.P.; Hawley, S.A.; Woods, A.; Carling, D.; Haystead, T.A.; Hardie, D.G. Purification of the AMP-activated protein kinase on ATP-γ-sepharose and analysis of its subunit structure. *Eur. J. Biochem.* **1994**, *223*, 351–357. [CrossRef] [PubMed]
21. Dasgupta, B.; Chhipa, R.R. Evolving lessons on the complex role of AMPK in normal physiology and cancer. *Trends Pharmacol. Sci.* **2016**, *37*, 192–206. [CrossRef] [PubMed]
22. Ross, F.A.; Jensen, T.E.; Hardie, D.G. Differential regulation by AMP and ADP of AMPK complexes containing different subunit isoforms. *Biochem. J.* **2016**, *473*, 189–199. [CrossRef] [PubMed]
23. Mauvezin, C.; Orpinell, M.; Francis, V.A.; Mansilla, F.; Duran, J.; Ribas, V.; Palacín, M.; Boya, P.; Teleman, A.A.; Zorzano, A. *The Nuclear Cofactor DOR Regulates Autophagy in Mammalian and Drosophila Cells*; EMBO Reports; Nature Publishing Group: London, UK, 2010.
24. Sala, D.; Ivanova, S.; Plana, N.; Ribas, V.; Duran, J.; Bach, D.; Turkseven, S.; Laville, M.; Vidal, H.; Karczewska-Kupczewska, M.; et al. Autophagy-regulating TP53INP2 mediates muscle wasting and is repressed in diabetes. *J. Clin. Investig.* **2014**, *124*, 1914–1927. [CrossRef] [PubMed]
25. Kirkin, V.; Lamark, T.; Sou, Y.S.; Bjørkøy, G.; Nunn, J.L.; Bruun, J.A.; Shvets, E.; McEwan, D.G.; Clausen, T.H.; Wild, P.; et al. A Role for NBR1 in Autophagosomal Degradation of Ubiquitinated Substrates. *Mol. Cell* **2009**, *33*, 505–516. [CrossRef] [PubMed]
26. Liang, J.; Xu, Z.X.; Ding, Z.; Lu, Y.; Yu, Q.; Werle, K.D.; Zhou, G.; Park, Y.Y.; Peng, G.; Gambello, M.J.; et al. Myristoylation confers noncanonical AMPK functions in autophagy selectivity and mitochondrial surveillance. *Nat. Commun.* **2015**, *6*, 7926. [CrossRef] [PubMed]
27. Jeon, S.M. Regulation and function of AMPK in physiology and diseases. *Exp. Mol. Med.* **2016**, *48*, e245. [CrossRef] [PubMed]
28. Kola, B.; Grossman, A.; Korbonits, M. The role of AMP-activated protein kinase in obesity. *Front. Horm Res.* **2008**, *36*, 198–211. [PubMed]
29. Goldman, S.; Zhang, Y.; Jin, S. Autophagy and adipogenesis: implications in obesity and type II diabetes. *Autophagy* **2010**, *6*, 179–181. [CrossRef] [PubMed]
30. Nassiri, I.; Lombardo, R.; Lauria, M.; Morine, M.J.; Moyseos, P.; Varma, V.; Nolen, G.T.; Knox, B.; Sloper, D.; Kaput, J.; et al. Systems view of adipogenesis via novel omics-driven and tissue-specific activity scoring of network functional modules. *Sci. Rep.* **2016**, *6*, 28851. [CrossRef] [PubMed]
31. Stuart, J.M.; Segal, E.; Koller, D.; Kim, S.K. A gene-coexpression network for global discovery of conserved genetic modules. *Science* **2003**, *302*, 249–255. [CrossRef] [PubMed]
32. Carter, S.L.; Brechbühler, C.M.; Griffin, M.; Bond, A.T. Gene co-expression network topology provides a framework for molecular characterization of cellular state. *Bioinformatics (Oxford England)* **2004**, *20*, 2242–2250. [CrossRef] [PubMed]
33. Ashburner, M.; Ball, C.A.; Blake, J.A.; Botstein, D.; Butler, H.; Cherry, J.M.; Davis, A.P.; Dolinski, K.; Dwight, S.S.; Eppig, J.T.; et al. Gene Ontology: Tool for the unification of biology. *Nat. Genet.* **2000**, *25*, 25–29. [CrossRef] [PubMed]
34. Carlson, M. *org.Mm.eg.db: Genome Wide Annotation for Mouse*; Bioconductor: Seattle, WA, USA, 2016.
35. Carlson, M. *GO.db: A Set of Annotation Maps Describing the Entire Gene Ontology*; R Package Version 3.2.2; Bioconductor: Seattle, WA, USA, 2015
36. Zhu, Y.; Davis, S.; Stephens, R.; Meltzer, P.S.; Chen, Y. GEOmetadb: Powerful alternative search engine for the Gene Expression Omnibus. *Bioinformatics* **2008**, *24*, 2798–2800. [CrossRef] [PubMed]

37. Sean, D.; Meltzer, P.S. GEOquery: A bridge between the Gene Expression Omnibus (GEO) and BioConductor. *Bioinformatics* **2007**, *23*, 1846–1847.
38. Szklarczyk, D.; Morris, J.H.; Cook, H.; Kuhn, M.; Wyder, S.; Simonovic, M.; Santos, A.; Doncheva, N.T.; Roth, A.; Bork, P.; et al. The STRING database in 2017: Quality-controlled protein–protein association networks, made broadly accessible. *Nucleic Acids Res.* **2017**, *45*, D362–D368. [CrossRef] [PubMed]
39. Langfelder, P.; Horvath, S. WGCNA: An R package for weighted correlation network analysis. *BMC Bioinform.* **2008**, *9*, 559. [CrossRef] [PubMed]
40. Langfelder, P.; Luo, R.; Oldham, M.C.; Horvath, S. Is my network module preserved and reproducible? *PLoS Comput. Biol.* **2011**. [CrossRef] [PubMed]
41. Xu, K.; Tang, C.; Tang, R.; Ali, G.; Zhu, J. A Comparative Study of Six Software Packages for Complex Network Research. In Proceedings of the 2010 Second International Conference on Communication Software and Networks, Singapore, 26–28 February 2010; pp. 350–354.
42. Ritchie, M.E.; Phipson, B.; Wu, D.; Hu, Y.; Law, C.W.; Shi, W.; Smyth, G.K. limma powers differential expression analyses for RNA-sequencing and microarray studies. *Nucleic Acids Res.* **2015**, *43*, e47. [CrossRef] [PubMed]
43. Yu, G.; Wang, L.G.; Han, Y.; He, Q.Y. clusterProfiler: An R Package for Comparing Biological Themes among Gene Clusters. *OMICS J. Integr.e Biol.* **2012**, *16*, 284–287. [CrossRef] [PubMed]
44. Ahmed, M.; Kim, D.R. pcr: An R package for quality assessment, analysis and testing of qPCR data. *PeerJ* **2018**, *6*, e4473. [CrossRef] [PubMed]
45. R Core Team. *R: A Language and Environment for Statistical Computing*; R Foundation for Statistical Computing: Vienna, Austria, 2017.
46. Huber, W.; Carey, V.J.; Gentleman, R.; Anders, S.; Carlson, M.; Carvalho, B.S.; Bravo, H.C.; Davis, S.; Gatto, L.; Girke, T.; et al. Orchestrating high-throughput genomic analysis with Bioconductor. *Nat. Methods* **2015**, *12*, 115–121. [CrossRef] [PubMed]
47. Merkel, D. Docker: Lightweight Linux containers for consistent development and deployment. *Linux J.* **2014**, *2014*, 2.

11

The Role of AMPK in the Regulation of Skeletal Muscle Size, Hypertrophy and Regeneration

David M. Thomson

Department of Physiology & Developmental Biology, Brigham Young University, Provo, UT 84602, USA; david_thomson@byu.edu

Abstract: AMPK (5′-adenosine monophosphate-activated protein kinase) is heavily involved in skeletal muscle metabolic control through its regulation of many downstream targets. Because of their effects on anabolic and catabolic cellular processes, AMPK plays an important role in the control of skeletal muscle development and growth. In this review, the effects of AMPK signaling, and those of its upstream activator, liver kinase B1 (LKB1), on skeletal muscle growth and atrophy are reviewed. The effect of AMPK activity on satellite cell-mediated muscle growth and regeneration after injury is also reviewed. Together, the current data indicate that AMPK does play an important role in regulating muscle mass and regeneration, with AMPKα1 playing a prominent role in stimulating anabolism and in regulating satellite cell dynamics during regeneration, and AMPKα2 playing a potentially more important role in regulating muscle degradation during atrophy.

Keywords: AMPK; LKB1; autophagy; proteasome; hypertrophy; atrophy; skeletal muscle; AICAR; mTOR; protein synthesis

---

1. Introduction

5′-adenosine monophosphate-activated protein kinase (AMPK) is an intracellular sensor of ATP consumption that emerged in the late 1990s as a key regulator of skeletal muscle metabolism [1–3]. Its role in the promotion of ATP-producing catabolic processes involved in glucose and fat oxidation is well characterized. Its general identity as a catabolic agent is further illustrated by its stimulation of protein degradation and autophagy [4–6]. Additionally, AMPK inhibits anabolic processes that consume ATP, such as protein synthesis [7]. Given these general actions, AMPK's potential negative effect on skeletal muscle growth has been well-studied over the past 20 years.

In this review, a very brief overview of AMPK structure and function will be presented. Then, AMPK's effect on cell processes that are relevant to the control of cell size, such as protein synthesis, protein degradation and autophagy, will be reviewed. Finally, the known experimental effects of AMPK modulation on skeletal muscle growth and regeneration will be presented.

2. AMPK and Its Activation

2.1. AMPK Structure and Activation

Many excellent sources are available in the literature that provide a thorough review of the molecular and mechanistic details of AMPK structure and activity (e.g., [3,8,9]). Only a brief summary is provided here. Active AMPK is a heterotrimer comprised of three subunits: α, β, and γ. The actual kinase domain is contained within the α subunit, along with the predominant regulatory phosphorylation site, Thr172, which must be phosphorylated to produce any significant activity. The α and γ subunits serve scaffolding and regulatory roles. The γ subunit confers AMP sensitivity to the enzyme through four cystathionine β-synthase (CBS) domains, which can bind AMP, ADP, or ATP.

This interaction with three of these nucleotides confers on AMPK its ability for effectively detecting cellular energy status. During energy stress, when ATP breakdown to ADP accelerates, AMP is generated through the action of adenylate kinase, which transfers a phosphate from one ADP molecule to another, resulting in the production of ATP and AMP. As AMP levels rise, it (and to some degree, ADP) activates AMPK by: (1) increasing AMPK phosphorylation by upstream kinases; (2) decreasing AMPK dephosphorylation by phosphatases; and (3) allosterically activating phosphorylated AMPK [8]. AMPK's response to the decrease in the ATP:AMP ratio are crucial for the cell's ability to maintain appropriate ATP levels because it promotes ATP-generating catabolic processes, while inhibiting ATP-consuming anabolic processes [3,8,9].

Different isoforms exist for each of the AMPK subunits. Two α (α1 and α2), two β (β1 and β2), and three γ (γ1, γ2, and γ3) isoforms result in the possibility of up to 12 distinct AMPK configurations. In human skeletal muscle, however, these configurations are likely limited to $\alpha2/\beta2/\gamma1$ (most abundant), $\alpha2/\beta2/\gamma3$, and $\alpha1/\beta2/\gamma1$ [10]. Of these three, $\alpha2/\beta2/\gamma3$ accounts for the majority of AMPK activation due to high-intensity exercise [11]. In contrast to human muscle, mouse muscle contains β1 trimers ($\alpha1\beta1\gamma1$ and $\alpha2\beta1\gamma1$), although these still only contribute slightly to the overall AMPK activity [12]. While some functional implications of these different configurations have been determined, a full understanding of the full impact of differing trimer contents in tissues and within muscles is still being worked out.

2.2. *Upstream AMPK Kinases*

In mature skeletal muscle, liver kinase B1 (LKB1) is generally considered the primary AMPK kinase since total AMPK activity is essentially eliminated by muscle-specific LKB1 knockout [13–16]. LKB1 seems, however, to play a more important role in AMPKα2 activation, since AMPKα1 activity is not heavily impacted in skeletal muscle by LKB1 knockout [13,14,17–19].

Calcium/calmodulin-dependent protein kinase (CamKK) [20,21], and transforming growth factor β-activated kinase-1 (TAK1) [19,22] likely also play important roles in the activation of AMPK in skeletal muscle under certain circumstances.

2.3. *AMPK Activators in Skeletal Muscle*

2.3.1. Exercise

As would be expected given its role as a cellular energy sensor, AMPK is strongly activated in skeletal muscle by repeated muscle contraction [23] and exercise [2,11,24] in both rodents and humans. Activation of AMPKα2-containing trimers by endurance exercise occurs within 5 min of the onset of exercise [25], and likely requires a relatively high intensity effort, usually somewhere above 50% of VO2max [24,26]. AMPK activity returns to baseline levels within 3 h after exercise [26].

While AMPKα2 activity is readily increased by exercise and muscle contraction in rodents [27], increases in AMPKα1 activity after exercise/contraction are less consistent. For example, AMPKα1 activity in mouse quadriceps muscle was approximately four times higher immediately after 90 min of treadmill running at 13–17 m/min [28], but was not activated at all after running at 10–15 m/min for 60 min [29]. Similarly, 30 min of treadmill running at 30% of maximum running capacity activated AMPKα2 in mouse skeletal muscle, but not AMPKα1, while running at 70% of maximum activated both isoforms [30]. In vitro contraction of the extensor digitorum longus (EDL) muscle for 25 min activated AMPKα1, while 20 min of in situ contraction of the tibialis anterior (TA) failed to do so [29]. The data from rodents is confirmed in human studies where cycling for 1 h at 50% and 70% VO2max failed to activate AMPKα1 [24], while a single 30 s sprint [31] or high intensity interval cycling (4×30 s bouts of cycle sprints) [32] activated both AMPKα1 and α2 isoforms. Thus, activation of AMPKα1 isoforms by exercise requires greater intensity work and/or duration than for the activation of AMPKα2. As discussed below, this has important implications in relation to AMPK's impact on muscle growth and repair, as AMPKα1 appears to be critical in the regulation of anabolism.

2.3.2. AICAR

The 5-amino-4-imidazolecarboxamide ribonucleoside (AICAR) has been used for nearly 25 years to activate AMPK in various tissues in the body [33], including skeletal muscle [1,34,35]. Upon administration, it is converted into ZMP (AICAR monophosphate), an AMP mimetic that activates AMPK without altering intracellular adenine nucleotide levels. Similar to relatively low-intensity exercise, intraperitoneal injection of AICAR activates AMPKα2 but not AMPKα1 in rat gastrocnemius [7]. Furthermore, AICAR-stimulated glucose uptake is eliminated in AMPKα2 knockout muscle, but not in AMPKα1 knockouts [36], suggesting that at least some of AICAR's metabolic effects are specifically AMPKα2 dependent. Nonetheless, AICAR can activate AMPKα1, since incubation of isolated rat epitrochlearis muscle with 2 mM AICAR activated AMPKα1, albeit to a lesser degree than AMPKα2 [27].

2.3.3. Metformin

Metformin has long been used as a front-line drug in the treatment of insulin resistance and diabetes because of its ability to improve hyperglycemia in an insulin-independent manner. Shortly after AMPK's metabolic actions began to be described, which are similar to those of metformin, it was discovered that at least some of metformin's effects are, indeed, AMPK-dependent, although some are not [37,38]. The activation of AMPK by metformin is mainly indirect, where metformin inhibits mitochondrial oxidative phosphorylation, thereby decreasing ATP production and generating an energetic stress on the cell [38]. Although the liver is considered the principal site of metformin's glucose-regulating effects, chronic, therapeutic dosing of metformin over a 10 week period does increase AMPKα2 (but not AMPKα1) activity in diabetic skeletal muscle [39]. However, it isn't known whether this is a direct effect of the metformin on skeletal muscle since the effect persisted after metformin withdrawal. In mice, a metformin injection modestly increased AMPKα1 and α2 activity, while treatment of isolated epitrochlearis and soleus (SOL) muscle ex vivo with 10 mM (but not 2 mM) metformin markedly activated AMPKα1 and α2 isoforms [40]. However, the relevance of these high concentrations to in vivo metformin action is questionable.

2.3.4. Small Molecule AMPK Activators

A-769662 was the first small molecule AMPK activator described in the literature [41]. It specifically targets β1-containing AMPK trimers, and in skeletal muscle only activates the scarcely-expressed $\alpha1\beta1$ complexes [42]. Several additional activators have subsequently been identified, with varying specificities for the different AMPK subunit isoforms. Of them, Ex229 (small molecule 991), PF-739, and MK-8722 have been demonstrated to activate AMPK in skeletal muscle [43], though effects on muscle growth, atrophy, and regeneration are unknown.

3. Regulation of Growth-Related Cell Processes by AMPK

Skeletal muscle growth, in essence, occurs when the rate of protein anabolism exceeds the rate of protein catabolism. Atrophy results when protein catabolism exceeds anabolism [44]. AMPK is known to regulate both processes.

3.1. Effect of AMPK on Protein Synthesis

The first indications that AMPK played a role in the regulation of protein metabolism came in 2002 when it was shown that the fractional rate of protein synthesis in skeletal muscle deceased approximately 45% 1 h after an injection of the AMPK-activating drug, AICAR [7]. This inhibitory effect of AMPK activation on protein synthesis was subsequently observed in cultured muscle cells [45] as well as hepatocytes/liver [46–48], cardiac myocytes [49,50], and cancer cells [51,52], among other cell types and tissues.

AMPK's inhibition of protein synthesis is mediated by regulation of protein translation through the mechanistic target of rapamycin, complex 1 (mTORC1) pathway. Regulation of mTORC1 activity is complex, as it serves as a signaling checkpoint for many environmental inputs including nutrients, energy status and mechanical strain. When activated, mTORC1 drives cell growth in part by stimulating protein synthesis through its phosphorylation of several downstream targets, the best characterized of which are the 70-kDa ribosomal protein S6 kinase (p70S6K1) and eukaryotic initiation factor 4E-binding protein 1 (4E-BP1).

AMPK has been shown to inhibit mTORC1 activity through multiple mechanisms. First, AMPK phosphorylates mTOR, a key component of the mTORC1 complex, at Thr2446 [53], which is thought to impair mTORC1 activity by preventing phosphorylation at Ser2448. This site (Ser2448) was initially thought to promote mTORC1 activity when phosphorylated. Since then, its relevance to mTORC1 activity has been reassessed, and it seems probable that phosphorylation of both sites (i.e., Thr2446 and Ser2448) is inhibitory on mTORC1 activity [54]. Nonetheless, AMPK also inhibits mTORC1 by phosphorylating tuberous sclerosis complex 2 (TSC2). Activation of mTORC1 occurs at the lysosomal membrane through interaction with GTP-bound Rheb [55]. TSC2 acts, in complex with binding partners tuberous sclerosis complex 1 and TBC1 domain family member 7 (TBC1D7) [56], as a GTPase activating protein that converts GTP to GDP, thereby greatly diminishing the ability of Rheb to promote mTOR activity. Finally, AMPK phosphorylates raptor, an mTOR binding partner that is essential for mTORC1 activity. This phosphorylation leads to sequestration of raptor by 14-3-3 proteins, and impaired mTORC1 activity [57].

In addition to its inhibitory action on mTORC1, AMPK also regulates protein synthesis through inhibition of eukaryotic elongation factor 2 (eEF2) activity. Phosphorylation of eEF2 at Thr56 inhibits binding of the elongation factor to the ribosome, thereby slowing elongation rate. Phosphorylation of eEF2 at this site is mediated by eEF2 kinase (eEF2K). AMPK impacts eEF2K activity in two ways. First, p70S6k phosphorylates and inhibits eEF2K (leading to eEF2 activation), and AMPK can prevent this by inhibiting the mTOR pathway, as described above. Secondly, AMPK directly phosphorylates and activates eEF2K, leading to eEF2 inactivation [47,58]. While translation initiation is often considered the rate-limiting step in protein synthesis, control of elongation can, under certain circumstances, be critical in protein synthetic rate [58,59]. For instance, inhibition of eEF2K partially blocks the acute inhibitory effect of contractions on protein synthesis, although this effect does not appear to be regulated by AMPK [60]. Thus, the capacity for eEF2 regulation by AMPK in skeletal muscle remains unclear.

3.2. Effect of AMPK on Catabolic Processes

3.2.1. AMPK and Autophagy

Defective cellular content (organelles, pathogens, etc.) is degraded and recycled through the process of autophagy under low-energy conditions such as nutrient deprivation and exercise. Autophagy involves several subprocesses including engulfment of the target components in an autophagosome, fusion of the autophagasome with a lysosome (forming an autophagolysosome), followed by degradation of the cargo. This is a complex process and a complete description will not be presented here (see reference [61] for an excellent review). However, several key points of autophagy regulation are important in the context of the current topic. Under low-energy conditions, uncoordinated 51-like kinase 1 (ULK1) phosphorylates and activates multiple downstream targets that promote the progression of autophagy, including several autophagy related (ATG) proteins and beclin-1. Under conditions of energy abundance mTORC1 inhibits ULK1 through phosphorylation at Ser757. This, along with its targeting of other autophagy components, leads to mTOR's inhibition of autophagy [61].

AMPK has long been known to regulate autophagy. Initial observations in rat hepatocytes suggested that AICAR-induced AMPK activation inhibited autophagy [62], but subsequent work demonstrated that the AMPK inhibitor, Compound C, and dominant negative AMPK expression also inhibited autophagy, suggesting that AICAR's effects might be AMPK-independent [63]. Since then, AMPK's role in the process remains complicated because its effect seems to be dependent on cell type and metabolic context. Nonetheless, it appears that AMPK generally supports and promotes autophagy [64,65], and this is true in skeletal muscle [66]. It does this through multiple mechanisms. As noted above, AMPK inhibits mTORC1 activity. This relieves mTORC1 inhibition of ULK1, and thereby promotes autophagic flux. Additionally, AMPK directly phosphorylates components of the autophagy regulatory machinery. AMPK phosphorylates ULK1 at several sites [67], and also targets autophagy related protein 9 (ATG9) [4] and beclin-1 [68] downstream of ULK1, promoting autophagy.

3.2.2. AMPK and Ubiquitin-Proteasome Mediated Catabolism

The 26S proteasome degrades proteins that have been tagged for destruction through the attachment of ubiquitin chains. The covalent attachment of ubiquitin to targeted proteins is catalyzed through the action of three enzymes (E1, E2, and E3). E3 actually refers to one of multiple ubiquitin ligases, each of which is specific for the degradation of particular proteins. In skeletal muscle, two E3 enzymes, Atrogin-1 and muscle ring finger-1 (MuRF-1), are known to play a prominent role in proteasomal protein breakdown during muscle atrophy [69].

Expression of the atrophy-related genes Atrogin-1 and MuRF-1 is regulated through members of the forkhead box (FoxO) transcription factors. Anabolic signaling that activates Akt (e.g., via nutritional and hormonal cues mediated by insulin and other growth factors) results in their cytoplasmic localization and subsequent degradation so that they do not induce atrogene transcription. Catabolic stimuli, such as oxidative stress and inflammation, increase MuRF-1 and Atrogin-1 expression in muscle through the mitogen-activated protein kinase (MAPK) p38 as well as nuclear factor-κB (NF-κB) [69].

AMPK stimulates FoxO activity. AICAR injection into mice increases FoxO1 and FoxO3 expression [28,70], although AICAR's upregulation of FoxO1 is not impacted by knockout of AMPKα2 [28]. Treatment of C2C12 myotubes with AICAR results in protein breakdown accompanied by increased expression of FOXO, Atrogin-1, MuRF-1, and two other FoxO target genes, microtubule-associated protein 1A/1B-light chain 3 (LC3), and Bnip3 [71], and these effects are Akt/mTOR independent [6]. AMPK also phosphorylates FoxO3a at a site known to activate the transcription factor and thereby induce generalized protein degradation but this may not necessarily affect its localization in the nucleus [72–74]. AICAR also increases Foxo3 binding to the MuRF-1 and Atrogin-1 promoters [6]. Furthermore, AMPK activation increases nicotinamide adenine dinucleotide (NAD^+) concentration which activates the sirtuin 1 (SIRT1) deacetylase. SIRT1-mediated deacetylation of FoxO proteins increases their transcriptional activity [75,76].

4. Influence of AMPK on Skeletal Muscle Size

4.1. AMPK Regulation of Basal Muscle Size

Initial observations in dominant-negative AMPK (AMPK-DN) transgenic mice in which a dominant negative AMPKα2 subunit was overexpressed under the muscle creatine kinase promoter (expressed in heart and skeletal muscle) showed that EDL muscles tended to be larger than in wild-type (WT) mice, suggesting that AMPK might negatively regulate basal muscle mass [77], as would be expected given AMPK's stimulation of catabolism and activation of anabolism. Those initial findings are consistent with later work in which skeletal muscle specific AMPKα1 and α2 double knockout (AMPKα1/α2 dKO) soleus muscles were larger by mass and fiber diameter compared to WT muscles [78]. Myotubes derived via primary muscle cell cultures from these AMPKα1/α2 dKO muscles were likewise larger than from those from WT muscles [78]. On the other hand, muscle-specific AMPKβ1/β2 double knockout (AMPKβ1/β2 dKO) SOL and EDL muscles were reportedly not different in size compared to WT muscles [79]. Why the double knockout of β isoforms did not lead to increased muscle size is not clear, but might be related to the promoter used to drive the Cre-mediated deletion of floxed AMPK. In the case of the AMPKβ1/β2 dKO mice, the muscle creatine kinase (MCK; cardiac and skeletal muscle specific) promoter was used, while in the AMPKα1/α2 dKO mice it was the human skeletal muscle actin (HSA; skeletal muscle specific) promoter. Alteration of AMPK activity in the heart in the AMPKβ1/β2 dKO mice could have influenced the response of skeletal muscle to the knockout since cardiac dysfunction is well-known to induce skeletal muscle atrophy. Thus, not all AMPK deficiency models support the notion that AMPK inhibits basal muscle mass, but these findings are usually derived from AMPK knockout models that are not specific to skeletal muscle. Plantaris muscles from germline AMPKα1 knockout mice are smaller than those from WT mice [80]. The lack of AMPKα1 in all tissues in this model, however, doesn't allow for conclusions regarding the role of AMPK in the muscle specifically, since the lack of AMPKα1 in other tissues may have impacted muscle size (e.g., by decreasing systemic growth factors or other humoral inputs). Furthermore, the lack of AMPKα1 was associated with compensatorily elevated AMPKα2 activity in the muscles, which could have resulted in decreased mass. In agreement with this interpretation, primary cultured myotubes derived from these cells (removed from the systemic environment of the mouse) were larger than WT muscles [80]. Another study found that muscle fibers, especially type IIb fibers, are smaller in whole-body AMPKβ2 knockout mice [81]. The TA muscles from the aforementioned cardiac/skeletal muscle AMPK β1/β2 dKO mice are also smaller vs. WT muscles [82]. Again, the smaller size of these muscle fibers in these studies could be secondary to systemic effects of altered function of other tissues (e.g., heart), though this has not been directly tested.

4.2. Role of AMPK in Skeletal Muscle Hypertrophy

Given AMPK's pro-catabolic and anti-anabolic actions, it was hypothesized that AMPK activity would block overload-induced muscle growth, and the available data generally support this. When comparing the hypertrophic response of rat muscle to synergist ablation-induced overload, AMPK phosphorylation in the hypertrophying muscle was associated with decreased muscle hypertrophy [83], and diminished mTOR pathway signaling [84]. Several subsequent studies have reported negative associations between AMPK phosphorylation or activity and skeletal muscle growth. Indeed, impaired overload hypertrophy in obese rats [85,86], attenuated mTOR phosphorylation in metabolic syndrome patients [87], myotube hypertrophy during differentiation [88], myostatin inhibition of eEF2 and protein synthesis in myotubes [89], and differences in hypertrophy with varied ladder-climbing protocols in rats [90] are all associated inversely with AMPK activity.

Direct pharmacological evidence showing that AMPK inhibits muscle growth has also been demonstrated. An AICAR injection 1 h prior to a bout of resistance exercise-mimicking contractions greatly attenuated the mTOR signaling response to the contraction bout [35], suggesting that AMPK activation would impair the normal increase in protein translation that occurs post resistance exercise. Likewise, continuous perfusion of overloaded plantaris muscles with AICAR after synergist ablation greatly attenuated muscle hypertrophy [91].

Genetic evidence for the inhibitory effect of AMPK on in vivo skeletal muscle hypertrophy was provided by Mounier, et al. [80], who performed synergist ablations on AMPKα1 knockout mice. After 7–21 days of overload, AMPKα1 expression and activity was significantly increased in WT mice (but not AMPKα1-KO mice, as expected). Despite lower basal muscle mass, whole muscle hypertrophy and muscle fiber hypertrophy at 7 and 21 days was greater for the AMPKα1 knockouts. In line with the hypertrophy measurements, mTOR pathway signaling, as assessed by p70S6k and 4E-BP1 phosphorylation was greater, while eEF2 phosphorylation was lower (corresponding to increased eEF2 activity) after overload in the AMPKα1-KO muscles. Importantly, this occurred despite a compensatory increase in AMPKα2 activity basally and at 7 and 21 days after overload in the KO muscles, demonstrating that AMPKα1 is likely the major isoform involved in regulation of overload-induced muscle growth.

On a related note, old age leads to the loss of muscle mass (sarcopenia) and a blunted anabolic response to hypertrophic stimuli. AMPKα2 activation by exercise and AICAR is typically blunted in old age [34,92]. However, AMPK phosphorylation in old overloaded muscle is elevated vs. young overloaded muscles and is negatively correlated to mTOR signaling and hypertrophy [83,84]. Similarly, AMPK phosphorylation 1–3 h after resistance exercise is elevated in old vs. young human muscle, and is associated with delayed mTOR pathway activation [93]. Interestingly, 10 min of continuous electrically stimulated muscle contractions resulted in increased AMPKα2 activity in muscles from young adult (8 months-old) and old (30 months-old) rats, but this response was attenuated by old age [34]. However, under this stimulation protocol, AMPKα1 activity increased after stimulation only in old muscle, and not young, suggesting that old muscle may be hypersensitive to exercise-induced AMPKα1 activation which perhaps contributes to sarcopenia.

4.3. Role of AMPK in the Regulation of Skeletal Muscle Atrophy

The response of the AMPK system to muscle atrophy is unclear. Disuse atrophy of rodent skeletal muscle after 1–4 weeks of hindlimb unloading (HU) has been reported to increase [94] or decrease AMPK phosphorylation [95–97], while others observed no effect of HU on AMPKα1 and AMPKα2 activity or on acetyl-CoA carboxylase (ACC; AMPK target and marker of AMPK activity) phosphorylation [98]. Similarly, AMPK activity is reported by some to increase after 4 and 7 days of denervation in mice and/or rats [99–102], while spinal cord transection does not alter AMPKα2 activity in muscle [103]. In the denervation model, these conflicting results may be due to timing since AMPK phosphorylation in denervated soleus muscles is decreased during at least the first 24 h post-denervation, is not different at 3 days, and is elevated by 7 days post-denervation [104] compared to control muscles. Thus, it appears that AMPK activity in general decreases initially, then increases later on during the adaptation to disuse, at least in the denervation model.

Consistent with a catabolic role for AMPK, HU-induced atrophy of the soleus muscle was partially attenuated in AMPK-DN soleus muscles, potentially through decreased ubiquitin-proteosome activity [98]. It should be noted, however, that in this model, AMPKα1 activity was only mildly decreased by dominant-negative (DN) expression in the transgenic muscles, so the anti-atrophy effect was mainly due to the loss of AMPKα2 activity. Similarly, atrophy in denervated TA muscles from AMPKα2-KO mice was partially blocked compared to WT muscles [101], and this was also associated with decreased autophagic markers, Atrogin-1/MuRF-1 expression and ubiquitination. Akt and 4E-BP1 phosphorylation were unaffected by AMPKα2-KO, suggesting that the attenuation of atrophy was due to decreased protein degradation rather than increased mTOR activity and synthesis. Together, these findings suggest that in contrast to AMPKα1's role in inhibiting skeletal muscle mTOR and hypertrophy, the presence of AMPKα2 plays a more pronounced role in supporting an atrophy response to disuse, and in promoting protein degradation through the ubiquitin-proteasome system.

While the lack of AMPKα2 attenuates atrophy, increased activation of AMPK above normal does not appear to accelerate the loss of muscle mass since daily AICAR injections during 3 days of tibial nerve denervation in rats did not significantly affect skeletal muscle atrophy in soleus and gastrocnemius muscles [105]. Furthermore, 4 weeks of AICAR treatment of mdx mice (a model for Duchenne muscular dystrophy) did not exacerbate atrophy associated with dystrophy, and actually improved muscle function, probably through enhanced autophagic clearing of damaged cell components [106], and/or promotion of a more oxidative muscle phenotype [107].

4.4. Effect of Disruption of LKB1 on Skeletal Muscle Size and Hypertrophy

LKB1 knockout in skeletal muscle results in a nearly complete elimination of basal, exercise, and AICAR-induced AMPKα2 activity [14,18,19] and overall AMPK phosphorylation [13,16,108–110], while it has little [13,14] to no [17,18] effect on AMPKα1 activity, which is an important consideration since AMPKα1 seems to be the major isoform regulating muscle growth [78,80]. LKB1 also phosphorylates several other AMPK family members, at least one of which, sucrose non-fermenting 1 AMPK related kinase (SNARK) is important in the maintenance of muscle mass [111].

The weight of muscles from relatively young mLKB1-KO mice is not statistically different from that of WT muscles [16,19]. However, after approximately 30 weeks of age, muscle mass begins to decline in mLKB1-KO muscles [15]. This atrophy is associated with the development of heart failure, however [15,112], and thus may be primarily due to cardiac cachexia. Consistent with that speculation, muscle weights were, for the most part, similar in a skeletal muscle specific dominant negative LKB1 model, though quadricep muscles were smaller, and diaphragms were larger [113].

Nonetheless, since LKB1 is a primary upstream activator of AMPK in skeletal muscle, McGee et al. [19] hypothesized that the lack of LKB1 in muscle would result in a greater hypertrophic response to overload. However, when plantaris muscles from conditional muscle-specific LKB1 knockout mice (mLKB1-KO; cardiac and skeletal muscle knockout) were overloaded via synergist ablation (of the gastrocnemius muscle), there was no significant difference in the degree of mTOR pathway activation or hypertrophy compared to WT muscles. Importantly, overload increased the activity of AMPKα1 (but not α2) in both WT and KO muscles, showing that this response is regulated at least in part in an LKB1-independent fashion, perhaps via CamKK or TAK1 signaling [19]. Therefore, based on the findings of Monier, et al., that it is the α1 subunit that regulates skeletal muscle mTOR signaling and size [80], the lack of difference in hypertrophic response in the mLKB1-KO muscles is not surprising.

However, when skmLKB1-KO muscles were subjected to an acute bout of intermittent contractions designed to mimic hypertrophy-inducing resistance exercise, mTOR signaling (p70S6k and ribosomal protein S6 phosphorylation) was elevated to a greater extent both basally and immediately post-contraction in knockout vs. WT muscles, as was protein synthesis at 8 h post contraction [110]. AMPK phosphorylation was increased with contractions in WT but not skmLKB1-KO muscles using this contraction protocol, suggesting that the increased mTOR signaling in the knockout muscle could be due to a lack of AMPK activation, but AMPKα1-specific activity was not measured. This suggests that LKB1 can exert catabolic effects under some circumstances, though evidence that this impacts gross muscle hypertrophy is lacking.

Potential effects of LKB1 on muscle atrophy during unloading or denervation are currently unknown.

4.5. Exercise-Induced AMPK Activation and Muscle Hypertrophy

That endurance training interferes with hypertrophy/strength gains has been well-established [114,115]. The accumulation of evidence demonstrating AMPK's anti-anabolic and pro-catabolic effects naturally leads to the question of whether its activation during exercise functionally impairs the ability of muscle to hypertrophy, which, if true, would mechanistically explain the conflict between endurance/hypertrophy responses. In support of this hypothesis, Atherton et al. [116] showed that tissue-autonomous differences in signaling pathway activation may contribute to the inherent differences in gross adaptation that is observed with endurance vs. resistance exercise training. Using in vitro electric stimulation protocols that mimic endurance (low frequency, continuous) and resistance (high-frequency, intermittent) exercise bouts in rat skeletal muscle, they showed that endurance-type stimulation (but not resistance-type stimulation) resulted in AMPK activation and accrual of peroxisome-proliferator-activated receptor γ coactivator-1 α, while resistance-type stimulation (but not endurance-type stimulation) increased phosphorylation of Akt, TSC2, mTOR, downstream mTOR targets, and increased protein synthesis.

In humans, however, the molecular responses to different exercise modalities is less clear and has generally been interpreted as not supporting the hypothesis that physiological AMPK activation (e.g., through endurance exercise training) significantly impacts mTOR signaling and/or protein synthesis [115]. Apró et al. showed in trained male subjects that activation AMPKα2 via 1 h of intense cycling did not significantly impair subsequent activation of mTOR pathway components or mixed muscle fractional protein synthesis after a resistance training bout [117]. However, AMPKα1 was not activated by either exercise bout in this case. Since AMPKα1 is the major AMPK isoform regulating skeletal muscle growth, at least in rodents [80], the lack of an effect of this endurance exercise bout on mTOR or protein synthesis would be expected and does not preclude an AMPK effect if the α1 subunit were actually activated (e.g., by more intense or prolonged exercise than that employed in this study).

Furthermore, while acute AMPK activation immediately after exercise is suppressed in endurance-trained muscle [118,119], chronic endurance exercise training increases basal AMPKα1 protein content and activity. Twelve weeks of treadmill training (90 min/day, 5 days/week) elevated both AMPKα1 and α2 protein content in rat muscle [120]. Similarly, in humans, AMPKα1 (but not AMPKα2) protein concentration [118,121,122] and basal AMPKα1 activity [122] is greater in endurance trained vs. untrained individuals. Thus, the question of whether or not resistance exercise-induced anabolic signaling and hypertrophy are impacted by AMPKα1 activation by endurance exercise training remains unresolved.

4.6. Does Pharmacological AMPK Activation Limit Skeletal Muscle Hypertrophy?

Data showing the effect of pharmacological AMPK activation on load-induced muscle hypertrophy is quite limited. As noted previously, AICAR activation of AMPK attenuates contraction-induced increases in mTOR signaling and overload-induced hypertrophy in rodent muscles [35,91].

Interestingly, metformin treatment of patients with severe burn injury at dosages (850–2550 mg/day for 8 days) previously shown to activate AMPKα2 but not AMPKα1 in skeletal muscle [39] led to a significant increase in protein synthesis [123]. Similarly, in tumor-bearing cachexic rat muscle, metformin treatment rescued protein synthesis and decreased protein degradation while activating AMPK, though isoform-specific activity measures were not taken [124]. This improvement in muscle anabolism may be attributable to the long-appreciated impact of metformin on insulin sensitivity. Improved insulin action at the skeletal muscle would not only improve glucose handling, but protein synthesis as well.

5. Influence of AMPK on Skeletal Muscle Regeneration after Injury

5.1. The Regenerative Process in Skeletal Muscle

Skeletal muscle regeneration after injury is dependent upon the action of muscle stem cells (MuSCs), primarily satellite cells (SCs) which, in uninjured muscle, reside underneath the basal lamina next to mature muscle fibers in a quiescent, mitotically inactive state. Upon muscle damage, these cells activate and proliferate, with their subsequent progeny either engaging in a process of self-renewal to maintain the MuSC pool, or differentiating into myoblasts that then fuse together with other myoblasts or existing myofibers, leading to repair or replacement of the damaged tissue. Many excellent reviews are available for more detail on these events (e.g., [125–127]).

Muscle regeneration is a precisely ordered process that is dependent on the actions and influence of many cellular players at or near the myogenic niche, including SCs, mature muscle fibers, immune cells, fibroblasts, fibroadipogenic progenitors (FAPs), and others [125,126]. Although AMPK likely plays an important role in the regulation of many of these cell types (in macrophages, for instance [128]), the discussion here will be limited to its role in SCs.

5.2. Effect of AMPK on Myogenesis in Culture

The C2C12 adult skeletal muscle myoblast cell line is frequently used as an in vitro culture model for studying the process of myogenesis. C2C12 myoblasts, prior to differentiation in low-serum media, express the α2, γ2, and γ3 AMPK isoforms, but minimal expression of α1, β1, β2, and γ1 isoforms. Differentiation of the myoblasts into myotubes by exposure to low-serum media, however, induces the expression of all isoforms except for γ1 [129]. Consistent with the lack of β isoforms in myoblasts, which should preclude AMPK activation, stimulation of the cells with oligomycin and serum withdrawal activated AMPK much more strongly in myotubes vs. myoblasts [129]. However, other findings show that AICAR is able to activate AMPK in undifferentiated myoblasts [130], suggesting that the lack of β isoforms and AMPK activity in myoblasts is not a generalizable finding.

Activation of AMPK impairs myoblast proliferation. When C2C12 myoblasts are cultured in low glucose conditions (≤5 mM), AMPK is activated leading to impaired differentiation into myotubes. The same phenomenon is true for primary myoblasts, but only at even lower glucose concentrations [131]. Pharmacological AMPK activation with AICAR, metformin and other drugs accomplishes the same impairment in differentiation [130,131].

AMPK also impairs myoblast differentiation in culture. Activation of AMPK with AICAR in differentiating C2C12 myoblasts decreased p21 expression (which normally increases dramatically during differentiation) and cell cycle transition, and decreased myotube formation and myosin heavy chain expression [130]. A similar inhibitory effect of AICAR on primary bovine myoblasts was also observed [132]. Furthermore, transfection of C2C12 myoblasts with CamKKβ, an established AMPK activator, resulted in AMPK activation in myoblasts, cell cycle arrest and impaired proliferation as well as impaired subsequent differentiation, and this effect on proliferation and differentiation was AMPK-dependent since it was blocked by dominant-negative AMPK expression [133]. Together, these in vitro findings suggest that hyperactivation of AMPK in myoblasts blocks muscle proliferation and differentiation.

5.3. Effect of AMPK on Muscle Regeneration In Vivo

Although hyperactivation of AMPK in culture impairs both proliferation and differentiation of myoblasts, the lack of AMPK in SCs in vivo blocks normal muscle regeneration after injury. AMPKα1 is the predominant catalytic isoform in quiescent, activated and differentiating satellite cells [134–136]. Regeneration of damaged muscle is impaired (vs. WT) in both constitutive AMPKα1-KO mice, as well as in mice with AMPKα1-KO induced just before injury [137], and this is associated with decreased satellite cell number and Pax7, myogenic factor 5 (Myf5), and myogenin expression in basal muscles. Furthermore, AMPKα1-KO satellite cells have diminished myogenic capacity when transplanted into WT muscles, showing that the defect in regeneration is mediated by the lack of AMPKα1 in the satellite cells themselves, rather than in other cells in the KO animals (such as fibroblasts, macrophages, etc.) [137].

A similar impairment of regeneration is demonstrated by satellite cell-specific AMPKα1-KO. Fu et al. reported that satellite cells lacking AMPKα1 activate and proliferate more slowly both in culture and in single fiber preparations, and result in a subsequent impairment of muscle regeneration after cardiotoxin injury [136]. SCs, with their scant mitochondria, depend heavily on glycolytic metabolism and, according to the findings of these authors, the lack of AMPK impairs SC activation and proliferation by decreasing Warburg-like glycolysis [136].

Theret et al. also reported that satellite cell-specific AMPKα1-KO impairs muscle regeneration [135]. They showed that when SCs from AMPKα1 knockout mice (but not AMPKα2 knockouts) were collected and differentiated, the lack of AMPKα1 resulted in increased self-renewal instead of differentiation [135]. Similarly, deletion of AMPKα1 in MuSCs in vivo resulted in decreased size of the regenerating fibers along with decreased differentiation and fusion, but increased proliferation of MuSCs. However, in contrast to the report of Fu et al., the impaired regeneration was attributed by these authors to increased lactate dehydrogenase activity and enhanced Warburg-like glycolysis in the

AMPKα1-KO SCs. The reason for this discrepancy is not clear, but could be due to different transgenic constructs. Regardless, both studies demonstrate the importance of SC AMPKα1 in allowing for proper regeneration through metabolic regulation. Together with the culture data, the available evidence indicates that AMPK activity must be kept within relatively tight bounds (not too high or too low) for optimal muscle regeneration.

5.4. LKB1's Role in Skeletal Muscle Regeneration

The content of the upstream AMPK kinase, LKB1, increases during myoblast differentiation [138]. Overexpression of LKB1 in C2C12 myoblasts enhances differentiation, while RNAi-mediated knockdown of LKB1 impairs differentiation [138]. While some of this effect is likely due to the action of LKB1 on other targets within the AMPK family, AMPK phosphorylation is also increased substantially during muscle differentiation [138].

The lack of LKB1 in SCs promotes proliferation and self-renewal of the satellite cell pool, but impairs myoblast differentiation [139,140]. The effect on self-renewal is due in part to the activation of the Notch signaling pathway in LKB1-deficient cells, leading to overexpression of Pax7 that appears to be dependent on the decreased AMPK activation in these cells [139]. Other findings indicate that AMPKα1 also regulates self-renewal in a LKB1-independent manner [135]. Furthermore, LKB1's role in SC differentiation is at least partly independent of AMPK through regulation of glycogen synthase kinase (GSK3)/Wnt signaling [140].

6. Conclusions and Future Perspectives

AMPK's role as a signaling nexus for cellular processes that control energy balance has been well established over recent decades. While it certainly is not the only player in the regulation of skeletal muscle development, size, and/or growth, it, and especially the AMPKα1 subunit, has emerged as a key factor that limits muscle size and capacity for hypertrophy. AMPKα2, on the other hand, may play a more substantial role in promoting muscle atrophy than AMPKα1 through its actions on autophagy and protein degradation (summarized in Figure 1). AMPK also limits myogenesis and regeneration after injury, although the loss of AMPKα1 also blocks these processes, showing that some (but not too much) AMPK activity is required for proper regenerative functioning. While many questions regarding AMPK's role in muscle growth and regeneration have been answered, others still remain unanswered. Does AMPKα1-specific activity after endurance exercise interfere with concomitant resistance-training adaptations? What cellular mediators control AMPK's effects on muscle growth and development? How does AMPK activity in neighboring accessory cells support or impair satellite cell function in muscle regeneration? Can pharmacological AMPK activation or inhibition be harnessed to improve hypertrophic and regenerative responses, especially in populations where these are impaired (aging, obesity, diabetes, myopathies, etc.)? What role do LKB1 and other AMPK family members play in these processes? Continuing work in this area will surely shed additional light on these and other important questions.

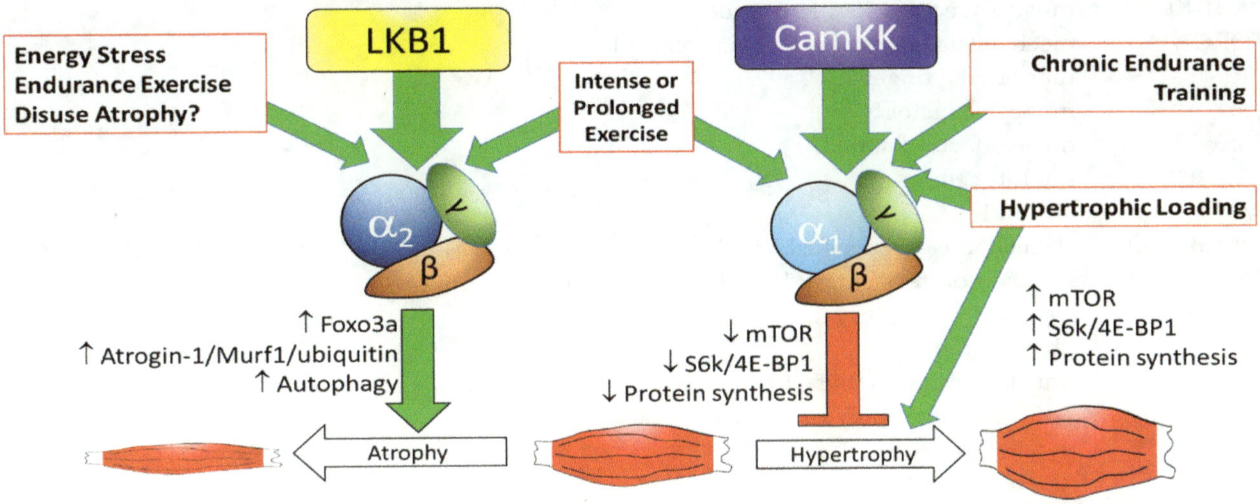

Figure 1. Proposed regulation of skeletal muscle size by 5′-adenosine monophosphate-activated protein kinase (AMPK). Energy stress (decreased ATP/AMP ratio; as in moderately intensive exercise) predominantly activates AMPKα2 via liver kinase B1 (LKB1), while AMPKα1 is only activated by highly-intense or prolonged exercise. Basal AMPKα1 content and activity is also increased by long-term endurance training, perhaps via Calcium/calmodulin-dependent protein kinase (CamKK) action, or other AMPK kinases. AMPKα2 stimulates catabolic processes by increasing Foxo3a, Atrogin-1 and MuRF-1 expression/activity and increasing autophagy, leading, under certain circumstance, to muscle atrophy, but has little effect on protein anabolism. AMPKα1 impairs mTOR signaling, slows protein synthesis, and blocks hypertrophy. Hypertrophic loading (i.e., resistance exercise) stimulates mechanistic target of rapamycin (mTOR) signaling, protein synthesis, and hypertrophy, but also activates AMPKα1 independent of LKB1 (perhaps via CamKK or other means), limiting the hypertrophic growth. ↑: increase expression or activity; ↓: decreased expression or activity.

Abbreviations

4E-BP1	eukaryotic initiation factor 4E binding protein 1
ACC	acetyl-CoA carboxylase
AICAR	5-amino-4-imidazolecarboxamide ribonucleoside
AMPK	AMP-activated protein kinase
CamKK	calcium/calmodulin-dependent protein kinase
CBS	cystathionine β-synthase
dKO	double knockout
DN	dominant negative
EDL	extensor digitorum longus
eEF2	eukaryotic elongation factor
eEF2K	eEF2 kinase
FAP	fibroadipogenic progenitor
FoxO	forkhead box
GSK3	glycogen synthase kinase 3
HU	hindlimb unloading
HSA	human skeletal muscle actin
KO	knockout
LC3	microtubule-associated protein 1A/1B-light chain 3
LKB1	liver kinase B1
MAPK	mitogen activated protein kinase
MCK	muscle creatine kinase

mLKB1-KO	muscle-specific LKB1 knockout
mTORC1	mechanistic target of rapamycin, complex 1
MuRF-1	muscle ring finger-1
Myf5	myogenic factor 5
NF-κB	nuclear factor kappa B
SC	satellite cell
SIRT-1	sirtuin 1
skmLKB1-KO	skeletal muscle-specific LKB1 knockout
SNARK	sucrose non-fermenting 1 AMPK related kinase
SOL	soleus
TA	tibialis anterior
TAK1	transforming growth factor β-activated protein kinase
TBC1D7	TBC1 domain family member 7
TGFβ	transforming growth factor β
TSC2	tuberous sclerosis complex 2
ULK1	uncoordinated 51-like kinase 1
Wnt	wingless/integrated
WT	wild-type

References

1. Merrill, G.F.; Kurth, E.J.; Hardie, D.G.; Winder, W.W. Aica riboside increases amp-activated protein kinase, fatty acid oxidation, and glucose uptake in rat muscle. *Am. J. Physiol.* **1997**, *273*, E1107–E1112. [CrossRef] [PubMed]
2. Winder, W.W.; Hardie, D.G. Inactivation of acetyl-coa carboxylase and activation of amp-activated protein kinase in muscle during exercise. *Am. J. Physiol.* **1996**, *270*, E299–E304. [CrossRef] [PubMed]
3. Winder, W.W.; Thomson, D.M. Cellular energy sensing and signaling by amp-activated protein kinase. *Cell. Biochem. Biophys.* **2007**, *47*, 332–347. [CrossRef] [PubMed]
4. Weerasekara, V.K.; Panek, D.J.; Broadbent, D.G.; Mortenson, J.B.; Mathis, A.D.; Logan, G.N.; Prince, J.T.; Thomson, D.M.; Thompson, J.W.; Andersen, J.L. Metabolic-stress-induced rearrangement of the 14-3-3zeta interactome promotes autophagy via a ulk1- and ampk-regulated 14-3-3zeta interaction with phosphorylated atg9. *Mol. Cell. Biol.* **2014**, *34*, 4379–4388. [CrossRef] [PubMed]
5. Sanchez, A.M.; Csibi, A.; Raibon, A.; Cornille, K.; Gay, S.; Bernardi, H.; Candau, R. Ampk promotes skeletal muscle autophagy through activation of forkhead foxo3a and interaction with ulk1. *J. Cell. Biochem.* **2012**, *113*, 695–710. [CrossRef] [PubMed]
6. Romanello, V.; Guadagnin, E.; Gomes, L.; Roder, I.; Sandri, C.; Petersen, Y.; Milan, G.; Masiero, E.; Del Piccolo, P.; Foretz, M.; et al. Mitochondrial fission and remodelling contributes to muscle atrophy. *EMBO J.* **2010**, *29*, 1774–1785. [CrossRef] [PubMed]
7. Bolster, D.R.; Crozier, S.J.; Kimball, S.R.; Jefferson, L.S. Amp-activated protein kinase suppresses protein synthesis in rat skeletal muscle through down-regulated mammalian target of rapamycin (mtor) signaling. *J. Biol. Chem.* **2002**, *277*, 23977–23980. [CrossRef] [PubMed]
8. Hardie, D.G. Keeping the home fires burning: Amp-activated protein kinase. *J. R. Soc. Interface* **2018**, *15*. [CrossRef] [PubMed]
9. Carling, D. Ampk signalling in health and disease. *Curr. Opin. Cell Biol.* **2017**, *45*, 31–37. [CrossRef] [PubMed]
10. Wojtaszewski, J.F.; Birk, J.B.; Frosig, C.; Holten, M.; Pilegaard, H.; Dela, F. 5′amp activated protein kinase expression in human skeletal muscle: Effects of strength training and type 2 diabetes. *J. Physiol.* **2005**, *564*, 563–573. [CrossRef] [PubMed]
11. Birk, J.B.; Wojtaszewski, J.F. Predominant α2/β2/γ3 ampk activation during exercise in human skeletal muscle. *J. Physiol.* **2006**, *577*, 1021–1032. [CrossRef] [PubMed]
12. Treebak, J.T.; Birk, J.B.; Hansen, B.F.; Olsen, G.S.; Wojtaszewski, J.F. A-769662 activates ampk β1-containing complexes but induces glucose uptake through a pi3-kinase-dependent pathway in mouse skeletal muscle. *Am. J. Physiol. Cell Physiol.* **2009**, *297*, C1041–C1052. [CrossRef] [PubMed]

13. Koh, H.J.; Arnolds, D.E.; Fujii, N.; Tran, T.T.; Rogers, M.J.; Jessen, N.; Li, Y.; Liew, C.W.; Ho, R.C.; Hirshman, M.F.; et al. Skeletal muscle-selective knockout of lkb1 increases insulin sensitivity, improves glucose homeostasis, and decreases trb3. *Mol. Cell. Biol.* **2006**, *26*, 8217–8227. [CrossRef] [PubMed]
14. Sakamoto, K.; McCarthy, A.; Smith, D.; Green, K.A.; Grahame Hardie, D.; Ashworth, A.; Alessi, D.R. Deficiency of lkb1 in skeletal muscle prevents ampk activation and glucose uptake during contraction. *EMBO J.* **2005**, *24*, 1810–1820. [CrossRef] [PubMed]
15. Thomson, D.M.; Hancock, C.R.; Evanson, B.G.; Kenney, S.G.; Malan, B.B.; Mongillo, A.D.; Brown, J.D.; Hepworth, S.; Fillmore, N.; Parcell, A.C.; et al. Skeletal muscle dysfunction in muscle-specific lkb1 knockout mice. *J. Appl. Physiol. (1985)* **2010**, *108*, 1775–1785. [CrossRef] [PubMed]
16. Thomson, D.M.; Porter, B.B.; Tall, J.H.; Kim, H.J.; Barrow, J.R.; Winder, W.W. Skeletal muscle and heart lkb1 deficiency causes decreased voluntary running and reduced muscle mitochondrial marker enzyme expression in mice. *Am. J. Physiol. Endocrinol. Metab.* **2007**, *292*, E196–E202. [CrossRef] [PubMed]
17. Tanner, C.B.; Madsen, S.R.; Hallowell, D.M.; Goring, D.M.; Moore, T.M.; Hardman, S.E.; Heninger, M.R.; Atwood, D.R.; Thomson, D.M. Mitochondrial and performance adaptations to exercise training in mice lacking skeletal muscle lkb1. *Am. J. Physiol. Endocrinol. Metab.* **2013**, *305*, E1018–E1029. [CrossRef] [PubMed]
18. Jeppesen, J.; Maarbjerg, S.J.; Jordy, A.B.; Fritzen, A.M.; Pehmoller, C.; Sylow, L.; Serup, A.K.; Jessen, N.; Thorsen, K.; Prats, C.; et al. Lkb1 regulates lipid oxidation during exercise independently of ampk. *Diabetes* **2013**, *62*, 1490–1499. [CrossRef] [PubMed]
19. McGee, S.L.; Mustard, K.J.; Hardie, D.G.; Baar, K. Normal hypertrophy accompanied by phosphoryation and activation of amp-activated protein kinase α1 following overload in lkb1 knockout mice. *J. Physiol.* **2008**, *586*, 1731–1741. [CrossRef] [PubMed]
20. Jensen, T.E.; Rose, A.J.; Jorgensen, S.B.; Brandt, N.; Schjerling, P.; Wojtaszewski, J.F.; Richter, E.A. Possible camkk-dependent regulation of ampk phosphorylation and glucose uptake at the onset of mild tetanic skeletal muscle contraction. *Am. J. Physiol. Endocrinol. Metab.* **2007**, *292*, E1308–E1317. [CrossRef] [PubMed]
21. Ferey, J.L.; Brault, J.J.; Smith, C.A.; Witczak, C.A. Constitutive activation of camkkα signaling is sufficient but not necessary for mtorc1 activation and growth in mouse skeletal muscle. *Am. J. Physiol. Endocrinol. Metab.* **2014**, *307*, E686–E694. [CrossRef] [PubMed]
22. Hindi, S.M.; Sato, S.; Xiong, G.; Bohnert, K.R.; Gibb, A.A.; Gallot, Y.S.; McMillan, J.D.; Hill, B.G.; Uchida, S.; Kumar, A. Tak1 regulates skeletal muscle mass and mitochondrial function. *JCI Insight* **2018**, *3*, e98441. [CrossRef] [PubMed]
23. Vavvas, D.; Apazidis, A.; Saha, A.K.; Gamble, J.; Patel, A.; Kemp, B.E.; Witters, L.A.; Ruderman, N.B. Contraction-induced changes in acetyl-coa carboxylase and 5′-amp-activated kinase in skeletal muscle. *J. Biol. Chem.* **1997**, *272*, 13255–13261. [CrossRef] [PubMed]
24. Fujii, N.; Hayashi, T.; Hirshman, M.F.; Smith, J.T.; Habinowski, S.A.; Kaijser, L.; Mu, J.; Ljungqvist, O.; Birnbaum, M.J.; Witters, L.A.; et al. Exercise induces isoform-specific increase in 5′amp-activated protein kinase activity in human skeletal muscle. *Biochem. Biophys. Res. Commun.* **2000**, *273*, 1150–1155. [CrossRef] [PubMed]
25. Stephens, T.J.; Chen, Z.P.; Canny, B.J.; Michell, B.J.; Kemp, B.E.; McConell, G.K. Progressive increase in human skeletal muscle ampkα2 activity and acc phosphorylation during exercise. *Am. J. Physiol. Endocrinol. Metab.* **2002**, *282*, E688–E694. [CrossRef] [PubMed]
26. Wojtaszewski, J.F.; Nielsen, P.; Hansen, B.F.; Richter, E.A.; Kiens, B. Isoform-specific and exercise intensity-dependent activation of 5′-amp-activated protein kinase in human skeletal muscle. *J. Physiol.* **2000**, *528*, 221–226. [CrossRef] [PubMed]
27. Musi, N.; Hayashi, T.; Fujii, N.; Hirshman, M.F.; Witters, L.A.; Goodyear, L.J. Amp-activated protein kinase activity and glucose uptake in rat skeletal muscle. *Am. J. Physiol. Endocrinol. Metab.* **2001**, *280*, E677–E684. [CrossRef] [PubMed]
28. Jorgensen, S.B.; Wojtaszewski, J.F.; Viollet, B.; Andreelli, F.; Birk, J.B.; Hellsten, Y.; Schjerling, P.; Vaulont, S.; Neufer, P.D.; Richter, E.A.; et al. Effects of α-ampk knockout on exercise-induced gene activation in mouse skeletal muscle. *FASEB J.* **2005**, *19*, 1146–1148. [CrossRef] [PubMed]
29. Dzamko, N.; Schertzer, J.D.; Ryall, J.G.; Steel, R.; Macaulay, S.L.; Wee, S.; Chen, Z.P.; Michell, B.J.; Oakhill, J.S.; Watt, M.J.; et al. Ampk-independent pathways regulate skeletal muscle fatty acid oxidation. *J. Physiol.* **2008**, *586*, 5819–5831. [CrossRef] [PubMed]

30. Maarbjerg, S.J.; Jorgensen, S.B.; Rose, A.J.; Jeppesen, J.; Jensen, T.E.; Treebak, J.T.; Birk, J.B.; Schjerling, P.; Wojtaszewski, J.F.; Richter, E.A. Genetic impairment of ampkα2 signaling does not reduce muscle glucose uptake during treadmill exercise in mice. *Am. J. Physiol. Endocrinol. Metab.* **2009**, *297*, E924–E934. [CrossRef] [PubMed]
31. Chen, Z.P.; McConell, G.K.; Michell, B.J.; Snow, R.J.; Canny, B.J.; Kemp, B.E. Ampk signaling in contracting human skeletal muscle: Acetyl-coa carboxylase and no synthase phosphorylation. *Am. J. Physiol. Endocrinol. Metab.* **2000**, *279*, E1202–E1206. [CrossRef] [PubMed]
32. Gibala, M.J.; McGee, S.L.; Garnham, A.P.; Howlett, K.F.; Snow, R.J.; Hargreaves, M. Brief intense interval exercise activates ampk and p38 mapk signaling and increases the expression of pgc-1α in human skeletal muscle. *J. Appl. Physiol. (1985)* **2009**, *106*, 929–934. [CrossRef] [PubMed]
33. Corton, J.M.; Gillespie, J.G.; Hawley, S.A.; Hardie, D.G. 5-aminoimidazole-4-carboxamide ribonucleoside. A specific method for activating amp-activated protein kinase in intact cells? *Eur. J. Biochem.* **1995**, *229*, 558–565. [CrossRef] [PubMed]
34. Hardman, S.E.; Hall, D.E.; Cabrera, A.J.; Hancock, C.R.; Thomson, D.M. The effects of age and muscle contraction on ampk activity and heterotrimer composition. *Exp. Gerontol.* **2014**, *55*, 120–128. [CrossRef] [PubMed]
35. Thomson, D.M.; Fick, C.A.; Gordon, S.E. Ampk activation attenuates s6k1, 4e-bp1, and eef2 signaling responses to high-frequency electrically stimulated skeletal muscle contractions. *J. Appl. Physiol. (1985)* **2008**, *104*, 625–632. [CrossRef] [PubMed]
36. Jorgensen, S.B.; Viollet, B.; Andreelli, F.; Frosig, C.; Birk, J.B.; Schjerling, P.; Vaulont, S.; Richter, E.A.; Wojtaszewski, J.F. Knockout of the α2 but not α1 5′-amp-activated protein kinase isoform abolishes 5-aminoimidazole-4-carboxamide-1-β-4-ribofuranosidebut not contraction-induced glucose uptake in skeletal muscle. *J. Biol. Chem.* **2004**, *279*, 1070–1079. [CrossRef] [PubMed]
37. Zhou, G.; Myers, R.; Li, Y.; Chen, Y.; Shen, X.; Fenyk-Melody, J.; Wu, M.; Ventre, J.; Doebber, T.; Fujii, N.; et al. Role of amp-activated protein kinase in mechanism of metformin action. *J. Clin. Investig.* **2001**, *108*, 1167–1174. [CrossRef] [PubMed]
38. Rena, G.; Hardie, D.G.; Pearson, E.R. The mechanisms of action of metformin. *Diabetologia* **2017**, *60*, 1577–1585. [CrossRef] [PubMed]
39. Musi, N.; Hirshman, M.F.; Nygren, J.; Svanfeldt, M.; Bavenholm, P.; Rooyackers, O.; Zhou, G.; Williamson, J.M.; Ljunqvist, O.; Efendic, S.; et al. Metformin increases amp-activated protein kinase activity in skeletal muscle of subjects with type 2 diabetes. *Diabetes* **2002**, *51*, 2074–2081. [CrossRef] [PubMed]
40. Oshima, R.; Yamada, M.; Kurogi, E.; Ogino, Y.; Serizawa, Y.; Tsuda, S.; Ma, X.; Egawa, T.; Hayashi, T. Evidence for organic cation transporter-mediated metformin transport and 5′-adenosine monophosphate-activated protein kinase activation in rat skeletal muscles. *Metabolism* **2015**, *64*, 296–304. [CrossRef] [PubMed]
41. Cool, B.; Zinker, B.; Chiou, W.; Kifle, L.; Cao, N.; Perham, M.; Dickinson, R.; Adler, A.; Gagne, G.; Iyengar, R.; et al. Identification and characterization of a small molecule ampk activator that treats key components of type 2 diabetes and the metabolic syndrome. *Cell Metab.* **2006**, *3*, 403–416. [CrossRef] [PubMed]
42. Scott, J.W.; van Denderen, B.J.; Jorgensen, S.B.; Honeyman, J.E.; Steinberg, G.R.; Oakhill, J.S.; Iseli, T.J.; Koay, A.; Gooley, P.R.; Stapleton, D.; et al. Thienopyridone drugs are selective activators of amp-activated protein kinase β1-containing complexes. *Chem. Biol.* **2008**, *15*, 1220–1230. [CrossRef] [PubMed]
43. Olivier, S.; Foretz, M.; Viollet, B. Promise and challenges for direct small molecule ampk activators. *Biochem. Pharmacol.* **2018**, *153*, 147–158. [CrossRef] [PubMed]
44. McGlory, C.; van Vliet, S.; Stokes, T.; Mittendorfer, B.; Phillips, S.M. The impact of exercise and nutrition on the regulation of skeletal muscle mass. *J. Physiol.* **2018**. [CrossRef] [PubMed]
45. Williamson, D.L.; Bolster, D.R.; Kimball, S.R.; Jefferson, L.S. Time course changes in signaling pathways and protein synthesis in c2c12 myotubes following ampk activation by aicar. *Am. J. Physiol. Endocrinol. Metab.* **2006**, *291*, E80–E89. [CrossRef] [PubMed]
46. Dubbelhuis, P.F.; Meijer, A.J. Hepatic amino acid-dependent signaling is under the control of amp-dependent protein kinase. *FEBS Lett.* **2002**, *521*, 39–42. [CrossRef]
47. Horman, S.; Browne, G.; Krause, U.; Patel, J.; Vertommen, D.; Bertrand, L.; Lavoinne, A.; Hue, L.; Proud, C.; Rider, M. Activation of amp-activated protein kinase leads to the phosphorylation of elongation factor 2 and an inhibition of protein synthesis. *Curr. Biol.* **2002**, *12*, 1419–1423. [CrossRef]

48. Reiter, A.K.; Bolster, D.R.; Crozier, S.J.; Kimball, S.R.; Jefferson, L.S. Repression of protein synthesis and mtor signaling in rat liver mediated by the ampk activator aminoimidazole carboxamide ribonucleoside. *Am. J. Physiol. Endocrinol. Metab.* **2005**, *288*, E980–E988. [CrossRef] [PubMed]
49. McLeod, L.E.; Proud, C.G. Atp depletion increases phosphorylation of elongation factor eef2 in adult cardiomyocytes independently of inhibition of mtor signalling. *FEBS Lett.* **2002**, *531*, 448–452. [CrossRef]
50. Chan, A.Y.; Soltys, C.L.; Young, M.E.; Proud, C.G.; Dyck, J.R. Activation of amp-activated protein kinase inhibits protein synthesis associated with hypertrophy in the cardiac myocyte. *J. Biol. Chem.* **2004**, *279*, 32771–32779. [CrossRef] [PubMed]
51. Xiang, X.; Saha, A.K.; Wen, R.; Ruderman, N.B.; Luo, Z. Amp-activated protein kinase activators can inhibit the growth of prostate cancer cells by multiple mechanisms. *Biochem. Biophys. Res. Commun.* **2004**, *321*, 161–167. [CrossRef] [PubMed]
52. Fay, J.R.; Steele, V.; Crowell, J.A. Energy homeostasis and cancer prevention: The amp-activated protein kinase. *Cancer Prev. Res. (Phila)* **2009**, *2*, 301–309. [CrossRef] [PubMed]
53. Cheng, S.W.; Fryer, L.G.; Carling, D.; Shepherd, P.R. Thr2446 is a novel mammalian target of rapamycin (mtor) phosphorylation site regulated by nutrient status. *J. Biol. Chem.* **2004**, *279*, 15719–15722. [CrossRef] [PubMed]
54. Figueiredo, V.C.; Markworth, J.F.; Cameron-Smith, D. Considerations on mtor regulation at serine 2448: Implications for muscle metabolism studies. *Cell. Mol. Life Sci.* **2017**, *74*, 2537–2545. [CrossRef] [PubMed]
55. Long, X.; Lin, Y.; Ortiz-Vega, S.; Yonezawa, K.; Avruch, J. Rheb binds and regulates the mtor kinase. *Curr. Biol.* **2005**, *15*, 702–713. [CrossRef] [PubMed]
56. Dibble, C.C.; Elis, W.; Menon, S.; Qin, W.; Klekota, J.; Asara, J.M.; Finan, P.M.; Kwiatkowski, D.J.; Murphy, L.O.; Manning, B.D. Tbc1d7 is a third subunit of the tsc1-tsc2 complex upstream of mtorc1. *Mol. Cell* **2012**, *47*, 535–546. [CrossRef] [PubMed]
57. Gwinn, D.M.; Shackelford, D.B.; Egan, D.F.; Mihaylova, M.M.; Mery, A.; Vasquez, D.S.; Turk, B.E.; Shaw, R.J. Ampk phosphorylation of raptor mediates a metabolic checkpoint. *Mol. Cell* **2008**, *30*, 214–226. [CrossRef] [PubMed]
58. Rose, A.J.; Richter, E.A. Regulatory mechanisms of skeletal muscle protein turnover during exercise. *J. Appl. Physiol. (1985)* **2009**, *106*, 1702–1711. [CrossRef] [PubMed]
59. Nielsen, P.J.; McConkey, E.H. Evidence for control of protein synthesis in hela cells via the elongation rate. *J. Cell. Physiol.* **1980**, *104*, 269–281. [CrossRef] [PubMed]
60. Rose, A.J.; Alsted, T.J.; Jensen, T.E.; Kobbero, J.B.; Maarbjerg, S.J.; Jensen, J.; Richter, E.A. A ca(2+)-calmodulin-eef2k-eef2 signalling cascade, but not ampk, contributes to the suppression of skeletal muscle protein synthesis during contractions. *J. Physiol.* **2009**, *587*, 1547–1563. [CrossRef] [PubMed]
61. Bento, C.F.; Renna, M.; Ghislat, G.; Puri, C.; Ashkenazi, A.; Vicinanza, M.; Menzies, F.M.; Rubinsztein, D.C. Mammalian autophagy: How does it work? *Annu. Rev. Biochem.* **2016**, *85*, 685–713. [CrossRef] [PubMed]
62. Samari, H.R.; Seglen, P.O. Inhibition of hepatocytic autophagy by adenosine, aminoimidazole-4-carboxamide riboside, and n6-mercaptopurine riboside. Evidence for involvement of amp-activated protein kinase. *J. Biol. Chem.* **1998**, *273*, 23758–23763. [CrossRef] [PubMed]
63. Meley, D.; Bauvy, C.; Houben-Weerts, J.H.; Dubbelhuis, P.F.; Helmond, M.T.; Codogno, P.; Meijer, A.J. Amp-activated protein kinase and the regulation of autophagic proteolysis. *J. Biol. Chem.* **2006**, *281*, 34870–34879. [CrossRef] [PubMed]
64. Herzig, S.; Shaw, R.J. Ampk: Guardian of metabolism and mitochondrial homeostasis. *Nat. Rev. Mol. Cell Biol.* **2018**, *19*, 121–135. [CrossRef] [PubMed]
65. Lee, J.W.; Park, S.; Takahashi, Y.; Wang, H.G. The association of ampk with ulk1 regulates autophagy. *PLoS ONE* **2010**, *5*, e15394. [CrossRef] [PubMed]
66. Martin-Rincon, M.; Morales-Alamo, D.; Calbet, J.A.L. Exercise-mediated modulation of autophagy in skeletal muscle. *Scand. J. Med. Sci. Sports* **2018**, *28*, 772–781. [CrossRef] [PubMed]
67. Kim, J.; Kundu, M.; Viollet, B.; Guan, K.L. Ampk and mtor regulate autophagy through direct phosphorylation of ulk1. *Nat. Cell Biol.* **2011**, *13*, 132–141. [CrossRef] [PubMed]
68. Zhang, D.; Wang, W.; Sun, X.; Xu, D.; Wang, C.; Zhang, Q.; Wang, H.; Luo, W.; Chen, Y.; Chen, H.; et al. Ampk regulates autophagy by phosphorylating becn1 at threonine 388. *Autophagy* **2016**, *12*, 1447–1459. [CrossRef] [PubMed]

69. Rom, O.; Reznick, A.Z. The role of e3 ubiquitin-ligases murf-1 and mafbx in loss of skeletal muscle mass. *Free Radic. Biol. Med.* **2016**, *98*, 218–230. [CrossRef] [PubMed]
70. Nystrom, G.J.; Lang, C.H. Sepsis and ampk activation by aicar differentially regulate foxo-1, -3 and -4 mrna in striated muscle. *Int J. Clin. Exp. Med.* **2008**, *1*, 50–63. [PubMed]
71. Nakashima, K.; Yakabe, Y. Ampk activation stimulates myofibrillar protein degradation and expression of atrophy-related ubiquitin ligases by increasing foxo transcription factors in c2c12 myotubes. *Biosci. Biotechnol. Biochem.* **2007**, *71*, 1650–1656. [CrossRef] [PubMed]
72. Greer, E.L.; Dowlatshahi, D.; Banko, M.R.; Villen, J.; Hoang, K.; Blanchard, D.; Gygi, S.P.; Brunet, A. An ampk-foxo pathway mediates longevity induced by a novel method of dietary restriction in *C. elegans*. *Curr. Biol.* **2007**, *17*, 1646–1656. [CrossRef] [PubMed]
73. Sanchez, A.M.; Candau, R.B.; Csibi, A.; Pagano, A.F.; Raibon, A.; Bernardi, H. The role of amp-activated protein kinase in the coordination of skeletal muscle turnover and energy homeostasis. *Am. J. Physiol. Cell Physiol.* **2012**, *303*, C475–C485. [CrossRef] [PubMed]
74. Greer, E.L.; Oskoui, P.R.; Banko, M.R.; Maniar, J.M.; Gygi, M.P.; Gygi, S.P.; Brunet, A. The energy sensor amp-activated protein kinase directly regulates the mammalian foxo3 transcription factor. *J. Biol. Chem.* **2007**, *282*, 30107–30119. [CrossRef] [PubMed]
75. Canto, C.; Gerhart-Hines, Z.; Feige, J.N.; Lagouge, M.; Noriega, L.; Milne, J.C.; Elliott, P.J.; Puigserver, P.; Auwerx, J. Ampk regulates energy expenditure by modulating nad+ metabolism and sirt1 activity. *Nature* **2009**, *458*, 1056–1060. [CrossRef] [PubMed]
76. Canto, C.; Jiang, L.Q.; Deshmukh, A.S.; Mataki, C.; Coste, A.; Lagouge, M.; Zierath, J.R.; Auwerx, J. Interdependence of ampk and sirt1 for metabolic adaptation to fasting and exercise in skeletal muscle. *Cell. Metab.* **2010**, *11*, 213–219. [CrossRef] [PubMed]
77. Mu, J.; Barton, E.R.; Birnbaum, M.J. Selective suppression of amp-activated protein kinase in skeletal muscle: Update on 'lazy mice'. *Biochem. Soc. Trans.* **2003**, *31*, 236–241. [CrossRef] [PubMed]
78. Lantier, L.; Mounier, R.; Leclerc, J.; Pende, M.; Foretz, M.; Viollet, B. Coordinated maintenance of muscle cell size control by amp-activated protein kinase. *FASEB J.* **2010**, *24*, 3555–3561. [CrossRef] [PubMed]
79. O'Neill, H.M.; Maarbjerg, S.J.; Crane, J.D.; Jeppesen, J.; Jorgensen, S.B.; Schertzer, J.D.; Shyroka, O.; Kiens, B.; van Denderen, B.J.; Tarnopolsky, M.A.; et al. Amp-activated protein kinase (ampk) $\beta1\beta2$ muscle null mice reveal an essential role for ampk in maintaining mitochondrial content and glucose uptake during exercise. *Proc. Natl. Acad. Sci. USA* **2011**, *108*, 16092–16097. [CrossRef] [PubMed]
80. Mounier, R.; Lantier, L.; Leclerc, J.; Sotiropoulos, A.; Pende, M.; Daegelen, D.; Sakamoto, K.; Foretz, M.; Viollet, B. Important role for ampkα1 in limiting skeletal muscle cell hypertrophy. *FASEB J.* **2009**, *23*, 2264–2273. [CrossRef] [PubMed]
81. Steinberg, G.R.; O'Neill, H.M.; Dzamko, N.L.; Galic, S.; Naim, T.; Koopman, R.; Jorgensen, S.B.; Honeyman, J.; Hewitt, K.; Chen, Z.P.; et al. Whole body deletion of amp-activated protein kinase β2 reduces muscle ampk activity and exercise capacity. *J. Biol. Chem.* **2010**, *285*, 37198–37209. [CrossRef] [PubMed]
82. Thomas, M.M.; Wang, D.C.; D'Souza, D.M.; Krause, M.P.; Layne, A.S.; Criswell, D.S.; O'Neill, H.M.; Connor, M.K.; Anderson, J.E.; Kemp, B.E.; et al. Muscle-specific ampk $\beta1\beta2$-null mice display a myopathy due to loss of capillary density in nonpostural muscles. *FASEB J.* **2014**, *28*, 2098–2107. [CrossRef] [PubMed]
83. Thomson, D.M.; Gordon, S.E. Diminished overload-induced hypertrophy in aged fast-twitch skeletal muscle is associated with ampk hyperphosphorylation. *J. Appl. Physiol. (1985)* **2005**, *98*, 557–564. [CrossRef] [PubMed]
84. Thomson, D.M.; Gordon, S.E. Impaired overload-induced muscle growth is associated with diminished translational signalling in aged rat fast-twitch skeletal muscle. *J. Physiol.* **2006**, *574*, 291–305. [CrossRef] [PubMed]
85. Paturi, S.; Gutta, A.K.; Kakarla, S.K.; Katta, A.; Arnold, E.C.; Wu, M.; Rice, K.M.; Blough, E.R. Impaired overload-induced hypertrophy in obese zucker rat slow-twitch skeletal muscle. *J. Appl. Physiol. (1985)* **2010**, *108*, 7–13. [CrossRef] [PubMed]
86. Katta, A.; Kakarla, S.K.; Manne, N.D.; Wu, M.; Kundla, S.; Kolli, M.B.; Nalabotu, S.K.; Blough, E.R. Diminished muscle growth in the obese zucker rat following overload is associated with hyperphosphorylation of ampk and dsrna-dependent protein kinase. *J. Appl. Physiol. (1985)* **2012**, *113*, 377–384. [CrossRef] [PubMed]

87. Layne, A.S.; Nasrallah, S.; South, M.A.; Howell, M.E.; McCurry, M.P.; Ramsey, M.W.; Stone, M.H.; Stuart, C.A. Impaired muscle ampk activation in the metabolic syndrome may attenuate improved insulin action after exercise training. *J. Clin. Endocrinol. Metab.* **2011**, *96*, 1815–1826. [CrossRef] [PubMed]
88. Egawa, T.; Ohno, Y.; Goto, A.; Ikuta, A.; Suzuki, M.; Ohira, T.; Yokoyama, S.; Sugiura, T.; Ohira, Y.; Yoshioka, T.; et al. Aicar-induced activation of ampk negatively regulates myotube hypertrophy through the hsp72-mediated pathway in c2c12 skeletal muscle cells. *Am. J. Physiol. Endocrinol. Metab.* **2014**, *306*, E344–E354. [CrossRef] [PubMed]
89. Deng, Z.; Luo, P.; Lai, W.; Song, T.; Peng, J.; Wei, H.K. Myostatin inhibits eef2k-eef2 by regulating ampk to suppress protein synthesis. *Biochem. Biophys. Res. Commun.* **2017**, *494*, 278–284. [CrossRef] [PubMed]
90. Luciano, T.F.; Marques, S.O.; Pieri, B.L.; de Souza, D.R.; Araujo, L.V.; Nesi, R.T.; Scheffer, D.L.; Comin, V.H.; Pinho, R.A.; Muller, A.P.; et al. Responses of skeletal muscle hypertrophy in wistar rats to different resistance exercise models. *Physiol. Res.* **2017**, *66*, 317–323. [PubMed]
91. Gordon, S.E.; Lake, J.A.; Westerkamp, C.M.; Thomson, D.M. Does amp-activated protein kinase negatively mediate aged fast-twitch skeletal muscle mass? *Exerc. Sport Sci. Rev.* **2008**, *36*, 179–186. [CrossRef] [PubMed]
92. Reznick, R.M.; Zong, H.; Li, J.; Morino, K.; Moore, I.K.; Yu, H.J.; Liu, Z.X.; Dong, J.; Mustard, K.J.; Hawley, S.A.; et al. Aging-associated reductions in amp-activated protein kinase activity and mitochondrial biogenesis. *Cell Metab.* **2007**, *5*, 151–156. [CrossRef] [PubMed]
93. Drummond, M.J.; Dreyer, H.C.; Pennings, B.; Fry, C.S.; Dhanani, S.; Dillon, E.L.; Sheffield-Moore, M.; Volpi, E.; Rasmussen, B.B. Skeletal muscle protein anabolic response to resistance exercise and essential amino acids is delayed with aging. *J. Appl. Physiol. (1985)* **2008**, *104*, 1452–1461. [CrossRef] [PubMed]
94. Hilder, T.L.; Baer, L.A.; Fuller, P.M.; Fuller, C.A.; Grindeland, R.E.; Wade, C.E.; Graves, L.M. Insulin-independent pathways mediating glucose uptake in hindlimb-suspended skeletal muscle. *J. Appl. Physiol. (1985)* **2005**, *99*, 2181–2188. [CrossRef] [PubMed]
95. Han, B.; Zhu, M.J.; Ma, C.; Du, M. Rat hindlimb unloading down-regulates insulin like growth factor-1 signaling and amp-activated protein kinase, and leads to severe atrophy of the soleus muscle. *Appl. Physiol. Nutr. Metab.* **2007**, *32*, 1115–1123. [CrossRef] [PubMed]
96. Liu, J.; Peng, Y.; Cui, Z.; Wu, Z.; Qian, A.; Shang, P.; Qu, L.; Li, Y.; Liu, J.; Long, J. Depressed mitochondrial biogenesis and dynamic remodeling in mouse tibialis anterior and gastrocnemius induced by 4-week hindlimb unloading. *IUBMB Life* **2012**, *64*, 901–910. [CrossRef] [PubMed]
97. Cannavino, J.; Brocca, L.; Sandri, M.; Grassi, B.; Bottinelli, R.; Pellegrino, M.A. The role of alterations in mitochondrial dynamics and pgc-1α over-expression in fast muscle atrophy following hindlimb unloading. *J. Physiol.* **2015**, *593*, 1981–1995. [CrossRef] [PubMed]
98. Egawa, T.; Goto, A.; Ohno, Y.; Yokoyama, S.; Ikuta, A.; Suzuki, M.; Sugiura, T.; Ohira, Y.; Yoshioka, T.; Hayashi, T.; et al. Involvement of ampk in regulating slow-twitch muscle atrophy during hindlimb unloading in mice. *Am. J. Physiol. Endocrinol. Metab.* **2015**, *309*, E651–E662. [CrossRef] [PubMed]
99. Gao, H.; Li, Y.F. Distinct signal transductions in fast- and slow-twitch muscles upon denervation. *Physiol. Rep.* **2018**, *6*. [CrossRef] [PubMed]
100. Paul, P.K.; Gupta, S.K.; Bhatnagar, S.; Panguluri, S.K.; Darnay, B.G.; Choi, Y.; Kumar, A. Targeted ablation of traf6 inhibits skeletal muscle wasting in mice. *J. Cell Biol.* **2010**, *191*, 1395–1411. [CrossRef] [PubMed]
101. Guo, Y.; Meng, J.; Tang, Y.; Wang, T.; Wei, B.; Feng, R.; Gong, B.; Wang, H.; Ji, G.; Lu, Z. Amp-activated kinase α2 deficiency protects mice from denervation-induced skeletal muscle atrophy. *Arch. Biochem. Biophys.* **2016**, *600*, 56–60. [CrossRef] [PubMed]
102. Ribeiro, C.B.; Christofoletti, D.C.; Pezolato, V.A.; de Cassia Marqueti Durigan, R.; Prestes, J.; Tibana, R.A.; Pereira, E.C.; de Sousa Neto, I.V.; Durigan, J.L.; da Silva, C.A. Leucine minimizes denervation-induced skeletal muscle atrophy of rats through akt/mtor signaling pathways. *Front. Physiol.* **2015**, *6*, 73. [CrossRef] [PubMed]
103. Dreyer, H.C.; Glynn, E.L.; Lujan, H.L.; Fry, C.S.; DiCarlo, S.E.; Rasmussen, B.B. Chronic paraplegia-induced muscle atrophy downregulates the mtor/s6k1 signaling pathway. *J. Appl. Physiol. (1985)* **2008**, *104*, 27–33. [CrossRef] [PubMed]
104. Stouth, D.W.; Manta, A.; Ljubicic, V. Protein arginine methyltransferase expression, localization, and activity during disuse-induced skeletal muscle plasticity. *Am. J. Physiol. Cell Physiol.* **2018**, *314*, C177–C190. [CrossRef] [PubMed]

105. Paulsen, S.R.; Rubink, D.S.; Winder, W.W. Amp-activated protein kinase activation prevents denervation-induced decline in gastrocnemius glut-4. *J. Appl. Physiol. (1985)* **2001**, *91*, 2102–2108. [CrossRef] [PubMed]
106. Pauly, M.; Daussin, F.; Burelle, Y.; Li, T.; Godin, R.; Fauconnier, J.; Koechlin-Ramonatxo, C.; Hugon, G.; Lacampagne, A.; Coisy-Quivy, M.; et al. Ampk activation stimulates autophagy and ameliorates muscular dystrophy in the mdx mouse diaphragm. *Am. J. Pathol.* **2012**, *181*, 583–592. [CrossRef] [PubMed]
107. Ljubicic, V.; Jasmin, B.J. Amp-activated protein kinase at the nexus of therapeutic skeletal muscle plasticity in duchenne muscular dystrophy. *Trends Mol. Med.* **2013**, *19*, 614–624. [CrossRef] [PubMed]
108. Thomson, D.M.; Brown, J.D.; Fillmore, N.; Condon, B.M.; Kim, H.J.; Barrow, J.R.; Winder, W.W. Lkb1 and the regulation of malonyl-coa and fatty acid oxidation in muscle. *Am. J. Physiol. Endocrinol. Metab.* **2007**, *293*, E1572–E1579. [CrossRef] [PubMed]
109. Brown, J.D.; Hancock, C.R.; Mongillo, A.D.; Benjamin Barton, J.; DiGiovanni, R.A.; Parcell, A.C.; Winder, W.W.; Thomson, D.M. Effect of lkb1 deficiency on mitochondrial content, fibre type and muscle performance in the mouse diaphragm. *Acta Physiol. (Oxf.)* **2011**, *201*, 457–466. [CrossRef] [PubMed]
110. Moore, T.M.; Mortensen, X.M.; Ashby, C.K.; Harris, A.M.; Kump, K.J.; Laird, D.W.; Adams, A.J.; Bray, J.K.; Chen, T.; Thomson, D.M. The effect of caffeine on skeletal muscle anabolic signaling and hypertrophy. *Appl. Physiol. Nutr. Metab.* **2017**, *42*, 621–629. [CrossRef] [PubMed]
111. Lessard, S.J.; Rivas, D.A.; So, K.; Koh, H.J.; Queiroz, A.L.; Hirshman, M.F.; Fielding, R.A.; Goodyear, L.J. The ampk-related kinase snark regulates muscle mass and myocyte survival. *J. Clin. Investig.* **2016**, *126*, 560–570. [CrossRef] [PubMed]
112. Jessen, N.; Koh, H.J.; Folmes, C.D.; Wagg, C.; Fujii, N.; Lofgren, B.; Wolf, C.M.; Berul, C.I.; Hirshman, M.F.; Lopaschuk, G.D.; et al. Ablation of lkb1 in the heart leads to energy deprivation and impaired cardiac function. *Biochim. Biophys. Acta* **2010**, *1802*, 593–600. [CrossRef] [PubMed]
113. Miura, S.; Kai, Y.; Tadaishi, M.; Tokutake, Y.; Sakamoto, K.; Bruce, C.R.; Febbraio, M.A.; Kita, K.; Chohnan, S.; Ezaki, O. Marked phenotypic differences of endurance performance and exercise-induced oxygen consumption between ampk and lkb1 deficiency in mouse skeletal muscle: Changes occurring in the diaphragm. *Am. J. Physiol. Endocrinol. Metab.* **2013**, *305*, E213–E229. [CrossRef] [PubMed]
114. Hickson, R.C. Interference of strength development by simultaneously training for strength and endurance. *Eur. J. Appl. Physiol. Occup. Physiol.* **1980**, *45*, 255–263. [CrossRef] [PubMed]
115. Coffey, V.G.; Hawley, J.A. Concurrent exercise training: Do opposites distract? *J. Physiol.* **2017**, *595*, 2883–2896. [CrossRef] [PubMed]
116. Atherton, P.J.; Babraj, J.; Smith, K.; Singh, J.; Rennie, M.J.; Wackerhage, H. Selective activation of ampk-pgc-1α or pkb-tsc2-mtor signaling can explain specific adaptive responses to endurance or resistance training-like electrical muscle stimulation. *FASEB J.* **2005**, *19*, 786–788. [CrossRef] [PubMed]
117. Apro, W.; Moberg, M.; Hamilton, D.L.; Ekblom, B.; van Hall, G.; Holmberg, H.C.; Blomstrand, E. Resistance exercise-induced s6k1 kinase activity is not inhibited in human skeletal muscle despite prior activation of ampk by high-intensity interval cycling. *Am. J. Physiol. Endocrinol. Metab.* **2015**, *308*, E470–E481. [CrossRef] [PubMed]
118. Nielsen, J.N.; Mustard, K.J.; Graham, D.A.; Yu, H.; MacDonald, C.S.; Pilegaard, H.; Goodyear, L.J.; Hardie, D.G.; Richter, E.A.; Wojtaszewski, J.F. 5′-amp-activated protein kinase activity and subunit expression in exercise-trained human skeletal muscle. *J. Appl. Physiol. (1985)* **2003**, *94*, 631–641. [CrossRef] [PubMed]
119. McConell, G.K.; Lee-Young, R.S.; Chen, Z.P.; Stepto, N.K.; Huynh, N.N.; Stephens, T.J.; Canny, B.J.; Kemp, B.E. Short-term exercise training in humans reduces ampk signalling during prolonged exercise independent of muscle glycogen. *J. Physiol.* **2005**, *568*, 665–676. [CrossRef] [PubMed]
120. Tarini, V.A.; Carnevali, L.C., Jr.; Arida, R.M.; Cunha, C.A.; Alves, E.S.; Seeleander, M.C.; Schmidt, B.; Faloppa, F. Effect of exhaustive ultra-endurance exercise in muscular glycogen and both α1 and α2 ampk protein expression in trained rats. *J. Physiol. Biochem.* **2013**, *69*, 429–440. [CrossRef] [PubMed]
121. Langfort, J.; Viese, M.; Ploug, T.; Dela, F. Time course of glut4 and ampk protein expression in human skeletal muscle during one month of physical training. *Scand. J. Med. Sci. Sports* **2003**, *13*, 169–174. [CrossRef] [PubMed]
122. Frosig, C.; Jorgensen, S.B.; Hardie, D.G.; Richter, E.A.; Wojtaszewski, J.F. 5′-amp-activated protein kinase activity and protein expression are regulated by endurance training in human skeletal muscle. *Am. J. Physiol. Endocrinol. Metab.* **2004**, *286*, E411–E417. [CrossRef] [PubMed]

123. Gore, D.C.; Wolf, S.E.; Sanford, A.; Herndon, D.N.; Wolfe, R.R. Influence of metformin on glucose intolerance and muscle catabolism following severe burn injury. *Ann. Surg.* **2005**, *241*, 334–342. [CrossRef] [PubMed]
124. Oliveira, A.G.; Gomes-Marcondes, M.C. Metformin treatment modulates the tumour-induced wasting effects in muscle protein metabolism minimising the cachexia in tumour-bearing rats. *BMC Cancer* **2016**, *16*, 418. [CrossRef] [PubMed]
125. Wosczyna, M.N.; Rando, T.A. A muscle stem cell support group: Coordinated cellular responses in muscle regeneration. *Dev. Cell* **2018**, *46*, 135–143. [CrossRef] [PubMed]
126. Deyhle, M.R.; Hyldahl, R.D. The role of t lymphocytes in skeletal muscle repair from traumatic and contraction-induced injury. *Front. Physiol.* **2018**, *9*, 768. [CrossRef] [PubMed]
127. Cornelison, D. "Known unknowns": Current questions in muscle satellite cell biology. *Curr. Top. Dev. Biol.* **2018**, *126*, 205–233. [PubMed]
128. Mounier, R.; Theret, M.; Arnold, L.; Cuvellier, S.; Bultot, L.; Goransson, O.; Sanz, N.; Ferry, A.; Sakamoto, K.; Foretz, M.; et al. Ampkα1 regulates macrophage skewing at the time of resolution of inflammation during skeletal muscle regeneration. *Cell Metab.* **2013**, *18*, 251–264. [CrossRef] [PubMed]
129. Niesler, C.U.; Myburgh, K.H.; Moore, F. The changing ampk expression profile in differentiating mouse skeletal muscle myoblast cells helps confer increasing resistance to apoptosis. *Exp. Physiol.* **2007**, *92*, 207–217. [CrossRef] [PubMed]
130. Williamson, D.L.; Butler, D.C.; Alway, S.E. Ampk inhibits myoblast differentiation through a pgc-1α-dependent mechanism. *Am. J. Physiol. Endocrinol. Metab.* **2009**, *297*, E304–E314. [CrossRef] [PubMed]
131. Fulco, M.; Cen, Y.; Zhao, P.; Hoffman, E.P.; McBurney, M.W.; Sauve, A.A.; Sartorelli, V. Glucose restriction inhibits skeletal myoblast differentiation by activating sirt1 through ampk-mediated regulation of nampt. *Dev. Cell* **2008**, *14*, 661–673. [CrossRef] [PubMed]
132. Miyake, M.; Takahashi, H.; Kitagawa, E.; Watanabe, H.; Sakurada, T.; Aso, H.; Yamaguchi, T. Ampk activation by aicar inhibits myogenic differentiation and myostatin expression in cattle. *Cell Tissue Res.* **2012**, *349*, 615–623. [CrossRef] [PubMed]
133. Ye, C.; Zhang, D.; Zhao, L.; Li, Y.; Yao, X.; Wang, H.; Zhang, S.; Liu, W.; Cao, H.; Yu, S.; et al. Camkk2 suppresses muscle regeneration through the inhibition of myoblast proliferation and differentiation. *Int. J. Mol. Sci.* **2016**, *17*. [CrossRef] [PubMed]
134. Fu, X.; Zhao, J.X.; Zhu, M.J.; Foretz, M.; Viollet, B.; Dodson, M.V.; Du, M. Amp-activated protein kinase α1 but not α2 catalytic subunit potentiates myogenin expression and myogenesis. *Mol. Cell. Biol.* **2013**, *33*, 4517–4525. [CrossRef] [PubMed]
135. Theret, M.; Gsaier, L.; Schaffer, B.; Juban, G.; Ben Larbi, S.; Weiss-Gayet, M.; Bultot, L.; Collodet, C.; Foretz, M.; Desplanches, D.; et al. Ampkα1-ldh pathway regulates muscle stem cell self-renewal by controlling metabolic homeostasis. *EMBO J.* **2017**, *36*, 1946–1962. [CrossRef] [PubMed]
136. Fu, X.; Zhu, M.J.; Dodson, M.V.; Du, M. Amp-activated protein kinase stimulates warburg-like glycolysis and activation of satellite cells during muscle regeneration. *J. Biol. Chem.* **2015**, *290*, 26445–26456. [CrossRef] [PubMed]
137. Fu, X.; Zhu, M.; Zhang, S.; Foretz, M.; Viollet, B.; Du, M. Obesity impairs skeletal muscle regeneration through inhibition of ampk. *Diabetes* **2016**, *65*, 188–200. [PubMed]
138. Mian, I.; Pierre-Louis, W.S.; Dole, N.; Gilberti, R.M.; Dodge-Kafka, K.; Tirnauer, J.S. Lkb1 destabilizes microtubules in myoblasts and contributes to myoblast differentiation. *PLoS ONE* **2012**, *7*, e31583. [CrossRef] [PubMed]
139. Shan, T.; Zhang, P.; Xiong, Y.; Wang, Y.; Kuang, S. Lkb1 deletion upregulates pax7 expression through activating notch signaling pathway in myoblasts. *Int. J. Biochem. Cell Biol.* **2016**, *76*, 31–38. [CrossRef] [PubMed]
140. Shan, T.; Zhang, P.; Liang, X.; Bi, P.; Yue, F.; Kuang, S. Lkb1 is indispensable for skeletal muscle development, regeneration, and satellite cell homeostasis. *Stem Cells* **2014**, *32*, 2893–2907. [CrossRef] [PubMed]

12

Loss of *AMPKα2* Impairs Hedgehog-Driven Medulloblastoma Tumorigenesis

Honglai Zhang [1], Rork Kuick [2], Sung-Soo Park [1], Claire Peabody [1], Justin Yoon [1], Ester Calvo Fernández [1], Junying Wang [3], Dafydd Thomas [1], Benoit Viollet [4,5,6], Ken Inoki [3], Sandra Camelo-Piragua [1,*] and Jean-François Rual [1,*]

1 Department of Pathology, University of Michigan Medical School, Ann Arbor, MI 48109, USA; honglaiz@med.umich.edu (H.Z.); parksung@med.umich.edu (S.-S.P.); peabodyc@umich.edu (C.P.); jxy673@case.edu (J.Y.); esfernan@med.umich.edu (E.C.F.); thomasda@med.umich.edu (D.T.)
2 Department of Biostatistics, School of Public Health, University of Michigan, Ann Arbor, MI 48109, USA; rork@umich.edu
3 Life Sciences Institute, University of Michigan, Ann Arbor, MI 48109, USA; wangjoy03@gmail.com (J.W.); inokik@umich.edu (K.I.)
4 Inserm, U1016, Institut Cochin, 75014 Paris, France; benoit.viollet@inserm.fr
5 CNRS, UMR8104, 75014 Paris, France
6 Université Paris Descartes, Sorbonne Paris cité, 75014 Paris, France
* Correspondence: sandraca@med.umich.edu (S.C.-P.); jrual@med.umich.edu (J.-F.R.)

Abstract: The AMP-activated protein kinase (AMPK) is a sensor of cellular energy status that has a dual role in cancer, i.e., pro- or anti-tumorigenic, depending on the context. In medulloblastoma, the most frequent malignant pediatric brain tumor, several in vitro studies previously showed that AMPK suppresses tumor cell growth. The role of AMPK in this disease context remains to be tested in vivo. Here, we investigate loss of *AMPKα2* in a genetically engineered mouse model of sonic hedgehog (SHH)-medulloblastoma. In contrast to previous reports, our study reveals that *AMPKα2* KO impairs SHH medulloblastoma tumorigenesis. Moreover, we performed complementary molecular and genomic analyses that support the hypothesis of a pro-tumorigenic SHH/AMPK/CNBP axis in medulloblastoma. In conclusion, our observations further underline the context-dependent role of AMPK in cancer, and caution is warranted for the previously proposed hypothesis that AMPK agonists may have therapeutic benefits in medulloblastoma patients. Note: an abstract describing the project was previously submitted to the American Society for Investigative Pathology PISA 2018 conference and appears in *The American Journal of Pathology* (Volume 188, Issue 10, October 2018, Page 2433).

Keywords: medulloblastoma; sonic hedgehog; AMPK

---

1. Introduction

Medulloblastoma is an embryonal tumor of the cerebellum and is the most common malignant brain tumor of childhood [1,2]. Genomics applied to medulloblastoma defined four medulloblastoma subgroups, each characterized by a distinct molecular/genetic signature, distinct patient demographics, and a distinct clinical profile (WNT, Sonic Hedgehog or SHH, Groups #3 and #4) [3,4]. More recently, integrative genomic analyses further underscored the highly heterogeneous and complex nature of medulloblastoma with a large spectrum of molecularly distinct consensus subgroups and subtypes within them [5–7]. Current therapeutic approaches to medulloblastoma are based on surgery, radiation, and non-targeted chemotherapy and are indistinguishably applied to all medulloblastoma subgroups. These therapies have led to significant improvements, with a 73% survival rate [1], but these results are achieved at a high cost to quality of life, e.g., neurocognitive or hormonal deficiencies [8,9]. Alternative therapeutic approaches are needed.

The AMP-activated protein kinase (AMPK) is a crucial energy sensor that controls cell metabolism and growth in response to low energy levels by phosphorylating a variety of substrates [10]. AMPK forms a heterotrimeric protein complex composed of three subunits that exist as multiple isoforms: a catalytic subunit ($\alpha 1$ or $\alpha 2$), a scaffolding subunit ($\beta 1$ or $\beta 2$), and an AMP-sensing subunit ($\gamma 1$, $\gamma 2$, or $\gamma 3$) [10]. AMPK is allosterically activated by an increased AMP/ATP ratio when cells are metabolically starved. Upon metabolic stress, AMPK activation results in both the inhibition of anabolic, energy-consuming mechanisms (e.g., biosynthetic pathways and the cell cycle) and the activation of catabolic processes (e.g., the promotion of mitochondrial biogenesis and glycolysis) [10]. As the master metabolic guardian, AMPK modulates numerous targets, including both oncogenes and tumor suppressors [11,12]. Thus, AMPK has a dual role in cancer, tumor-suppressive, or pro-oncogenic, depending on the cellular or tissue context [13–15]. For example, as an inhibitor of the cell cycle and as a target of the LKB1 kinase [16–18], AMPK has long been considered an essential mediator of LKB1's tumor-suppressive effect in cancer [13,19]. The tumor suppressor role of AMPK is also exemplified by the observation that inactivation of *AMPKα1* in murine B-cell lineages promotes *Myc*-driven lymphomagenesis [20] and that *AMPKα2* suppresses murine embryonic fibroblast transformation and tumorigenesis [21]. Yet, loss of *AMPKα* impairs tumor growth in vivo in various carcinogenic contexts [22–27]. Indeed, maintenance of the metabolic balance by AMPK is likely a critical process for survival during the metabolic stress that can occur in the tumor microenvironment [13,14,26].

In SHH-driven medulloblastoma tumorigenesis, several studies support a tumor-suppressive role for AMPK [28–33]. First, in medulloblastoma cells, AMPK phosphorylates the SHH pathway transcription factor GLI1, promoting its proteasomal degradation, thus inhibiting SHH signaling and SHH-driven medulloblastoma [28–30,33]. Second, a potential tumor-suppressive role for AMPK in SHH-driven medulloblastoma was also inferred by the fact that the increased survival observed upon *Hk2* KO in a mouse model of medulloblastoma is accompanied by a gain of AMPK activity [31,32]. In light of these observations, AMPK agonists could have therapeutic value for the treatment of SHH medulloblastoma patients. As noted above, though, the multifaceted role of AMPK in cancer warrants caution [13–15]. As a matter of fact, D'amico et al. recently proposed that a non-canonical SHH/AMPK axis promotes medulloblastoma via activation of CCHC type nucleic acid binding protein (CNBP), ornithine decarboxylase 1 (ODC1), and polyamine metabolism [34]. Though indirectly inferred, this study supports a putative pro-tumorigenic role for AMPK in SHH medulloblastoma [34]. Importantly, in all of the aforementioned studies, the role of AMPK in medulloblastoma was only investigated in vitro in medulloblastoma cells.

The potential role of AMPK as a promoter or as a suppressor of medulloblastoma tumorigenesis warrants further investigation and remains to be tested in vivo. Here, we describe the analysis of loss of *AMPKα2* in a genetically engineered mouse model of SHH-driven medulloblastoma. Remarkably, in disagreement with the previous studies supporting a tumor-suppressive role for AMPK in medulloblastoma tumorigenesis [28–33], our analysis reveals that loss of *AMPKα2* impairs SHH-driven medulloblastoma tumorigenesis.

2. Results

2.1. *AMPKα2* Is Required for SHH-Driven Medulloblastoma In Vivo

We investigated the role of *AMPKα2* in vivo in medulloblastoma using the [*GFAP-tTA;TRE-SmoA1*] genetically engineered mouse model of SHH-driven medulloblastoma [35,36] in combination with the *AMPKα2* KO mouse [37]. We note that, as previously described [37], *AMPKα2* KO mice are born with the expected Mendelian ratio, are fertile, appear indistinguishable from their WT littermates, and have a normal lifespan. Accordingly, *AMPKα2* KO mice exhibit normal cerebellar development with properly tri-laminated cerebellar cortex (molecular, Purkinje cell and internal granular cell layers).

Three groups of C57BL/6 mice were analyzed: (a) *TRE-SmoA1* mice: the oncogenic gain-of-function allele of *Smo*, i.e., *SmoA1*, is not expressed in the absence of *GFAP-tTA* in these negative control group mice; (b) [*GFAP-tTA;TRE-SmoA1*] mice: all mice develop medulloblastoma in this positive control group; (c) [*GFAP-tTA;TRE-SmoA1;AMPKα2*$^{-/-}$] mice: this test group allows us to assess the extent to which *AMPKα2* KO modulates tumor incidence. Genotyping data for each group is shown in Figure S1. For each study group, we performed histological (Figure 1) and survival (Figure 2) analyses.

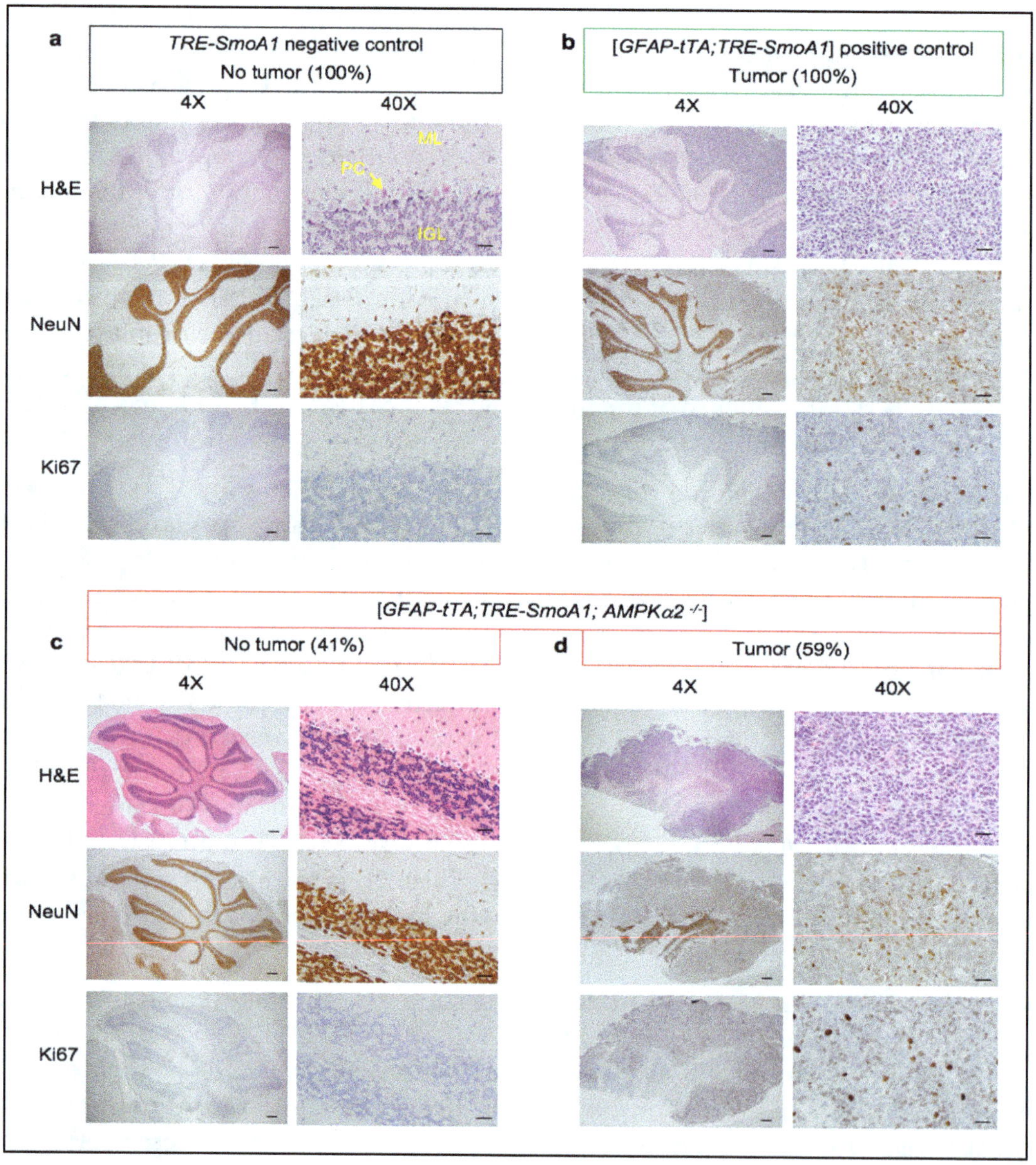

Figure 1. Loss of *AMPKα* impairs SHH medulloblastoma tumorigenesis in vivo. Histopathological examination of the cerebellums of (**a**) a Postnatal Day 63 (P63) *TRE-SmoA1* mouse, (**b**) a P56 [*GFAP-tTA;TRE-SmoA1*] mouse, (**c**) a one-year-old [*GFAP-tTA;TRE-SmoA1;AMPKα2*$^{-/-}$] mouse representative of the 7/17 (41%) [*GFAP-tTA;TRE-SmoA1;AMPKα2*$^{-/-}$] mice that do not show any evidence of tumor upon histological analysis, and (**d**) a P30 [*GFAP-tTA;TRE-SmoA1;AMPKα2*$^{-/-}$] mouse representative of the 10/17 (59%) [*GFAP-tTA;TRE-SmoA1;AMPKα2*$^{-/-}$] mice that show a tumor. The histological analysis includes H&E staining as well as Ki67 (marker of proliferation) and NeuN (marker of neuronal differentiation) immunohistochemistry (magnification: 4× and 40×). The different cell layers of the cerebellum, i.e., the molecular (ML), Purkinje cell (PC), and internal granular cell (IGL) layers are labeled in the 40× *TRE-SmoA1* control H&E picture. Scale bars: 200 μm (4×) or 25 μm (40×).

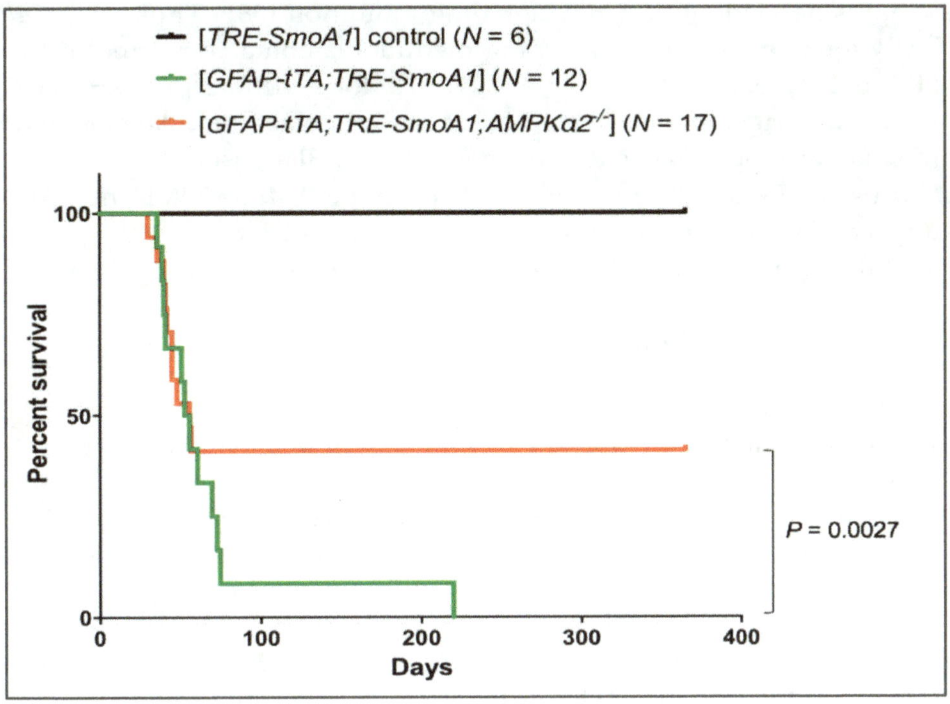

Figure 2. Loss of *AMPKα* increases survival in *SmoA1* mouse model of SHH medulloblastoma. Survival analysis of *AMPKα2* KO in *SmoA1*-driven medulloblastoma. Three groups of *C57BL/6* mice were assessed for survival: (i) *TRE-SmoA1*, (ii) [*GFAP-tTA;TRE-SmoA1*], and (iii) [*GFAP-tTA;TRE-SmoA1;AMPKα2*$^{-/-}$]. The *AMPKα2* KO mice have a significantly lower chance than the *AMPKα2* WT mice of developing medulloblastoma [59% (10/17) for *AMPKα2* KO versus 100% (12/12) for *AMPKα2* WT; $p = 0.0027$, likelihood ratio chi-square test]. For the log-rank test, which is designed to be most powerful when hazards are proportional, we obtained $p = 0.108$. Indeed, the *AMPKα2* KO mice that develop medulloblastoma succumb to the tumor at the same pace as WT mice in the first 60 days.

Cerebellar development is normal in the *TRE-SmoA1* control mice. Indeed, as expected for adult mice, we observe properly tri-laminated molecular, Purkinje cell, and internal granular cell layers, but no external granular cell layer. The strong staining for the neuronal marker NeuN indicates the presence of normal cerebellar neurons in the internal granular layer. The absence of staining for Ki-67 indicates that cells are in a quiescent, non-proliferative stage (Figure 1a). As previously described [35,36], all [*GFAP-tTA;TRE-SMOA1*] mice develop large medulloblastoma tumors that can extend along the entire rostral-caudal length of the cerebellum, account for more than a quarter of the total cerebellar volume, express NeuN, indicative of neuronal origin, and show marked proliferative activity (positive for Ki67) (Figure 1b). The vast majority (11/12) of the [*GFAP-tTA;TRE-SMOA1*] mice succumb to the tumor within 80 days (Figure 2), as previously described [35,36]. One [*GFAP-tTA;TRE-SMOA1*] mouse survived beyond 200 days with a very large tumor mass identified post-mortem (Figure S2). We note that, apart from medulloblastoma formation, no other types of tumors can be detected in the brain or the other organs of [*GFAP-tTA;TRE-SMOA1*] mice.

In contrast, 41% (7/17) of the [*GFAP-tTA;TRE-SmoA1;AMPKα2*$^{-/-}$] mice survived beyond one year (Figure 2) and do not show any evidence of tumor upon histological analysis (Figure 1c). Two of these seven mice show focal, microscopic remnants of external granular cell (EGC) neurons in groups no larger than 10 cells, which were non-proliferative and non-neoplastic (Figure S3). These cells likely correspond to granular cell neurons that have failed to properly migrate into the internal granular layer, a process that is dependent on SHH signaling inhibition [38]. Thus, the *AMPKα2* KO mice have a significantly lower chance of developing medulloblastoma compared to the *AMPKα2* WT mice [59% (10/17) for *AMPKα2* KO versus 100% (12/12) for *AMPKα2* WT; $p = 0.0027$]. Our results demonstrate the pro-tumorigenic role of *AMPKα2* in medulloblastoma. We note that the number of pyknotic cells undergoing apoptosis in either the normal cerebellar tissues (very rare occurrence, <1%) or the tumorigenic tissues (~10–20%) remains the same in the *AMPKα2* WT and KO mice. Thus, the effect of *AMPKα2* on apoptosis is not a likely to be a major factor underlying the pro-tumorigenic role of *AMPKα2* in medulloblastoma. Below, we explore the potential effect of *AMPKα2* KO on CNBP.

2.2. Loss of *AMPKα2* Results in the Decreased Expression of the CNBP Protein

In response to SHH signaling, AMPK phosphorylates CNBP, thus promoting its interaction with SUFU and its stabilization [34]. Subsequently, the increased level of CNBP protein expression results in the increased translation of ODC1 and associated activation of polyamine metabolism, which is essential for the SHH-dependent proliferation of medulloblastoma cells [34]. In their study, D'amico et al. also showed that targeting CNBP or ODC1 impairs SHH-driven medulloblastoma and thus indirectly inferred that AMPK may be required for tumor formation in this context [34]. Our observation that *AMPKα2* KO impairs tumor progression in the *SmoA1* mouse model of medulloblastoma directly verifies this hypothesis (Figures 1 and 2). To further assess the relevance of this SHH/AMPK/CNBP axis in SHH medulloblastoma, we investigated the level of protein expression of CNBP in both normal and tumorigenic cerebellums in the presence or in the absence of *AMPKα2*. We observed that, in both physiological (*SmoA1*−) and pathophysiological (*SmoA1*+) contexts, the loss of *AMPKα2* resulted in lower levels of the CNBP protein (Figure 3a). In conclusion, notwithstanding that other AMPK substrates [11,12] may also contribute to the pro-tumorigenic role of *AMPKα2* in medulloblastoma, these results further support the hypothesis of a critical role for the SHH/AMPK/CNBP axis in SHH-driven medulloblastoma.

Loss of *AMPKα2* can be associated with a compensatory increase in AMPKα1 protein, e.g., in the muscle [39]. We investigated this effect in the cerebellum. While the level of expression of AMPKα1 is higher in the *SmoA1*+ mice than in the *SmoA1*− mice, it remains the same in the control and the *AMPKα2* KO mice in both the *SmoA1− and SmoA1*+ contexts (Figure 3b). Thus, in both normal and tumorigenic cerebellums, the deficiency in AMPKα2 was not compensated for by AMPKα1 overexpression.

Similarly, we investigated the level of expression of phosphorylated acetyl-CoA carboxylase alpha (pACC1) in both normal and tumorigenic cerebellums in the presence or in the absence of *AMPKα2*. ACC1 is a key enzyme in the fatty acid synthesis pathway that mediates the conversion of acetyl-CoA to malonyl-CoA. ACC1 phosphorylation by AMPK results in its inactivation, thus inhibiting lipogenesis [40,41]. We observed that, in the *SmoA1*+ context, the loss of *AMPKα2* resulted in lower levels of the pACC1 protein (Figure 3c). This analysis of pACC1 suggests a potential impact associated with *AMPKα2* KO on lipogenesis. Interestingly, while a pro-tumorigenic role of ACC1 has been previously described in various contexts [42], a tumor-suppressive role has also been suggested. Indeed, ACC1 inhibition by AMPK is required to maintain NADPH levels and in vivo growth of lung and breast tumor xenografts [23]. Similarly, impaired inhibition of ACC1 upon *AMPKα2* KO may result in low NADPH levels in medulloblastoma cells and thus may contribute to the pro-tumorigenic role of AMPKα2 in medulloblastoma.

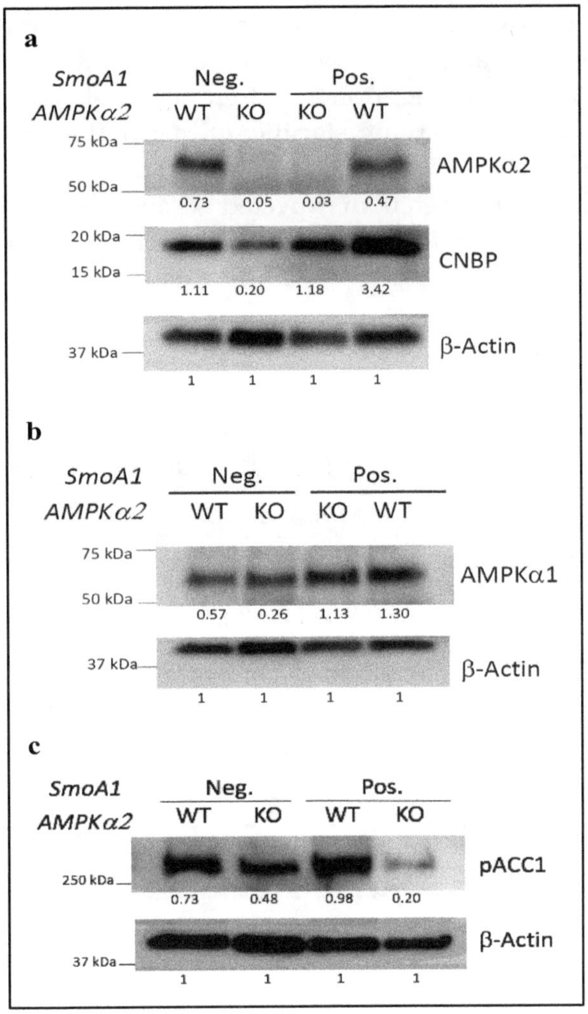

Figure 3. Loss of *AMPKα2* results in lower level of expression of the CNBP and pACC1 proteins in both physiological (*SmoA1−*) and pathophysiological (*SmoA1+*) contexts. Western blot analyses of the AMPKα2 (**a**), AMPKα1 (**b**), CNBP (**a**), pACC1 (**c**), and β-Actin proteins in whole cerebellum protein extracts obtained from WT, *AMPKα2*$^{-/-}$, [*GFAP-tTA;TRE-SmoA1; AMPKα2*$^{-/-}$], and [*GFAP-tTA;TRE-SmoA1*] Postnatal Day 16 (P16) littermate mice. Below each Western blot band, we show the ratio of the band intensity of the targeted protein over the band intensity of β-Actin.

2.3. Frequent Copy Number Gains for the AMPK, CNBP, and ODC1 Genes in SHH and Group #3 Medulloblastoma Patients

We analyzed the publicly available copy number data for 345 medulloblastoma samples [5] (Figure 4). Notably, *AMPK* genes do not have frequent copy number losses in medulloblastoma, the sole exception being the loss of *AMPKγ3* in 6% of Group #4 tumors. In contrast, *AMPK* genes tend to have frequent copy number gains in medulloblastoma in general, and particularly in Group #3 tumors, which have frequent gains for the *AMPKα1* (30%), *AMPKβ2* (38%), and *AMPKγ2* (43%) genes. *AMPKα2* also has frequent copy number gains in Group #3 tumors (11%) (Figure 4a,b). These observations resonate with the previous report that Group #3 medulloblastoma has frequent chromosome 1 gains [43]. In SHH medulloblastoma patients, we observed frequent copy number gains for *AMPKβ2* (6%), *AMPKγ2* (5%) and *AMPKγ3* (10%). Interestingly, strong copy number gains are also observed for *CNBP* and *ODC1* in SHH medulloblastoma (19% and 11%, respectively), as well as in Group #3 medulloblastoma (10% and 22%, respectively). These observations draw yet another interesting parallel with the D'amico et al. study in that their immunohistochemistry analysis of CNBP and ODC1 in a cohort of 42 medulloblastoma patients revealed high levels of expression of both proteins in both SHH and Group

#3 medulloblastoma [34]. The extent to which genetic events co-occur or not in patients can reveal functional relationships between genes, e.g., dependence or redundancy [44]. We found that the copy numbers for the nine genes under investigation are often positively correlated, with a remarkable 21 out of 33 non-syntenic pairs of genes having significant ($p < 0.01$) positive correlations (Figure 4c). In other words, not only are *AMPK*, *CNBP*, and *ODC1* genes frequently gained in medulloblastoma, but these gains tend to co-occur more often than expected at random, suggesting that the co-occurrence of these gains may provide a synergistic advantage. Together, these observations are in agreement with a pro-tumorigenic role for the SHH/AMPK/CNBP axis in medulloblastoma.

a

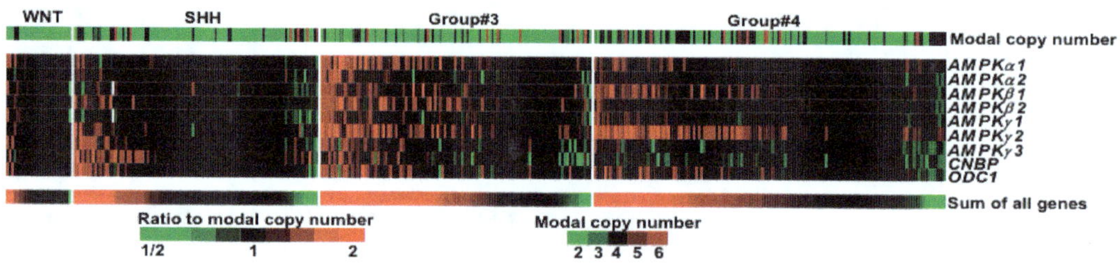

b

			Percentage of tumors with gains				Percentage of tumors with losses				P-values from Liklihood Ratio Chi-Square tests asking if frequency of losses and gains are the same					P-values from Likelihood Ratio Chi-Square tests asking if the proportions of tumors with gains are equal in the four subgroups
Chromosome	Location	Gene	WNT (N=26)	SHH (N=98)	G3 (N=82)	G4 (N=139)	WNT (N=26)	SHH (N=98)	G3 (N=82)	G4 (N=139)	WNT (N=26)	SHH (N=98)	G3 (N=82)	G4 (N=139)	All (N=345)	All (N=345)
5	40,770,000	*AMPKα1*	12%	2%	30%	5%	0%	1%	0%	0%	0.041	0.56	3.9E-09	0.002	4.4E-11	7.8E-09
1	57,150,000	*AMPKα2*	0%	0%	11%	0%	0%	1%	5%	2%	1	0.24	0.16	0.041	0.81	7.0E-06
12	120,110,000	*AMPKβ1*	4%	1%	15%	14%	4%	1%	0%	1%	1	1	4.5E-05	4.4E-06	3.4E-08	2.2E-04
1	146,630,000	*AMPKβ2*	0%	6%	38%	1%	0%	0%	1%	1%	1	0.004	2.6E-09	1	3.2E-10	7.4E-16
12	49,400,000	*AMPKγ1*	4%	0%	7%	1%	4%	3%	1%	1%	1	0.041	0.047	0.56	0.44	0.012
7	151,350,000	*AMPKγ2*	0%	5%	43%	37%	0%	1%	0%	0%	1	0.088	3.3E-12	4.2E-17	3.7E-27	8.3E-14
2	219,690,000	*AMPKγ3*	8%	10%	11%	1%	0%	1%	5%	6%	0.096	0.003	0.16	0.013	0.13	9.5E-04
3	128,900,000	*CNBP*	4%	19%	10%	0%	0%	0%	13%	9%	0.24	2.9E-07	0.49	4.5E-05	0.48	4.6E-08
2	10,585,000	*ODC1*	8%	11%	22%	9%	0%	2%	2%	3%	0.096	0.009	1.2E-04	0.041	2.8E-07	0.036

c

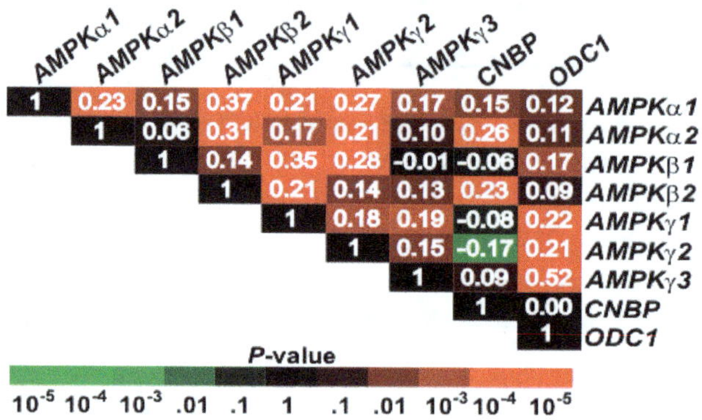

Figure 4. Frequent copy number gains for the AMPK, CNBP, and ODC1 genes in SHH and Group #3 medulloblastoma patients. (**a**) Estimated ratio of copy number to modal copy number for each tumor for 345 medulloblastomas, including 26 WNT, 98 SHH, 82 Group #3, and 139 Group #4 samples. Modal copy number estimates are shown in the first row and the rows for individual genes show copy number estimates normalized to the modal copy number. The "sum of all genes" row was computed from the sum of the log-transformed ratios for the nine genes and was used to determine the sorting of samples in each group. (**b**) The percentage of tumors within each of the four subgroups with copy number gains [$\log2(\frac{copy\ number}{modal\ copy\ number}) > 0.4$] or losses [$\log2(\frac{copy\ number}{modal\ copy\ number}) < -0.4$]. Gains or losses in greater than 5% of the tumors in a given group are highlighted with red or green background, respectively. Similarly, p values < 0.01 are highlighted. (**c**) Analysis of the correlations of log-transformed normalized copy number estimates shows that positive correlations are frequent. Pearson correlations are shown as numbers in white, while their significance is indicated by the color of the background in each cell.

3. Discussion

In this study, we demonstrate that loss of AMPKα2 impairs SHH-driven medulloblastoma tumorigenesis. This is a surprising observation given the previous reports supporting a tumor-suppressive role for AMPK in medulloblastoma tumorigenesis [28–33] and the fact that the subunit AMPKα2 alone can suppress tumorigenesis in other contexts [21,45,46]. In fact, together with a previous analysis of AMPKα2 in mammary carcinoma [25], our analysis represents the first in vivo evidence that the loss of AMPKα2 alone can impair oncogenesis.

Several reasons may underlie this apparent AMPK paradox, both in cancer in general and in medulloblastoma in particular. As previously reviewed, AMPK modulates numerous targets both to inhibit anabolism and to promote catabolism. As such, AMPK has a dual role in cancer, tumor-suppressive or pro-tumorigenic, depending on the cellular or tissue context [13–15]. For example, on the one hand, as an inhibitor of the mTOR complex 1 and the cell cycle, AMPK can be tumor-suppressive [19,20]. On the other hand, maintenance of the metabolic balance by AMPK may be critical for survival during metabolic stress that can occur in the tumor microenvironment—hence the requirement for AMPK activity in some cancer cells [22–27]. Similarly, in medulloblastoma, the role of AMPK as both an inhibitor of GLI1 and SHH signaling [28–30,33] (tumor-suppressive effect) as well as a promoter of CNBP and polyamine metabolism (pro-tumorigenic effect) supposes a dual role for AMPK in this disease context. In agreement with the observation that CNBP protein levels are decreased in the absence of AMPKα2, we observed that the pro-tumorigenic function of AMPKα2 prevails in the context of the highly penetrant [GFAP-tTA;TRE-SMOA1] mouse model of medulloblastoma. It would be interesting, though, to assess the loss of AMPKα2 in other mouse models of medulloblastoma with lower incidence/longer latency [47,48], in order to investigate which of AMPKα2's roles, tumor-suppressive or pro-tumorigenic, prevails in these contexts.

The apparent discrepancy between our results observed in mouse and the previously described tumor-suppressive role of AMPK in human medulloblastoma cell lines [28–30,33] could be due to the species-specific role of AMPK in the cerebellum. While all studies in which the role of AMPK has been directly tested genetically in human medulloblastoma cells agree, i.e., AMPK suppresses SHH-driven cell growth, the role of AMPK in mouse cell lines is debated [28–30,33]. For example, several studies by the Yang laboratory showed that AMPK phosphorylates GLI1 at multiple sites and targets it for proteasomal degradation in murine cells [28,29,33]. Yet, Di Magno et al. argue that AMPK phosphorylates GLI1 at the unique residue Ser408, which is conserved only in primates but not in other species [30]. If the regulation of GLI1 by AMPK does not occur in mouse, as suggested by Di Magno et al., this species-specific difference could explain why our observations obtained in mouse contradict the previous reports that AMPK suppresses growth in human SHH medulloblastoma [28–33]. Further investigation of the species-specific effect of AMPK on GLI1 and medulloblastoma is warranted.

In some contexts, AMPKα1 and AMPKα2 mediate specific, non-redundant functions, e.g., AMPK$\alpha 2^{-/-}$ but not AMPK$\alpha 1^{-/-}$, mice are resistant to hypoglycemic AICAR effects [37,49]. In other contexts, AMPKα1 and AMPKα2 are genetically redundant, e.g., while both AMPKα1 and AMPKα2 are dispensable for development, the combined loss of both genes is embryonic lethal at E10.5 [50]. While we cannot exclude the possibility that a complete KO of AMPKα may have an even more dramatic effect on medulloblastoma tumorigenesis and survival, our results suggest that AMPKα2 has a specific function in this context. This lack of genetic redundancy may underlie various mechanisms, as previously reviewed [51,52]. Could this be due to different biochemical function for AMPKα1- and AMPKα2-containing heterotrimeric complexes? AMPKα2 may be the target of different regulators and/or AMPKα2 may target different substrates. In that regard, it would be interesting to assess whether CNBP is specifically targeted by AMPKα2- but not by AMPKα1-containing heterotrimeric complexes. Could this be due to variations in the spatio-temporal expression for AMPKα1 and AMPKα2? Cell types may express one gene, but not the other. In agreement with an AMPKα2-specific function in SHH medulloblastoma and the fact that the cerebellar granular neuron progenitor cell

(CGNP) is the "cell of origin" in SHH-driven medulloblastoma [53], AMPKα2 is expressed at higher level than AMPKα1 in cerebellar granule neurons [54].

Genomics applied to medulloblastoma defined four medulloblastoma subgroups (WNT, SHH, and Groups #3 and #4) [3,4] and, more recently, different subtypes within them [5–7]. D'amico et al. previously showed that targeting CNBP or ODC1 impairs SHH-driven medulloblastoma and inferred that the SHH/AMPK/CNBP axis may be pro-tumorigenic in this context [34]. Notwithstanding the debate on the nature of the effect (pro- versus anti-tumorigenic) [28–34], our observation that AMPKα2 impairs tumor progression in the SmoA1 mouse model of medulloblastoma validates this hypothesis. Is AMPK functionally relevant in other medulloblastoma subgroups? We observed high copy numbers for the AMPK, CNBP, and ODC1 genes not only in SHH but also in Group #3 medulloblastoma. Taken together with the complementary report that both SHH and Group #3 medulloblastomas show a high level of expression of CNBP and ODC1 proteins [34], the SHH/AMPK/CNBP axis could then have a critical role in both the SHH and Group #3 subgroups. Group #3 medulloblastomas account for ~25% of all medulloblastomas, are characterized by a transcriptional signature associated with photoreceptors and gamma aminobutyric acid–secreting (GABAergic) neurons, and are associated with a poor prognosis [3]. Though differences in the metabolic profiles between medulloblastoma subgroups have not yet been comprehensively examined, the higher degree of aggressiveness associated with Group #3 medulloblastomas may translate to a higher need for energy and thus a higher sensibility to the proper maintenance of the metabolic balance by AMPK under metabolic stress conditions. Investigation of AMPK KO in mouse models of Group #3 medulloblastoma, when they become available, is warranted.

Our study reveals a pro-tumorigenic role for AMPKα2 in SHH medulloblastoma. In light of the previous reports on the tumor-suppressive role of AMPK in SHH medulloblastoma, this observation further underscores the multifaceted role of AMPK in cancer. It also warrants caution to the previously proposed use of AMPK agonists for the treatment of cancer patients in general, and in medulloblastoma patients in particular.

4. Materials and Methods

4.1. [GFAP-tTA;TRE-SmoA1] Mouse Model

We studied medulloblastoma tumorigenesis in the absence or presence of AMPKα2 in the previously published bitransgenic [GFAP-tTA;TRE-SmoA1] model [35,36], where the expression of the tetracycline-regulated transactivator (tTA) is driven by a GFAP promoter and the expression of oncogenic SmoA1 is under the control of the tetracycline responsive element (TRE). The experimental breeders used in this study were an AMPKα2 KO mouse [37], a TRE-SmoA1 mouse [35], and a GFAP-tTA mouse [55], all of which were maintained on a C57/BL6 background for at least five generations prior to initiating experiments. TRE-SmoA1, [GFAP-tTA;TRE-SmoA1], and [GFAP-tTA;TRE-SmoA1;AMPKα2$^{-/-}$] mouse littermates were generated by crossing [GFAP-tTA;AMPKα2$^{+/-}$] mice with [TRE-SmoA1;AMPKα2$^{+/-}$] mice. Animals that meet the guidelines for end-stage illness and/or found to be at protocol endpoint (evidence of large tumor formation, enlarged dome head, and/or severe neurological dysfunction) were humanely euthanized in accordance with the institutional guidelines for the welfare of experimental animals. Maintenance of mouse colonies and experimental procedures were approved by the University of Michigan Committee on the Use and Care of Animals.

4.2. Mouse Genotyping

The mouse genotyping experiments were performed as previously described [36], with minor modifications. Genotyping experiments were performed by PCR analysis using tail genomic DNA obtained from pups at Postnatal Days 10 (P10) and 14 (P14). Genotyping of *AMPKα2* was determined by PCR using the following oligonucleotides: forward: 5′-gcttagcacgttaccctggat-3′ and WT reverse: 5′-gtcttcactgaaatacatagca-3′ or mutant reverse: 5′-gcattgaaccacagtccttcctc-3′.

Genotyping of the *GFAP-tTA* transgene was determined by PCR using the following oligonucleotides: forward: 5′-ctcgcccagaagctaggtgt-3′ and reverse: 5′-ccatcgcgatgacttagt-3′. Genotyping of the *TRE-SmoA1* transgene was determined by PCR using the following oligonucleotides: forward: 5′-ggaactgatgaatgggagca-3′ and reverse: 5′-gggaggtgtgggaggttt-3′. For internal control genotyping, we used the following primers (forward: 5′-caaatgttgctgtctggtg-3′ and reverse: 5′-gtcagtcgagtgcacagttt-3′).

4.3. Immunohistochemistry

Immunohistochemistry (IHC) experiments were performed as previously described [36].

4.4. Western Blot Analyses

Whole cerebellum protein samples were prepared using RIPA buffer, sonicated three times on ice, and supernatants were collected after centrifugation. The resulting cerebellum protein extracts were separated on acrylamide gels, transferred to PVDF membranes, and proteins were detected using standard immunoblotting techniques. The following antibodies were used: AMPKα1 (Bethyl Laboratory®, Montgomery, TX, USA; Cat# A300-507A), AMPKα2 (Bethyl Laboratory®, Cat# A300-508A), pACC1 (Cell Signaling®, Danvers, MA, USA; Cat# 11818), β-actin (Cell Signaling®, Cat# 5125), and goat α-rabbit IgG (Jackson Immunoresearch Laboratory®, West Grove, PA, USA; Cat# 111-035-045). The CNBP antibody [34] was a gift from Dr. Gianluca Canettieri, Sapienza University of Rome, Italy. The intensities of Western blot signal bands were quantified using Gel Analysis in ImageJ.

4.5. Copy Number Variations in Medulloblastoma Patients

We analyzed the publicly available copy number data for medulloblastoma samples estimated from short-read DNA sequencing [5]. We downloaded a single segment file holding estimated $\log_2(\frac{copy\ number}{2})$ data from PedcBioportal (http://pedcbioportal.org/). Following the methods of the original paper [5], we excluded 16 samples and analyzed the remaining 345. Many tumors had chromosome doublings, so we normalized our data by first computing the average of the $\log_2(\frac{copy\ number}{2})$ estimates for each chromosome and then counted how many chromosomes gave estimates within 0.1 unit of each possible log-transformed integer copy number. We then selected the most common integer chromosomal copy number for each tumor, after excluding those chromosomes with the most frequent copy number changes (Chr. 6, 7, 8, and 17, X and Y), and referred to this as the modal copy number. Among the 345 tumors, we observed 212 tumors with a 2-modal copy number, 19 tumors with a 3-modal copy number, 91 tumors with a 4-modal copy number, 2 tumors with a 5-modal copy number, and 21 tumors with a 6-modal copy number. Our final estimates of the relative gene copy numbers were of $\log_2(\frac{copy\ number}{modal\ copy\ number})$.

Author Contributions: J.-F.R. conceived and directed the project. H.Z., C.P., J.Y., E.C.F., B.V., and J.-F.R. designed and performed the in vivo mouse studies. H.Z., S.-S.P., J.W., D.T., K.I., and S.C.-P. performed and analyzed the molecular studies and the IHC experiments. R.K. performed the copy number variation and statistical analyses. J.-F.R. wrote the manuscript, with contributions from other co-authors.

Acknowledgments: We thank the University of Michigan Rogel Cancer Center Tissue Core for technical support with the immunohistochemistry experiments; Gianluca Canettieri, Sapienza University of Rome, Rome, Italy, for the CNBP antibody; Tao Xu for his comments and suggestions in the early development of the project.

Abbreviations

AMPK	AMP-activated protein kinase
CGNPs	Cerebellar granular neuron progenitor cells
Chr.	Chromosome
DAB	Diaminobenzidine

GABAergic	Gamma aminobutyric acid–secreting
H&E	Hematoxylin and eosin
HRP	Horseradish peroxidase
IHC	Immunohistochemistry
IGL	Internal granular cell
ML	Molecular
P10	Postnatal day 10
pACC1	phosphorylated acetyl-CoA carboxylase alpha
PC	Purkinje cell
SHH	Sonic Hedgehog
TRE	Tetracycline responsive element
WT	Wild type

References

1. Ostrom, Q.T.; Gittleman, H.; Liao, P.; Vecchione-Koval, T.; Wolinsky, Y.; Kruchko, C.; Barnholtz-Sloan, J.S. CBTRUS Statistical Report: Primary brain and other central nervous system tumors diagnosed in the United States in 2010–2014. *Neuro Oncol.* **2017**, *19*, v1–v88. [CrossRef] [PubMed]
2. Liu, K.W.; Pajtler, K.W.; Worst, B.C.; Pfister, S.M.; Wechsler-Reya, R.J. Molecular mechanisms and therapeutic targets in pediatric brain tumors. *Sci. Signal.* **2017**, *10*. [CrossRef] [PubMed]
3. Taylor, M.D.; Northcott, P.A.; Korshunov, A.; Remke, M.; Cho, Y.J.; Clifford, S.C.; Eberhart, C.G.; Parsons, D.W.; Rutkowski, S.; Gajjar, A.; et al. Molecular subgroups of medulloblastoma: The current consensus. *Acta Neuropathol.* **2012**, *123*, 465–472. [CrossRef] [PubMed]
4. Northcott, P.A.; Jones, D.T.; Kool, M.; Robinson, G.W.; Gilbertson, R.J.; Cho, Y.J.; Pomeroy, S.L.; Korshunov, A.; Lichter, P.; Taylor, M.D.; et al. Medulloblastomics: The end of the beginning. *Nat. Rev. Cancer* **2012**, *12*, 818–834. [CrossRef] [PubMed]
5. Northcott, P.A.; Buchhalter, I.; Morrissy, A.S.; Hovestadt, V.; Weischenfeldt, J.; Ehrenberger, T.; Grobner, S.; Segura-Wang, M.; Zichner, T.; Rudneva, V.A.; et al. The whole-genome landscape of medulloblastoma subtypes. *Nature* **2017**, *547*, 311–317. [CrossRef] [PubMed]
6. Lin, C.Y.; Erkek, S.; Tong, Y.; Yin, L.; Federation, A.J.; Zapatka, M.; Haldipur, P.; Kawauchi, D.; Risch, T.; Warnatz, H.J.; et al. Active medulloblastoma enhancers reveal subgroup-specific cellular origins. *Nature* **2016**, *530*, 57–62. [CrossRef] [PubMed]
7. Cavalli, F.M.G.; Remke, M.; Rampasek, L.; Peacock, J.; Shih, D.J.H.; Luu, B.; Garzia, L.; Torchia, J.; Nor, C.; Morrissy, A.S.; et al. Intertumoral Heterogeneity within Medulloblastoma Subgroups. *Cancer Cell* **2017**, *31*, 737–754. [CrossRef] [PubMed]
8. Mulhern, R.K.; Merchant, T.E.; Gajjar, A.; Reddick, W.E.; Kun, L.E. Late neurocognitive sequelae in survivors of brain tumours in childhood. *Lancet Oncol.* **2004**, *5*, 399–408. [CrossRef]
9. Laughton, S.J.; Merchant, T.E.; Sklar, C.A.; Kun, L.E.; Fouladi, M.; Broniscer, A.; Morris, E.B.; Sanders, R.P.; Krasin, M.J.; Shelso, J.; et al. Endocrine outcomes for children with embryonal brain tumors after risk-adapted craniospinal and conformal primary-site irradiation and high-dose chemotherapy with stem-cell rescue on the SJMB-96 trial. *J. Clin. Oncol.* **2008**, *26*, 1112–1118. [CrossRef] [PubMed]
10. Hardie, D.G.; Ross, F.A.; Hawley, S.A. AMPK: A nutrient and energy sensor that maintains energy homeostasis. *Nat. Rev. Mol. Cell Biol.* **2012**, *13*, 251–262. [CrossRef] [PubMed]
11. Hardie, D.G.; Schaffer, B.E.; Brunet, A. AMPK: An Energy-Sensing Pathway with Multiple Inputs and Outputs. *Trends Cell Biol.* **2016**, *26*, 190–201. [CrossRef] [PubMed]
12. Cheng, J.; Zhang, T.; Ji, H.; Tao, K.; Guo, J.; Wei, W. Functional characterization of AMP-activated protein kinase signaling in tumorigenesis. *Biochim. Biophys. Acta* **2016**, *1866*, 232–251. [CrossRef] [PubMed]
13. Liang, J.; Mills, G.B. AMPK: A contextual oncogene or tumor suppressor? *Cancer Res.* **2013**, *73*, 2929–2935. [CrossRef] [PubMed]
14. Zadra, G.; Batista, J.L.; Loda, M. Dissecting the Dual Role of AMPK in Cancer: From Experimental to Human Studies. *Mol. Cancer Res.* **2015**, *13*, 1059–1072. [CrossRef] [PubMed]
15. Dasgupta, B.; Chhipa, R.R. Evolving Lessons on the Complex Role of AMPK in Normal Physiology and Cancer. *Trends Pharmacol. Sci.* **2016** *37*, 192–206. [CrossRef] [PubMed]

16. Hawley, S.A.; Boudeau, J.; Reid, J.L.; Mustard, K.J.; Udd, L.; Makela, T.P.; Alessi, D.R.; Hardie, D.G. Complexes between the LKB1 tumor suppressor, STRAD alpha/beta and MO25 alpha/beta are upstream kinases in the AMP-activated protein kinase cascade. *J. Biol.* **2003**, *2*, 28. [CrossRef] [PubMed]
17. Woods, A.; Johnstone, S.R.; Dickerson, K.; Leiper, F.C.; Fryer, L.G.; Neumann, D.; Schlattner, U.; Wallimann, T.; Carlson, M.; Carling, D. LKB1 is the upstream kinase in the AMP-activated protein kinase cascade. *Curr. Biol.* **2003**, *13*, 2004–2008. [CrossRef] [PubMed]
18. Shaw, R.J.; Kosmatka, M.; Bardeesy, N.; Hurley, R.L.; Witters, L.A.; DePinho, R.A.; Cantley, L.C. The tumor suppressor LKB1 kinase directly activates AMP-activated kinase and regulates apoptosis in response to energy stress. *Proc. Natl. Acad. Sci. USA* **2004**, *101*, 3329–3335. [CrossRef] [PubMed]
19. Shackelford, D.B.; Shaw, R.J. The LKB1-AMPK pathway: Metabolism and growth control in tumour suppression. *Nat. Rev. Cancer* **2009**, *9*, 563–575. [CrossRef] [PubMed]
20. Faubert, B.; Boily, G.; Izreig, S.; Griss, T.; Samborska, B.; Dong, Z.; Dupuy, F.; Chambers, C.; Fuerth, B.J.; Viollet, B.; et al. AMPK is a negative regulator of the Warburg effect and suppresses tumor growth in vivo. *Cell Metab.* **2013**, *17*, 113–124. [CrossRef] [PubMed]
21. Phoenix, K.N.; Devarakonda, C.V.; Fox, M.M.; Stevens, L.E.; Claffey, K.P. AMPKalpha2 Suppresses Murine Embryonic Fibroblast Transformation and Tumorigenesis. *Genes Cancer* **2012**, *3*, 51–62. [CrossRef] [PubMed]
22. Laderoute, K.R.; Amin, K.; Calaoagan, J.M.; Knapp, M.; Le, T.; Orduna, J.; Foretz, M.; Viollet, B. 5′-AMP-activated protein kinase (AMPK) is induced by low-oxygen and glucose deprivation conditions found in solid-tumor microenvironments. *Mol. Cell. Biol.* **2006**, *26*, 5336–5347. [CrossRef] [PubMed]
23. Jeon, S.M.; Chandel, N.S.; Hay, N. AMPK regulates NADPH homeostasis to promote tumour cell survival during energy stress. *Nature* **2012**, *485*, 661–665. [CrossRef] [PubMed]
24. Rios, M.; Foretz, M.; Viollet, B.; Prieto, A.; Fraga, M.; Costoya, J.A.; Senaris, R. AMPK activation by oncogenesis is required to maintain cancer cell proliferation in astrocytic tumors. *Cancer Res.* **2013**, *73*, 2628–2638. [CrossRef] [PubMed]
25. Hindupur, S.K.; Balaji, S.A.; Saxena, M.; Pandey, S.; Sravan, G.S.; Heda, N.; Kumar, M.V.; Mukherjee, G.; Dey, D.; Rangarajan, A. Identification of a novel AMPK-PEA15 axis in the anoikis-resistant growth of mammary cells. *Breast Cancer Res.* **2014**, *16*, 420. [CrossRef] [PubMed]
26. Saito, Y.; Chapple, R.H.; Lin, A.; Kitano, A.; Nakada, D. AMPK Protects Leukemia-Initiating Cells in Myeloid Leukemias from Metabolic Stress in the Bone Marrow. *Cell Stem Cell* **2015**, *17*, 585–596. [CrossRef] [PubMed]
27. Chhipa, R.R.; Fan, Q.; Anderson, J.; Muraleedharan, R.; Huang, Y.; Ciraolo, G.; Chen, X.; Waclaw, R.; Chow, L.M.; Khuchua, Z.; et al. AMP kinase promotes glioblastoma bioenergetics and tumour growth. *Nat. Cell Biol.* **2018**, *20*, 823–835. [CrossRef] [PubMed]
28. Li, Y.H.; Luo, J.; Mosley, Y.Y.; Hedrick, V.E.; Paul, L.N.; Chang, J.; Zhang, G.; Wang, Y.K.; Banko, M.R.; Brunet, A.; et al. AMP-Activated Protein Kinase Directly Phosphorylates and Destabilizes Hedgehog Pathway Transcription Factor GLI1 in Medulloblastoma. *Cell Rep.* **2015**, *12*, 599–609. [CrossRef] [PubMed]
29. Zhang, R.; Huang, S.Y.; Ka-Wai Li, K.; Li, Y.H.; Hsu, W.H.; Zhang, G.J.; Chang, C.J.; Yang, J.Y. Dual degradation signals destruct GLI1: AMPK inhibits GLI1 through beta-TrCP-mediated proteasome degradation. *Oncotarget* **2017**, *8*, 49869–49881. [CrossRef] [PubMed]
30. Di Magno, L.; Basile, A.; Coni, S.; Manni, S.; Sdruscia, G.; D'Amico, D.; Antonucci, L.; Infante, P.; De Smaele, E.; Cucchi, D.; et al. The energy sensor AMPK regulates Hedgehog signaling in human cells through a unique Gli1 metabolic checkpoint. *Oncotarget* **2016**, *7*, 9538–9549. [CrossRef] [PubMed]
31. Gershon, T.R.; Crowther, A.J.; Tikunov, A.; Garcia, I.; Annis, R.; Yuan, H.; Miller, C.R.; Macdonald, J.; Olson, J.; Deshmukh, M. Hexokinase-2-mediated aerobic glycolysis is integral to cerebellar neurogenesis and pathogenesis of medulloblastoma. *Cancer Metab.* **2013**, *1*, 2. [CrossRef] [PubMed]
32. Tech, K.; Gershon, T.R. Energy metabolism in neurodevelopment and medulloblastoma. *Transl. Pediatr* **2015**, *4*, 12–19. [CrossRef] [PubMed]
33. Huang, S.Y.; Chen, S.K.; Yang, J.Y. Activation of AMPK inhibits medulloblastoma cell growth and Gli1 activity. *Cancer Rep. Rev.* **2017**. [CrossRef]
34. D'Amico, D.; Antonucci, L.; Di Magno, L.; Coni, S.; Sdruscia, G.; Macone, A.; Miele, E.; Infante, P.; Di Marcotullio, L.; De Smaele, E.; et al. Non-canonical Hedgehog/AMPK-Mediated Control of Polyamine Metabolism Supports Neuronal and Medulloblastoma Cell Growth. *Dev. Cell* **2015**, *35*, 21–35. [CrossRef] [PubMed]

35. Michael, L.E.; Westerman, B.A.; Ermilov, A.N.; Wang, A.; Ferris, J.; Liu, J.; Blom, M.; Ellison, D.W.; van Lohuizen, M.; Dlugosz, A.A. Bmi1 is required for Hedgehog pathway-driven medulloblastoma expansion. *Neoplasia* **2008**, *10*, 1343–1349. [CrossRef] [PubMed]
36. Xu, T.; Zhang, H.; Park, S.S.; Venneti, S.; Kuick, R.; Ha, K.; Michael, L.E.; Santi, M.; Uchida, C.; Uchida, T.; et al. Loss of Pin1 Suppresses Hedgehog-Driven Medulloblastoma Tumorigenesis. *Neoplasia* **2017**, *19*, 216–225. [CrossRef] [PubMed]
37. Viollet, B.; Andreelli, F.; Jorgensen, S.B.; Perrin, C.; Geloen, A.; Flamez, D.; Mu, J.; Lenzner, C.; Baud, O.; Bennoun, M.; et al. The AMP-activated protein kinase alpha2 catalytic subunit controls whole-body insulin sensitivity. *J. Clin. Investig.* **2003**, *111*, 91–98. [CrossRef] [PubMed]
38. De Luca, A.; Cerrato, V.; Fuca, E.; Parmigiani, E.; Buffo, A.; Leto, K. Sonic hedgehog patterning during cerebellar development. *Cell. Mol. Life Sci.* **2016**, *73*, 291–303. [CrossRef] [PubMed]
39. Jorgensen, S.B.; Treebak, J.T.; Viollet, B.; Schjerling, P.; Vaulont, S.; Wojtaszewski, J.F.; Richter, E.A. Role of AMPKalpha2 in basal, training-, and AICAR-induced GLUT4, hexokinase II, and mitochondrial protein expression in mouse muscle. *Am. J. Physiol. Endocrinol. Metab.* **2007**, *292*, E331–E339. [CrossRef] [PubMed]
40. Carling, D.; Zammit, V.A.; Hardie, D.G. A common bicyclic protein kinase cascade inactivates the regulatory enzymes of fatty acid and cholesterol biosynthesis. *FEBS Lett.* **1987**, *223*, 217–222. [CrossRef]
41. Fullerton, M.D.; Galic, S.; Marcinko, K.; Sikkema, S.; Pulinilkunnil, T.; Chen, Z.P.; O'Neill, H.M.; Ford, R.J.; Palanivel, R.; O'Brien, M.; et al. Single phosphorylation sites in Acc1 and Acc2 regulate lipid homeostasis and the insulin-sensitizing effects of metformin. *Nat. Med.* **2013**, *19*, 1649–1654. [CrossRef] [PubMed]
42. Wang, C.; Ma, J.; Zhang, N.; Yang, Q.; Jin, Y.; Wang, Y. The acetyl-CoA carboxylase enzyme: A target for cancer therapy? *Expert Rev. Anticancer Ther.* **2015**, *15*, 667–676. [CrossRef] [PubMed]
43. Northcott, P.A.; Shih, D.J.; Peacock, J.; Garzia, L.; Morrissy, A.S.; Zichner, T.; Stutz, A.M.; Korshunov, A.; Reimand, J.; Schumacher, S.E.; et al. Subgroup-specific structural variation across 1000 medulloblastoma genomes. *Nature* **2012**, *488*, 49–56. [CrossRef] [PubMed]
44. Ashworth, A.; Lord, C.J.; Reis-Filho, J.S. Genetic interactions in cancer progression and treatment. *Cell* **2011**, *145*, 30–38. [CrossRef] [PubMed]
45. Fox, M.M.; Phoenix, K.N.; Kopsiaftis, S.G.; Claffey, K.P. AMP-Activated Protein Kinase alpha 2 Isoform Suppression in Primary Breast Cancer Alters AMPK Growth Control and Apoptotic Signaling. *Genes Cancer* **2013**, *4*, 3–14. [CrossRef] [PubMed]
46. Vila, I.K.; Yao, Y.; Kim, G.; Xia, W.; Kim, H.; Kim, S.J.; Park, M.K.; Hwang, J.P.; Billalabeitia, E.G.; Hung, M.C.; et al. A UBE2O-AMPKα2 Axis That Promotes Tumor Initiation and Progression Offers Opportunities for Therapy. *Cancer Cell* **2017**, *31*, 208–224. [CrossRef] [PubMed]
47. Goodrich, L.V.; Milenkovic, L.; Higgins, K.M.; Scott, M.P. Altered neural cell fates and medulloblastoma in mouse patched mutants. *Science* **1997**, *277*, 1109–1113. [CrossRef] [PubMed]
48. Hallahan, A.R.; Pritchard, J.I.; Hansen, S.; Benson, M.; Stoeck, J.; Hatton, B.A.; Russell, T.L.; Ellenbogen, R.G.; Bernstein, I.D.; Beachy, P.A.; et al. The SmoA1 mouse model reveals that notch signaling is critical for the growth and survival of sonic hedgehog-induced medulloblastomas. *Cancer Res.* **2004**, *64*, 7794–7800. [CrossRef] [PubMed]
49. Jorgensen, S.B.; Viollet, B.; Andreelli, F.; Frosig, C.; Birk, J.B.; Schjerling, P.; Vaulont, S.; Richter, E.A.; Wojtaszewski, J.F. Knockout of the alpha2 but not alpha1 5′-AMP-activated protein kinase isoform abolishes 5-aminoimidazole-4-carboxamide-1-beta-4-ribofuranosidebut not contraction-induced glucose uptake in skeletal muscle. *J. Biol. Chem.* **2004**, *279*, 1070–1079. [CrossRef] [PubMed]
50. Viollet, B.; Athea, Y.; Mounier, R.; Guigas, B.; Zarrinpashneh, E.; Horman, S.; Lantier, L.; Hebrard, S.; Devin-Leclerc, J.; Beauloye, C.; et al. AMPK: Lessons from transgenic and knockout animals. *Front. Biosci.* **2009**, *14*, 19–44. [CrossRef]
51. Ross, F.A.; MacKintosh, C.; Hardie, D.G. AMP-activated protein kinase: A cellular energy sensor that comes in 12 flavours. *FEBS J.* **2016**, *283*, 2987–3001. [CrossRef] [PubMed]
52. Herzig, S.; Shaw, R.J. AMPK: Guardian of metabolism and mitochondrial homeostasis. *Nat. Rev. Mol. Cell Biol.* **2018**, *19*, 121–135. [CrossRef] [PubMed]
53. Yang, Z.J.; Ellis, T.; Markant, S.L.; Read, T.A.; Kessler, J.D.; Bourboulas, M.; Schuller, U.; Machold, R.; Fishell, G.; Rowitch, D.H.; et al. Medulloblastoma can be initiated by deletion of Patched in lineage-restricted progenitors or stem cells. *Cancer Cell* **2008**, *14*, 135–145. [CrossRef] [PubMed]

54. Turnley, A.M.; Stapleton, D.; Mann, R.J.; Witters, L.A.; Kemp, B.E.; Bartlett, P.F. Cellular distribution and developmental expression of AMP-activated protein kinase isoforms in mouse central nervous system. *J. Neurochem.* **1999**, *72*, 1707–1716. [CrossRef] [PubMed]
55. Lin, W.; Kemper, A.; McCarthy, K.D.; Pytel, P.; Wang, J.P.; Campbell, I.L.; Utset, M.F.; Popko, B. Interferon-gamma induced medulloblastoma in the developing cerebellum. *J. Neurosci.* **2004**, *24*, 10074–10083. [CrossRef] [PubMed]

13

Nutritional Modulation of AMPK-Impact upon Metabolic-Inflammation

Claire L. Lyons [1,2] and Helen M. Roche [2,3,*]

[1] Unit of Molecular Metabolism, Lund University Diabetes Center, Clinical Research Center, Lund University, 205 02 Malmö, Sweden; claire.lyons@med.lu.se
[2] Nutrigenomics Research Group, UCD Institute of Food and Health, UCD Conway Institute of Biomolecular and Biomedical Research, University College Dublin, Belfield, 4 Dublin, Ireland
[3] Institute of Global Food Security, Queen's University Belfast BT7 1NN, Northern Ireland, UK
* Correspondence: helen.roche@ucd.ie

Abstract: Nutritional status provides metabolic substrates to activate AMP-Activated Protein Kinase (AMPK), the energy sensor that regulates metabolism. Recent evidence has demonstrated that AMPK has wider functions with respect to regulating immune cell metabolism and function. One such example is the regulatory role that AMPK has on NLRP3-inlflammasome and IL-1β biology. This in turn can result in subsequent negative downstream effects on glucose, lipid and insulin metabolism. Nutrient stress in the form of obesity can impact AMPK and whole-body metabolism, leading to complications such as type 2 diabetes and cancer risk. There is a lack of data regarding the nature and extent that nutrient status has on AMPK and metabolic-inflammation. However, emerging work elucidates to a direct role of individual nutrients on AMPK and metabolic-inflammation, as a possible means of modulating AMPK activity. The posit being to use such nutritional agents to re-configure metabolic-inflammation towards more oxidative phosphorylation and promote the resolution of inflammation. The complex paradigm will be discussed within the context of if/how dietary components, nutrients including fatty acids and non-nutrient food components, such as resveratrol, berberine, curcumin and the flavonoid genistein, modulate AMPK dependent processes relating to inflammation and metabolism.

Keywords: AMPK; IL-1β; NLRP3; nutrition; dietary fatty acids; metabolic-inflammation; nutrigenomics

---

1. Introduction

AMP-activated protein kinase (AMPK) is a serine/threonine kinase (once thought of as solely an energy sensor) and has since been assigned widespread roles in metabolism [1], inflammation [2] and Type 2 Diabetes (T2D) [3]. It is responsible for adapting cellular metabolism in response to nutritional and environmental variations. This involves activating pathways to produce energy whilst also inhibiting energy-consuming pathways. The body uses adenosine triphosphate (ATP) as an energy source, which is broken down to adenosine diphosphate (ADP) and then adenosine monophosphate (AMP). When ATP levels are low, the ADP:AMP ratio increases and in turn activates AMPK to activate pathways to replenish ATP levels and restore energy. AMPK is a heterotrimer composed of a catalytic α (α) (α1 and α2) subunit and regulatory beta (β) (β1 and β2) and γ (γ) (γ1, γ2 and γ3) subunits. AMPK can be activated through both allosteric and phosphorylation (p) means. The threonine-172 (Thr-172) residue within the activation loop of the kinase domain on the α-subunit is the main phosphorylation site for AMPK activation [4]. The upstream protein kinases, liver kinase B1 (LKB1), Ca^{2+}/calmodulin-dependent protein kinase kinases (CaMKK) [5] and AMPK kinase (AMPKK) [4], are responsible for AMPK activation by phosphorylation. The allosteric activation of AMPK occurs through AMP and helps to prevent the dephosphorylation of AMPK [6]. AMP has also been shown to directly increase the CaMKKβ- and LKB1-mediated α-Thr^{172} phosphorylation [7]. AMPK activation, in response to low energy status, re-configures glucose, lipid and mitochondrial metabolism towards adenosine triphosphate (ATP) production and decreases anabolic pathways that would otherwise further deplete ATP levels (Figure 1). Insulin can inhibit AMPK activity through phosphorylation of the $Ser^{485/491}$ in the α1/α2 site in multiple tissues, including skeletal muscle and liver, without affecting Thr^{172} phosphorylation, through the Akt pathway [8]. One of the main mechanisms of AMPK activation is the prevention of its dephosphorylation by AMP, especially within the context of metabolic disease [6]. The main pathways activated are those that are involved in growth and metabolism [9]. It activates fatty acid oxidation (FAO) to generate ATP and inhibits unnecessary pathways, such as fatty acid synthesis [1]. Alternatively, AMPK affects the cell cycle and neuronal membrane excitability, as a way to regulate ATP levels [10]. In this review, we will discuss how AMPK can be activated by nutrients, such as glucose and lipids, but is mainly determined by nutrient status. The widespread effects of AMPK on metabolism, ranging from nutrient to mitochondrial metabolism, will be covered. In times of stress, such as obesity or direct influence by fatty acids, AMPK activation becomes altered and leads to dysregulated metabolism. Furthermore, there is an increasing amount of research pointing to the effect of metabolic signals on immune function and inflammation, and how a reciprocal relationship exists. The literature to date has focused on the role of nutritional status on AMP activation and how this affects cellular processes and metabolism. However, whilst AMPK-mediated metabolic-inflammation is a critical biological interaction, there is a paucity of data in relation to the nature and the extent to which individual nutrients affects cellular process and metabolism, thus this will be a focus within the review.

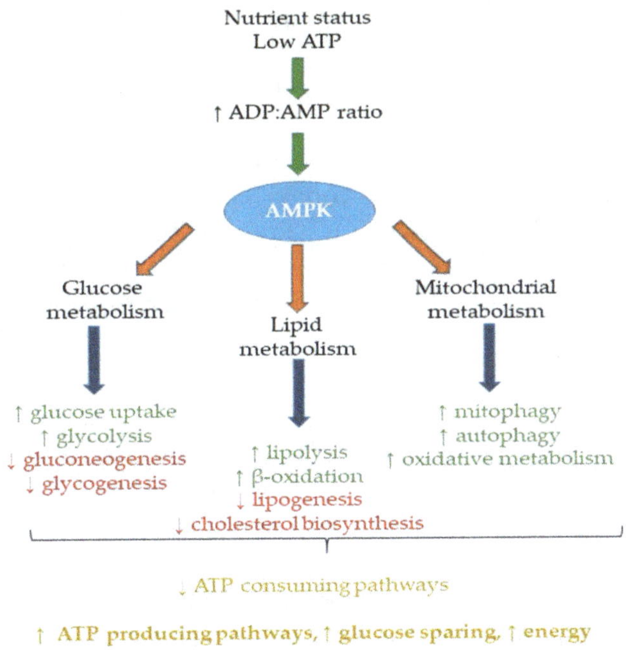

Figure 1. The role of AMP-Activated Protein Kinase (AMPK) on whole-body metabolism. AMPK is a nutrient sensor, which is activated in response to low adenosine triphosphate (ATP) levels, and an increased adenosine diphosphate: adenosine monophosphate (ADP:AMP) ratio. As a result, it activates pathways that produce ATP through glucose, lipid and mitochondrial metabolism pathways, thus increasing ATP levels. Conversely, pathways that deplete ATP are inhibited by AMPK. An ↑ arrow represents an upregulation of the process and ↓ represents a downregulation of the process. AMPK = AMP-Activated Protein Kinase, ATP = Adenosine triphosphate, ADP = Adenosine diphosphate, AMP = Adenosine monophosphate.

2. AMPK Activation, Metabolism and Nutrient Status

Metabolism is fueled by the nutrients we consume and is a tightly regulated process. Nutrient status therefore has a direct effect on the energy status of the organism. On the cellular level, AMPK activation is dependent upon energy status in the form of low ATP levels, and an increased ADP:AMP ratio. It is activated by energetic stress characterized by low levels of ATP. AMPK activation re-configures cellular metabolism, switching on catabolic pathways to generate ATP and switching off anabolic pathways that would otherwise deplete ATP. The metabolic impact of AMPK activation/de-activation has already been thoroughly reviewed by Herzig and Shaw [11]. Briefly, as illustrated in Figure 1, lipid and glucose metabolism are re-configured to supply energy; whilst protein metabolism, particularly protein synthesis is shut down. AMPK activation promotes glucose uptake and glycolysis, and activates lipolysis and oxidation, which is associated with significant upregulation of mitochondrial metabolism, mitophagy and autophagy. Conversely, gluconeogenesis and glycogenesis, as well as fatty acid synthesis/lipogenesis and cholesterol biosynthesis, are attenuated. In essence, AMPK activation promotes glucose sparing and oxidative metabolism to generate maximal ATP from cellular energy substrates. This process is used by most quiescent cells, as opposed to the glucose dependent, glycolytic metabolism relied upon by activated immune and proliferating cells. The current review will focus on the effects of AMPK on cellular metabolism in the context of inflammation and whole-body responses to obesity. For more detailed reviews about AMPK and the effects on lipid, glucose and mitochondrial metabolism, the readers are directed to the following reviews [9,12,13]. Recent work suggests that glucose metabolites mediate AMPK activation via non-canonical energy sensing mechanisms. Whilst it is well acknowledged that cellular glucose status activates AMPK as described above, more recent work suggests that glucose status affects AMPK activation, independent of increasing AMP:ATP and ADP ratios. It was initially presumed that low glucose status activated AMPK via reduced glucose catabolism leading to ATP depletion via the canonical energy sensing mechanism. But Zhang and colleagues recently showed that AMPK was activated when mouse embryonic fibroblasts were transferred from high to low (below 5 mM) glucose concentrations without any changes in cellular AMP/ATP or ADP/ATP ratios [14]. It was demonstrated that during glycolysis, glucose is converted to fructose-1, 6-bisphosphate (FBP), which is then processed or sensed by FBP aldolases. Such low glucose status reduces FBP-bound aldolase, which in turn activates AMPK via LKB1 phosphorylation. Thus, metabolites such as FBP indicative of poor glucose availability modulate FBP aldolases, which in turn sense low FBP and activate AMPK.

2.1. Nutrient Status and Impaired AMPK Action

On the whole-body level, energy status also affects AMPK activation. Obesity, in its simplest form, is caused by a chronic energy imbalance, wherein caloric intake exceeds caloric expenditure. This represents a metabolic stress event and hence affects AMPK activation, as illustrated in Figure 2. Obesity is associated with reduced AMPK activation, concomitant with alterations in glycolysis, insulin sensitivity, hepatic lipid metabolism and inflammation. The AMP metabolic stress signal results in β-myristoylation, allowing AMPK membrane association and facilitating the phosphorylation required for activation [7]. In man AMPK activity, as determined by its phosphorylation state, the P-AMPK/T-AMPK ratio, was reduced in visceral adipose tissue (VAT), rather than subcutaneous adipose tissue (SAT), of obese humans [15]. Obesity also affects immune cell AMPK status. Macrophage phosphorylated AMPK (pAMPK) expression was 33% lower in mouse models of genetic obesity compared to their lean counterparts [16]. Rats on a high fat diet (HFD) also show reduced renal AMPK activity [17]. Obesity arises as a result of excess energy intake, usually excess fat with or without surplus simple carbohydrates/sugars. Later in the review, we will deal with the impact of different dietary components within the context of obesity. It is important to note that high fat diets are usually preferentially enriched in saturated fatty acid (SFA), and initial data would suggest that SFA enriched high fat diets are particularly potent with respect to reducing pAMPK expression in adipose tissue and bone marrow derived macrophages (BMDM) [2]. Interestingly, recent data suggests that altered AMPK activation is not just a function of obesity but may play a role in energy homeostasis. Chronic activation of AMPK, through a mutation in the $\gamma 2$ subunit, was involved in hyperphagia, obesity and impaired pancreatic function, which was observed in both mice and humans [18]. Therefore, long-term, non-discriminate activation of AMPK should also be taken with caution. It is possible that the adverse effect of obesity on pAMPK is mediated via adiponectin, the adipokine responsible for regulating glucose levels and fatty acid synthesis, which is decreased in obesity [19]. Adiponectin can activate AMPK via LKB1, but this is impaired in the skeletal muscle of *ob/ob* mice [20]. AMPK can be pharmacologically induced through treatment with the agonist, 5′-aminoimidazole-4-carboxamide ribonucleotide (AICAR). Mice given AICAR in conjunction with an HFD showed increased plasma adiponectin levels, reduced macrophage infiltration into the kidney, and reduced kidney hypertrophy, but the impact on adipose tissue biology was not investigated in this study [21].

2.2. *The Involvement of AMPK in Insulin Resistance*

AMPK expression is altered in insulin resistant obese individuals, with reduced expression compared to insulin sensitive body mass index-matched counterparts [15]. AMPK$\beta 1^{-/-}$ contributes to insulin resistance with decreased phosphorylation of protein kinase B (Akt), increased adipose non-esterified fatty acids, hyperglycemia and hyperinsulinemia [16]. Serum leptin levels increased, and adiponectin reduced, indicating the negative effect of AMPK signaling disruption on adipose biology. Furthermore, in a co-culture system of AMPK$\alpha 1^{-/-}$ macrophages and 3T3-L1 adipocytes, insulin stimulated phosphorylation of the insulin receptor and subsequent insulin stimulated glucose uptake, were both decreased with the deletion of the AMPK subunit. This provides confirmation that AMPKs anti-inflammatory effect in macrophages leads to a positive effect on adipose biology [22]. Adipocyte-specific deletion of AMPKβ subunits had a deleterious effect on glucose tolerance and insulin sensitivity, in response to an HFD. Further analysis demonstrated that reduced energy expenditure in brown adipose tissue and hepatic steatosis, but not white adipose tissue inflammation, was responsible for the insulin resistant phenotype [23].

Leclerc and colleagues demonstrated that glucose could regulate and reduce AMPK activity in both human and rodent islets, which resulted in decreased insulin secretion [24]. AMPK can regulate insulin-induced gene expression of L-type pyruvate kinase and pre-proinsulin promotor in islets [25]. A study by Mottillo and colleagues utilized an inducible model for deletion of the two AMPK β subunits in adipocytes (i$\beta 1\beta 2$AKO) [26], whereby deletion of AMPK exacerbated the insulin resistant phenotype in terms of hepatic steatosis and glucose tolerance in response to an HFD [23]. AMPK knockout models are associated with differing phenotypes, based on which subunit is targeted. Those with the $\alpha 1/2$ subunit deficiency show increased lipolysis [27], whereas $\beta 1/2$ deficiency show reduced oxidative metabolism [23]. AMPK activation has been shown to inhibit the first phase of adipogenesis [28] but also increase peroxisome proliferator-activated receptor γ (PPAR-γ) expression [29] in other studies. Despite these confounding results, AMPK$\alpha 1$- and AMPK$\alpha 2$-deficient mice display hypertrophic adiposity following an HFD [27,28,30], but these were conducted in whole body knock-out models. Global AMPK$\alpha 1^{-/-}$ mice display increased adiposity, systemic insulin resistance and increased inflammation. When the bone marrow cells were deficient in AMPK$\alpha 1$, the mice had insulin resistance but no obesity, whereas adipocyte AMPK$\alpha 1$ deficiency had the same phenotype as the global knockout [31]. In genetically obese *ob/ob* mice, improved insulin sensitivity was a consequence of increased AMPK levels and glucose production in the liver, but no effects were observed in the skeletal muscle of these mice [20]. When AMPK levels were increased with metformin in HFD-rats switched to a chow diet, the metformin group had reduced weight gain and plasma glucose and triglyceride levels compared to a diet switch alone [17]. The authors speculated that metformin was increasing renal FAO by modulating AMPK/acetyl-CoA carboxylase (ACC) pathway and thereby reducing renal lipotoxicity. Metformin treatment in HFD mice also improves glucose metabolism and decreases the adipocyte size compared to HFD fed mice alone [32], thereby illustrating the multitude of effects that AMPK plays in whole body metabolism.

2.3. AMPK and Its Link to Cancer

As in obesity, AMPK is downregulated in cancer. AMPK has been implicated in cancer due to its effects on cellular growth and metabolism [33,34]. Tumor cells downregulate AMPK, and thus re-configure cellular metabolism towards glycolytic metabolism to enhance cell growth and proliferation. The mechanistic basis of which is complex and may be related to LKB1 dependent AMPK Thr^{172} phosphorylation to downregulate AMPK-mediated metabolism, over-signalling of the insulin/IGF1-regulated protein kinase Akt/PKB pathway, and/or other mechanisms [35]. LKB, the upstream kinase of AMPK activation, is a tumor suppressor but is mutated in cancer cells. Mammalian target of rapamycin (mTOR) controls cell growth and proliferation and is a target of AMPK. When LKB1 is deficient in cancer cells, mTOR activity inhibition is lost and thus cell proliferation is increased [36]. Cell growth is an ATP-consuming pathway and as such, AMPK directly inhibits mTOR activity. Fatty acid synthesis is increased in many cancers and through AMPKs ability to inhibit this process, AMPK is thought to have anti-cancer roles. Pharmacological AMPK activation, with metformin or salicylate, may protect against cancer initiation and development. Aspirin treated patients with CVD and metformin treated patients with type 2 diabetes have a lower incidence of cancer [37,38]. Whilst these studies seem promising, it is important to note that whether AMPK is a tumor suppressor or potential oncogene, at different cancer stages, is highly controversial and ripe for investigation [35,39,40]. Given the important regulatory role AMPK mediates within the context of metabolism, the next section will deal with the extent to which metabolism may alter inflammation. Also, the impact of nutrient-derived compounds is detailed in Section 4.

3. The Intersection between Metabolism and Inflammation

It is now appreciated that there is a complex and dynamic inter-relationship between metabolism and inflammation. Thus, rather than investigating simple pathways/biological processes, there is an intense interest in identifying potential hubs that co-regulate metabolism and inflammation, and in this context, AMPK is an interesting candidate. As stated previously, obesity can affect AMPK activity, but it is difficult to know if the HFD obesogenic effect reflects energy overload or a direct impact of dietary fatty acid (FA) composition. Interestingly, there is an increasing body of evidence that suggests that the nature of fatty acids may co-regulate inflammation and pAMPK expression. Attenuated pAMPK expression is associated with an inverse increase in inflammatory markers in both mouse models of obesity [16] and in obese humans [15]. In nutrient-rich conditions, and with inflammatory stimuli, AMPK phosphorylation and the activity of the α 1 subunit of AMPK is reduced. However, when AMPK is activated there is a subsequent reduction in both nuclear factor kappa B (NF-κB) and tumour necrosis factor α (TNF-α) secretion in macrophages [22].

3.1. AMPK and the NLRP3 Inflammasome

The nucleotide-binding domain, leucine-rich-containing family, pyrin domain-containing-3 (NLPR3) inflammasome, mediates interleukin-1β (IL-1β) production, but requires a metabolic product, reactive oxygen species (ROS), for its complete activation. NLRP3 has been shown to play a downstream role in metabolism, with $Nlrp3^{-/-}$ mice on an HFD displaying increased FAO [41]. Caspase-1 is required to process pro-IL-1β to IL-1β activation and *ob/ob* mice administered a caspase-1 inhibitor also display increased FAO [42]. AMPK can no longer reduce FAO when the regulatory β-1 subunit is deleted, thereby altering the tetrameric formation of AMPK. This occurs through changes in ACC phosphorylation and altered mitochondrial content [16]. AMPK can inhibit inducible nitric oxide synthase production as a way to reduce inflammation [43], while AMPK inhibition blocks autophagy increasing mitochondrial ROS production, an instrumental step in NLRP3 inflammasome activation [44]. ROS and nicotinamide adenine dinucleotide phosphate inhibition, as well as AMPK activation by AICAR, can all prevent the lipopolysaccharide (LPS)- and palmitic acid (PA)- induction of pro-IL-1β and mature IL-1β production [44]. NLRP3 inflammasome components and the activation of caspase-1 and IL-1β were significantly upregulated in monocyte-derived macrophages from T2D patients and were reduced following treatment with metformin, an insulin sensitizing drug that activates AMPK. The ability of metformin to reduce NLPR3 inflammasome activation was attributed to a reduction in mitochondrial ROS, a known activator of the inflammasome [45].

Metformin has been widely used in the treatment of T2D [3], wherein it induces not only metabolic but also positive anti-inflammatory effects. Metformin indirectly activates AMPK by inhibiting mitochondrial ATP synthesis [46], while decreasing the secretion of inflammatory markers, cyclooxygenase-2 and IL-1β, in the kidney of HFD-fed rats [17]. Citrate is involved in FA synthesis and has also been noted as being a potent inflammatory stimulus [47]. Accumulation of citrate can inhibit phosphofructokinase and back up the glycolytic pathway. The treatment of BMDM with citrate increased NLRP3-dependent IL-1β production and caspase-1 cleavage [48]. The isocitrate/pyruvate cycle has also been implicated in increasing glucose-stimulated insulin secretion, but this area of research warrants further investigation [47]. Treatment of BMDM with α-ketoglutarate has been shown to increase interleukin-10 (IL-10) but also significantly enhance IL-1β expression compared to LPS alone [49]. Conflicting reports however demonstrate that LPS-primed BMDM stimulated with α-ketoglutarate derivative attenuated the LPS-induced IL-1β mRNA and IL-1β protein, through reduced activity of prolyl hydroxylases, which is responsible for stabilising hypoxia-inducible factor 1-α (HIF-1α) protein [50].

3.2. Cellular Metabolism and Its Effect on Immune Cell Function

Early work suggested that sub-acute chronic inflammation that typified diet related diseases including obesity, T2D and atherosclerosis disrupted metabolism [51–53], wherein cytokines impeded a range of metabolic pathways such as insulin signaling, and reverse cholesterol transport [51–55]. However, it is now evident that there is a much more dynamic reciprocal regulatory relationship between metabolism and inflammation, wherein the nature of cellular metabolism determines immune cells functionality and response. At the simplest level, the cellular balance or switch between oxidative phosphorylation (OxPHOS) versus glucose dependent glycolytic metabolism, defines the functional nature of inactive versus activated immune and proliferating cells. Interestingly immune cells preferentially utilize different metabolic pathways depending on their pro-inflammatory, anti-inflammatory or resolving nature. The metabolic pathway selected reflects the functionality and requirements of the immune cell; whether it is a rapid increase in energy to carry out an innate immune response, or a prolonged availability of energy to enable adaptive responses, such as healing and repair [56]. An elegant combined transcriptomic and metabolomic analysis of BMDM functionality identified important transcriptional and metabolic rewiring during the macrophage polarization [57]. Inactive BMDM rely on the tricarboxylic acid (TCA) cycle and OxPHOS for cellular metabolism, wherein fatty acids and glutamine, but to a lesser extent glucose, are fully oxidized for maximal ATP generation. AMPK plays a key role in modulating metabolism and the balance between OxPHOS and glycolytic metabolism [11]. Upon activation, pro-inflammatory immune cells shift from utilizing the TCA cycle and generation of ATP via OxPHOS to anaerobic glycolysis [57]; this metabolic switch favors the use of glucose as the energy substrate [58]. The pro-inflammatory gram-negative bacterial product, LPS, is often used to induce a classically activated M1 macrophage M1 pro-inflammatory response, as it increases glycolysis and decreases oxygen consumption [50]. This results in a decrease in the AMP:ADP ratio, and a depletion of ATP. When this occurs, there is a shift in ATP production from OxPHOS to glycolysis, as glycolysis produces less ATP. As a result, AMPK would thus be activated to replenish the ATP levels [49]. Increased glycolysis is facilitated by a switch in the phosphofructokinase isoenzymes to its' ubiquitous isoform, leading to increased phosphorylation of fructose-6-phosphate, the rate limiting step in glycolysis [56].

3.3. Metabolic Reprogramming of T Cells

Metabolic reprogramming occurs in T cells, with regulatory T cells favoring OxPHOS and T effector cells switching to glycolysis upon activation. Treg cells have increased AMPK activation and in turn increased lipid oxidation, while T effector cells activate the opposing mTOR pathway [59]. T effector cells have significantly increased glucose transporter 1 (GLUT1) expression compared to T regulatory cells, to deal with the increased need for glycolysis. T regulatory cells require lipid metabolism for their differentiation which is mediated by AMPK where metformin administration to mice increases the number of T reg cells [60]. T effector cells can switch their metabolism under low glucose conditions by decreasing glycolysis and maintaining ATP levels by employing glutamine-dependent OxPHOS [61]. Specifically, LKB1 is critical for multiple T cell functions including development, viability, metabolism and activation. LKB1 was found to be involved in glucose metabolism as T cells lacking LKB1 had increased glycolysis and glucose uptake. A consequence of the disrupted metabolism in LKB1-deficient T cells was an increase in inflammatory $CD4^+$ and $CD8^+$ cells. In vivo studies in mice lacking T-cell specific AMPKα1 confirmed that AMPK activation was required for these processes [62]. AMPKα1 is required for T effector cells to maintain metabolic flexibility in times of nutrient stress and for T helper cell development in response to infection [61]. When T effector cells need to revert back to metabolically quiescent memory cells, AMPKα1null CD8 T cells were unable to make this transition, further confirming the role that AMPK plays in T cell metabolic switching [63]. Pozanski and colleagues have reviewed T cell and natural kill cell immune-metabolism, and recent emerging evidence suggests that AMPK may also play a role in the metabolic phenotype of natural killer cells [64]. Given the important inter-relationship between metabolism and inflammation, which is partly attributable to AMPK, we now address the extent to which this is sensitive to nutritional status and/or nutritional interventions. Thus, to explore the extent to which AMPK metabolic-inflammation may be either up- or down-regulated by different nutritional components, and how metabolic switching can also lead to beneficial changes in the cell [65]

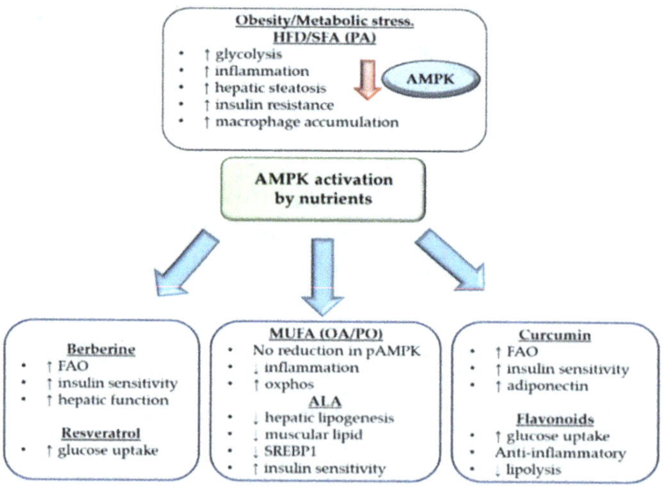

Figure 2. Whole body AMPK modulation and the impact of different nutrients on AMPK activation. Excess nutrient consumption through high fat diet (HFD) and obesity can downregulate AMP-Activated Protein Kinase (AMPK) expression and cause dysregulated metabolism, inflammation and insulin resistance. Nutrients, including monounsaturated fatty acids (MUFA), α linoleic acid (ALA), berberine, resveratrol, curcumin and flavonoids can all activate AMPK and downstream positive effects in relation to improved mitochondrial metabolism, improved liver function and reduced inflammation. Therefore, modulation of AMPK through nutrient intervention can improve whole body metabolism. An $\uparrow$ arrow represents an upregulation of the process or increased expression and $\downarrow$ represents a downregulation of the process or decreased expression. AMPK = AMP-Activated Protein Kinase, MUFA = Monounsaturated fatty acids, OA = Oleic acid, PO = Palmitoleic acid, ALA = α linoleic acid, SREBP1 = Sterol regulatory element-binding protein 1c, FAO = fatty acid oxidation.

4. Modulation of AMPK Activation by Nutrients

4.1. Fatty Acids Differentially Affect AMPK Function

One of the main drivers of obesity is the excessive consumption of dietary fat. Dietary fat can have widespread negative effects in relation to inflammation, insulin resistance and metabolism, depending on the nature of the fatty acids. Briefly, fatty acids can be either saturated or unsaturated, wherein the latter are sub-divided into monounsaturated (MUFA) or polyunsaturated (PUFA) fatty acids depending on the number of double bonds in the carbon chain. The inclusion of double bonds and the degree of unsaturation has very different effects on metabolism and inflammation. From the experimental point of view, rodent high fat diets are usually enriched with lard which is rich in the saturated fatty acid (SFA) lauric acid (C12:0) or palm oil which is enriched in another SFA palmitic acid (PA) (C16:0). Whilst it has long been acknowledged that obesity is associated with reduced pAMPK expression and a pro-inflammatory phenotype, the true nature of the apparent reciprocal regulation and putative impact of different dietary components is ill defined.

In vitro studies suggest that different fatty acids modulate AMPK mediated inflammation. AMPK$\beta1^{-/-}$ BMDM display a predominant M1 profile, which was further exacerbated with two saturated fatty acids, PA and stearic acid (C18:0). Hematopoietic deletion of AMPK$\beta1^{-/-}$ was sufficient to induce systemic inflammation in an HFD setting [16]. PA-treated macrophages display increased inflammatory markers, but this is reduced with metformin co-treatment, a phenomenon mimicked in vivo in HFD mice treated with metformin. When AMPK was inhibited with compound C (CC) and included with metformin in vitro, inflammatory markers were increased, thus confirming that metformin is mediating its anti-inflammatory effect through AMPK activation [32]. Interestingly the AMPK activator, AICAR impeded LPS- and FA-induced cytokine response, in part through NF-κB inhibition [16]. However, AICAR has its non-specific limitations, it can inhibit NF-κB DNA binding in human macrophages without AMPK activation [66].

Unsaturated fatty acid may have less of an adverse effect on AMPK mediated metabolism and inflammation (Figure 2). Two groups have compared the impact of palmitic acid and oleic acid (OA) on AMPK. In vitro, direct stimulation with PA reduces pAMPK expression in macrophages. In contrast two MUFA, both palmitoleic acid (PO) (C16:1) and OA (C18:1) do not [2,67]. In vivo, feeding SFA enriched HFD amplifies insulin resistance, inflammation and reduces pAMPK expression. In contrast, feeding an HFD derived from the MUFA, OA did not reduce adipose pAMPK, despite obesity, compared to an SFA enriched HFD [2]. In order to define if the apparent co-regulation of IL-1β inflammation and AMPK was fatty acid dependent, a series of in vitro experiments using CC and AICAR demonstrated that when AMPK was inhibited with CC, IL-1β was increased in BMDM. Conversely, AMPK activation with AICAR in conjunction with PA could inhibit IL-1β secretion [2]. When AMPK is activated pharmacologically through AICAR or by overexpression, it reduces TNF-α- and PA- induced increases in NF-κB expression in endothelial cells [68]. Alpha-linoleic acid (ALA) is n-3 polyunsaturated fatty acid (n-3 PUFA) (C18:3 n-3 *cis*-Δ^3) which acts as a naturally occurring antioxidant and is a cofactor for mitochondrial respiratory enzymes [69]. It is important to note that α- linoleic acid is not a fatty acid per se, but an organosulfur compound derived from the fatty acid caprylic acid (C8:0). ALA was shown to improve both insulin-stimulated glucose uptake and FAO in the skeletal muscle of diabetes prone, obese rats. This was attributed to AMPK activation and reduced lipid accumulation, as dominant-negative AMPK prevented the positive effects of ALA [70]. Subsequent studies found an improvement in whole body glucose tolerance in HFD + ALA rats but did not attribute this to an AMPK mediated lowering of intramuscular lipid [71]. Within the liver of HFD rats, ALA can reduce hepatic lipogenesis through reduced SREBP-1 expression. As a result, hepatic steatosis was attenuated, and this was mediated in part by AMPK activation [72]. ALA can directly increase AMPK in hepatocytes but its lipid lowering and beneficial effects on hepatic metabolism were not AMPK-dependent [73]. In terms of understanding the relative impact of fatty acids versus glucose, Kratz and colleagues identified a distinct population of metabolically activated macrophages (MMe), following a palmitate, glucose and insulin challenge within the adipose tissue. MMe macrophages display attributes of both alternatively activated M2 anti-inflammatory markers (lipid metabolism) but secreted M1-associted pro-inflammatory cytokines [74]. It was noteworthy that PA was more potent in this system compared to glucose and insulin metabolic challenges.

4.2. Reversing Metabolic Inflammation through AMPK

The concept of reversing HFD induced metabolic-inflammation is intriguing and involves AMPK. Bone marrow derived macrophages (BMDM) derived from HFD mice retained a "dietary memory" with increased mRNA levels of TNF-α, interleukin-6 (IL-6) and nitric oxide synthase 2, compared to those of a low-fat diet [75], which could be reversed with incubation of the n-6 MUFA, *cis*-PO (C16:1 *cis*-Δ^9). This PO mediated anti-inflammatory effect was AMPK-dependent. Both OA and PO maintain AMPK expression, and this is thought to mediate their anti-inflammatory effects [75]. Work in Thp-1 human macrophages containing a knockdown of AMPK showed the same inflammatory response by OA when AMPK could not function [2]. Co-incubation of PO with PA, restored pAMPK levels to that of PO alone in BMDM. AMPK inhibition results in PO-induced increases in nitric oxide expression [75]. PA, in a similar manner to the inflammatory stimulus LPS, can increase the cells use of glycolysis in BMDM, while exposure to PO increased OxPHOS. Furthermore, when PO was co-incubated with PA, they displayed similar oxygen consumption levels compared to PO alone. This was thought to reflect the increased M2 phenotype induced through PO incubation [75]. Anti-inflammatory M2 macrophages employ the OxPHOS pathway, which is capable of producing more ATP than glycolysis [57]. M2 macrophages rely on FA as an energy source [58]. The inhibition of various steps of the mitochondrial OxPHOS pathway were able to reduce the expression of arginine 1, a common M2 marker, and abrogate the anti-inflammatory potential of these cells, but had no effect on classically activated macrophages [58]. Activation of pyruvate kinase isozymes M2 enhanced IL-10 secretion in LPS-stimulated BMDM, favouring an M2 phenotype and reduced M1 polarisation [65]. Macrophage M1 polarisation was reduced through AMPK activation by fibronectin type III domain-containing 5 (FNDC5), and worsened when FNDC5 was inhibited, thus lowering AMPK levels in the adipose tissue of HFD-fed mice [76]. Palmitate-treated macrophages treated with AICAR was able to reduce the number of M1 macrophages while subsequently increasing the M2 macrophage population [32]. Microarray analysis of adipose tissue following an 8wk SFA diet showed upregulation of pathways involved in immune function and inflammation. Consumption of a MUFA diet showed no change or downregulation of the same pathways [77]. MUFA-HFD mice display increased energy expenditure compared to SFA-HFD mice, and given the role of AMPK as an energy sensor, the involvement of AMPK in regulating this process cannot be ruled out [2].

4.3. AMPK Regulation by Resveratrol, Berberine and Curcumin

Resveratrol, a natural phytochemical, increases glucose uptake in insulin-resistant 3T3-L1 adipocytes by increasing the phosphorylation of pAkt and downstream AMPK activation. A similar effect was noted in vivo in resveratrol treated KKAy mice who demonstrated increased pAkt and elevated AMPK expression [78]. Resveratrol can inhibit cell proliferation in the HT-29 cancer cell line, which was AMPK-dependent. The ability of resveratrol to increase ROS production was elucidated as the mechanism of AMPK activation [79]. Berberine is a naturally occurring plant-derived compound that can indirectly activate AMPK. Berberine can increase energy expenditure and improve liver function in obese mice by stimulating fatty acid oxidation and inhibiting lipogenic genes [80]. Berberine inhibits complex I of the mitochondria to reduce respiration in cultured myotubes and muscle mitochondria. Furthermore, a derivative of berberine demonstrated improved insulin sensitivity and reduced adiposity in vivo in HFD rats [81], owing to the possibility of nutrient intervention treatments. Berberine has positive effects on cancer by inhibiting their metastatic potential and reducing the inflammatory COX-2 pathway by increasing AMPK phosphorylation through increased ROS production [82]. Curcumin is a naturally occurring polyphenol and is the active constituent of turmeric. In a review by Shehzad and colleagues detailing curcumin studies over a 10-year period, this polyphenol was found to have multiple anti-inflammatory and insulin sensitizing effects, involving the suppression of anti-inflammatory transcription factors, upregulation of adiponectin and interactions with glucose and insulin signal transduction pathways [83]. HFD supplemented with 0.15% curcumin reduced body weight, adiposity, serum lipid and glucose levels, and insulin resistance in mice compared to HFD alone controls. Furthermore, beneficial effects on hepatic lipid accumulation and metabolism were noted with reductions in ACC and fatty acid synthase but increased FAO [84]. Curcumin can also increase FAO in muscle which was mediated in part by AMPK and improved palmitate-induced insulin resistance in L6-myotubes [85]. Curcumin demonstrates chemo preventative properties by inducing cell death through p53 phosphorylation, but this is prevented if AMPK action is inhibited with CC treatment in an ovarian cancer cell line [86].

4.4. AMPK Regulation by Flavonoids

Plant flavonoids are plant secondary metabolites known to have beneficial health effects [87]. Genistein, a soy isoflavone, can reduce blood glucose in a KK-Ay/Ta Jcl T2D mouse model. When used in vitro in L6 myotubes, genistein stimulates phosphorylation of AMPK and increases glucose uptake by increasing GLUT4 translocation to the plasma membrane, which was prevented when AMPK was inhibited with CC [88]. Genistein has anti-inflammatory properties in macrophages through inhibition of NF-κB mediated by AMPK [89]. In a diabetic rat model (Zucker *fa/fa*) fed soy protein, AMPK and FAO were increased in the skeletal muscle and associated with decreased weight gain and improved glucose and triglyceride levels, demonstrating the positive effect that flavonoids can have on overall metabolism [90]. Apigenin, another member of the flavonoid family, dose-dependently increases AMPK expression, while inhibiting adipogenic gene expression and lipolysis in 3T3-L1 adipocytes [91]. In vivo administration of apigenin to HFD-mice reduced the HFD-induced weight gain, increased serum glucose and muscle pro-inflammatory cytokine expression. The apigenin treated mice displayed increased OxPHOS and citrate synthase activity compared to the muscle HFD alone mice, through improvements in mitochondrial biogenesis and mitochondrial function [92]. Similar improvements were observed by Jung and colleagues, with additional benefits observed in relation to improved hepatic steatosis, reduced plasma FFA, and downregulation of lipogenic genes in the liver of apigenin + HFD mice [93]. Flavonoids have been ascribed cancer preventative roles through the induction of apoptotic pathways mediated through AMPK signaling pathways [94]. The ability of FA to undergo metabolic reconfiguration, combined with the anti-oxidant and anti-inflammatory properties of nutrients and how these are involved in the diet-induced modulation of inflammation therefore requires further investigation. FA are an important source of metabolites and no doubt their metabolism will feed more into the immuno-metabolism circuit, however to date the nature and extent to which dietary FA and/or nutritional status can modulate this dynamic two-way process requires definition.

5. Conclusions and Future Perspectives

AMPK is an exciting target given its co-regulatory role with respect to metabolism and inflammation. It is evident that Western dietary elements, which are obesogenic in nature, attenuate AMPK, the metabolic impact of which probably explains the sub-acute pro-inflammatory phenotype that typifies obesity, T2D and CVD. The challenge remains to discern if/how certain nutrients either up- or down-regulate cellular AMPK status. It is evident that feeding SFA are particularly deleterious, and therefore negatively impact upon metabolism and sub-acute chronic inflammation that typifies obesity T2D. Speculatively it is probable that high-fructose diets will also have deleterious effect on AMPK, given the fact that feeding fructose increases endogenous synthesis of SFA. Conversely, there is some evidence to suggest that feeding high fat diets rich in MUFA and PUFA, or supplemented with natural AMPK agonists, such as resveratrol and berberine, do not attenuate AMPK. Furthermore, sirtuin 1 (SIRT) is another nutrient sensor that has widespread effects on metabolism in response to caloric restriction and may also be involved in regulating metabolism and inflammation [95]. Whilst the paradigm of nutritional enhancement of AMPK activity is worthy of investigation in light of the obesity epidemic, to date the impact of these nutritional interventions have been observed in mice. Therefore, we need to define if/how this translates to humans.

Author Contributions: C.L.L. wrote the review. H.M.R. advised on content, wrote and critically evaluated the manuscript. All authors approved the final submission.

Abbreviations

ACC	Acetyl-CoA carboxylase
ADP	Adenosine diphosphate
AICAR	5′-aminoimidazole-4-carboxamide ribonucleotide
Akt	Protein kinase B
ALA	A-linoleic acid
AMP	Adenosine monophosphate
AMPK	AMP-Activated Protein Kinase
AMPKK	AMPK kinase
ATP	Adenosine triphosphate
BMDM	Bone marrow derived macrophages
CaMKK	Ca^{2+}/calmodulin-dependent protein kinase kinases
CC	Compound C
CVD	Cardiovascular disease
FA	Fatty acid
FAO	Fatty acid oxidation
FBP	Fructose-1,6-bisphosphate
FNDC5	Fibronectin type III domain-containing 5
Glut1	Glucose transporter 1
HIF-1α	Hypoxia-inducible factor 1-α
HFD	High fat diet
IL-1β	Interleukin-1β
IL-6	Interleukin 6
IL-10	Interleukin 10
LKB1	Liver kinase B1
LPS	Lipopolysaccharide
Mme	Metabolically activated macrophages
mTOR	Mammalian target of rapamycin
MUFA	Monounsaturated fatty acids
OA	Oleic acid
OxPHOS	Oxidative phosphorylation
P	Phosphorylation
PA	Palmitic acid
PO	Palmitoleic acid
PPAR-γ	Peroxisome proliferator-activated receptor γ
PUFA	Polyunsaturated fatty acid

ROS	Reactive oxygen species
SAT	Subcutaneous adipose tissue
SFA	Saturated fatty acid
SIRT1	Sirtuin
SREBP-1	Sterol regulatory element-binding protein 1
T2D	Type 2 diabetes
TCA	Tricarboxylic acid
Thr	Threonine
TNF-α	Tumour necrosis factor α
VAT	Visceral adipose tissue
α	Alpha
β	Beta
γ	Gamma

References

1. Hardie, D.G. The AMP-activated protein kinase pathway—New players upstream and downstream. *J. Cell Sci.* **2004**, *117*, 5479–5487. [CrossRef] [PubMed]
2. Finucane, O.M.; Lyons, C.L.; Murphy, A.M.; Reynolds, C.M.; Klinger, R.; Healy, N.P.; Cooke, A.A.; Coll, R.C.; McAllan, L.; Nilaweera, K.N.; et al. Monounsaturated fatty acid-enriched high-fat diets impede adipose NLRP3 inflammasome-mediated IL-1? Secretion and insulin resistance despite obesity. *Diabetes* **2015**, *64*, 2116–2128. [CrossRef] [PubMed]
3. Knowler, W.C.; Connor, E.B.; Fowler, S.E.; Hamman, R.F.; Lachin, J.M.; Walker, E.A.; Nathan, D.M. Reduction in the incidence of type 2 diabetes with lifestyle intervention or metformin. *N. Engl. J. Med.* **2002**, *346*, 393–403. [PubMed]
4. Stein, S.C.; Woods, A.; Jones, J.A.; Davison, M.D.; Carling, D. The regulation of AMP-activated protein kinase by phosphorylation. *Biochem. J.* **2000**, *345*, 437–443. [CrossRef] [PubMed]
5. Hawley, S.A.; Davison, M.; Woods, A.; Davies, S.P.; Beri, R.K.; Carling, D.; Hardie, D.G. Characterization of the AMP-activated protein kinase kinase from rat liver and identification of threonine 172 as the major site at which it phosphorylates AMP-activated protein kinase. *J. Biol. Chem.* **1996**, *271*, 27879–27887. [CrossRef]
6. Xiao, B.; Sanders, M.J.; Underwood, E.; Heath, R.; Mayer, F.V.; Carmena, D.; Jing, C.; Walker, P.A.; Eccleston, J.F.; Haire, L.F.; et al. Structure of mammalian AMPK and its regulation by ADP. *Nature* **2011**, *472*, 230–233. [CrossRef] [PubMed]
7. Oakhill, J.S.; Chen, Z.P.; Scott, J.W.; Steel, R.; Castelli, L.A.; Ling, N.; Macaulay, S.L.; Kemp, B.E. β-Subunit myristoylation is the gatekeeper for initiating metabolic stress sensing by AMP-activated protein kinase (AMPK). *Proc. Natl. Acad. Sci. USA* **2010**, *107*, 19237–19241. [CrossRef] [PubMed]
8. Valentine, R.J.; Coughlan, K.A.; Ruderman, N.B.; Saha, A.K. Insulin inhibits AMPK activity and phosphorylates AMPK Ser485/491 through Akt in hepatocytes, myotubes and incubated rat skeletal muscle. *Arch. Biochem. Biophys.* **2014**, *562*, 62–69. [CrossRef] [PubMed]
9. Mihaylova, M.M.; Shaw, R.J. The AMPK signalling pathway coordinates cell growth, autophagy and metabolism. *Nat. Cell Biol.* **2011**, *13*, 1016–1023. [CrossRef] [PubMed]
10. Hardie, D.G.; Ross, F.A.; Hawley, S.A. AMPK: A nutrient and energy sensor that maintains energy homeostasis. *Nat. Rev. Mol. Cell Biol.* **2012**, *13*, 251–262. [CrossRef] [PubMed]
11. Herzig, S.; Shaw, R.J. AMPK: Guardian of metabolism and mitochondrial homeostasis. *Nat. Rev. Mol. Cell Biol.* **2017**, *19*, 121–135. [CrossRef] [PubMed]
12. O'Neill, L.A.J.; Hardie, D.G. Metabolism of inflammation limited by AMPK and pseudo-starvation. *Nature* **2013**, *493*, 346–355. [CrossRef] [PubMed]
13. O'Neill, H.M.; Holloway, G.P. AMPK regulation of fatty acid metabolism and mitochondrial biogenesis: Implications for obesity. *Mol. Cell. Endocrinol.* **2013**, *366*, 135–151. [CrossRef] [PubMed]
14. Zhang, C.S.; Hawley, S.A.; Zong, Y.; Li, M.; Wang, Z.; Gray, A.; Ma, T.; Cui, J.; Feng, J.W.; Zhu, M.; et al. Fructose-1,6-bisphosphate and aldolase mediate glucose sensing by AMPK. *Nature* **2017**, *548*, 112–116. [CrossRef] [PubMed]

15. Gauthier, M.S.; O'Brien, E.L.; Bigornia, S.; Mott, M.; Cacicedo, J.M.; Xu, X.J.; Gokce, N.; Apovian, C.; Ruderman, N. Decreased AMP-activated protein kinase activity is associated with increased inflammation in visceral adipose tissue and with whole-body insulin resistance in morbidly obese humans. *Biochem. Biophys. Res. Commun.* **2011**, *404*, 382–387. [CrossRef] [PubMed]
16. Galic, S.; Fullerton, M.D.; Schertzer, J.D.; Sikkema, S.; Marcinko, K.; Walkley, C.R.; Izon, D.; Honeyman, J.; Chen, Z.P.; van Denderen, B.J.; et al. Hematopoietic AMPK β1 reduces mouse adipose tissue macrophage inflammation and insulin resistance in obesity. *J. Clin. Investig.* **2011**, *121*, 4903–4915. [CrossRef] [PubMed]
17. Tikoo, K.; Sharma, E.; Amara, V.R.; Pamulapati, H.; Dhawale, V.S. Metformin improves metabolic memory in high fat diet (HFD)-induced renal dysfunction. *J. Biol. Chem.* **2016**, *291*, 21848–21856. [CrossRef] [PubMed]
18. Yavari, A.; Stocker, C.J.; Ghaffari, S.; Wargent, E.T.; Steeples, V.; Czibik, G.; Pinter, K.; Bellahcene, M.; Woods, A.; Martínez de Morentin, P.B.; et al. Chronic Activation of γ2 AMPK Induces Obesity and Reduces β Cell Function. *Cell Metab.* **2016**, *23*, 821–836. [CrossRef] [PubMed]
19. Arita, Y.; Kihara, S.; Ouchi, N.; Takahashi, M.; Maeda, K.; Miyagawa, J.; Hotta, K.; Shimomura, I.; Nakamura, T.; Miyaoka, K.; et al. Paradoxical Decrease of an Adipose-Specific Protein, Adiponectin, in Obesity. *Biochem. Biophys. Res. Commun.* **1999**, *257*, 79–83. [CrossRef] [PubMed]
20. Kadowaki, T.; Yamauchi, T.; Kubota, N.; Hara, K.; Ueki, K.; Tobe, K. Adiponectin and adiponectin receptors in insulin resistance, diabetes, and the metabolic syndrome. *J. Clin. Investig.* **2006**, *116*, 1784–1792. [CrossRef] [PubMed]
21. Declèves, A.-E.; Mathew, A.V.; Cunard, R.; Sharma, K. AMPK mediates the initiation of kidney disease induced by a high-fat diet. *J. Am. Soc. Nephrol.* **2011**, *22*, 1846–1855. [CrossRef] [PubMed]
22. Yang, Z.; Kahn, B.B.; Shi, H.; Xue, B.Z. Macrophage α1 AMP-activated protein kinase (α1 AMPK) antagonizes fatty acid-induced inflammation through SIRT1. *J. Biol. Chem.* **2010**, *285*, 19051–19059. [CrossRef] [PubMed]
23. Mottillo, E.P.; Desjardins, E.M.; Crane, J.D.; Smith, B.K.; Green, A.E.; Ducommun, S.; Henriksen, T.I.; Rebalka, I.A.; Razi, A.; Sakamoto, K.; et al. Lack of Adipocyte AMPK Exacerbates Insulin Resistance and Hepatic Steatosis through Brown and Beige Adipose Tissue Function. *Cell Metab.* **2016**, *24*, 118–129. [CrossRef] [PubMed]
24. Leclerc, I.; Woltersdorf, W.W.; da Silva Xavier, G.; Rowe, R.L.; Cross, S.E.; Korbutt, G.S.; Rajotte, R.V.; Smith, R.; Rutter, G.A. Metformin, but not leptin, regulates AMP-activated protein kinase in pancreatic islets: Impact on glucose-stimulated insulin secretion. *Am. J. Physiol. Endocrinol. Metab.* **2004**, *286*, E1023–E1031. [CrossRef] [PubMed]
25. Da Silva Xavier, G.; Leclerc, I.; Varadi, A.; Tsuboi, T.; Moule, S.K.; Rutter, G.A. Role of AMP-activated protein kinase in the regulation by glucose of islet β cell gene expression. *Proc. Natl. Acad. Sci. USA* **2000**, *97*, 4023–4028. [CrossRef] [PubMed]
26. Mottillo, E.P.; Balasubramanian, P.; Lee, Y.-H.; Weng, C.; Kershaw, E.E.; Granneman, J.G. Coupling of lipolysis and de novo lipogenesis in brown, beige, and white adipose tissues during chronic β3-adrenergic receptor activation. *J. Lipid Res.* **2014**, *55*, 2276–2286. [CrossRef] [PubMed]
27. Kim, S.J.; Tang, T.; Abbott, M.; Viscarra, J.A.; Wang, Y.; Sul, H.S. AMPK phosphorylates desnutrin/ATGL and HSL to regulate lipolysis and fatty acid oxidation within adipose tissue. *Mol. Cell. Biol.* **2016**, *36*, 1961–1976. [CrossRef] [PubMed]
28. Habinowski, S.A.; Witters, L.A. The Effects of AICAR on Adipocyte Differentiation of 3T3-L1 Cells. *Biochem. Biophys. Res. Commun.* **2001**, *286*, 852–856. [CrossRef] [PubMed]
29. Ceddia, R.B. The role of AMP-activated protein kinase in regulating white adipose tissue metabolism. *Mol. Cell. Endocrinol.* **2013**, *366*, 194–203. [CrossRef] [PubMed]
30. Villena, J.A.; Viollet, B.; Andreelli, F.; Kahn, A.; Vaulont, S.; Sul, H.S. Induced Adiposity and Adipocyte Hypertrophy in Mice Lacking the AMP-Activated Protein Kinase-alpha2 Subunit. *Diabetes* **2004**, *53*, 2242–2249. [CrossRef] [PubMed]
31. Zhang, W.; Zhang, X.; Wang, H.; Guo, X.; Li, H.; Wang, Y.; Xu, X.; Tan, L.; Mashek, M.T.; Zhang, C.; et al. AMP-activated protein kinase α1 protects against diet-induced insulin resistance and obesity. *Diabetes* **2012**, *61*, 3114–3125. [PubMed]
32. Jing, Y.; Wu, F.; Li, D.; Yang, L.; Li, Q.; Li, R. Metformin improves obesity-associated inflammation by altering macrophages polarization. *Mol. Cell. Endocrinol.* **2018**, *461*, 256–264. [CrossRef] [PubMed]

33. Li, W.; Saud, S.M.; Young, M.R.; Chen, G.; Hua, B. Targeting AMPK for cancer prevention and treatment. *Oncotarget* **2015**, *6*, 7365–7378. [CrossRef] [PubMed]
34. Motoshima, H.; Goldstein, B.J.; Igata, M.; Araki, E. AMPK and cell proliferation—AMPK as a therapeutic target for atherosclerosis and cancer. *J. Physiol.* **2006**, *574*, 63–71. [CrossRef] [PubMed]
35. Hardie, D.G. Molecular Pathways: Is AMPK a Friend or a Foe in Cancer? *Clin. Cancer Res.* **2015**, *21*, 3836–3840. [CrossRef] [PubMed]
36. Carretero, J.; Medina, P.P.; Blanco, R.; Smit, L.; Tang, M.; Roncador, G.; Maestre, L.; Conde, E.; Lopez-Rios, F.; Clevers, H.C.; et al. Dysfunctional AMPK activity, signalling through mTOR and survival in response to energetic stress in LKB1-deficient lung cancer. *Oncogene* **2007**, *26*, 1616–1625. [CrossRef] [PubMed]
37. Rothwell, P.M.; Fowkes, F.G.; Belch, J.F.; Ogawa, H.; Warlow, C.P.; Meade, T.W. Effect of daily aspirin on long-term risk of death due to cancer: Analysis of individual patient data from randomised trials. *Lancet* **2011**, *377*, 31–41. [CrossRef]
38. Evans, J.M.M.; Donnelly, L.A.; Emslie-Smith, A.M.; Alessi, D.R.; Morris, A.D. Metformin and reduced risk of cancer in diabetic patients. *BMJ* **2005**, *330*, 1304–1305. [CrossRef] [PubMed]
39. Liang, J.; Mills, G.B. AMPK: A Contextual Oncogene or Tumor Suppressor? *Cancer Res.* **2013**, *73*, 2929–2935. [CrossRef] [PubMed]
40. Yang, Y. Metformin for cancer prevention. *Front. Med.* **2011**, *5*, 115–117. [CrossRef] [PubMed]
41. Vandanmagsar, B.; Youm, Y.H.; Ravussin, A.; Galgani, J.E.; Stadler, K.; Mynatt, R.L.; Ravussin, E.; Stephens, J.M.; Dixit, V.D. The NLRP3 inflammasome instigates obesity-induced inflammation and insulin resistance. *Nat. Med.* **2011**, *17*, 179–188. [CrossRef] [PubMed]
42. Stienstra, R.; Joosten, L.A.; Koenen, T.; van Tits, B.; van Diepen, J.A.; van den Berg, S.A.; Rensen, P.C.; Voshol, P.J.; Fantuzzi, G.; Hijmans, A.; et al. The inflammasome-mediated caspase-1 activation controls adipocyte differentiation and insulin sensitivity. *Cell Metab.* **2010**, *12*, 593–605. [CrossRef] [PubMed]
43. Pilon, G.; Dallaire, P.; Marette, A. Inhibition of Inducible Nitric-oxide Synthase by Activators of AMP-activated Protein Kinase. *J. Biol. Chem.* **2004**, *279*, 20767. [CrossRef] [PubMed]
44. Wen, H.; Gris, D.; Lei, Y.; Jha, S.; Zhang, L.; Huang, M.T.; Brickey, W.J.; Ting, J.P. Fatty acid-induced NLRP3-ASC inflammasome activation interferes with insulin signaling. *Nat. Immunol.* **2011**, *12*, 408–415. [CrossRef] [PubMed]
45. Lee, H.M.; Kim, J.J.; Kim, H.J.; Shong, M.; Ku, B.J.; Jo, E.K. Upregulated NLRP3 inflammasome activation in patients with type 2 diabetes. *Diabetes* **2013**, *62*, 194–204. [CrossRef] [PubMed]
46. Hawley, S.A.; Ross, F.A.; Chevtzoff, C.; Green, K.A.; Evans, A.; Fogarty, S.; Towler, M.C.; Brown, L.J.; Ogunbayo, O.A.; Evans, A.M.; et al. Use of cells expressing γ subunit variants to identify diverse mechanisms of AMPK activation. *Cell Metab.* **2010**, *11*, 554–565. [CrossRef] [PubMed]
47. Iacobazzi, V.; Infantino, V. Citrate—New functions for an old metabolite. *Biol. Chem.* **2014**, *395*, 387–399. [CrossRef] [PubMed]
48. Wolf, A.J.; Reyes, C.N.; Liang, W.; Becker, C.; Shimada, K.; Wheeler, M.L.; Cho, H.C.; Popescu, N.I.; Coggeshall, K.M.; Arditi, M.; et al. Hexokinase Is an Innate Immune Receptor for the Detection of Bacterial Peptidoglycan. *Cell* **2016**, *166*, 624–636. [CrossRef] [PubMed]
49. Mills, E.L.; Kelly, B.; Logan, A.; Costa, A.S.H.; Varma, M.; Bryant, C.E.; Tourlomousis, P.; Däbritz, J.H.M.; Gottlieb, E.; Latorre, I.; et al. Succinate Dehydrogenase Supports Metabolic Repurposing of Mitochondria to Drive Inflammatory Macrophages. *Cell* **2016**, *167*, 457–470. [CrossRef] [PubMed]
50. Tannahill, G.M.; Curtis, A.M.; Adamik, J.; Palsson-McDermott, E.M.; McGettrick, A.F.; Goel, G.; Frezza, C.; Bernard, N.J.; Kelly, B.; Foley, N.H.; et al. Succinate is an inflammatory signal that induces IL-1β through HIF-1α. *Nature* **2013**, *496*, 238–242. [CrossRef] [PubMed]
51. McArdle, M.A.; Finucane, O.M.; Connaughton, R.M.; McMorrow, A.M.; Roche, H.M. Mechanisms of obesity-induced inflammation and insulin resistance: Insights into the emerging role of nutritional strategies. *Front. Endocrinol.* **2013**, *4*, 52. [CrossRef] [PubMed]
52. Cooke, A.A.; Connaughton, R.M.; Lyons, C.L.; McMorrow, A.M.; Roche, H.M. Fatty acids and chronic low grade inflammation associated with obesity and the metabolic syndrome. *Eur. J. Pharmacol.* **2016**, *785*, 207–214. [CrossRef] [PubMed]
53. Murphy, A.M.; Lyons, C.L.; Finucane, O.M.; Roche, H.M. Interactions between differential fatty acids and inflammatory stressors-impact on metabolic health. *Prostaglandins Leukot. Essent. Fat. Acids* **2015**, *92*, 49–55. [CrossRef] [PubMed]

54. Reynolds, C.M.; McGillicuddy, F.C.; Harford, K.A.; Finucane, O.M.; Mills, K.H.; Roche, H.M. Dietary saturated fatty acids prime the NLRP3 inflammasome via TLR4 in dendritic cells-implications for diet-induced insulin resistance. *Mol. Nutr. Food Res.* **2012**, *6*, 1212–1222. [CrossRef] [PubMed]
55. O'Reilly, M.; Dillon, E.; Guo, W.; Finucane, O.; McMorrow, A.; Murphy, A.; Lyons, C.; Jones, D.; Ryan, M.; Gibney, M.; et al. High-Density Lipoprotein Proteomic Composition, and not Efflux Capacity, Reflects Differential Modulation of Reverse Cholesterol Transport by Saturated and Monounsaturated Fat Diets. *Circulation* **2016**, *133*, 1838–1850. [CrossRef] [PubMed]
56. Rodríguez-Prados, J.C.; Través, P.G.; Cuenca, J.; Rico, D.; Aragonés, J.; Martín-Sanz, P.; Cascante, M.; Boscá, L. Substrate fate in activated macrophages: A comparison between innate, classic, and alternative activation. *J. Immunol.* **2010**, *185*, 605–614. [CrossRef] [PubMed]
57. Jha, A.K.; Huang, S.C.; Sergushichev, A.; Lampropoulou, V.; Ivanova, Y.; Loginicheva, E.; Chmielewski, K.; Stewart, K.M.; Ashall, J.; Everts, B.; et al. Network Integration of Parallel Metabolic and Transcriptional Data Reveals Metabolic Modules that Regulate Macrophage Polarization. *Immunity* **2015**, *42*, 419–430. [CrossRef] [PubMed]
58. Vats, D.; Mukundan, L.; Odegaard, J.I.; Zhang, L.; Smith, K.L.; Morel, C.R.; Wagner, R.A.; Greaves, D.R.; Murray, P.J.; Chawla, A. Oxidative metabolism and PGC-1beta attenuate macrophage-mediated inflammation. *Cell Metab.* **2006**, *4*, 13–24. [CrossRef] [PubMed]
59. MacIver, N.J.; Michalek, R.D.; Rathmell, J.C. Metabolic regulation of T lymphocytes. *Annu. Rev. Immunol.* **2013**, *31*, 259–283. [CrossRef] [PubMed]
60. Michalek, R.D.; Gerriets, V.A.; Jacobs, S.R.; Macintyre, A.N.; MacIver, N.J.; Mason, E.F.; Sullivan, S.A.; Nichols, A.G.; Rathmell, J.C. Cutting edge: Distinct glycolytic and lipid oxidative metabolic programs are essential for effector and regulatory CD4+ T cell subsets. *J. Immunol.* **2011**, *186*, 3299–3303. [CrossRef] [PubMed]
61. Blagih, J.; Coulombe, F.; Vincent, E.E.; Dupuy, F.; Galicia-Vázquez, G.; Yurchenko, E.; Raissi, T.C.; van der Windt, G.J.; Viollet, B.; Pearce, E.L.; et al. The Energy Sensor AMPK Regulates T Cell Metabolic Adaptation and Effector Responses In Vivo. *Immunity* **2015**, *42*, 41–54. [CrossRef] [PubMed]
62. MacIver, N.J.; Blagih, J.; Saucillo, D.C.; Tonelli, L.; Griss, T.; Rathmell, J.C.; Jones, R.G. The Liver Kinase B1 Is a Central Regulator of T Cell Development, Activation, and Metabolism. *J. Immunol.* **2011**, *187*, 4187–4198. [CrossRef] [PubMed]
63. Rolf, J.; Zarrouk, M.; Finlay, D.K.; Foretz, M.; Viollet, B.; Cantrell, D.A. AMPKα1: A glucose sensor that controls CD8 T-cell memory. *Eur. J. Immunol.* **2013**, *43*, 889–896. [CrossRef] [PubMed]
64. Poznanski, S.M.; Barra, N.G.; Ashkar, A.A.; Schertzer, J.D. Immunometabolism of T cells and NK cells: Metabolic control of effector and regulatory function. *Inflamm. Res.* **2018**, *67*, 813–828. [CrossRef] [PubMed]
65. Palsson-McDermott, E.M.; Curtis, A.M.; Goel, G.; Lauterbach, M.A.R.; Sheedy, F.J.; Gleeson, L.E.; van den Bosch, M.W.M.; Quinn, S.R.; Domingo-Fernandez, R.; Johnston, D.G.W.; et al. Pyruvate kinase M2 regulates Hif-1α activity and IL-1β induction and is a critical determinant of the warburg effect in LPS-activated macrophages. *Cell Metab.* **2015**, *21*, 65–80. [CrossRef] [PubMed]
66. Kirchner, J.; Brüne, B.; Namgaladze, D. AICAR inhibits NFκB DNA binding independently of AMPK to attenuate LPS-triggered inflammatory responses in human macrophages. *Sci. Rep.* **2018**, *8*, 7081. [CrossRef] [PubMed]
67. Moon, J.S.; Nakahira, K.; Choi, A.M. Fatty acid synthesis and NLRP3-inflammasome. *Oncotarget* **2015**, *6*, 21765–21766. [CrossRef] [PubMed]
68. Cacicedo, J.M.; Yagihashi, N.; Keaney, J.F., Jr; Ruderman, N.B.; Ido, Y. AMPK inhibits fatty acid-induced increases in NF-kappaB transactivation in cultured human umbilical vein endothelial cells. *Biochem. Biophys. Res. Commun.* **2004**, *324*, 1204–1209. [CrossRef] [PubMed]
69. Dicter, N.; Madar, Z.; Tirosh, O. α-Lipoic Acid Inhibits Glycogen Synthesis in Rat Soleus Muscle via Its Oxidative Activity and the Uncoupling of Mitochondria. *J. Nutr.* **2002**, *132*, 3001–3006. [CrossRef] [PubMed]
70. Lee, W.J.; Song, K.H.; Koh, E.H.; Won, J.C.; Kim, H.S.; Park, H.S.; Kim, M.S.; Kim, S.W.; Lee, K.U.; Park, J.Y. α-Lipoic acid increases insulin sensitivity by activating AMPK in skeletal muscle. *Biochem. Biophys. Res. Commun.* **2005**, *332*, 885–891. [CrossRef] [PubMed]
71. Timmers, S.; de Vogel-van den Bosch, J.; Towler, M.C.; Schaart, G.; Moonen-Kornips, E.; Mensink, R.P.; Hesselink, M.K.; Hardie, D.G.; Schrauwen, P. Prevention of high-fat diet-induced muscular lipid accumulation in rats by α lipoic acid is not mediated by AMPK activation. *J. Lipid Res.* **2010**, *51*, 352–359. [CrossRef] [PubMed]

72. Park, K.G.; Min, A.K.; Koh, E.H.; Kim, H.S.; Kim, M.O.; Park, H.S.; Kim, Y.D.; Yoon, T.S.; Jang, B.K.; Hwang, J.S.; et al. Alpha-lipoic acid decreases hepatic lipogenesis through adenosine monophosphate-activated protein kinase (AMPK)-dependent and AMPK-independent pathways. *Hepatology* **2008**, *48*, 1477–1486. [CrossRef] [PubMed]

73. Yang, Y.; Li, W.; Liu, Y.; Sun, Y.; Li, Y.; Yao, Q.; Li, J.; Zhang, Q.; Gao, Y.; Gao, L.; et al. Alpha-lipoic acid improves high-fat diet-induced hepatic steatosis by modulating the transcription factors SREBP-1, FoxO1 and Nrf2 via the SIRT1/LKB1/AMPK pathway. *J. Nutr. Biochem.* **2014**, *25*, 1207–1217. [CrossRef] [PubMed]

74. Kratz, M.; Coats, B.R.; Hisert, K.B.; Hagman, D.; Mutskov, V.; Peris, E.; Schoenfelt, K.Q.; Kuzma, J.N.; Larson, I.; Billing, P.S.; et al. Metabolic dysfunction drives a mechanistically distinct proinflammatory phenotype in adipose tissue macrophages. *Cell Metab.* **2014**, *20*, 614–625. [CrossRef] [PubMed]

75. Chan, K.L.; Pillon, N.J.; Sivaloganathan, D.M.; Costford, S.R.; Liu, Z.; Théret, M.; Chazaud, B.; Klip, A. Palmitoleate Reverses High Fat-induced Proinflammatory Macrophage Polarization via AMP-activated Protein Kinase (AMPK). *J. Biol. Chem.* **2015**, *290*, 16979–16988. [CrossRef] [PubMed]

76. Xiong, X.Q.; Geng, Z.; Zhou, B.; Zhang, F.; Han, Y.; Zhou, Y.B.; Wang, J.J.; Gao, X.Y.; Chen, Q.; Li, Y.H.; et al. FNDC5 attenuates adipose tissue inflammation and insulin resistance via AMPK-mediated macrophage polarization in obesity. *Metabolism* **2018**, *83*, 31–41. [CrossRef] [PubMed]

77. Van Dijk, S.J.; Feskens, E.J.; Bos, M.B.; Hoelen, D.W.; Heijligenberg, R.; Bromhaar, M.G.; de Groot, L.C.; de Vries, J.H.; Müller, M.; Afman, L.A. A saturated fatty acid—Rich diet induces an obesity-linked proinflammatory gene expression profile in adipose tissue of subjects at risk of metabolic syndrome. *Am. J. Clin. Nutr.* **2009**, *90*, 1656–1664. [CrossRef] [PubMed]

78. Chen, S.; Zhao, Z.; Ke, L.; Li, Z.; Li, W.; Zhang, Z.; Zhou, Y.; Feng, X.; Zhu, W. Resveratrol improves glucose uptake in insulin-resistant adipocytes via Sirt1. *J. Nutr. Biochem.* **2018**, *55*, 209–218. [CrossRef] [PubMed]

79. Hwang, J.T.; Kwak, D.W.; Lin, S.K.; Kim, H.M.; Kim, Y.M.; Park, O.J. Resveratrol Induces Apoptosis in Chemoresistant Cancer Cells via Modulation of AMPK Signaling Pathway. *Ann. N. Y. Acad. Sci.* **2007**, *1095*, 441–448. [CrossRef] [PubMed]

80. Kim, W.S.; Lee, Y.S.; Cha, S.H.; Jeong, H.W.; Choe, S.S.; Lee, M.R.; Oh, G.T.; Park, H.S.; Lee, K.U.; Lane, M.D.; et al. Berberine improves lipid dysregulation in obesity by controlling central and peripheral AMPK activity. *Am. J. Physiol. Endocrinol. Metab.* **2009**, *296*, 812–819. [CrossRef] [PubMed]

81. Turner, N.; Li, J.Y.; Gosby, A.; To, S.W.; Cheng, Z.; Miyoshi, H.; Taketo, M.M.; Cooney, G.J.; Kraegen, E.W.; James, D.E.; et al. Berberine and its more biologically available derivative, dihydroberberine, inhibit mitochondrial respiratory complex I: A mechanism for the action of berberine to activate AMP-activated protein kinase and improve insulin action. *Diabetes* **2008**, *57*, 1414–1418. [CrossRef] [PubMed]

82. Kim, H.S.; Kim, M.J.; Kim, E.J.; Yang, Y.; Lee, M.S.; Lim, J.S. Berberine-induced AMPK activation inhibits the metastatic potential of melanoma cells via reduction of ERK activity and COX-2 protein expression. *Biochem. Pharmacol.* **2011**, *83*, 385–394. [CrossRef] [PubMed]

83. Shehzad, A.; Ha, T.; Subhan, F.; Lee, Y.S. New mechanisms and the anti-inflammatory role of curcumin in obesity and obesity-related metabolic diseases. *Eur. J. Nutr.* **2011**, *50*, 151–161. [CrossRef] [PubMed]

84. Um, M.Y.; Hwang, K.H.; Ahn, J.; Ha, T.Y. Curcumin Attenuates Diet-Induced Hepatic Steatosis by Activating AMP-Activated Protein Kinase. *Basic Clin. Pharmacol. Toxicol.* **2013**, *113*, 152–157. [CrossRef] [PubMed]

85. Na, L.X.; Zhang, Y.L.; Li, Y.; Liu, L.Y.; Li, R.; Kong, T.; Sun, C.H. Curcumin improves insulin resistance in skeletal muscle of rats. *Nutr. Metab. Cardiovasc. Dis.* **2011**, *21*, 526–533. [CrossRef] [PubMed]

86. Pan, W.; Yang, H.; Cao, C.; Song, X.; Wallin, B.; Kivlin, R.; Lu, S.; Hu, G.; Di, W.; Wan, Y. AMPK mediates curcumin-induced cell death in CaOV3 ovarian cancer cells. *Oncol. Rep.* **1994**, *20*, 1553–1559.

87. Panche, A.N.; Diwan, A.D.; Chandra, S.R. Flavonoids: An overview. *J. Nutr. Sci.* **2016**, *5*, e47. [CrossRef] [PubMed]

88. Ha, B.G.; Nagaoka, M.; Yonezawa, T.; Tanabe, R.; Woo, J.T.; Kato, H.; Chung, U.I.; Yagasaki, K. Regulatory mechanism for the stimulatory action of genistein on glucose uptake in vitro and in vivo. *J. Nutr. Biochem.* **2012**, *23*, 501–519. [CrossRef] [PubMed]

89. Ji, G.; Zhang, Y.; Yang, Q.; Cheng, S.; Hao, J.; Zhao, X.; Jiang, Z. Genistein Suppresses LPS-Induced Inflammatory Response through Inhibiting NF-κB following AMP Kinase Activation in RAW 264.7 Macrophages. *PLoS ONE* **2012**, *7*, e53101. [CrossRef] [PubMed]

90. Palacios-González, B.; Zarain-Herzberg, A.; Flores-Galicia, I.; Noriega, L.G.; Alemán-Escondrillas, G.; Zariñan, T.; Ulloa-Aguirre, A.; Torres, N.; Tovar, A.R. Genistein stimulates fatty acid oxidation in a leptin

receptor-independent manner through the JAK2-mediated phosphorylation and activation of AMPK in skeletal muscle. *Biochim. Biophys. Acta Mol. Cell Biol. Lipids* **2014**, *1841*, 132–140. [CrossRef] [PubMed]

91. Ono, M.; Fujimori, K. Antiadipogenic Effect of Dietary Apigenin through Activation of AMPK in 3T3-L1 Cells. *J. Agric. Food Chem.* **2011**, *59*, 13346–13352. [CrossRef] [PubMed]
92. Choi, W.H.; Son, H.J.; Jang, Y.J.; Ahn, J.; Jung, C.H.; Ha, T.Y. Apigenin Ameliorates the Obesity-Induced Skeletal Muscle Atrophy by Attenuating Mitochondrial Dysfunction in the Muscle of Obese Mice. *Mol. Nutr. Food Res.* **2017**, *61*, 1700218. [CrossRef] [PubMed]
93. Jung, U.J.; Cho, Y.Y.; Choi, M.S. Apigenin Ameliorates Dyslipidemia, Hepatic Steatosis and Insulin Resistance by Modulating Metabolic and Transcriptional Profiles in the Liver of High-Fat Diet-Induced Obese Mice. *Nutrients* **2016**, *8*, 305. [CrossRef] [PubMed]
94. Sung, B.; Chung, H.Y.; Kim, N.D. Role of Apigenin in Cancer Prevention via the Induction of Apoptosis and Autophagy. *J. Cancer Prev.* **2016**, *21*, 216–226. [CrossRef] [PubMed]
95. Chang, H.C.; Guarente, L. SIRT1 and other sirtuins in metabolism. *Trends Endocrinol. Metab.* **2014**, *25*, 138–145. [CrossRef] [PubMed]

PERMISSIONS

All chapters in this book were first published in MDPI; hereby published with permission under the Creative Commons Attribution License or equivalent. Every chapter published in this book has been scrutinized by our experts. Their significance has been extensively debated. The topics covered herein carry significant findings which will fuel the growth of the discipline. They may even be implemented as practical applications or may be referred to as a beginning point for another development.

The contributors of this book come from diverse backgrounds, making this book a truly international effort. This book will bring forth new frontiers with its revolutionizing research information and detailed analysis of the nascent developments around the world.

We would like to thank all the contributing authors for lending their expertise to make the book truly unique. They have played a crucial role in the development of this book. Without their invaluable contributions this book wouldn't have been possible. They have made vital efforts to compile up to date information on the varied aspects of this subject to make this book a valuable addition to the collection of many professionals and students.

This book was conceptualized with the vision of imparting up-to-date information and advanced data in this field. To ensure the same, a matchless editorial board was set up. Every individual on the board went through rigorous rounds of assessment to prove their worth. After which they invested a large part of their time researching and compiling the most relevant data for our readers.

The editorial board has been involved in producing this book since its inception. They have spent rigorous hours researching and exploring the diverse topics which have resulted in the successful publishing of this book. They have passed on their knowledge of decades through this book. To expedite this challenging task, the publisher supported the team at every step. A small team of assistant editors was also appointed to further simplify the editing procedure and attain best results for the readers.

Apart from the editorial board, the designing team has also invested a significant amount of their time in understanding the subject and creating the most relevant covers. They scrutinized every image to scout for the most suitable representation of the subject and create an appropriate cover for the book.

The publishing team has been an ardent support to the editorial, designing and production team. Their endless efforts to recruit the best for this project, has resulted in the accomplishment of this book. They are a veteran in the field of academics and their pool of knowledge is as vast as their experience in printing. Their expertise and guidance has proved useful at every step. Their uncompromising quality standards have made this book an exceptional effort. Their encouragement from time to time has been an inspiration for everyone.

The publisher and the editorial board hope that this book will prove to be a valuable piece of knowledge for researchers, students, practitioners and scholars across the globe.

LIST OF CONTRIBUTORS

Michael Forster, Xiaojun Julia Liang and Matthias Gehringer
Department of Pharmaceutical/Medicinal Chemistry, Institute of Pharmaceutical Sciences, Faculty of Sciences, University of Tübingen, 72076 Tübingen, Germany
Cluster of Excellence iFIT (EXC 2180) 'Image-Guided & Functionally Instructed Tumor Therapies', University of Tübingen, 72076 Tübingen, Germany

Martin Schröder and Apirat Chaikuad
Structural Genomics Consortium, Goethe University Frankfurt, Buchmann Institute for Molecular Life Sciences, Max-von-Laue-Straße 15, 60438 Frankfurt am Main, Germany
Institute of Pharmaceutical Chemistry, Goethe University Frankfurt, Buchmann Institute for Molecular Life Sciences, Max-von-Laue-Straße 9, 60438 Frankfurt am Main, Germany

Stefan Gerstenecker
Department of Pharmaceutical/Medicinal Chemistry, Institute of Pharmaceutical Sciences, Faculty of Sciences, University of Tübingen, 72076 Tübingen, Germany

Stefan Knapp
Structural Genomics Consortium, Goethe University Frankfurt, Buchmann Institute for Molecular Life Sciences, Max-von-Laue-Straße 15, 60438 Frankfurt am Main, Germany
Institute of Pharmaceutical Chemistry, Goethe University Frankfurt, Buchmann Institute for Molecular Life Sciences, Max-von-Laue-Straße 9, 60438 Frankfurt am Main, Germany
Frankfurt Cancer Institute (FCI) and German Translational Cancer Network (DKTK) Site Frankfurt/ Mainz, 60438 Frankfurt am Main, Germany

Stefan Laufer
Department of Pharmaceutical/Medicinal Chemistry, Institute of Pharmaceutical Sciences, Faculty of Sciences, University of Tübingen, 72076 Tübingen, Germany
Cluster of Excellence iFIT (EXC 2180) 'Image-Guided & Functionally Instructed Tumor Therapies', University of Tübingen, 72076 Tübingen, Germany
Tübingen Center for Academic Drug Discovery (TüCAD2), 72076 Tübingen, Germany

Veronika Obsilova
Department of Structural Biology of Signaling Proteins, Division BIOCEV, Institute of Physiology of the Czech Academy of Sciences, 25250 Vestec, Czech Republic

Tomas Obsil
Department of Structural Biology of Signaling Proteins, Division BIOCEV, Institute of Physiology of the Czech Academy of Sciences, 25250 Vestec, Czech Republic
Department of Physical and Macromolecular Chemistry, Faculty of Science, Charles University, 12843 Prague, Czech Republic

Anastasiya Strembitska, Sarah J. Mancini, George S. Baillie and Ian P. Salt
Institute of Cardiovascular and Medical Sciences, College of Medical, Veterinary and Life Sciences, University of Glasgow, Glasgow G12 8QQ, UK

Jonathan M. Gamwell
Institute of Cardiovascular and Medical Sciences, College of Medical, Veterinary and Life Sciences, University of Glasgow, Glasgow G12 8QQ, UK
Radcliffe Department of Medicine, University of Oxford, Oxford OX3 9DU, UK

Timothy M. Palmer
Institute of Cardiovascular and Medical Sciences, College of Medical, Veterinary and Life Sciences, University of Glasgow, Glasgow G12 8QQ, UK
Centre for Atherothrombosis and Metabolic Disease, Hull York Medical School, University of Hull, Hull HU6 7RX, UK

Pascal Rowart
Groupe Interdisciplinaire de Génoprotéomique Appliquée (GIGA), Cardiovascular Sciences, University of Liège (ULiège), Avenue de L'Hôpital 11, 4000 Liège, Belgium

Jingshing Wu and Michael J. Caplan
Department of Cellular and Molecular Physiology, Yale School of Medicine, New Haven, CT 06520, USA

François Jouret
Groupe Interdisciplinaire de Génoprotéomique Appliquée (GIGA), Cardiovascular Sciences, University of Liège (ULiège), Avenue de L'Hôpital 11, 4000 Liège, Belgium
Division of Nephrology, Centre Hospitalier Universitaire de Liège (CHU of Liège), University of Liège (CHU ULiège), 13-B4000 Liège, Belgium

Brendan Gongol
Department of Medicine, University of California, San Diego, CA 92093, USA
Department of Cardiopulmonary Sciences, School of Allied Health Professions, Loma Linda University, Loma Linda, CA 92350, USA

Indah Sari, Tiffany Bryant and Geraldine Rosete
Department of Cardiopulmonary Sciences, School of Allied Health Professions, Loma Linda University, Loma Linda, CA 92350, USA

Traci Marin
Department of Medicine, University of California, San Diego, CA 92093, USA
Department of Health Sciences, Victor Valley College, Victorville, CA 92395, USA

Arnaud Jacquel, Frederic Luciano, Guillaume Robert and Patrick Auberger
Université Côte d'Azur, C3M Inserm U1065, 06204 Nice, France
Equipe Labellisée par la Fondation ARC, 94803 Villejuif, France

David Martin-Hidalgo
Research Group of Intracellular Signaling and Technology of Reproduction (SINTREP), Institute of Biotechnology in Agriculture and Livestock (INBIO G+C), University of Extremadura, 10003 Cáceres, Spain
Unit for Multidisciplinary Research in Biomedicine (UMIB), Laboratory of Cell Biology, Department of Microscopy, Institute of Biomedical Sciences Abel Salazar (ICBAS), University of Porto, 40050-313 Porto, Portugal

Ana Hurtado de Llera
Research Group of Intracellular Signaling and Technology of Reproduction (SINTREP), Institute of Biotechnology in Agriculture and Livestock (INBIO G+C), University of Extremadura, 10003 Cáceres, Spain
Hormones and Metabolism Research Group, Faculty of Health Sciences, University of Beira Interior, 6200-506 Covilhã, Portugal

Violeta Calle-Guisado, Lauro Gonzalez-Fernandez, Luis Garcia-Marin and M. Julia Bragado
Research Group of Intracellular Signaling and Technology of Reproduction (SINTREP), Institute of Biotechnology in Agriculture and Livestock (INBIO G+C), University of Extremadura, 10003 Cáceres, Spain

Dietbert Neumann
Department of Pathology, CARIM School for Cardiovascular Diseases, Faculty of Health, Medicine and Life Sciences, Maastricht University, 6200 MD Maastricht, The Netherlands

You-Lin Tain
Departments of Pediatrics, Kaohsiung Chang Gung Memorial Hospital and Chang Gung University College of Medicine, Kaohsiung 833, Taiwan
Institute for Translational Research in Biomedicine, Kaohsiung Chang Gung Memorial Hospital and Chang Gung University College of Medicine, Kaohsiung 833, Taiwan

Chien-Ning Hsu
Department of Pharmacy, Kaohsiung Chang Gung Memorial Hospital, Kaohsiung 833, Taiwan

Mahmoud Ahmed, Jin Seok Hwang, Trang Huyen Lai, Sahib Zada, Huynh Quoc Nguyen, Trang Min Pham and Deok Ryong Kim
Department of Biochemistry and Convergence Medical Sciences and Institute of Health Sciences, Gyeongsang National University School of Medicine, Jinju 527-27, Korea

Miyong Yun
Department of Bioindustry and Bioresource Engineering, College of Life Sciences, Sejong University, Seoul 05006, Korea

David M. Thomson
Department of Physiology & Developmental Biology, Brigham Young University, Provo, UT 84602, USA

Honglai Zhang, Sung-Soo Park, Claire Peabody, Justin Yoon, Ester Calvo Fernández, Dafydd Thomas, Sandra Camelo-Piragua and Jean-François Rual
Department of Pathology, University of Michigan Medical School, Ann Arbor, MI 48109, USA

Rork Kuick
Department of Biostatistics, School of Public Health, University of Michigan, Ann Arbor, MI 48109, USA

Junying Wang and Ken Inoki
Life Sciences Institute, University of Michigan, Ann Arbor, MI 48109, USA

Benoit Viollet
Inserm, U1016, Institut Cochin, 75014 Paris, France
CNRS, UMR8104, 75014 Paris, France
Université Paris Descartes, Sorbonne Paris cité, 75014 Paris, France

Claire L. Lyons
Unit of Molecular Metabolism, Lund University Diabetes Center, Clinical Research Center, Lund University, 205 02 Malmö, Sweden
Nutrigenomics Research Group, UCD Institute of Food and Health, UCD Conway Institute of Biomolecular and Biomedical Research, University College Dublin, Belfield, 4 Dublin, Ireland

Helen M. Roche
Nutrigenomics Research Group, UCD Institute of Food and Health, UCD Conway Institute of Biomolecular and Biomedical Research, University College Dublin, Belfield, 4 Dublin, Ireland
Institute of Global Food Security, Queen's University Belfast BT7 1NN, Northern Ireland, UK

Index

www.ingramcontent.com/pod-product-compliance
Lightning Source LLC
LaVergne TN
LVHW080951210926
848485LV00004B/816